Nanocrystals

edited by
Yoshitake Masuda

Nanocrystals
Edited by Yoshitake Masuda

Published by InTech

Janeza Trdine 9, 51000 Rijeka, Croatia

Publishing Process Manager Iva Lipovic

Technical Editor Martina Peric

Cover Designer InTech Design Team

Additional hard copies can be obtained from orders@intechopen.com

Nanocrystals, Edited by Yoshitake Masuda
p. cm.
ISBN 978-953-307-126-8

For EU product safety concerns:
IN TECH d.o.o., Prolaz Marije Krucifikse Kozulić 3, 51000 Rijeka, Croatia,
info@intechopen.com or visit our website at intechopen.com.

Meet the editor

Dr. Yoshitake Masuda is a senior research scientist at the National Institute of Advanced Industrial Science and Technology (AIST), Japan. He graduated from Tsukuba University in 1994, and received his Master of Engineering degree from Tsukuba University in 1996. He was an engineer at NGK Spark Plug Co. Ltd., from 1996 to 1998, and an assistant professor at Nagoya University from 2000 to 2006. Dr. Masuda received his Doctor of Engineering degree from Nagoya University in 2004, and his PhD thesis was on the Patterning of TiO2 Thin Films and Particles using Self-assembled Monolayers.

Contents

Preface

Nanocrystals have attracted much attention for the next generation science and technology. They cover many areas of materials science such as inorganic materials, organic materials, hybrid materials, biomaterials, and relate to various research fields, such as chemistry, physics, biology, electronic devices, optic devices, etc.

The nanocrystals provide excellent novel functions by control of crystal growth, morphology, size, crystallinity, anisotropy, orientation, self-assembly, integration, etc. Each crystal has a different crystal structure. Thus, they have anisotropy of crystal faces and crystal axis. Morphology control of nanocrystals can be realized by the use of several crystal growth modes through the change of crystal growth conditions. For instance, variety of crystal growth modes can be used for morphology control of nanocrystals in aqueous solution processes by the change of solution temperature, ion concentration, pH, additives, etc. Novel nanocrystals that have unique structures and functions were developed using these techniques. Additionally, self-assembly, orientation, integration and patterning of nanocrystals provide novel functions and values. Self-assembly and 2D-patterning of nanocrystals, for instance, were developed for photonic crystals. Self-assembly brought out a novel function from the nanocrystals.

Nanocrystal is one of the most important research fields for the next generation science and technologies. Nanocrystal science is a cross-cutting research and includes various materials and research areas. It has ripple effects throughout the materials science. The scientific knowledge and technologies in this book will contribute to the development of nanocrystal research and related future applications.

In closing, I wish to express my sincere sense of gratitude to the authors, book manager Ms. Iva Lipovic, and the publishing staff. I dedicate this book to my parents, Mr. Toshio Masuda and Ms. Nobuko Masuda, my sisters, Ms. Shinobu Horita and Ms. Satoe Amaya, my children, Ms. Yuuka Masuda, Ms. Arisa Masuda and Mr. Ikuto Masuda, and my wife, Ms. Yumi Masuda.

Editor

Yoshitake Masuda
National Institute of Advanced Industrial Science and Technology (AIST)
Japan

1

Morphology control, self-assembly and site-selective deposition of metal oxide nanocrystals

Yoshitake Masuda
National Institute of Advanced Industrial Science and Technology (AIST), 2266-98 Anagahora, Shimoshidami, Moriyama-ku, Nagoya 463-8560, Japan

Keywords

Nanocrystals, Metal Oxide, Morphology Control, Self-assembly, Site-selective Deposition, Self-assembled Monolayer, Liquid Phase Process, Aqueous Phase Process, Aqueous Solution Process

1. Introduction

Nanocrystals of metal oxides have been attracted much attention for future science and technology. Morphology control, self-assembly and site-selective deposition of nanocrystals will open a new frontier in materials science. They were realized in this chapter using aqueous solution processes for metal oxide nanocrystals.

Site-selective deposition (SSD) of metal oxide thin films was developed to fabricate nano/microstructures of metal oxide such as TiO_2, Fe_3O_4, ZnO, Y_2O_3:Eu, etc.. Several conceptual processes for SSD using self-assembled monolayers (SAMs) as templates were proposed, and nano/micropatterns of ceramic thin films were successfully fabricated. Molecular recognition of SAMs was effectively used to achieve high site-selectivity. These processes can be used for the fabrication of various metal oxide devices under environment-friendly conditions.

We also developed a self-assembly process of particles to fabricate desired patterns of colloidal crystals. A micropattern of colloidal methanol prepared on a SAM in hexane was used as a mold for particle patterning, and slow dissolution of methanol into hexane caused shrinkage of molds to form micropatterns of close-packed particle assemblies. This result is a step toward the realization of nano/micro periodic structures for next-generation photonic devices by a self-assembly process.

Furthermore, metal oxides were synthesized in aqueous solutions to form anisotropic nanostructures. Stand-alone ZnO self-assembled films were, for instance, prepared using air-liquid interfaces. The ZnO films had sufficiently high strength to free-stand-alone and showed high *c*-axis orientation. The films can be pasted onto desired substrates. ZnO particles having a hexagonal cylinder shape, long ellipse shape or hexagonal symmetry radial whiskers were also prepared in aqueous solutions. The morphology was controlled by changing the supersaturation degree. Anatase TiO_2 particles with high surface area of 270 m^2/g were prepared at 50°C. The particles were assemblies of nano TiO_2 crystals covered with nanorelief surface structures. The crystals grew anisotropically along the *c*-axis to form

acicular crystals. TiO_2 films consisted of anisotropic acicular crystals were also prepared. The films showed high *c*-axis orientation. Acicular $BaTiO_3$ particles were prepared using morphology control of $BaC_2O_4 \bullet 0.5H_2O$. They were prepared in aqueous solutions and annealed with co-precipitated amorphous phase to form acicular $BaTiO_3$ particles.
In this chapter, we will mainly focus on "liquid phase morphology control of metal oxides nanocrystals" of ZnO[1], TiO_2[2] and $BaTiO_3$ particles[3], and "liquid phase site-selective deposition of metal oxide nanocrystals" of TiO_2[4] and Eu:Y_2O_3[5].

2. Liquid Phase Morphology Control of Metal Oxide Nanocrystals

2.1. Morphology Control of Stand-alone ZnO Self-assembled Film[1]

Stand-alone ZnO films were fabricated using aqueous solutions. The films were assemblies of sheet shaped nanocrystals. They had gradient structure, high *c*-axis orientation, high surface area and unique morphology. They were formed at the top of solutions without substrate. Air-liquid interface was used as a template in this process.
Zinc nitrate hexahydrate ($Zn(NO_3)_2$ $6H_2O$) (15 mM) was dissolved in distilled water at 60°C. Ethylenediamine ($H_2NCH_2CH_2NH_2$) (15 mM) was added to the solution to induce the formation of ZnO[1]. The solution was kept at 60°C using a water bath for 6 h with no stirring. The solution was then left to cool for 42 h in the bath. Polyethylene terephthalate (PET) film, glass (S-1225, Matsunami Glass Ind., Ltd.) and an Si wafer (p-type Si [100], NK Platz Co., Ltd.) were used as substrates.
The solution color was changed from transparent to white after the addition of ethylenediamine. ZnO particles were formed by the homogeneous nucleation and growth. The solution became transparent again after 6h. The supersaturation degree of the solution was high at the initial stage of the reaction for the first 1 h and decreased as the color of the solution changed. Ethylenediamine accelerated deposition of ZnO. Zinc-ethylenediamine complex forms in the solution as shown by eq. 1[6].

$$Zn^{2+} + 3H_2NCH_2CH_2NH_2 \rightleftharpoons [Zn(H_2NCH_2CH_2NH_2)_3]^{2+} \quad (1)$$

The chemical equilibrium in eq. 1 moves to the left and the zinc-ethylenediamine complex decomposes to increase the concentration of Zn^{2+} at elevated temperature.
OH^- concentration increases by the hydrolysis of ethylenediamine as shown by eq. 2.

$$H_2NCH_2CH_2NH_2 + 2H_2O \rightleftharpoons H_3NCH_2CH_2NH_3^{2+} + 2OH^- \quad (2)$$

ZnO and $Zn(OH)_2$ are thus formed in the aqueous solution as shown by eq. 3.

$$Zn^{2+} + 2OH^- \rightarrow Zn(OH)_2 \rightarrow ZnO + H_2O \quad (3)$$

Films were formed at the top of the solution. Air-liquid interface was used as template. The films had sufficiently high strength to be obtained as stand-alone films. Additionally, a film was scooped to past onto a desired substrate such PET film, Si wafer, glass plate or paper, and the pasted ZnO film was then dried to bond it to the substrate. Both sides of the film can be pasted on substrate. The film physically adhered to the substrate. The film maintained its adhesion during immersion in lightly ultrasonicated water, however, it can be easily peeled off again by strong ultrasonication. The film can be handled easily from substrate to other substrate. It also

can be attached strongly to substrate by annealing or addition of chemical regents such as silane coupling agent to form chemical bonds between the film and the substrate.
The film grew to a thickness of about 5 μm after 48 h, i.e., 60°C for 6 h, and was left to cool for 42 h.
The air side of the stand-alone film had a smooth surface over a wide area due to the flat air-liquid interface (Fig. 1-a1), whereas the liquid side of the film had a rough surface (Fig. 1-b1). The films consisted of ZnO nano-sheets were clearly observed from the liquid side (Fig. 1-b2) and the fracture edge-on profile of the film (Fig. 1-c1, 1-c2). The nano-sheets had a thickness of 5-10 nm and were 1-5 μm in size. They mainly grew forward to the bottom of the solution, i.e., perpendicular to the air-liquid interface, such that the sheets stood perpendicular to the air-liquid interface. Thus, the liquid side of the film had many ultra-fine spaces surrounded by nano-sheet and had a high specific surface area. The air side of the film, on the other hand, had a flat surface that followed the flat shape of the air-liquid interface. The air-liquid interface was thus effectively utilized to form the flat surface of the film. This flatness would contribute to the strong adhesion strength to substrates for pasting of the film. The air-side surface prepared for 48 h had holes of 100-500 nm in diameter (Fig. 1-a2), and were hexagonal, rounded hexagonal or round in shape. The air-side surface prepared for 6 h, in contrast, had no holes on the surface. The air-side surface was well crystallized to form a dense surface and ZnO crystals would partially grow to a hexagonal shape because of the hexagonal crystal structure. Well-crystallized ZnO hexagons were then etched to form holes on the surface by decrease in pH. The growth face of the film would be liquid side. ZnO nano-sheets would grow to form a large ZnO film by Zn ion supply from the aqueous solution. Further investigation of the formation mechanism would contribute to the development of crystallography in the solution system and the creation of novel ZnO fine structures.
The film showed a very strong 0002 x-ray diffraction peak of hexagonal ZnO at $2\theta = 34.04°$ and weak 0004 diffraction peak at $2\theta = 72.16°$ with no other diffractions of ZnO (Fig. 2). (0002) planes and (0004) planes were perpendicular to the *c*-axis, and the diffraction peak only from (0002) and (0004) planes indicates high *c*-axis orientation of ZnO film. The inset figure shows that the crystal structure of hexagonal ZnO stands on a substrate to make the *c*-axis perpendicular to the substrate. Crystallite size parallel to (0002) planes was estimated from the half-maximum full-width of the 0002 peak to 43 nm. This is similar to the threshold limit value of our XRD equipment and thus the crystallite size parallel to (0002) planes is estimated to be greater than or equal to 43 nm. Diffraction peaks from a silicon substrate were observed at $2\theta = 68.9°$ and $2\theta = 32.43°$. Weak diffractions at $2\theta = 12.5°$, 24.0°, 27.6°, 30.5°, 32.4° and 57.6° were assigned to co-precipitated zinc carbonate hydroxide ($Zn_5(CO_3)_2(OH)_6$, JCPDS No. 19-1458).
Stand-alone ZnO film was further evaluated by TEM and electron diffraction. The film was crushed to sheets and dispersed in an acetone. The sheets at the air-liquid interface were skimmed by a cupper grid with a carbon supporting film. The sheets were shown to have uniform thickness (Fig. 2a). They were dense polycrystalline films constructed of ZnO nanoparticles (Fig. 2b). Lattice image was clearly observed to show high crystallinity of the particles. The film was shown to be single phase of ZnO by electron diffraction pattern. These observations were consistent with XRD and SEM evaluations.

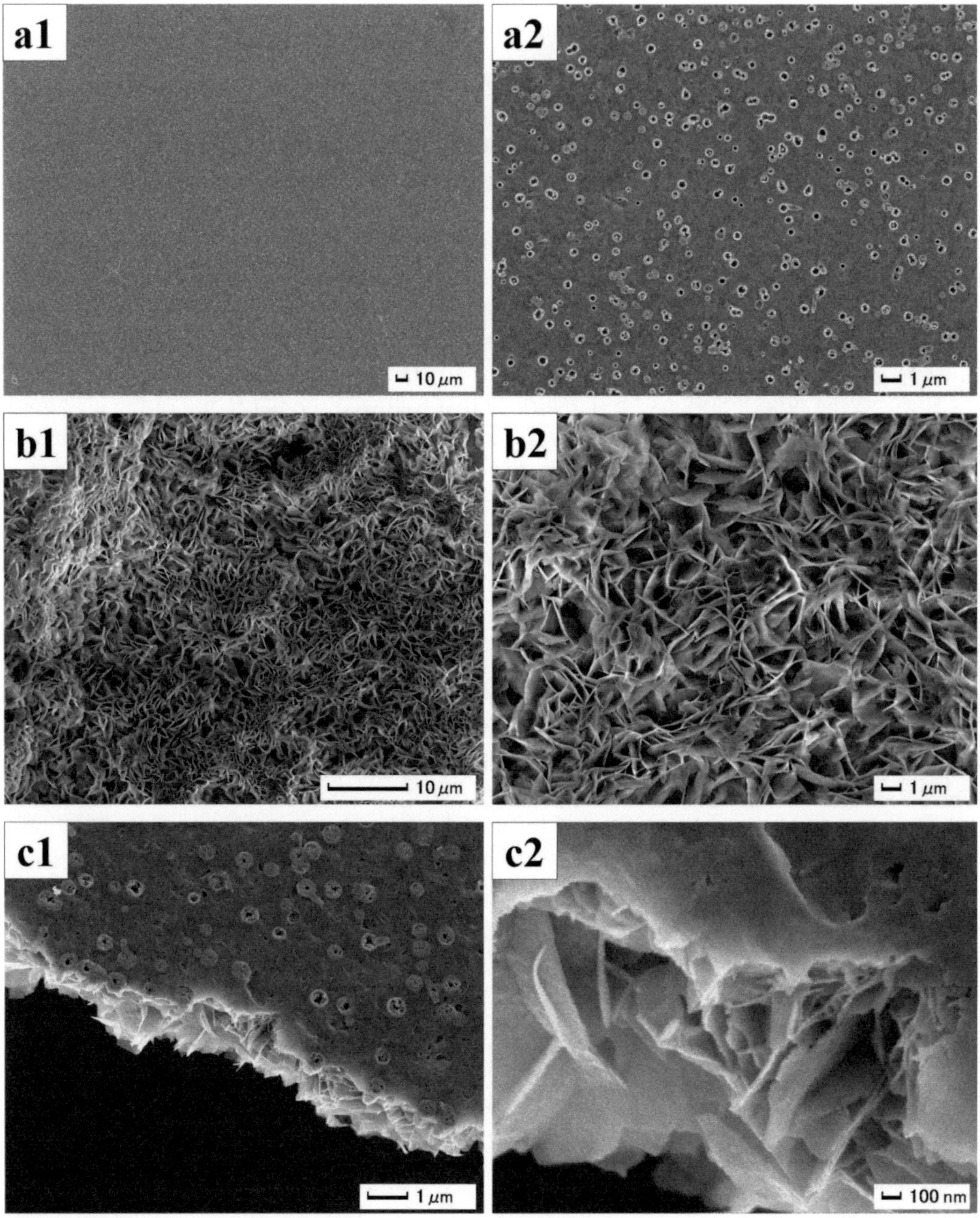

Fig. 1. SEM micrographs of high *c*-axis oriented stand-alone ZnO self-assembled film. (a1) Air-side surface of ZnO film. (a2) Magnified area of (a1). (b1) Liquid-side surface of ZnO film. (b2) Magnified area of (b2). (c1) Fracture cross section of ZnO film from air side. (c2) Magnified area of (c1).

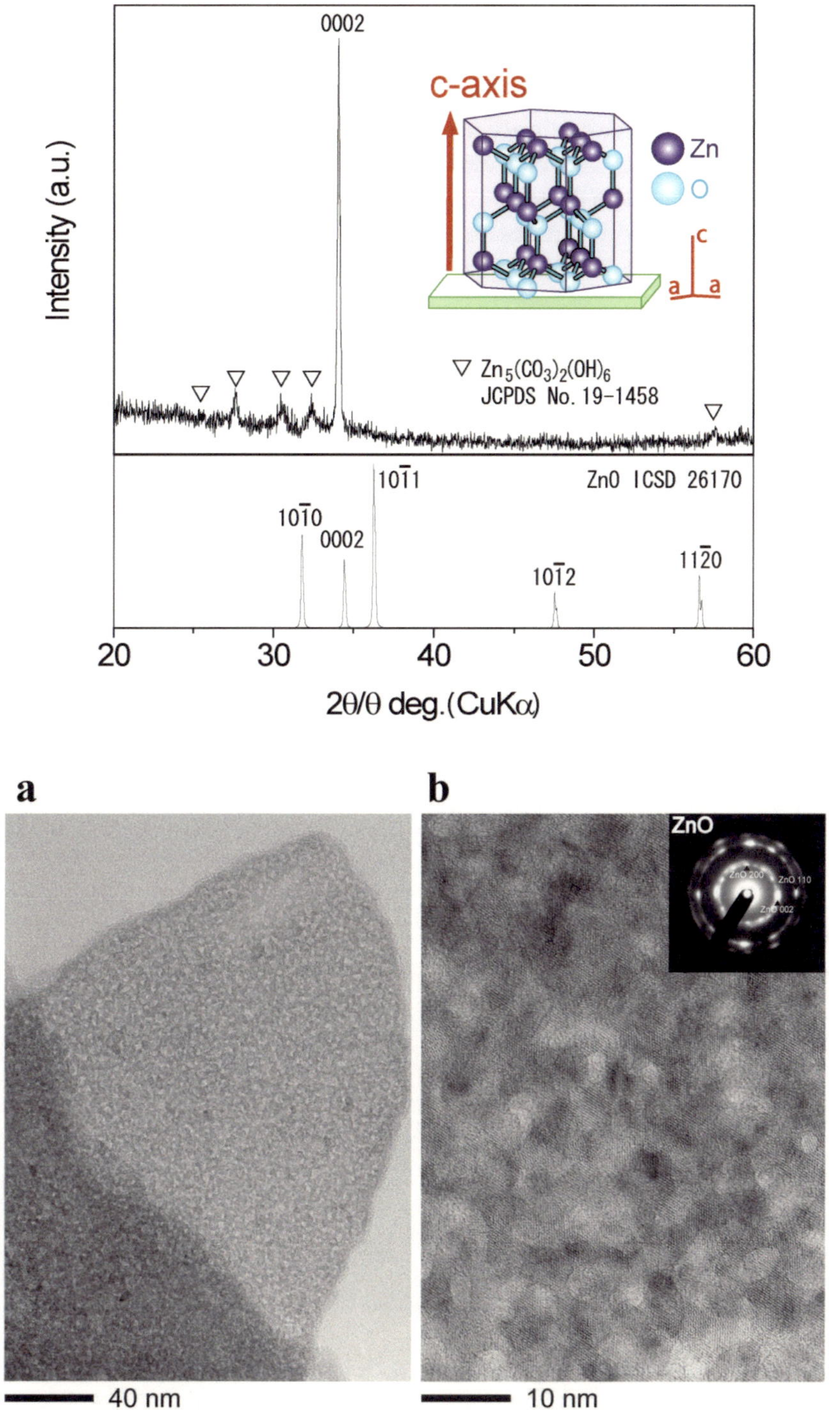

Fig. 2. XRD diffraction pattern of high *c*-axis oriented stand-alone ZnO self-assembled film. (a) TEM micrograph of ZnO nano-sheets. (b) Magnified area of (a). (Insertion) Electron diffraction pattern of ZnO.

The film pasted on a silicon wafer was annealed at 500°C for 1 h in air to evaluate the details of the films. ZnO film maintained its structure during the annealing (Fig. 3). The air side of the film showed a smooth surface (Fig. 3-a1) and the liquid side showed a relief structure having a high specific surface area (Fig. 3-b1, 3-b2). The air side showed the film consisted of dense packing of small ZnO nanosheets and the size of sheets increased toward the liquid-side surface (Fig. 3-a2). ZnO sheets would grow from the air side to the liquid side, i.e., the sheets would nucleate at the liquid-air interface and grow down toward the bottom of the solution by the supply of Zn ions from the solution. Annealed film showed X-ray diffractions of ZnO and Si substrate with no additional phases. As-deposited ZnO nano-sheets were shown to be crystalline ZnO because the sheets maintained their fine structure during the annealing without any phase transition. High *c*-axis orientation was also maintained during the annealing, showing a very strong 0002 diffraction peak.

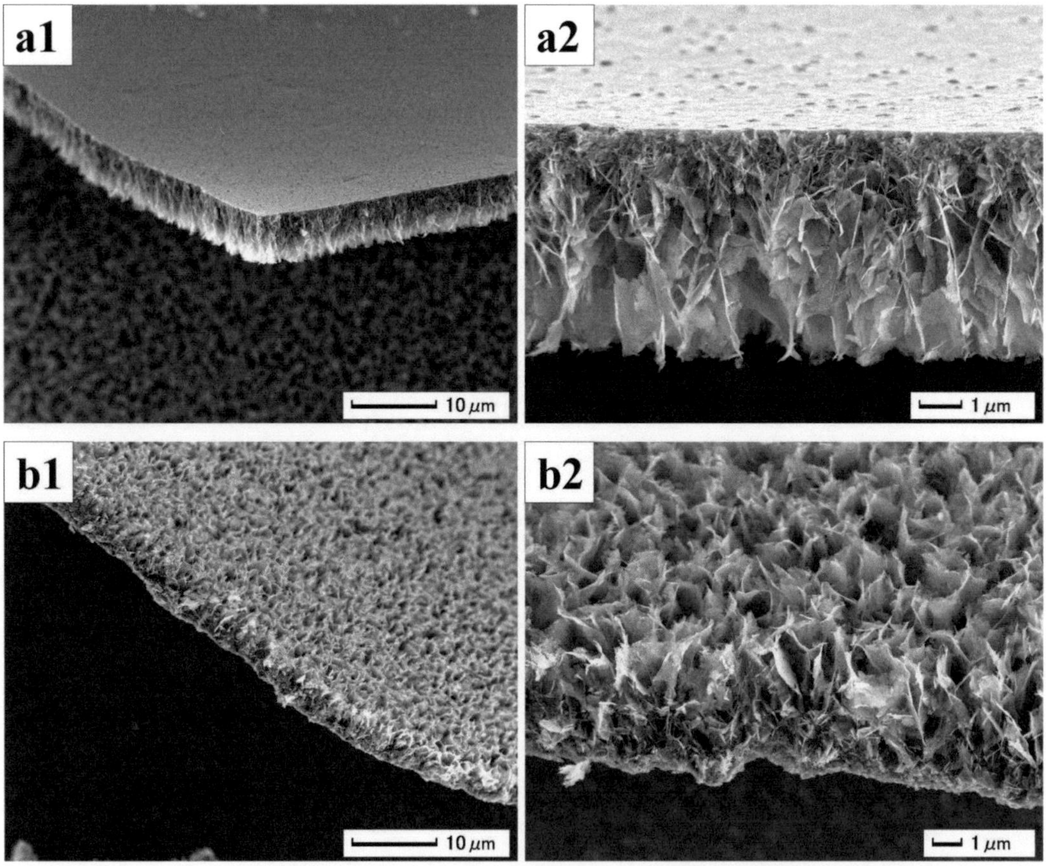

Fig. 3. SEM micrographs of high *c*-axis oriented stand-alone ZnO self-assembled film annealed at 500°C for 1 h in air. (a1) Fracture edge-on profile of ZnO film from air side. (a2) Cross-section profile of ZnO film from air side. (b1) Fracture edge-on profile of ZnO film from liquid side. (b2) Cross-section profile of ZnO film from liquid side.

The solution was further kept at 25°C for 1 month to evaluate the details of the deposition mechanism. The film prepared at the air-liquid interface for 1 month was not hexagonal ZnO. The film showed strong X-ray diffractions of zinc carbonate hydroxide single phase. ZnO would be dissolved by decrease in pH. ZnO would be crystallized at the initial reaction stage for the first 48 h. ZnO was then gradually etched and dissolved by nitric acid and zinc carbonate hydroxide was crystallized using Zn ions which were supplied by the dissolution of crystalline ZnO.

In summary, nano-sheet assembled stand-alone ZnO film was successfully fabricated using a simple solution process. Air-liquid interface was used as a template to form the films. The film had high *c*-axis orientation and showed a strong 0002 diffraction peak and weak 0004 peak. The air side of the film had a flat surface, whereas the liquid side had a rough surface having many ultra-fine spaces surrounded by ZnO nano-sheets. The rough surface of the liquid side was suitable for sensors or dye-sensitized solar cells. The film was also pasted on a desired substrate such as PET films, Si substrate or glass plates. The surface of low heat-resistant flexible polymer film was modified with high *c*-axis oriented crystalline ZnO film without heat treatment. This low-cost, low-temperature technique can be used for a wide range of applications including sensors, solar cells, electrical devices and optical devices using the various properties of high *c*-axis oriented crystalline ZnO.

2.2 Morphology Control of Nanocrystal Assembled TiO_2 Particles[2]

TiO_2 particles were prepared in aqueous solutions at ordinary temperature. The particles were assemblies of nanocrystals that had acicular shape. They had high surface area of 270 m^2/g and unique morphology. They are candidate material for dye-sensitized solar cells and photo catalyst.

Ammonium hexafluorotitanate (12.372 g) and boric acid (11.1852 g) were dissolved in deionized water (600 mL) at 50°C[2]. Concentration of them were 0.15 and 0.05 M, respectively. The solution was kept at 50°C for 30 min using a water bath with no stirring. The solution was centrifuged at 4000 rpm for 10 min (Model 8920, Kubota Corp.). Precipitated particles were dried at 60°C for 12 h after removal of supernatant solution.

The solution became clouded about 10 min after mixing ammonium hexafluorotitanate solution and boric acid solution. The particles were homogeneously nucleated in the solution, turning the solution white.

X-ray diffraction analysis indicated that the particles were single phase of anatase TiO_2. The peaks were observed at 2θ = 25.1, 37.9, 47.6, 54.2, 62.4, 69.3, 75.1, 82.5 and 94.0°. They were assigned to the 101, 004, 200, 105 + 211, 204, 116 + 220, 215, 303 + 224 + 312 and 305 + 321 diffraction peaks of anatase TiO_2 (JCPSD No. 21-1272, ICSD No. 9852) (Fig. 4).

The 004 diffraction intensity of randomly oriented particles is usually 0.2 times the 101 diffraction intensity as shown in JCPDS data (No. 21-1272). However, the 004 diffraction intensity of the particles deposited in our process was 0.36 times the 101 diffraction intensity. Additionally, the integral intensity of the 004 diffraction was 0.18 times the 101 diffraction intensity, indicating the *c*-axis orientation of the particles. Particles were not oriented on the glass holder for XRD measurement. Therefore, TiO_2 crystals would be an anisotropic shape in which the crystals were elongated along the *c*-axis. The crystals would have a large number of stacks of c planes such as (001) planes compared to stacks of (101) planes. The diffraction intensity from the (004) planes would be enhanced compared to that from the (101) planes.

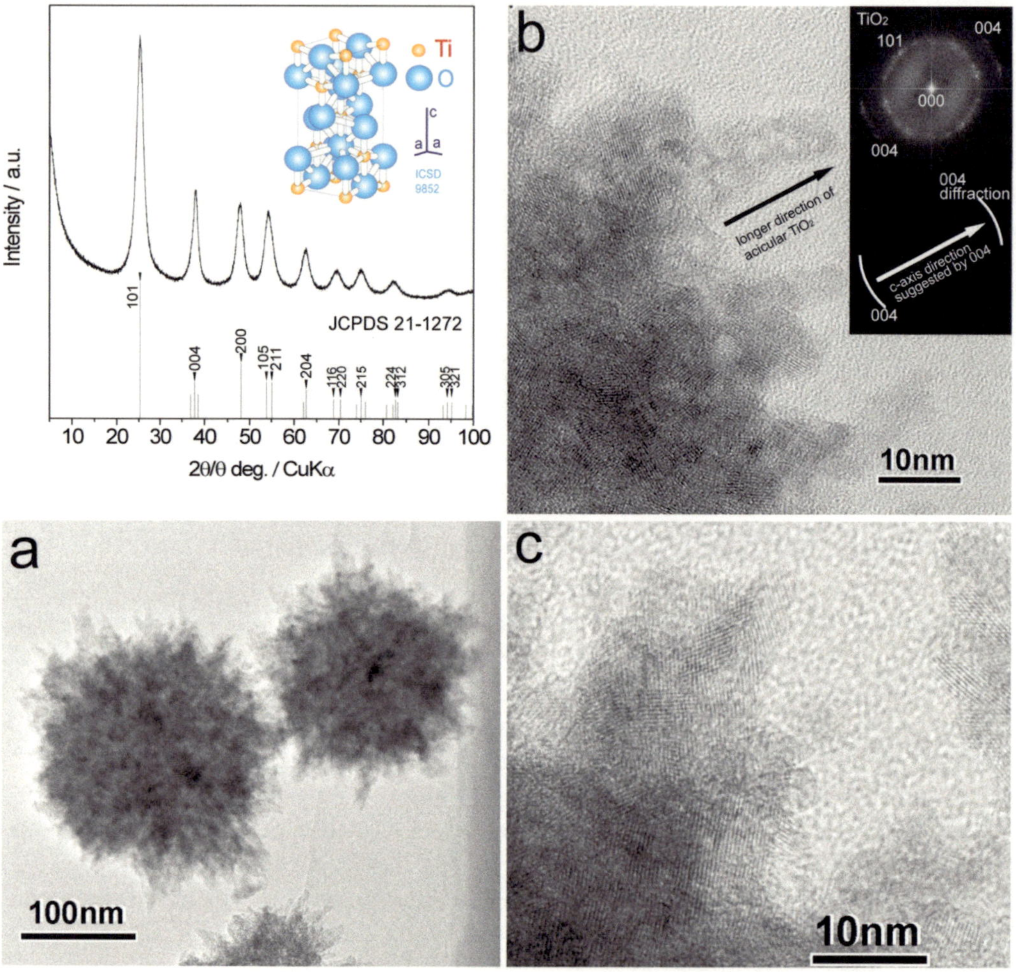

Fig. 4. XRD diffraction pattern of anatase TiO_2 particles. (a): TEM micrograph of anatase TiO_2 particles. (b): Magnified area of (a) showing morphology of acicular crystals. Insertion in (b): FFT image of (b) anatase TiO_2. (c): Magnified area of (a) showing lattice images of anatase TiO_2.

Crystallite size perpendicular to the (101) or (004) planes was estimated from the full-width half-maximum of the 101 or 004 peak to be 3.9 nm or 6.3 nm, respectively. Elongation of crystals in the *c*-axis direction was also suggested by the difference in crystallite size.
The particles were shown to be assemblies of nano TiO_2 crystals (Fig. 4a). Particle diameter was estimated to be 100–200 nm. Relief structures had formed on the surfaces and open pores had formed inside because the particles were porous assemblies of nanocrystals.

Nanocrystals were shown to have acicular shapes (Fig. 4b). They were about 5–10 nm in length. The longer direction of acicular TiO_2 is indicated by the black arrow. The inserted FFT image shows the 101 and 004 diffractions of anatase TiO_2. Nanocrystals are assigned to the single phase of anatase TiO_2. It is notable that the diffraction from the (101) planes has a ring shape due to random orientation but that from the (004) planes was observed only in the upper right region and lower left region in the FFT image. Anisotropic 004 diffractions indicated the direction of the *c*-axis, which was perpendicular to the (004) planes, as shown by the white arrow. It was roughly parallel to the longer direction of acicular TiO_2. These results suggest that acicular TiO_2 grew along the *c*-axis to enhance the diffraction intensity from the (004) planes. Crystal growth of anatase TiO_2 along the *c*-axis was previously observed in TiO_2 films[7]. Anisotropic crystal growth is one of the features of liquid phase crystal deposition.

Acicular nanocrystals showed lattice images of anatase TiO_2 (Fig. 4c). They were constructed of anatase TiO_2 crystals without amorphous or additional phases. Anatase crystals were not covered with amorphous or additional phases even at the tips. Bare anatase crystal with nanosized structure is important to achieve high performance for catalysts and devices.

Crystallization of TiO_2 was effectively utilized to form assemblies of acicular nanocrystals in the process. Open pores and surface relief structures were successfully formed on the particles.

The dried particles were dispersed in water to evaluate zeta potential and particle size distribution after evaluation of N_2 adsorption. The particles had positive zeta potential of 30.2 mV at pH 3.1, which decreased to 5.0, −0.6, −11.3 and −36.3 mV at pH 5.0, 7.0, 9.0 and 11.1, respectively. The isoelectric point was estimated to be pH 6.7, slightly higher than that of anatase TiO_2 (pH 2.7–6.0)[8]. Zeta potential is very sensitive to the particle surface conditions, ions adsorbed on the particle surfaces, and the kind and concentration of ions in the solution. The variations in zeta potential were likely caused by the difference in the surface conditions of TiO_2 particles, affected by the interaction between particles and ions in the solution.

Mean particle size was estimated to be ~550 nm in diameter with a standard deviation (STD) of 220 nm at pH 3.1. This was larger than that observed by TEM. Slight aggregation occurred at pH 3 because the particles were dried completely prior to measurement. Particle size increased with pH and showed a maximum of near the isoelectric point (550 nm at pH 3.1, 3150 nm at pH 5, 4300 nm at pH 7, 5500 nm at pH 9 or 2400 nm at pH 11.1). Strong aggregation resulted from the lack of repulsion force between particles near the isoelectric point.

The particles were generated in the solution at pH 3.8 in this study. It would be suitable to obtain repulsion force between particles for crystallization without strong aggregation.

TiO_2 particles exhibited N_2 adsorption-desorption isotherms of Type IV (Fig. 5a). The desorption isotherm differed from adsorption isotherm in the relative pressure (P/P_0) range from 0.4 to 0.7, showing mesopores in the particles. BET surface area of the particles was estimated to be 270 m^2/g (Fig. 5b). This is higher than that of TiO_2 nanoparticles such as Aeroxide P25 (BET 50 m^2/g, 21 nm in diameter, anatase 80% + rutile 20%, Degussa), Aeroxide P90 (BET 90–100 m^2/g, 14 nm in diameter, anatase 90% + rutile 10%, Degussa), MT-01 (BET 60 m^2/g, 10 nm in diameter, rutile, Tayca Corp.) and Altair TiNano (BET 50 m^2/g, 30–50 nm in diameter, Altair Nanotechnologies Inc.)[9]. A high BET surface area cannot be obtained from particles having a smooth surface even if the particle size is less than 100 nm. A high BET surface area would be realized by the unique morphology of TiO_2 particles constructed of nanocrystal assemblies.

Total pore volume and average pore diameter were estimated from pores smaller than 230 nm at P/Po = 0.99–0.431 cc/g and 6.4 nm, respectively. They were estimated to be 0.212 cc/g and 3.1 nm, respectively, from pores smaller than 11 nm at P/Po = 0.80. Total pore volume was also estimated by the BJH method from pores smaller than 154 nm to be 0.428 cc/g. Average pore diameter was estimated to be 6.3 nm using BET surface area.

Pore size distribution was calculated by the BJH method using adsorption isotherms (Fig. 5c). It showed a pore size distribution curve having a peak at ~2.8 nm and pores larger than 10 nm. TiO_2 particles would have mesopores of ~2.8 nm surrounded by nanocrystals. Pores larger than 10 nm are considered to be interparticle spaces. The pore size distribution also suggested the existence of micropores smaller than 1 nm.

Pore size distribution was further calculated by the DFT/Monte-Carlo method. The model was in fair agreement with adsorption isotherms (Fig. 5d). Pore size distribution showed a peak at ~3.6 nm that indicated the existence of mesopores of ~3.6 nm (Fig. 5e). The pore size calculated by the DFT/Monte-Carlo method was slightly larger than that calculated from the BJH method because the latter method is considered to have produced an underestimation[10-12]. The pore size distribution also suggested the existence of micropores of ~1 nm, probably resulting from microspaces surrounded by nanocrystals and the uneven surface structure of nanocrystals.

The particles were shown to have a large surface area as well as micropores of ~1 nm, mesopores of ~2.8–3.6 nm and pores larger than 10 nm, by N_2 adsorption characteristics. Assembly of acicular nanocrystals resulted in unique features and high surface area.

TiO_2 particles were generated in the solutions at 90°C for 1h using an oil bath with no stirring for comparison. The solutions became clouded after the addition of boric acid solutions into ammonium hexafluorotitanate solutions. High temperature accelerated crystal growth of TiO_2. Hydrogen chloride of 0.6 ml was added into the solutions of 200ml to decrease crystallization speed of TiO_2. The pH of the solutions was 2.4 one hour after mixing the solutions. BET surface area of the particles was estimated to 18 m^2/g. This is much lower than that of the particles prepared at 50°C and slightly lower than that prepared at 90°C for 8 min in our previous work (44 m^2/g)[13]. Formation of TiO_2 was accelerated at high temperature and it decreased surface area. The particles grew in the solutions to decrease surface area as function of time. Crystallization of TiO_2 was shown to be strongly affected by growth conditions such as solution temperature and growth time.

In summary, anatase TiO_2 particles, 100–200 nm in diameter, were successfully fabricated in aqueous solution. They were assemblies of nanocrystals 5–10 nm that grew anisotropically along the *c*-axis to form acicular shapes. The particles thus had nanorelief surface structures constructed of acicular crystals. They showed *c*-axis orientation due to high-intensity X-ray diffraction from the (004) crystal planes. The particles had a high BET surface area of 270 m^2/g. Total pore volume and average pore diameter were estimated from pores smaller than 230 nm at P/Po = 0.99–0.43 cc/g and 6.4 nm, respectively. They were also estimated from pores smaller than 11 nm at P/Po = 0.80–0.21 cc/g and 3.1 nm, respectively. BJH and DFT/Monte-Carlo analysis of adsorption isotherm indicated the existence of pores ~2.8 and ~3.6 nm, respectively. Additionally, the analyses suggested the existence of micropores of ~1 nm. Crystallization and self-assembly of nano TiO_2 were effectively utilized to fabricate nanocrystal assembled TiO_2 particles having high surface area and nanorelief surface structure.

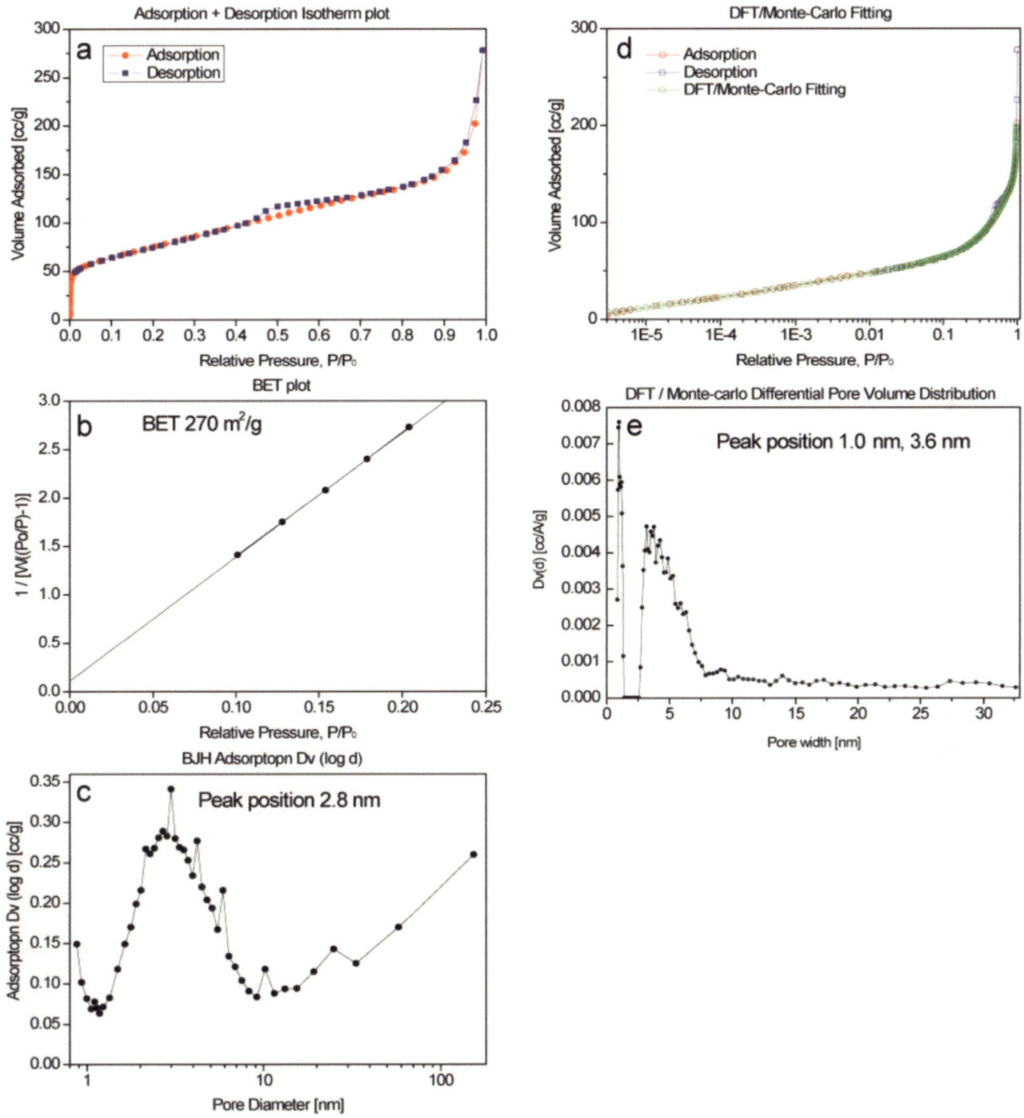

Fig. 5. (a): N_2 adsorption-desorption isotherm of anatase TiO_2 particles. (b): BET surface area of anatase TiO_2 particles. (c): Pore size distribution calculated from N_2 adsorption data of anatase TiO_2 particles using BJH equation. (d): N_2 adsorption-desorption isotherm and DFT/Monte-Carlo fitting curve of anatase TiO_2 particles. (e): Pore size distribution calculated from N_2 adsorption data of anatase TiO_2 particles using DFT/Monte-Carlo equation.

2.3. Morphology Control of Acicular $BaTiO_3$ Particles[3]

Acicular $BaTiO_3$ crystals were fabricated using solution processes. Morphology control of them was realized precise control of crystallization in the solutions.

Oxalic acid (252 mg) was dissolved into isopropyl alcohol (4 ml)[3]. Butyl titanate monomer (0.122 ml) was mixed with the oxalic acid solution, and the solution was then mixed with distilled water (100 ml). The pH of the solution was increased to pH = 7 by adding NaOH (1 M) and distilled water, while the volume of the solution was adjusted to 150 ml by these additions. The aqueous solution (50 ml) with barium acetate (39.3 mg) was mixed with the oxalic acid solution. The mixed solution containing barium acetate (0.77 mM), butyl titanate monomer (2 mM) and oxalic acid (10 mM) was kept at room temperature for several hours with no stirring, and the solution gradually became cloudy. Stirring causes the collision of homogeneously nucleated particles and destruction of large grown particles, and so was avoided in this process. The size of the precipitate was easily controlled from nanometer order to micrometer order by changing the growth period. Large particles were grown by immersion for several hours to evaluate the morphology and crystallinity in detail.

Oxalate ions ($C_2O_4^{2-}$) react with barium ions (Ba^{2+}) to form barium oxalate (BaC_2O_4 $0.5H_2O$). BaC_2O_4 $0.5H_2O$ is dissolved in weak acetate acid provided by barium acetate ($(CH_3COO)_2Ba$), however, it can be deposited at pH 7 which is adjusted by adding NaOH. $BaC_2O_4 \cdot 0.5H_2O$ was thus successfully precipitated from the solution.

Acicular particles were homogeneously nucleated and precipitated from the solution (Fig. 6a). They were on average 23 μm (ranging from 19 to 27 μm) in width and 167 μm (ranging from 144 to 189 μm) in length, giving a high aspect ratio of 7.2. They had sharp edges and clear crystal faces, indicating high crystallinity. A gel-like solid was also coprecipitated from the solution as a second phase.

XRD diffraction patterns for the mixture of acicular particles and gel-like solid showed sharp diffraction peaks of crystalline $BaC_2O_4 \cdot 0.5H_2O$ with no additional phase. Acicular particles were crystalline $BaC_2O_4 \cdot 0.5H_2O$ and the gel-like solid would be an amorphous phase.

Fortunately, $BaC_2O_4 \cdot 0.5H_2O$ has a triclinic crystal structure as shown by the model calculated from structure data [14] (Fig. 6b XRD first step) and thus anisotropic crystal growth was allowed to proceed to produce an acicular shape. Each crystal face has a different surface energy and surface nature such as zeta potential and surface groups. Anisotropic crystal growth is induced by minimizing the total surface energy in ideal crystal growth. Additionally, site-selective adsorption of ions or molecules on specific crystal faces suppresses crystal growth perpendicular to the faces and so induces anisotropic crystal growth. These factors would cause anisotropic crystal growth of $BaC_2O_4 \cdot 0.5H_2O$ and hence allow us to control morphology and fabricate acicular $BaC_2O_4 \cdot 0.5H_2O$ particles. The positions of diffraction peaks corresponded with that of JCPDS No. 20-0134 (Fig. 6b XRD third step) and that calculated from crystal structure data [14] (Fig. 6b XRD second step), however, several diffraction peaks, especially 320 and 201, were enhanced strongly compared to their relative intensity. The enhancement of diffraction intensity from specific crystal faces would be related to anisotropic crystal growth; a large crystal size in a specific crystal orientation increases the x-ray diffraction intensity for the crystal face perpendicular to the crystal orientation.

EDX elemental analysis indicated the chemical ratio of the precipitate, which included acicular particles and gel-like solid, to be about Ba / Ti = 1 to 1.5. The chemical ratio indicated that the coprecipitated amorphous gel contained Ti ions. Additional Ba ions can be transformed into $BaCO_3$ by annealing and removed by HCl treatment in the next step. The ratio was thus controlled to slightly above Ba / Ti = 1 by adjusting the volume ratio of acicular particles and gel-like solid. Consequently, acicular particles of crystalline $BaC_2O_4 \cdot 0.5H_2O$ with Ti-containing gel-like solid were successfully fabricated in an aqueous solution process.

In comparison, isotropic particles of barium titanyl oxalate ($BaTiO(C_2O_4)_2 \cdot 4H_2O$) were precipitated at pH 2. $TiOC_2O_4$ was formed by the following reaction in which the reaction of oxalic acid ($H_2C_2O_4 \cdot 2H_2O$) with butyl titanate monomer ($(C_4H_9O)_4Ti$) and hydrolysis can take place simultaneously [15].

$$(C_4H_9O)_4Ti + H_2C_2O_4 \cdot 2H_2O \rightarrow TiOC_2O_4 + 4C_4H_9OH + H_2O \text{..................(a)}$$

$TiO(C_2O_4)$ was then converted to oxalotitanic acid ($H_2TiO(C_2O_4)_2$) by the reaction:

$$TiO(C_2O_4) + H_2C_2O_4 \cdot 2H_2O \rightarrow H_2TiO(C_2O_4)_2 + 2H_2O \text{..............(b)}$$

Alcoholic solution containing oxalotitanic acid ($H_2TiO(C_2O_4)_2$) formed by reaction (b) was subjected to the following cation exchange reaction by rapidly adding an aqueous solution of barium acetate at room temperature:

$$H_2TiO(C_2O_4)_2 + Ba(CH_3COO)_2 \rightarrow BaTiO(C_2O_4)_2\downarrow + 2CH_3COOH \quad (c)$$

$BaTiO(C_2O_4)_2$ isotropic particles were formed by reaction (c).

On the other hand, neither $BaC_2O_4 \cdot 0.5H_2O$ nor $BaTiO(C_2O_4)_2$ was precipitated at pH 3 to pH 6. Gel-like solid was formed in the solution and their XRD spectra showed no diffraction peaks. The amorphous gel that precipitated at pH = 3 to 6 would be the same as the amorphous gel coprecipitated at pH 7.

These comparisons show that the crystal growth and morphology control of $BaC_2O_4 \cdot 0.5H_2O$ are sensitive to the solution conditions.

The precipitate was annealed at 750 °C for 5 h in air. Acicular $BaC_2O_4 \cdot 0.5H_2O$ particles were reacted with Ti-containing amorphous gel to introduce Ti ions to transform into crystalline $BaTiO_3$. X-ray diffraction of the annealed precipitate showed crystalline $BaTiO_3$ and an additional barium carbonate phase ($BaCO_3$). Excess precipitation of $BaC_2O_4 \cdot 0.5H_2O$ caused the generation of barium carbonate phase ($BaCO_3$) as expected.

The annealed precipitate was further immersed in HCl solution (1 M) to dissolve barium carbonate ($BaCO_3$). Acicular particles of crystalline $BaTiO_3$ were successfully fabricated with no additional phase. Particles showed acicular shape with 2.8×10×50 μm and x-ray diffraction of single-phase crystalline $BaTiO_3$ (Fig. 6c). The high aspect ratio of the particles (17.8 = 50 / 2.8) would be provided by that of $BaC_2O_4 \cdot 0.5H_2O$ particles. The particle size of acicular $BaTiO_3$ can be easily controlled by the growth period and solution concentration for $BaC_2O_4 \cdot 0.5H_2O$ precipitation which decides the particle size of $BaC_2O_4 \cdot 0.5H_2O$.

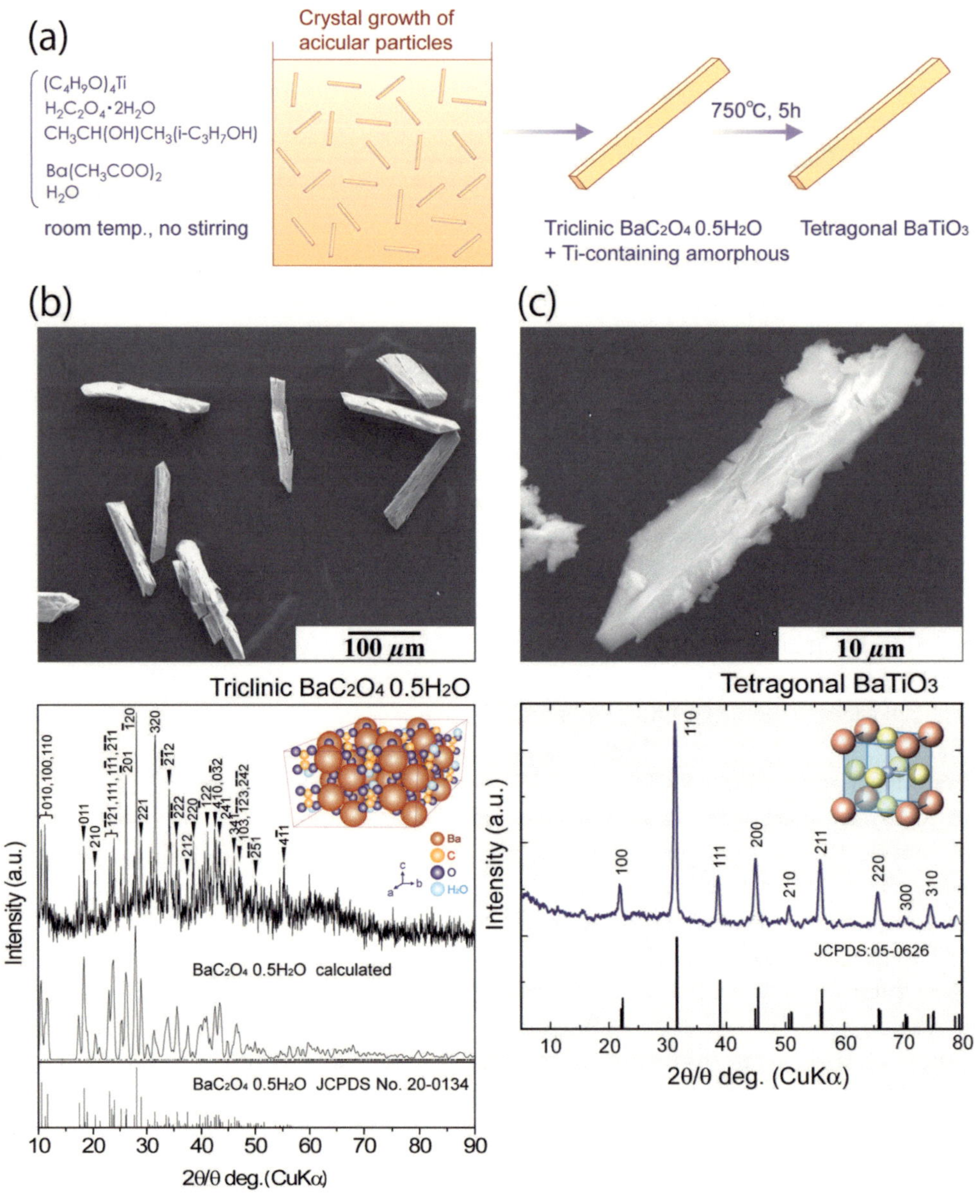

Fig. 6. (a) Conceptual process for fabricating acicular $BaTiO_3$ particles. Morphology control of $BaC_2O_4 \cdot 0.5H_2O$ particles and phase transition to $BaTiO_3$. (b) SEM micrograph and XRD diffraction pattern of acicular $BaC_2O_4 \cdot 0.5H_2O$ particles precipitated from an aqueous solution at pH = 7. XRD diffraction measurement data (first step), XRD pattern calculated from crystal structure data[16] (second step) and XRD pattern of JCPDS No. 20-134 (third step) are shown for triclinic $BaC_2O_4 \cdot 0.5H_2O$. (c) SEM micrograph and XRD diffraction pattern of acicular $BaTiO_3$ particles after annealing at 750 °C for 5 h and HCl treatment. XRD diffraction measurement data (first step) and XRD pattern of JCPDS No. 05-0626 (second step) are shown for tetragonal $BaTiO_3$.

$BaTiO_3$ has a cubic crystal structure at high temperature above phase transition and has a tetragonal crystal structure at room temperature. The cubic crystal structure is completely isotropic and the tetragonal crystal structure results from stretching a cubic lattice along one of its lattice vectors. For both of the crystal structures it is difficult to control anisotropic crystal growth, however, with our newly developed process we could successfully control the morphology and fabricate acicular particles. This was achieved by controlling the morphology of triclinic $BaC_2O_4 \cdot 0.5H_2O$ to acicular shape and the phase transition to $BaTiO_3$ by introducing Ti ions from the coprecipitated amorphous phase. The novel concept can be applied to a wide variety of morphology control and crystal growth control for advanced electronic devices composed of crystalline materials.

In summary, a novel process to fabricate acicular $BaTiO_3$ particles was developed. Morphology control of crystalline $BaC_2O_4 \cdot 0.5H_2O$ to acicular shape was realized in an aqueous solution. The particles were then transformed into crystalline $BaTiO_3$ by introducing Ti ions from the coprecipitated amorphous gel phase during the annealing process. Consequently, acicular particles of tetragonal $BaTiO_3$ were produced by combining several key technologies. Morphology control in this system has high scientific value for crystal growth, and acicular particles of crystalline $BaTiO_3$ may have a great impact on ultra-thin MLCC in future.

3. Liquid Phase Site-selective Deposition of Metal Oxide Nanocrystals

3.1. Site-selective Deposition of Anatase TiO_2[4]

Site-selective deposition of anatase TiO_2 nanocrystals was achieved in aqueous solutions. Nucleation and crystal growth of TiO_2 were accelerated on super hydrophilic surfaces. It allowed us to form micro-patterns of TiO_2 nanocrystals.

Transparent conductive substrate of F doped SnO_2 (FTO, SnO_2: F, Asahi Glass Co., Ltd., 9.3-9.7 Ω/□, 26 × 50 × 1.1 mm) was blown by air to remove dust and was exposed to ultraviolet light (low-pressure mercury lamp PL16-110, air flow, 100 V, 200 W, SEN Lights Co.) for 10 min through a photomask (Test-chart-No.1-N type, quartz substrate, 1.524 mm thickness, Toppan Printing Co., Ltd.) (Fig. 7)[4]. The initial SnO_2: F substrate showed a water contact angle of 96°. The UV-irradiated surface was, however, wetted completely (contact angle 0–1°). The contact angle decreased with irradiation time (96°, 70°, 54°, 35°, 14°, 5° and 0° for 0 min, 0.5 min, 1 min, 2 min, 3 min, 4 min and 5 min, respectively). This suggests that a small amount of adsorbed molecules on the SnO_2: F substrate was removed completely by UV irradiation. The surface of the SnO_2: F substrate would be covered by hydrophilic OH groups after irradiation. Consequently, the SnO_2: F substrate was modified to have a patterned surface with hydrophobic regions and super-hydrophilic regions.

Ammonium hexafluorotitanate ($[NH_4]_2TiF_6$) (2.0620 g) and boric acid (H_3BO_3) (1.8642 g) were separately dissolved in deionized water (100 mL) at 50°C. Boric acid solution was added to ammonium hexafluorotitanate solution at concentrations of 0.15 M and 0.05 M, respectively. The SnO_2: F substrate having a patterned surface with hydrophobic regions and super-hydrophilic regions was covered by a silicon rubber sponge sheet (Silico-sheet, SR-SG-S 5mmt RA grade, Shin-etsu Finetech Co., Ltd.) to suppress deposition of TiO_2 at the initial stage. The substrate was immersed perpendicularly in the middle of the solution

(Fig. 7). The solution was kept at 50°C with no stirring. The silicon rubber sponge sheet was removed from the SnO_2: F substrate after 25 h, then the substrate was kept for a further 2 h at 50°C. The substrate was covered by the sheet instead of immersion of substrate at 25h to avoid agitation of the solution.

The solution became clouded in about 10 min after the mixing of ammonium hexafluorotitanate solution and boric acid solution. The particles were homogeneously nucleated in the solution and made the solution white. They then gradually precipitated and fell to the bottom of the vessel, so the solution became transparent over a period of hours. Ti ions were consumed for crystallization of TiO_2 particles, thus decreasing the ion concentration in the solution. The super-saturation degree of the solution was sufficiently low to realize slow heterogeneous nucleation without homogeneous nucleation which forms TiO_2 particles. The silicon rubber sponge sheet was removed from the SnO_2: F substrate after 25 h, then the substrate was kept for a further 2 h at 50°C. Consequently, the patterned surface on the SnO_2: F substrate was exposed to the transparent solution including Ti ions at a low concentration for 2 h. Heterogeneous nucleation and slow crystallization of TiO_2 progressed only on the substrate.

Deposition of anatase TiO_2 proceeds by the following mechanisms[7]:

$$TiF_6^{2-} + 2H_2O \rightleftarrows TiO_2 + 4H^+ + 6F^- \quad \cdots(a)$$

$$BO_3^{3-} + 4F^- + 6H^+ \longrightarrow BF_4^- + 3H_2O \quad \cdots(b)$$

Equation (a) is described in detail by the following two equations:

$$TiF_6^{2-} \xrightarrow{nOH^-} TiF_{6-n}(OH)_n^{2-} + nF^- \xrightarrow{(6-n)OH^-} Ti(OH)_6^{2-} + 6F^- \quad \cdots(c)$$

$$Ti(OH)_6^{2-} \longrightarrow TiO_2 + 2H_2O + 2OH^- \quad \cdots(d)$$

Fluorinated titanium complex ions gradually change into titanium hydroxide complex ions in an aqueous solution as shown in Eq. (c). The increase of F^- concentration displaces Eqs. (a) and (c) to the left, however, the produced F^- can be scavenged by H_3BO_3 (BO_3^{3-}) as shown in Eq. (b) to displace Eqs. (a) and (c) to the right. Anatase TiO_2 formed from titanium hydroxide complex ions ($Ti(OH)_6^{2-}$) in Eq. (d).

Liquid phase patterning was not realized in the initial solution but realized in the solution after 25 h. Solution condition was evaluated as function of time to clarify this reason. The solution was transparent immediately after the mixing of ammonium hexafluorotitanate solution and boric acid solution, became clouded after 0.5 h and showed maximum whiteness after 1 h. Anatase TiO_2 particles nucleated homogeneously in the solution and grew to form large particles, which gradually precipitated and made the bottom of the vessel white. The solution became slightly white after 5 h and transparent after 25 h. The solutions changed to transparent by the filtrations. Precipitated particles from the residual solution and particles from the supernatant solution trapped by filters were determined by XRD evaluation to be a single phase of anatase TiO_2.

Precipitated particles from the residual solution increased as 0 mg, 165.1 mg, 230.9 mg and 424.4 mg at 0 h, 2 h, 5 h and 25 h, respectively. The precipitation increased rapidly at the initial stage and moderately after 2 h, reflecting the decrease of crystal growth rate.
The weight of particles > 2.5 μm in diameter was estimated to be 2.8 mg, 49.7 mg, 9.5 mg, 0 mg and 0 mg at 0.5 h, 1 h, 2 h, 5 h and 25 h, respectively. Particles were formed at the initial stage and precipitated, making the bottom of the vessel white. This is consistent with the color change of the solution.
The precipitate from the filtrate collected by all of the filters was evaluated by XRD. The white powder contained anatase TiO_2, rutile TiO_2 and a large amount of boric acid. These were crystallized from ions in the filtrate during drying. The weight was estimated to be 1455 mg, 1429 mg, 1392 mg, 1345 mg and 1341 mg at 0.5 h, 1 h, 2 h, 5 h and 25 h, respectively. This indicated that the solution contained a high concentration of ions at the initial stage, which then decreased as a function of time. Ion concentration would decrease by the crystallization and precipitation of anatase TiO_2. This result is consistent with the weight variation of precipitated particles, weight variation of suspended particles and solution color change shown in the photographs.
Liquid phase patterning was not realized in the initial clouded solution but realized in the transparent solution after 25 h. Evaluation of solution condition as function of time showed the reason of this phenomenon. TiO_2 particles formed at the initial stage around 1 h and precipitated gradually. Ions were consumed for crystallization of TiO_2 and decreased as a function of time. Heterogeneous nucleation predominantly progressed after 5 h. Consequently, TiO_2 was formed on super hydrophilic regions selectively to realize liquid phase patterning.
FTO substrate was immersed in the solution for 25 h to form a thick film and ultrasonicated in water for 20 min. TiO_2 film was constructed of two layers. Under layer with 200 nm thickness was a polycrystalline film of anatase TiO_2. Upper layer with 300 nm thickness was an assembly of acicular TiO_2 crystals which grew perpendicular to the substrate. The film was shown by electron diffraction pattern to be a single phase of anatase TiO_2. Electron diffraction from the 004 plane was stronger than that of the 101, 200, 211 planes, etc to show anisotropic crystal growth along the *c*-axis. Additionally, 004 diffractions were strong perpendicular to the substrate, showing that the *c*-axis orientation of acicular crystals was perpendicular to the substrate. The FTO layer was shown to be a single phase of SnO_2 with high crystallinity. Acicular TiO_2 crystals had a long shape, being ~ 300 nm in length and 10 - 100 nm in diameter. A lattice image of anatase TiO_2 was observed from the crystals.
The film deposited on the substrate was evaluated by XRD analysis. Strong X-ray diffractions were observed for films deposited on FTO substrates and assigned to SnO_2 of FTO films. The 004 diffraction peak of anatase TiO_2 was not observed clearly for TiO_2 film on FTO substrates because both of the weak 004 diffraction peak of TiO_2 and the strong diffraction peak of FTO were observed at the same angle. Glass substrates with no FTO coating were immersed in the solution. Weak X-ray diffraction peaks were observed at 2θ = 25.3, 37.7, 48.0, 53.9, 55.1 and 62.7° for the films deposited on glass substrates. They were assigned to 101, 004, 200, 105, 211 and 204 diffraction peaks of anatase TiO_2 (ICSD No. 9852) (Fig. 7). A broad diffraction peak from the glass substrate was also observed at about 2θ = 25°.

The intensity of the 004 diffraction peak was stronger than that of the 101 diffraction peak for the film obtained by the liquid phase crystal deposition method, though the intensity of 101 was stronger than that of 004 for anatase TiO_2 powders with no orientation (ICSD No. 9852). The integral intensity or peak height of 004 was 2.6 times or 2.2 times that of 101, respectively, suggesting high *c*-axis orientation of anatase TiO_2 crystals. Crystallite size perpendicular to the 101 or 004 planes was estimated from the full-width half-maximum of the 101 or 004 peak to be 9 nm or 17 nm, respectively. Elongation of crystals in the *c*-axis direction was also suggested by the difference of crystallite size. These evaluations were consistent with high *c*-axis orientation observed by TEM and electron diffraction. Crystallite size estimated by XRD was similar to that in TiO_2 under layer rather than that of acicular crystals observed by TEM. TiO_2 thin film prepared on a glass would be constructed of not acicular crystals but polycrystals in under layer.

After having been immersed in the solution, the substrate was rinsed with distilled water and dried in air (Fig. 7). The initial FTO surface appeared to be blue-green under white light due to light diffracted from the FTO layer. On the other hand, TiO_2 films deposited on the super-hydrophilic surface appeared to be yellow-green. The color change would be caused by deposition of transparent TiO_2 film which influenced the wavelength of the diffracted light.

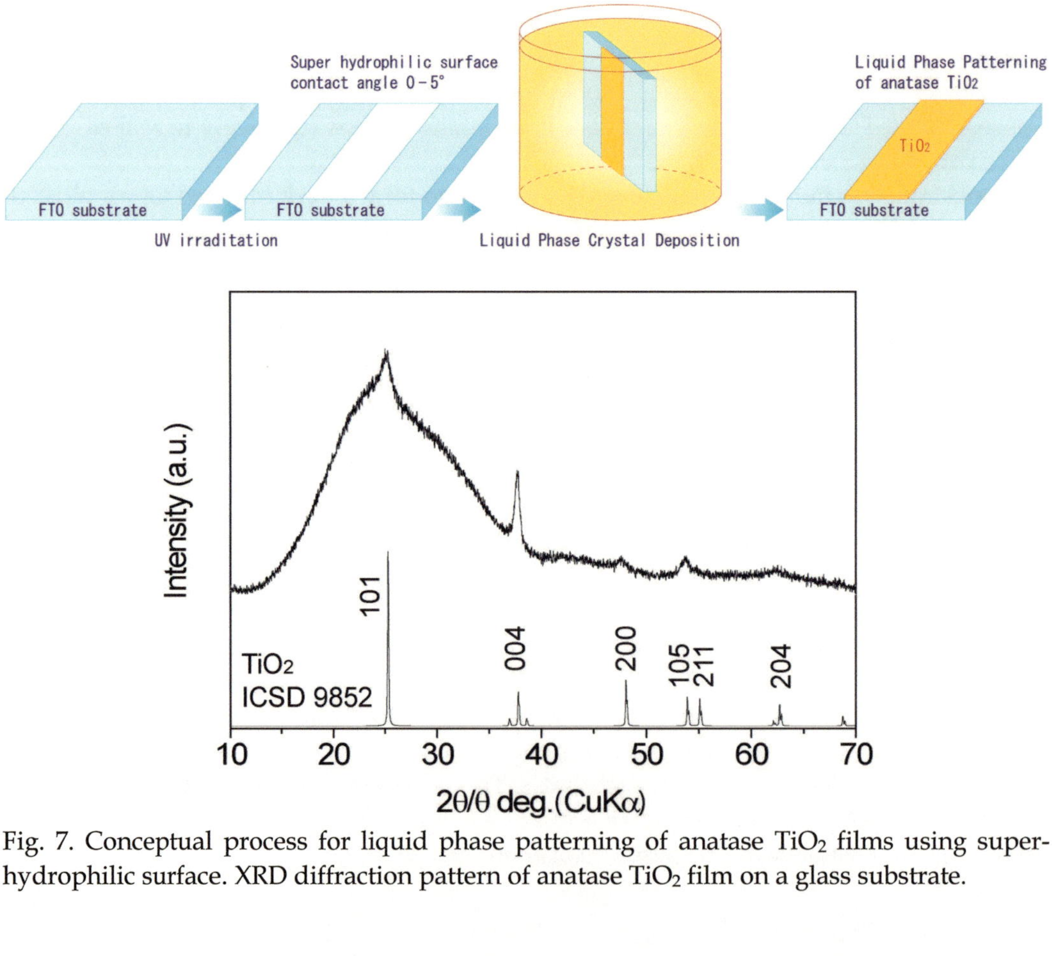

Fig. 7. Conceptual process for liquid phase patterning of anatase TiO_2 films using super-hydrophilic surface. XRD diffraction pattern of anatase TiO_2 film on a glass substrate.

The micropattern of TiO_2 was shown by SEM evaluation to be successfully fabricated (Fig. 8(Top)). TiO_2 deposited on super-hydrophilic regions showed black contrast, while the initial FTO regions without deposition showed white contrast in Fig. 8(top). The average line width in Fig. 8(top) is 55 μm. Line edge roughness[16], as measured by the standard deviation of the line width, is ~2.8 μm. This represents a ~5% variation (i.e., 2.8/55) in the nominal line width, similar to the usual 5% variation afforded by current electronics design rules. The minimum line width of the pattern depends on the resolution of the photomask and wavelength of irradiated light (184.9 nm). It would be improved to ~ 1 μm by using a high-resolution photomask.

The FTO layer was a particulate film having a rough surface (Fig. 8-b1, b2). Edged particles of 100 - 500 nm in diameter were observed on the surface. The micropattern of TiO_2 thin film was covered by an assembly of nano crystals of 10 - 30 nm in diameter (Fig. 8-a1, a2). The nano crystals would be anatase TiO_2 which grew anisotropically. The TiO_2 film also had large structural relief of 100 - 500 nm in diameter. As the thin TiO_2 film was deposited on the edged particulate surface of the FTO layer, the surface of TiO_2 had large structural relief.

The morphology of the TiO_2 layer and FTO layer was further observed by fracture cross section profiles (Fig. 9). The polycrystalline FTO layer prepared on a flat glass substrate was shown to have a thickness of ~ 900 nm, and a high roughness of 100 - 200 nm on the surface (Fig. 9a). Nano TiO_2 crystals were deposited on the super-hydrophilic FTO surface (Fig. 9a), whereas no deposition was observed on the initial FTO surface. The super-hydrophilic FTO surface was covered with an array of nano TiO_2 crystals (Fig. 9b, c), which had a long shape of ~ 150 nm in length and ~ 20 nm in diameter. These observations were consistent with TEM and XRD evaluations. Nano TiO_2 crystals would grow along the *c*-axis and thus enhance the 004 X-ray diffraction peak and 004 electron diffraction peak. They formed a long shape having a high aspect ratio of 7.5 (150 nm in length / 20 nm in diameter) as shown in the SEM fracture cross section profile (Fig. 9b, c) and TEM micrograph. The orientation of nano TiO_2 crystals with their long axis perpendicular to the FTO layer (Fig. 9b, c) would also enhance the 004 diffraction peak.

In summary, a micropattern of anatase TiO_2 thin film was successfully fabricated on an SnO_2: F substrate in an aqueous solution. Crystalline anatase TiO_2 was deposited by liquid phase crystal deposition at 50°C. Nucleation and crystal growth of TiO_2 were accelerated on the super-hydrophilic SnO_2: F surface, but were suppressed on the hydrophobic initial SnO_2: F surface. Consequently, liquid phase patterning of anatase TiO_2 was achieved on an SnO_2: F substrate. TiO_2 crystals were directly deposited on the SnO_2: F surface without any insulating layers which decrease the electrical conductivity between TiO_2 and the SnO_2: F substrate. The micropattern of anatase TiO_2 on the SnO_2: F substrates could be applied to electrodes of dye-sensitized solar cells or molecular sensors. Additionally, this process can be used to form a flexible micropattern of anatase TiO_2 electrodes on low-heat-resistant conductive polymer films. This process will contribute to the microfabrication of TiO_2 electrodes for dye-sensitized solar cells or molecular sensors.

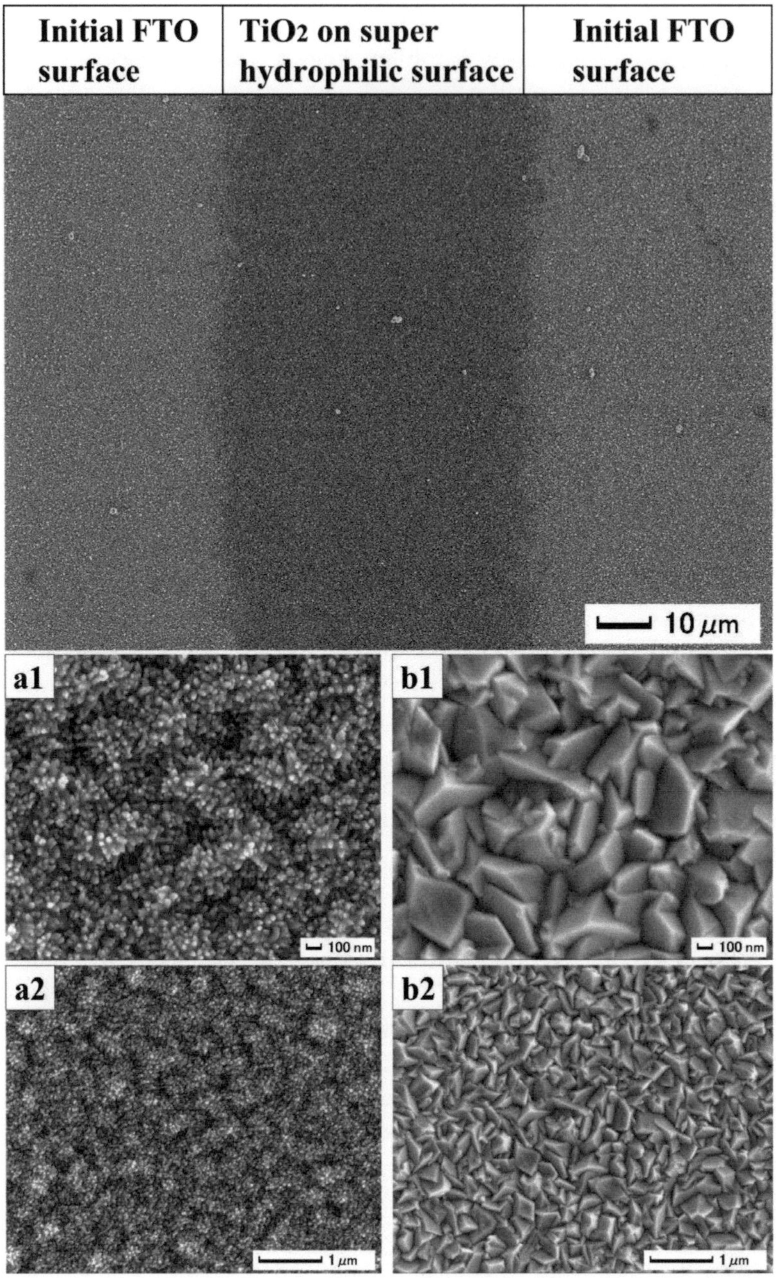

Fig. 8. SEM micrographs of a micropattern of anatase TiO_2 films on SnO_2: F substrates. (Top) Micropattern of anatase TiO_2 films. (a1) Surface of anatase TiO_2 films deposited on super-hydrophilic region. TiO_2 was formed on super-hydrophilic region which was cleaned by UV irradiation before the immersion. (a2) Magnified area of (a1) showing surface morphology of anatase TiO_2 film. (b1) Surface of SnO_2: F substrate without TiO_2 deposition. TiO_2 was not formed on non-cleaned region. (b2) Magnified area of (b1) showing surface morphology of SnO_2: F substrate.

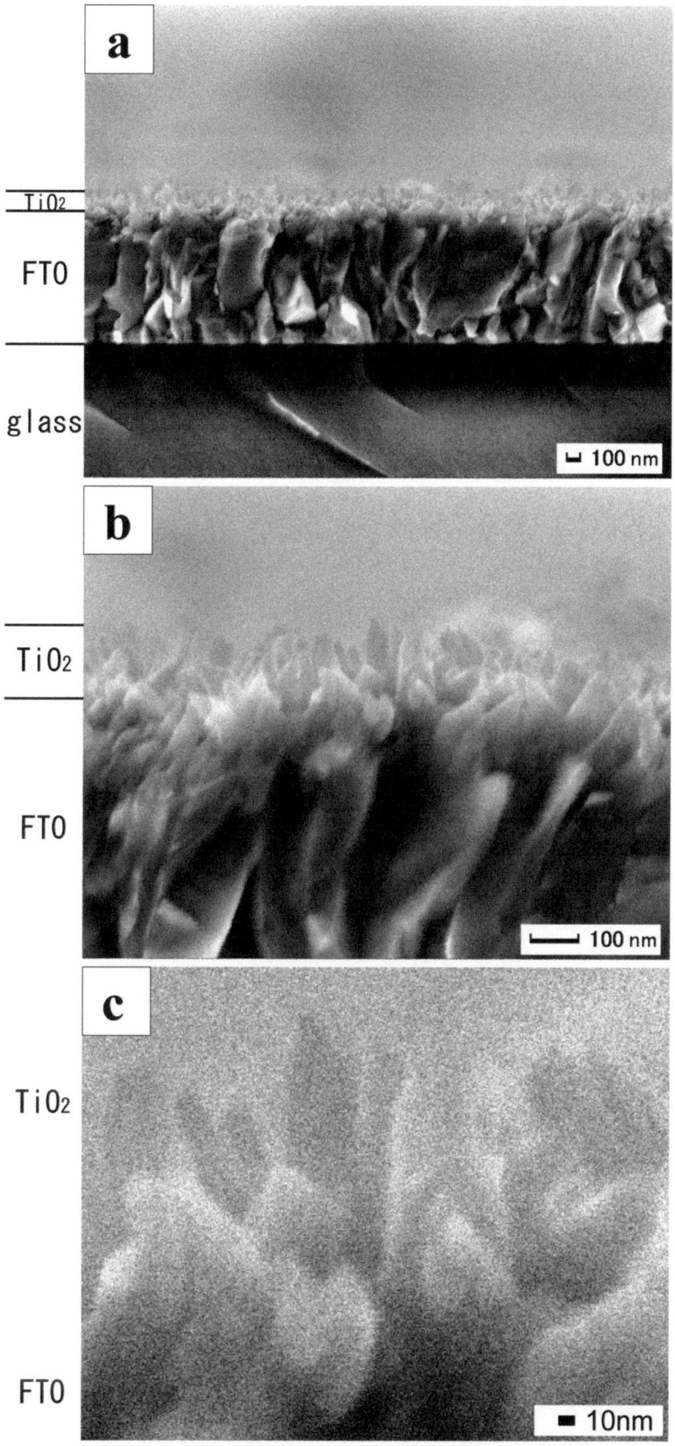

Fig. 9. SEM micrographs of anatase TiO_2 films on SnO_2: F substrates. (a) Fracture cross section of TiO_2 films. (b, c) Magnified area of (a) showing morphology of nano TiO_2 crystals.

3.2. Site-selective Deposition of Eu:Y_2O_3[5]

Micropatterning of Eu doped Y_2O_3 nanocrystals was also realized using site-selective deposition in the solutions. Self-assembled monolayer was utilized as template in this system. Si substrate (p-type [100], 1–50 Ωcm, Newwingo Co., Ltd.) was cleaned ultrasonically in acetone, ethanol and deionized water for 5 min, respectively, in this order and was exposed to ultraviolet light and ozone gas for 10 min to remove organic contamination by using a UV/ozone cleaner (184.9 nm and 253.7 nm) (low-pressure mercury lamp 200 W, PL21-200, SEN Lights Co., 18 mW/cm^2, distance from lamp 30 mm, 24 °C, humidity 73%, air flow 0.52 m^3/min, 100 V, 320 W)[17-20]. APTS (3-Aminopropyltriethoxysilane)-SAM was prepared by immersing the Si substrate in an anhydrous toluene solution containing 1 vol% APTS for 1 h in N_2 atmosphere. The substrate was rinsed with a fresh anhydrous toluene in N_2 atmosphere. The substrate with SAM was baked at 120 °C for 5 min to remove residual solvent and promote chemisorption of the SAM.

APTS-SAM was then irradiated by ultraviolet light (PL21-200) through a photomask (Test-chart-No.1-N type, quartz substrate, 1.524 mm thickness, guaranteed line width 2 μm ± 0.5 μm, Toppan Printing Co., Ltd.) for 10 min. UV irradiation modified an amino-terminated silane to a silanol forming a pattern of amino-terminated silane regions and silanol regions[17-20]. Patterned APTS-SAM having amino regions and silanol regions was used as a template for patterning of yttrium oxide. Initially deposited APTS-SAM showed water contact angles of 48°. The UV-irradiated surface of SAM was, however, wetted completely (contact angle <5°). This suggests that SAM of APTS was modified to hydrophilic OH group surfaces by UV irradiation.

The patterned APTS-SAM was immersed in an aqueous solution containing $Y(NO_3)_3 \cdot 6H_2O$ (4 mM), $Eu(NO_3)_3 \cdot 6H_2O$ (0.4 mM) and NH_2CONH_2 (50 mM) at 25 °C[5]. The solution was heated to 77 °C gradually as shown in Fig. 10 since urea (NH_2CONH_2) decomposes to form ammonium ions (NH_4^+) above 70 °C (Eq. (a)). The decomposition of urea at elevated temperature plays an essential role in the deposition of yttrium oxide. The aqueous solution of urea yields ammonium ions and cyanate ions (OCN^-) at temperatures above 70 °C[21] (Eq. (a)). Cyanate ions react rapidly according to Eq. (b). Yttrium ions are weakly hydrolyzed[22,23] in water to $YOH(H_2O)_n^{2+}$ (Eq. (c)). The resulting release of protons (H^+) and/or hydronium ions (H_3O^+) accelerates urea decomposition (Eq. (b)). The precipitation of the amorphous basic yttrium carbonate ($Y(OH)CO_3 \cdot xH_2O$, x=1) can take place through the reaction in Eq. (d)[24,25]. The controlled release of cyanate ions by urea decomposition causes deposition of basic yttrium carbonate once the critical supersaturation in terms of reacting component is achieved. Since the decomposition of urea is quite slow, the amount needed to reach supersaturation within a given period of time must be considerably higher than the stoichiometric amount of yttrium ions, as revealed by previous studies of lanthanide compounds[26].

The temperature of the solution increased gradually and reached 77 °C in about 80 min[5]. The solution was kept at ~ 77 °C during deposition. The pH of the solution increased from 5.2 to 5.8 in about 90 min and then gradually decreased to 5.6. Temperature and pH increased for the initial 90 min and became stable after 90 min. The average size of particles homogeneously nucleated in the solution at 100 min was about 227 nm and increased to 262 nm at 150 min, 282 nm at 180 min, 310 nm at 210 min, and 323 nm at 240 min (Fig. 10a). Particles nucleated and grew after the solution temperature exceeded 70 °C because urea decomposes above 70 °C to form carbonate ions[21] which causes deposition of basic yttrium carbonate[23-26]. The particles grew rapidly at the beginning of the growth period and then their growth rate decreased exponentially (Fig. 10a). The decrease in growth rate was caused by the decrease of supersaturation degree influenced by a decrease in solution concentration.

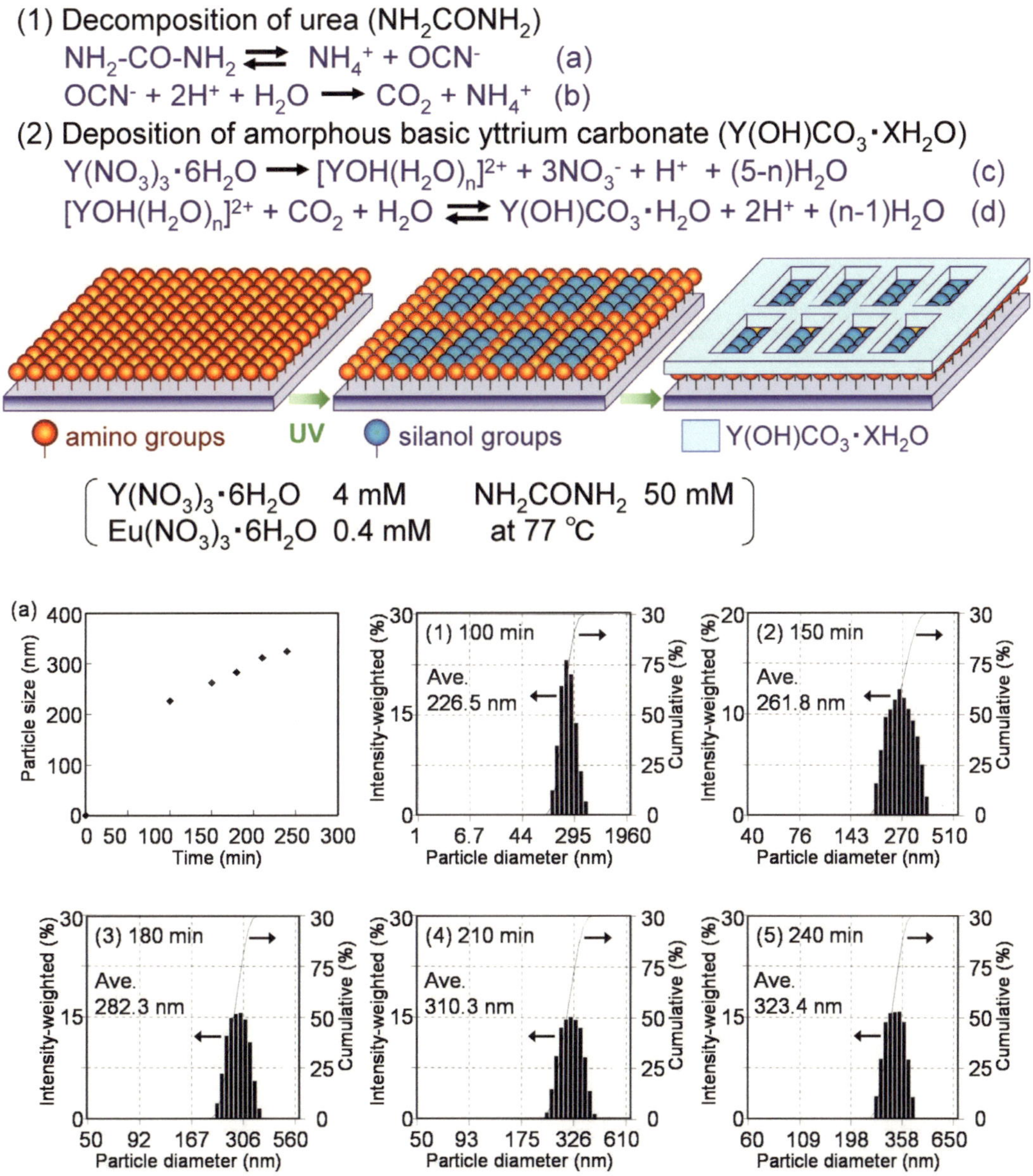

Fig. 10. Conceptual process for site-selective deposition of visible-light emitting Y_2O_3:Eu thin films using a self-assembled monolayer. (a) Time variation of particle size distribution. (1-5) Particle size distribution of yttrium carbonate particles at (1) 100 min, (2) 150 min, (3) 180 min, (4) 210 min or (5) 240 min.

Yttrium carbonate films were observed to deposit on amino regions of a patterned SAM after the immersion in an aqueous solution (Fig. 11A, B). Deposits showed white contrast, while silanol regions without deposition showed black contrast in SEM observation. Narrow lines of depositions having 10–50 μm width were successfully fabricated in an aqueous solution. Patterned APTS-SAM showed high ability for site-selective deposition of yttrium carbonate in solution systems.

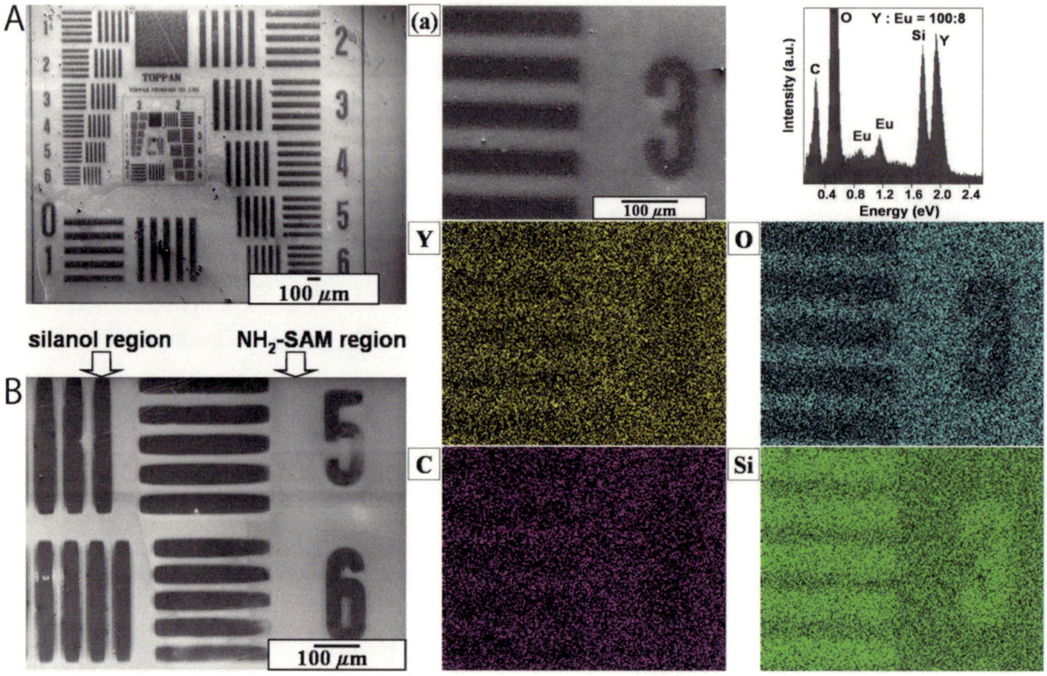

Fig. 11. (A) SEM micrograph of patterned Y_2O_3:Eu thin films and (B) magnified area of (A). (a) SEM micrograph of patterned Y_2O_3:Eu thin films. Characteristic X-ray images of Y, O, C and Si for (a) Y_2O_3:Eu thin films. (Top right) Elemental analysis of Y_2O_3:Eu thin films deposited for 90 min.

Yttrium carbonate films were also deposited on the hydrophobic octadecyl surface of OTS(octadecyltrichlorosilane)-SAM having water contact angle (WCA) of 116 ° and as-purchased silicon wafer having WCA of about 20-50 ° which was kept in a plastic case in air. On the other hand, the films were not deposited on UV irradiated silicon wafer having WCA < 5 °. The super hydrophilic surface of WCA < 5 ° suppressed film deposition, whereas the hydrophobic surface and medium surface of WCA > 20-30 ° accelerated film deposition possibly because of hydrophobic interaction between deposition and substrate surface. This is consistent with a former study[32]. Yttrium carbonate was deposited both on bare single crystal Si wafers, and on Si wafers coated with sulfonate-functionalized organic self-assembled monolayers.

Yttrium, europium, oxygen and carbon were observed from as-deposited thin films on amino regions, while silicon and oxygen were detected from non-covered silanol regions by

EDX (Fig. 11). The molecular ratio of yttrium to europium was determined to be 100 : 8 (Fig. 11). It was close to that of $Y(NO_3)_3 \cdot 6H_2O$ to $Eu(NO_3)_3 \cdot 6H_2O$, i.e., 100 : 10, in the solution because the chemistry of $Eu(NO_3)_3$ is similar to that of $Y(NO_3)_3$ to incorporate europium in the precipitation. The content of europium was in the range we had expected. Y_2O_3:Eu with atomic ratio Y : Eu = 100 : ~ 8 was reported to have strong photoluminescence[27,28]. Carbon was detected from yttrium carbonate. Silicon and oxygen were detected from silicon wafer covered with a natural oxide layer (amorphous SiO_2).
Amino regions were covered with thin films composed of many large particles (about 100–300 nm in diameter) and very high roughness (RMS 25.6 nm) (Fig. 12A). Silanol regions, on the other hand, showed only nano-sized small particles (about 10–50 nm in diameter) and very low roughness (RMS 1.7 nm). The high site-selectivity of deposition and the big difference in surface morphology and roughness were clearly shown by AFM observation. The thickness of the films was estimated from AFM scans across deposited and undeposited regions of the substrate. It increased with immersion time after 45 min (0 nm at 45 min, 60 nm at 70 min and 100 nm at 90 min (Fig. 12A)). The average growth rate (70 nm/h = 100 / 90 min) was higher than that previously reported (2 nm/h = 35 nm / 15 h)[25]. An amorphous yttrium basic carbonate film was deposited at 80 °C from aqueous solutions of $YNO_3 \bullet 5H_2O$ and urea on Si wafers coated with sulfonate-functionalized organic self-assembled monolayers in previous studies. The thickness was then evaluated by TEM after the treatment with ultrasonication for half an hour in distilled water. The difference of growth rate was caused mainly by the difference of the substrate treatment by ultrasonication. Additionally, the thickness of our film was smaller than the particle size in the solution shown in Fig. 10 (227 nm at 100 min). Heterogeneous nucleation and attachment of initial particles of yttrium carbonate occurred without the attachment of aggregated large particles shown in Fig. 10. The yttrium carbonate was then grown on the substrate to form a film of 100 nm thickness after immersion for 90 min. The particles of about 100 nm in height were removed by ultrasonication for 30 min and the film of several nm in height remained as reported[25].
Yttrium was not detected by XPS from the substrate immersed for 45 min, however, it was clearly observed from that immersed for 90 min. This indicates that the deposition began between 45 and 90 min after immersion. The solution temperature reached 70 °C in ~ 45 min and then the solution began to decompose and release carbonate ions, causing the deposition of basic yttrium carbonate. The deposition mechanism evaluated by XPS is consistent with the change of solution temperature, decomposition temperature of urea and chemical reaction of this system. The binding energy of Y 3d5/2 spectrum from the deposition (158.2 eV) was higher than that of metal yttrium (155.8 eV)[29]. The spectrum shifted to lower binding energy (156.7 eV) after annealing at 800 °C in air for 1 h and is similar to that of Y_2O_3 (157.0 eV)[30]. The binding energies of Y 3d5/2 spectra in as-deposited films and annealed films were higher than that of metal yttrium possibly due to the chemical bonds formed between yttrium ions and oxygen ions. The chemical shift of Y 3d5/2 binding energy by annealing is consistent with crystallization of as-deposited films to crystalline Y_2O_3. C 1s spectra were detected at 289.7 eV and 284.6 eV from as-deposited films. The C 1s spectrum at 289.7 eV then disappeared by the annealing. C 1s at 284.6 eV was assigned to surface contamination and C 1s at 289.7 eV was detected from as-deposited

yttrium carbonate. The disappearance of C 1s at 289.7 eV is consistent with the phase transition from yttrium carbonate to Y_2O_3.

As-deposited film was shown to be an amorphous phase (Fig. 12B-a) by XRD measurement. The film showed no diffraction peak after annealing at 400 °C for 1 h, however, it showed 222, 400 and 440 diffraction peaks of crystalline cubic Y_2O_3[31] without any additional phase after annealing at 600 °C for 1 h and the intensities of diffraction peaks increased further by annealing at 800 °C for 1 h (Fig. 12B-b). The film was shown to be a polycrystalline Y_2O_3 film constructed from randomly deposited Y_2O_3 particles without crystal-axis orientation. The crystal structure model and diffraction pattern of Y_2O_3 were calculated from the crystal structure data of ICSD #23811 as shown in Fig. 12B. The crystallization by annealing confirmed from XRD measurement is consistent with XPS evaluation.

We attempted to remove Y_2O_3 films from the silicon substrate by debonding with scotch tape or by ultrasonication for 5 min in water. However, the films maintained their bonds with the substrate, indicating that strong adhesion had formed between films and substrate.

The thin film annealed at 800 °C for 1 h, i.e., crystalline Y_2O_3:Eu thin film, was shown to be excited by 230–250 nm (center: 243 nm) and emit red light photoluminescence centered at 611 nm in the fluorescence excitation spectrum (Fig. 12C-a). Neither the as-deposited film nor the film annealed at 400 °C for 1 h showed photoluminescence, on the other hand, the films annealed at 600 °C or 800 °C for 1 h emitted light centered at 617 nm by 250 nm in fluorescence emission spectra (Fig. 12C-b). The fluorescence intensity of the film annealed at 800 °C was stronger than that of the film annealed at 600 °C. Fluorescence intensity increased by the phase transformation from amorphous yttrium carbonate to yttrium oxide and crystal growth by the heat treatments, and is consistent with the crystallization observed by XRD[32]. The spectra are described by the well-known $^5D_0$–^7F_J line emissions (J = 0, 1, 2, …) of the Eu^{3+} ion with the strongest emission for J = 2 at 612 nm. The thin film annealed at 800 °C produced visible red light photoluminescence by excitation from Nd: YAG laser (266 nm) (Fig. 12C, inset). The white square shows the edges of the Y_2O_3:Eu thin film and the red color shows visible red emission from the irradiated area on the substrate.

In summary, we have proposed a novel process for fabricating visible red light emitting Eu-doped Y_2O_3 and its micropattern using a self-assembled monolayer and an aqueous solution system. The patterned APTS-SAM having amino groups regions and silanol groups regions achieved site-selective deposition of yttrium oxide in an aqueous solution. The deposited films were crystallized by annealing at 600 °C or 800 °C for 1 h. Crystalline Y_2O_3:Eu produced visible red light photoluminescence centered at 611 nm by excitation from Nd: YAG laser (266 nm). This study showed the high potential of aqueous solution systems and self-assembled monolayers for the fabrication of functional metal oxide thin films and their micropatterns.

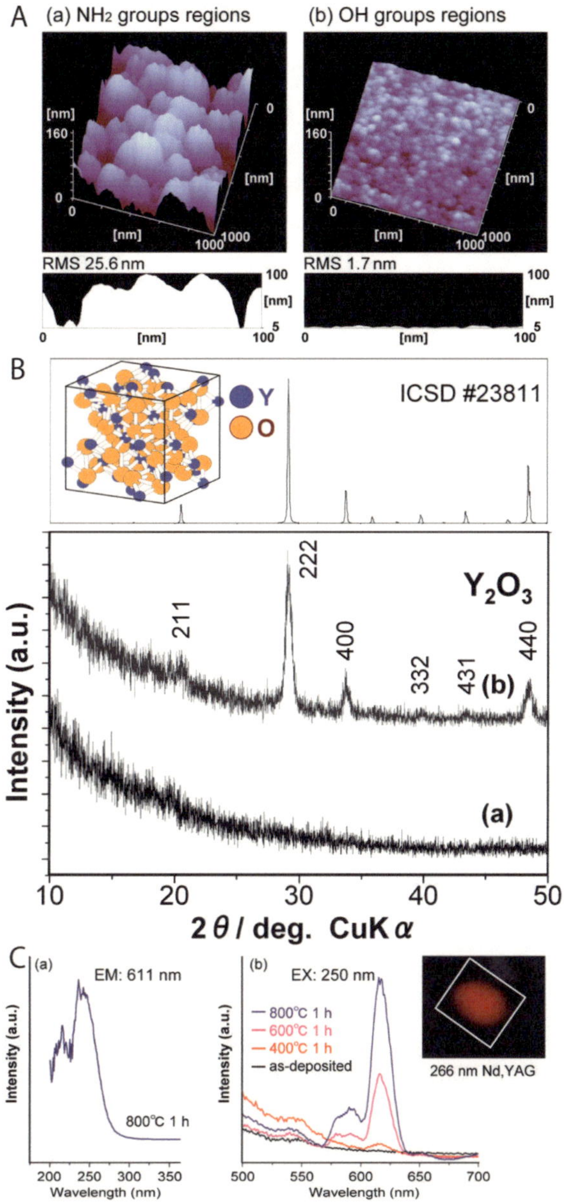

Fig. 12. (A) (a) AFM images and cross-section profile of Y_2O_3:Eu thin films on NH_2 groups regions. (b) AFM images and cross-section profile of Y_2O_3:Eu thin films on OH groups regions. (B) XRD patterns of Y_2O_3:Eu thin films (a) before and (b) after annealing at 800 °C for 1 h. (The upper picture) Crystal structure model and diffraction pattern of cubic Y_2O_3 calculated from crystal structure data of ICSD #23811. (C) (a) Fluorescence excitation spectrum (emission: 611 nm) for Y_2O_3:Eu thin film after annealing at 800 °C for 1 h. (b) Fluorescence emission spectra (excitation: 250 nm) for Y_2O_3:Eu thin films before and after annealing at 400, 600 or 800 °C for 1 h. Inset: Photoluminescence image for Y_2O_3:Eu thin film annealed at 800 °C for 1 h (excitation: 266 nm).

4. Summary

Morphology control, self-assembly and site-selective deposition of metal oxide nanocrystals were reported in this chapter. The novel nano/micro structures of nanocrystals will create the next generation of metal oxide devices. Additionally, they were realized in solutions at ordinary temperature and atmospheric pressure. These environmental-friendly processes will contribute to green innovations.

5. Acknowledgement

The author thanks Prof. Kunihito Koumoto of Nagoya University, Dr. Tatsuki Ohji and Dr. Kazumi Kato of AIST, Dr. Masako Ajimi, Dr. Makoto Bekki and Dr. Shuji Sonezaki of TOTO Ltd. Research Laboratory. Some of works that were reviewed in this chapter were partially supported by METI, Japan, as part of R&D for High Sensitivity Environment Sensor Components.

6. References

(1) Masuda, Y.; Kato, K. *Cryst. Growth Des.* **2008**, *8*, 275-279.
(2) Masuda, Y.; Kato, K. *Cryst. Growth Des.* **2008**, *8*, 3213-3218.
(3) Masuda, Y.; Yamada, T.; Koumoto, K. *Cryst. Growth Des.* **2008**, *8*, 169-171.
(4) Masuda, Y.; Kato, K. *Chem. Mater.* **2008**, *20*, 1057-1063.
(5) Masuda, Y.; Yamagishi, M.; Koumoto, K. *Chem. Mater.* **2007**, *19*, 1002-1008.
(6) Gao, M. D.; Li, M. M.; Yu, W. D. *J. Phys. Chem. B* **2005**, *109*, 1155-1161.
(7) Masuda, Y.; Sugiyama, T.; Seo, W. S.; Koumoto, K. *Chem. Mater.* **2003**, *15*, 2469-2476.
(8) Furlong, D. N.; Parfitt, G. D. *J. Colloid Interface Sci.* **1978**, *65*, 548-554.
(9) Wahi, R. K.; Liu, Y. P.; Falkner, J. C.; Colvin, V. L. *J. Colloid Interface Sci.* **2006**, *302*, 530-536.
(10) Kruk, M.; Jaroniec, M. *Chem. Mater.* **2001**, *13*, 3169-3183.
(11) Ravikovitch, P. I.; Odomhnaill, S. C.; Neimark, A. V.; Schuth, F.; Unger, K. K. *Langmuir* **1995**, *11*, 4765-4772.
(12) Lastoskie, C.; Gubbins, K. E.; Quirke, N. *J. Phys. Chem.* **1993**, *97*, 4786-4796.
(13) Katagiri, K.; Ohno, K.; Masuda, Y.; Koumoto, K. *J. Ceram. Soc. Japan* **2007**, *115*, 831-834.
(14) Mutin, J. C.; Dusausoy, Y.; Protas, J. *J. Solid State Chem.* **1981**, *36*, 356-364.
(15) Potdar, H. S.; Deshpande, S. B.; Date, S. K. *J. Am. Ceram. Soc.* **1996**, *79*, 2795-2797.
(16) Masuda, Y.; Sugiyama, T.; Lin, H.; Seo, W. S.; Koumoto, K. *Thin Solid Films* **2001**, *382*, 153-157.
(17) Dressick, W. J.; Calvert, J. M. *Jpn. J. Appl. Phys.* **1993**, *32*, 5829-5839.
(18) Collins, R. J.; Shin, H.; De Guire, M. R.; Heuer, A. H.; Sukenik, C. N. *Appl. Phys. Lett.* **1996**, *69*, 860-862.
(19) Masuda, Y.; Koumura, T.; Okawa, T.; Koumoto, K. *J. Colloid Interface Sci.* **2003**, *263*, 190-195.
(20) Masuda, Y.; Itoh, M.; Yonezawa, T.; Koumoto, K. *Langmuir* **2002**, *18*, 4155-4159.
(21) Shaw, W. H. R.; Bordeaux, J. J. *J. Am. Chem. Soc.* **1955**, *77*, 4729-4733.
(22) Ryabchikov, D. E.; Ryabukhin, V. A. *Analytical Chemistry of Yttrium and the Lanthanide Elements*; Humphrey Science: Ann Arbor, MI, 1970.

(23) Baes, C. F.; Mesmer, R. E. *The Hydrolysis of Captions*; Wiley: New York, 1976.
(24) Aiken, B.; Hsu, W. P.; Matijevic, E. *J. Am. Ceram. Soc.* **1988**, *71*, 845-853.
(25) Agarwal, M.; DeGuire, M. R.; Heuer, A. H. *Appl. Phys Lett.* **1997**, *71*, 891-893.
(26) Matijevic, E.; Hsu, W. P. *J. Colloid Interface Sci.* **1987**, *118*, 506-523.
(27) Sharma, P. K.; Jilavi, M. H.; Nass, R.; Schmidt, H. *J. Lumin.* **1999**, *82*, 187-193.
(28) Kwaka, M. G.; Parkb, J. H.; Shon, S. H. *Solid State Commun.* **2004**, *130*, 199-201.
(29) Fuggle, J. C.; Martensson, N. *J. Electron Spectrosc. Relat. Phenom.* **1980**, *21*, 275.
(30) Wagner, C. D. *Practical Surface Analysis*; 2 ed.; John Wiley, 1990; Vol. 1.
(31) Paton, M. G.; Maslen, E. N. *Acta Crystallographica* **1967**, *1*, 1948-23.

Figure Captions

Figure 1. SEM micrographs of high *c*-axis oriented stand-alone ZnO self-assembled film. (a1) Air-side surface of ZnO film. (a2) Magnified area of (a1). (b1) Liquid-side surface of ZnO film. (b2) Magnified area of (b2). (c1) Fracture cross section of ZnO film from air side. (c2) Magnified area of (c1). [Reprinted with permission from Ref.[1], Y. Masuda and K. Kato, Cryst. Growth Des. 8, 1, 275, 2008. Copyright @American Chemical Society (2008)]
Figure 2. XRD diffraction pattern of high *c*-axis oriented stand-alone ZnO self-assembled film. (a) TEM micrograph of ZnO nano-sheets. (b) Magnified area of (a). (Insertion) Electron diffraction pattern of ZnO. [Reprinted with permission from Ref.[1], Y. Masuda and K. Kato, Cryst. Growth Des. 8, 1, 275, 2008. Copyright @American Chemical Society (2008)]
Figure 3. SEM micrographs of high *c*-axis oriented stand-alone ZnO self-assembled film annealed at 500°C for 1 h in air. (a1) Fracture edge-on profile of ZnO film from air side. (a2) Cross-section profile of ZnO film from air side. (b1) Fracture edge-on profile of ZnO film from liquid side. (b2) Cross-section profile of ZnO film from liquid side. [Reprinted with permission from Ref.[1], Y. Masuda and K. Kato, Cryst. Growth Des. 8, 1, 275, 2008. Copyright @American Chemical Society (2008)]

Figure 4. XRD diffraction pattern of anatase TiO_2 particles. (a): TEM micrograph of anatase TiO_2 particles. (b): Magnified area of (a) showing morphology of acicular crystals. Insertion in (b): FFT image of (b) anatase TiO_2. (c): Magnified area of (a) showing lattice images of anatase TiO_2. [Reprinted with permission from Ref.[2], Y. Masuda and K. Kato, Crystal Growth & Design, 8, 9, 3213, 2008. Copyright @American Chemical Society (2008)]
Figure 5. (a): N_2 adsorption-desorption isotherm of anatase TiO_2 particles. (b): BET surface area of anatase TiO_2 particles. (c): Pore size distribution calculated from N_2 adsorption data of anatase TiO_2 particles using BJH equation. (d): N_2 adsorption-desorption isotherm and DFT/Monte-Carlo fitting curve of anatase TiO_2 particles. (e): Pore size distribution calculated from N_2 adsorption data of anatase TiO_2 particles using DFT/Monte-Carlo equation. [Reprinted with permission from Ref.[2], Y. Masuda and K. Kato, Crystal Growth & Design, 8, 9, 3213, 2008. Copyright @American Chemical Society (2008)]

Figure 6. (a) Conceptual process for fabricating acicular $BaTiO_3$ particles. Morphology control of $BaC_2O_4 \bullet 0.5H_2O$ particles and phase transition to $BaTiO_3$. (b) SEM micrograph and XRD diffraction pattern of acicular $BaC_2O_4 \bullet 0.5H_2O$ particles precipitated from an aqueous solution at pH = 7. XRD diffraction measurement data (first step), XRD pattern calculated from crystal structure data16 (second step) and XRD pattern of JCPDS No. 20-

134 (third step) are shown for triclinic BaC_2O_4 • $0.5H_2O$. (c) SEM micrograph and XRD diffraction pattern of acicular $BaTiO_3$ particles after annealing at 750 °C for 5 h and HCl treatment. XRD diffraction measurement data (first step) and XRD pattern of JCPDS No. 05-0626 (second step) are shown for tetragonal $BaTiO_3$. [Reprinted with permission from Ref. [3], Y. Masuda, T. Yamada and K. Koumoto, Cryst. Growth Des. 8, 169, 2008. Copyright @American Chemical Society (2008)]

Figure 7. Conceptual process for liquid phase patterning of anatase TiO_2 films using super-hydrophilic surface. XRD diffraction pattern of anatase TiO_2 film on a glass substrate. [Reprinted with permission from Ref.[4], Y. Masuda and K. Kato, Chem. Mater., 20, 3, 1057, 2008. Copyright @American Chemical Society (2008)]

Figure 8. SEM micrographs of a micropattern of anatase TiO_2 films on SnO_2: F substrates. (Top) Micropattern of anatase TiO_2 films. (a1) Surface of anatase TiO_2 films deposited on super-hydrophilic region. TiO_2 was formed on super-hydrophilic region which was cleaned by UV irradiation before the immersion. (a2) Magnified area of (a1) showing surface morphology of anatase TiO_2 film. (b1) Surface of SnO_2: F substrate without TiO_2 deposition. TiO_2 was not formed on non-cleaned region. (b2) Magnified area of (b1) showing surface morphology of SnO_2: F substrate. [Reprinted with permission from Ref.[4], Y. Masuda and K. Kato, Chem. Mater., 20, 3, 1057, 2008. Copyright @American Chemical Society (2008)]

Figure 9. SEM micrographs of anatase TiO_2 films on SnO_2: F substrates. (a) Fracture cross section of TiO_2 films. (b, c) Magnified area of (a) showing morphology of nano TiO_2 crystals. [Reprinted with permission from Ref.[4], Y. Masuda and K. Kato, Chem. Mater., 20, 3, 1057, 2008. Copyright @American Chemical Society (2008)]

Figure 10. Conceptual process for site-selective deposition of visible-light emitting Y_2O_3:Eu thin films using a self-assembled monolayer. (a) Time variation of particle size distribution. (1-5) Particle size distribution of yttrium carbonate particles at (1) 100 min, (2) 150 min, (3) 180 min, (4) 210 min or (5) 240 min. [Reprinted with permission from Ref.[5], Y. Masuda, M. Yamagishi, K. Koumoto, Chem. Mater., 19, 1002, 2007. Copyright @American Chemical Society (2007)]

Figure 11. (A) SEM micrograph of patterned Y_2O_3:Eu thin films and (B) magnified area of (A). (a) SEM micrograph of patterned Y_2O_3:Eu thin films. Characteristic X-ray images of Y, O, C and Si for (a) Y_2O_3:Eu thin films. (Top right) Elemental analysis of Y_2O_3:Eu thin films deposited for 90 min. [Reprinted with permission from Ref.[5], Y. Masuda, M. Yamagishi, K. Koumoto, Chem. Mater., 19, 1002, 2007. Copyright @American Chemical Society (2007)]

Figure 12. (A) (a) AFM images and cross-section profile of Y_2O_3:Eu thin films on NH_2 groups regions. (b) AFM images and cross-section profile of Y_2O_3:Eu thin films on OH groups regions. (B) XRD patterns of Y_2O_3:Eu thin films (a) before and (b) after annealing at 800 °C for 1 h. (The upper picture) Crystal structure model and diffraction pattern of cubic Y_2O_3 calculated from crystal structure data of ICSD #23811. (C) (a) Fluorescence excitation spectrum (emission: 611 nm) for Y_2O_3:Eu thin film after annealing at 800 °C for 1 h. (b) Fluorescence emission spectra (excitation: 250 nm) for Y_2O_3:Eu thin films before and after annealing at 400, 600 or 800 °C for 1 h. Inset: Photoluminescence image for Y_2O_3:Eu thin film annealed at 800 °C for 1 h (excitation: 266 nm). [Reprinted with permission from Ref.[5], Y. Masuda, M. Yamagishi, K. Koumoto, Chem. Mater., 19, 1002, 2007. Copyright @American Chemical Society (2007)]

2

Growth of undoped and metal doped ZnO nanostructures by solution growth

Rathinam Chandramohan, Thirukonda Anandamoorthy Vijayan
and Jagannathan Thirumalai
Department of Physics,Sree Sevugan Annamalai College,Devakottai - 630 303
India

1. Introduction

ZnO is one of the most studied materials of the II-VI oxide materials that derive continuous attention of the researchers worldwide since forties (Bunn 1935). Because of its current and possible applications in several novel devices, renewed interest has emerged and several reviews (Liu et al, 2005; Tsukazaki et al, 2005), and conference proceedings are published exclusively for ZnO nano crystallites similar systems at Singapore (2005), (2009) and Changchan, China (2006) to explore the feasibility of commercial application for future devices. Yet the ream of novel devices from this wonderful material is yet to be accomplished in full (Wellings, et al, 2008). With a wide band gap of 3.2 eV and a large exciton binding energy of 60 meV at room temperature, ZnO, line GaN, will be important for blue and ultraviolent optical devices. ZnO has several advantages over GaN in this applications range however, the most important being its longer exciton binding energy and the ability to grow single crystal substrates. Other favourable aspects of ZnO include its broad chemistry leading to many opportunities for wet chemical etching, low power threshold for optical pumping, radiation hardness and biocompatibility. Together, these properties of ZnO make it an ideal candidate for a variety of devices ranging from sensors through to ultra-violet laser devices and nanotechnology based devices such as displays. As fervent research into ZnO continues, difficulties such as the fabrication of p-type ZnO that have so far stated that the development of devices had over come (Yang etal, 2008). To give a quantitative report on the state of art of ZnO nanocrystals is quite difficult and an attempt has been made to survey the chemical growth of this system in this study. The chemical solution growth of ZnO nano thin films composed of nano crystallites using a two step double dip chemical deposition method has been discussed In detail in this chapter. The growth and characterization of nano structures of ZnO has been reported by Wang (Wang, 2004).

Mitra et al (1998) has prepared Zinc Oxide thin films using chemical deposition technique. The structural, morphological properties of the prepared films are characterized using X-ray diffraction and scanning electron microscope. They have used Zn salts as precursor and successfully synthesized ZnO films. The growth of highly textured Zinc oxide (ZnO) thin films with a preferred (101) orientation has been prepared by employing chemical bath deposition using a sodium zincate bath on glass substrates has been reported by (Ramamoorthy et al, 2004). The films were characterized by XRD, SEM, EDX, UV-Vis-NIR, FTIR and PL in order to justify the suitability for commercial device quality. (Natsume et al, 2000) have studied the d.c electrical conductivity and optical properties of zinc oxide film prepared by a sol-gel spin coating technique. The temperature dependence of the conductivity indicated that electron transport in the conduction band was due to thermal execution of donor electrons for temperatures from 250 to 300 K. (Chapparro et al, 2003) have proposed the spontaneous growth of ZnO thin films from aqueous solutions. An electroless - chemical process is proposed, consisting in the formation of the super oxide radical (O_2^-) followed by chemical reaction of two O_2^- with Zn $(NH_3)_4^{2+}$ cations. (Wellings et al, 2008) have deposited ZnO thin films from aqueous zinc nitrate solution at 80°C onto fluorine doped tin oxide (FTO) coated glass substrates. Structural analysis, surface morphology, optical studies and electrical conductivity were studied and thickness of the ZnO films was found to be 0.40 μm. (Walter Walter et al, 2007) have studied the characterization of strontium doped ZnO thin films on love wave filter applications. X-ray diffraction, scanning electron microscopy and atomic force microscopy studied the crystalline structure and surface morphology of films. The electrochemical coupling coefficient, dielectric constant, and temperature coefficient of frequency of filters were then determined using a network analyzer. (Vijayan et al, 2008, a, b) have reported the preparation conditions for undoped ZnO using double dip technique and used them for gas sensor applications. They have also reported the synthesis of Sr doped ZnO using double dip technique and used them for gas sensor applications. Recently (Chandramohan et al, 2010) have synthesized Mg doped ZnO thin films using double dip chemical growth and reported the ferromagnetic properties of the films. Tahir Saeed et al. (1995) have deposited thin films of mono phase crystalline hexagonal ZnO from solutions of zinc acetate in the presence of ethylenediamine and sodium hydroxide on to glass microscope studies. Two distinct morphologies of ZnO were observed by scanning electron microscopy. The deposited films were specular and adherent. (Cheng et al 2006) have fabricated thin films transistors (TFTs) with active channel layers of zinc oxide using a low - temperature chemical bath deposition. Current voltage (I-V) properties measured through the gate reveal that the ZnO channel is n-type. (Sadrnezhaad et al 2006) have studied the effect of addition of Tiron as a surfactant on the microstructure of chemically deposited zinc oxide. Addition of tiron charges the surface morphology and causes to form the fine - grained structure. The obtained results indicate that increasing the number of dipping carves to progress the deposition process. (Piticescu, et al 2007) have studied the influence of the synthesis parameters on the chemical and microstructural characteristics of nanophases synthesized in the two methods. 'Al' doping tends to a lower material density and to a smaller gown size. Zhou et al (2007) have studied microstructure electrical and optical properties of aluminium doped zinc oxide films. The ZnO:Al thin films are transparent (~ 90%) in near ultraviolet and visible region A. with the annealing temperature increasing from 300°C to 500°C. The film was oriented more preferentially along the (002) direction, the grain size of the film increased, the transmittance also became higher and the electrical resistivity decreased. Joseph et al (2006) have reported the structural, electrical and optical properties of Al-doped ZnO thin films prepared by chemical spray deposition. XRD studies and SEM studies revealed that the film was polycrystalline in nature with (002) preferred concentration and smooth conditions have exhibited a resistivity of 2.45×10^{-4} Ωm with an optical transmittance of 97% of 550 nm. Oral et al (2007) have studied microstructure and optical properties of monocrystalline ZnO and ZnO : Li /Al thin films. Crystallized films had a grain size under 50 nm and showed C-axis grain orientation. All films had a very smooth surface with RMS and surface roughness values between 0.23 and 0.35 nm. Peiro et al (2005) have reported microwave-activated chemical bath depositions of zinc oxide thin films. Scanning electron microscopic characterization suggested that both the shape of the crystals and the textures of the film were highly influenced by the chemical path composition. Composition of films grown on bone glass or fluorine - doped tin oxide (SnO_2: F) showed that heterogeneous deposition was favoured on conducting substrates due to the localized heating. Bulk ZnO is quite expensive and unavailable in large wafers. So, for the time being, thin films of ZnO are relatively a good choice. Usually, the doped ZnO films with optimum properties (perfect crystalline structure, good conducting properties, high transparency, high intensity of luminescence) are obtained when they are grown on heated substrates and annealed after deposition at high temperature in oxygen atmosphere (Peiro et al, 2005 Lokhande et al, 2000; Srinivasan et al, 2006; Chou et al, 2005). However, for an extensive use in the commercial applications pure and doped ZnO films must be prepared at much lower substrate temperatures. Therefore, it is necessary to develop a low-temperature deposition technology for the growth of ZnO films. Many works are seen in the low temperature growth of this interesting ZnO system both undoped and metal doped (Tang et al, 1998, Cracium et al, 1994; Gorla et al, 1999; Kotlyarchuk et al, 2005) thin films and nano thin films. Advantages are effectiveness and simplicity of the deposition equipment, high deposition rates, wide spectrum of deposition parameters for the control and the optimization of film properties, and film thickness. The sum of all these special features enables the growth of oxide thin films at low temperature substrates with perfect crystallinity. The present work is a preparation and characterization of undoped ZnO, Sr-doped ZnO (SZO) and Al- doped ZnO (AZO) thin films by chemical deposition technique. In which the influence of solution concentration, solution pH value, film thickness, annealing temperature and concentration of strontium and aluminium atoms of the grown films are investigated. In addition it demonstrates that any dopant can be used in principle along with the precursor to enable them to be included in the system. The technique can be tuned to get the desired morphology and nanocrystallites of desired sizes distributed over any type of substrate for various applications.

2. Substrate and its preparation

Thin film requires a substrate to support itself. The substrate provides the necessary mechanical strength and rigidity needed for the film and it has adequate thermal ability to ensure at room temperature and withstand at high temperature. The function of the substrate is to provide the base on to which the thin film circuits are fabricated and various thin film multilayers are deposited. To form the thin film with defined electrical parameters, the substrates must be smooth and flat otherwise electrical and optical properties may be affected. Therefore in choosing a suitable substrate, in addition to considering the need to provide the mechanical support to the deposits, due consideration must be given to the possible influence of the substrates on the properties of the deposits. Commonly used substrate materials for polycrystalline thin film circuits include alumina, glass, silicon and metals, beryllium oxide based ceramic, aluminium nitride. When the films are deposited into glass, electrical and optical measurements are not disturbed by an underlying layer and are thus easier to interpret. Of all these, glass is found to posses all the requirements and is economically and widely used. However any type of substrate may be used in this simple growth method. Substrate cleaning in thin film technology is an important step prior to deposition. It is necessary to remove the contaminants that would otherwise affect the properties of the film. Cleaning involves the removal of contaminants without damage to the substrate. While cleaning, the bond between the substrates is broken and contaminants are set free from the substrates. The properties that can be affected by the presence of contaminants include morphology, nucleation electronic properties and the substrate film interface. Expected contaminants include fingerprints, dust, oil, and lint particles. The proper cleaning technique depends on the nature of the substrate and nature of the contaminants. The composition, physical properties such as porosity, thermal expansion, melting point, conductivity and chemistry of the substrate should be carefully considered in designing the cleaning operation. The energy required to break those bond could be supported by chemical, salvation, thermal (or) mechanical process. As the other techniques this technique also involves rigorous cleaning of the substrates.

3. Growth of nanocrystallites of undoped and doped ZnO

Experimental

A schematic diagram of the shape-selective synthesis of doped and undoped metal oxide nanostructures *via* double dip technique is shown in Fig. 1. Preparation of undoped and doped ZnO nano thin films ZnO thin films were performed using a two-step chemical bath deposition technique using a solution comprising of high purity zinc sulphate, magnesium sulphate and sodium hydroxide with a pH value of 9 as first step and a dip in hot water kept near boiling point as the second step. Before deposition, the glass substrates were cleaned by chromic acid followed by cleaning with acetone. The well-cleaned substrates were immersed in the chemical bath for a known standardized time followed by immersion in hot water for the same time for hydrogenation.

Possible formation mechanism

The process of solution dip (step 1) followed by hot water dipping (step 2) is repeated for known number of times. According to the following equation, the complex layer deposited on the substrate during the dipping in sodium zincate bath will be decomposed to ZnO due to subsequent dipping in hot water. The proposed reaction mechanism for undoped ZnO is according to the following equations

$$ZnSO_4 + 2\,NaOH \rightarrow Na_2ZnO_2 + H2SO_4 \uparrow .... \quad (1)$$

$$Na_2ZnO_2 + H_2O \rightarrow ZnO + 2\,NaOH .... \quad (2)$$

Part of the ZnO so formed was deposited onto the substrate as a strongly adherent film and the remainder formed as a precipitate. The addition of Metal sulphate in the ratio of Zn: Metal as 100:1 in the first dip solution leads to the formation of MZO films.

ZnO thin films were prepared using double dip technique shown in Fig. 1 by varying deposition parameters such as solvent medium, solution pH, concentrations, temperature, number of dippings, etc., The effect of these parameters are studied using various characterizations and the optimized deposition parameters are arrived for undoped ZnO thin films. The 'Al 'and 'Sr' doping were carried out by adding the respective metallc salts in the solution bath at different proportions (Zn : M as 100 : 1 or 10 : 1 where M =' Sr' or 'Al'.

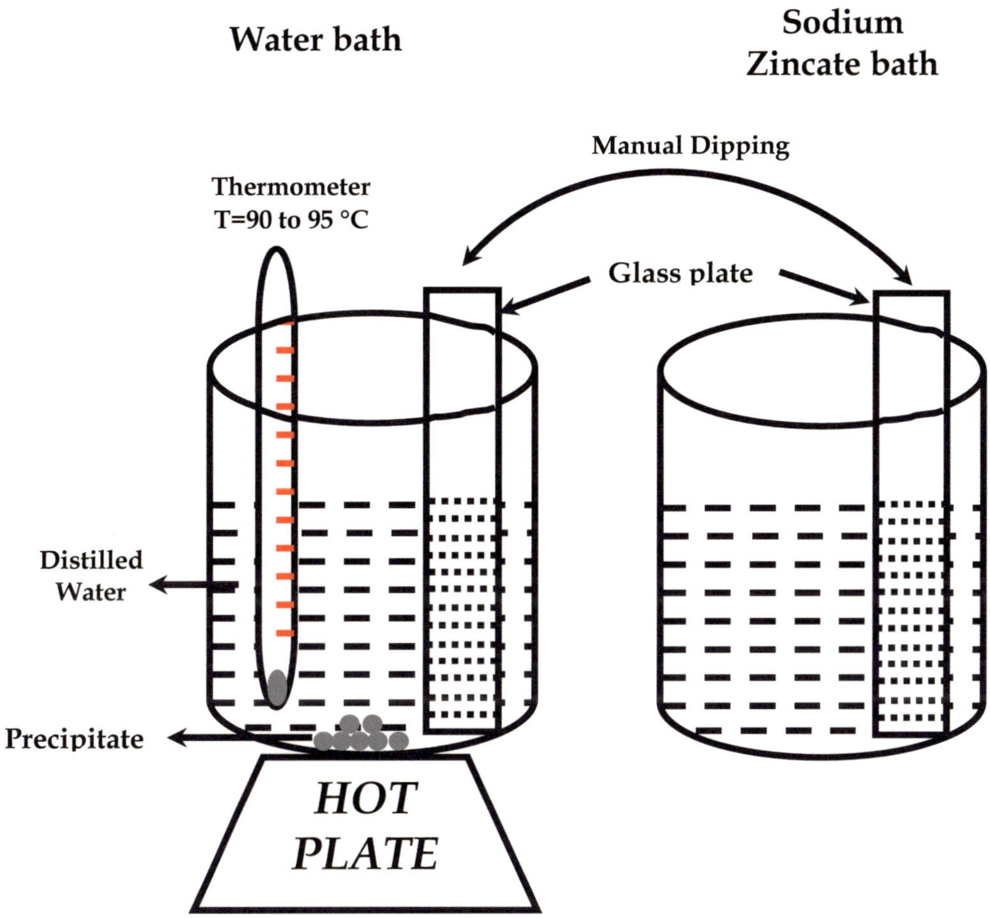

Fig. 1. Schematic representation of solution dip technique

Opimization of growth conditions

Undoped, Strontium doped ZnO (SZO) and Aluminium doped ZnO (AZO) thin films are prepared by solution grown double dip technique. The films are annealed in air to improve the crystallinity and grain sizes. All the synthesized films are characterized for their structural, optical, surface morphology, surface roughness, compositional analysis, X-ray photoemission spectroscopy analysis and electrical properties. The doping of Strontium and Aluminium (0.1mM and 1mM) concentration thin films by solution grown double dip technique is performed and investigated. The salient features of various studies carried out and the important findings are presented in this chapter. Zinc Oxide and doped (Sr and Al) thin films find interesting applications in the filed of gas sensor, solar cells, optoelectronic devices etc., Aluminium doped Zinc Oxide thin films are more suitable for gas sensor applications owing to their band gap and stability. This chapter describes widely the synthesis ZnO, Sr-doped ZnO (SZO) and Al-doped ZnO (AZO) thin films by solution grown double dip technique from aqueous solutions of $ZnSO_4$. The films prepared are found to be compact and homogeneous. The deposition conditions are optimized to obtain uniform, thin films suitable for gas sensor applications. The optimized deposition conditions to prepare ZnO films are

Bath composition and deposition conditions

Zinc Sulphate	: 0.1M
Sodium hydroxide	: 0.2M
Solution pH	: 9 ± 0.2
Bath temperature	: 90°C

SZO and AZO thin films are prepared at various molarities (0.1mM and 1mM) under optimized condition to obtain uniform, thin film suitable for gas sensor applications. The optimized deposition conditions to prepare SZO and AZO thin films are

Bath composition and deposition conditions

Strontium doped Zinc oxide (SZO) thin films

Zinc sulphate	: 0.1M
Sodium hydroxide	: 0.2M
Strontium Sulphate	: 0.1mM and 1mM
Solution pH	: 9 ± 0.2
Hot water temperature	: 90° C

Aluminium doped Zinc Oxide (AZO) thin films

Zinc sulphate	: 0.1M
Sodium hydroxide	: 0.2M
Aluminium Sulphate	: 0.1mM and 1mM
Solution pH	: 9 ± 0.2
Hot water temperature	: 90° C

The growth conditions have been optimized by us for various dopants including non metals. Several parameters involved in this technique offer wide range of selection of parameters. It is found that the films deposited at room temperature are found to be smooth and uniform and compatible with any physical or chemical techniques.

4. The structure and morphology of the nanostructures

To support the discussion on the optimization of growth conditions the SEM studies were carried out. The studies reveal that by altering the deposition conditions morphologies with minor variations can be obtained. Figure 2-4 shows the SEM micrographs of ZnO grown using double dip technique where 40, 80 and 100 respectively.

It is observed that improving the number of dipping yields the better morphology films. It is observed that the film is quite uniform up to a thickness in the range of few microns with the variation in the range of 10 nm. While the grains of the film surfaces are uniformly in the average range of 300 nm, the surface seems to be formed by the stacking of different nanorods or cylindrical grains whose size varied from 20 to 500 nm.

To support the discussion on the optimization of growth conditons the AFM studies were carried out. The studies reveal that by altering the deposition conditions morphologies with minor varitions can be obtained. Figures 5 (a - c) represents 3D AFM micrographs obtained for ZnO samples grown at room temperature under optimized conditions. It is observed that the stacking of different nanorods or tubular grains whose size varied from 20 to 500 nm makes the samples topography. Adjusting the deposition parameters may control the size of the grains. It is interesting that the morphology when further explored using 2D and 3D analysis supported by SEM investigations that morphology is also due to the nearly spherical nano grains.

Adjusting the deposition parameters gives a control on the growth of nanocrystallites with various sizes and shapes. It is interesting to know that the dip rate, interval between successive dippings, the variations in second dip bath, etc produces significant changes in the morphology inviting more number of researchers in this low cost nanocrystallites growth. For doping with other metals respective salts may be replaced in the place of aluminum sulphate. This method of growth facilitates fabricating excellent structures for future devices at low cost and low temperature. The technique is found to be highly reproducible and can be extended to large area and large scale fabrication systems.

In summary, the synthesis and optimization of undoped and doped ZnO systems have been reported. The Morphological studies through AFM and SEM reveal's excellent features associated with nanocrystallites of which the structure is made. The SEM reveals continuously stacked nanorods of diameter ranging from 20nm to few hundred nanometers. The AFM studies reveal the surface to be of minimum roughness composed of spherical and hexagonal shaped grains. They are uniformly distributed throughout the surface exhibiting the superiority of the films. Extensive characterizations on the structure, microstructure optical and electrical properties have been made and the exotic choice available in this simple method has paved way for the synthesis of many similar systems by our group like Fe, Mg and Mn doped ZnO thin films and other TCO systems like CdO, etc. Also the properties of these thin film nanocrystallites can be tailored to suit variety of applications like, phosphors, display panels, thermal conduction and opto electronic devices. The technique is easy for automation and anticorrosive coatings can be coated employing doped ZnO systems on to various mechanical spares. The potential of this technique is yet to be exploited in full by the industrial community. The crystallite shape and size control is also feasible in this excellent method.

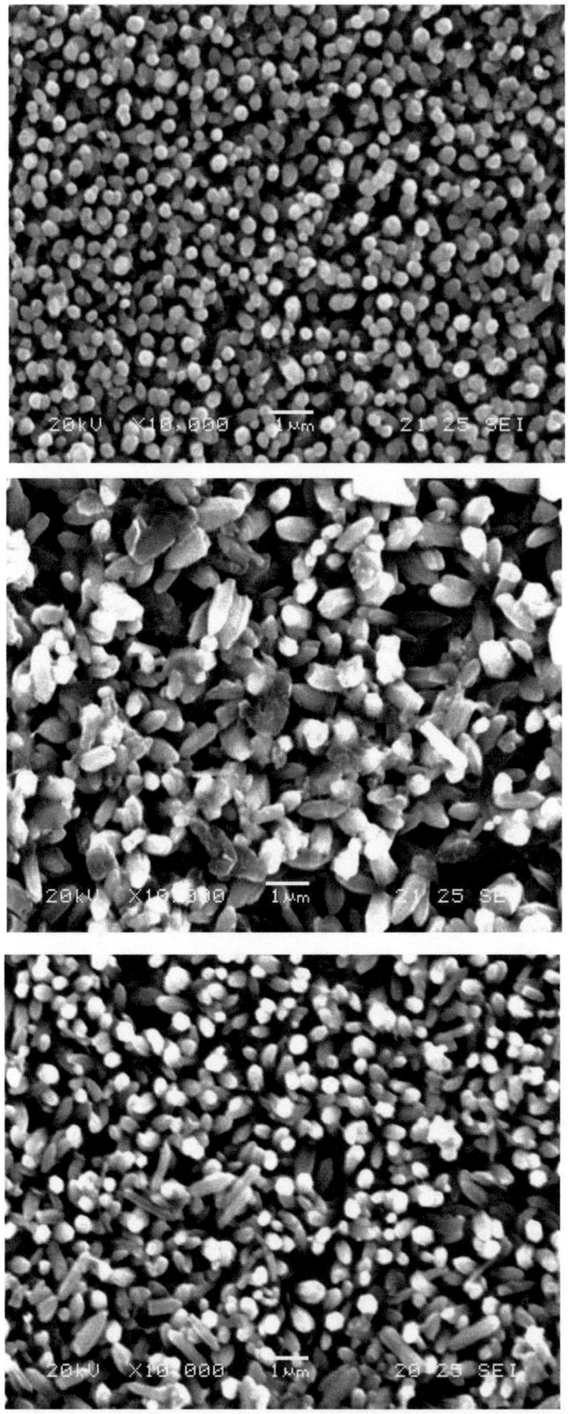

Fig. 2. Surface morphology on (a) pH =8, (b) pH =9 and (c) pH=10 of ZnO thin films.

(a)

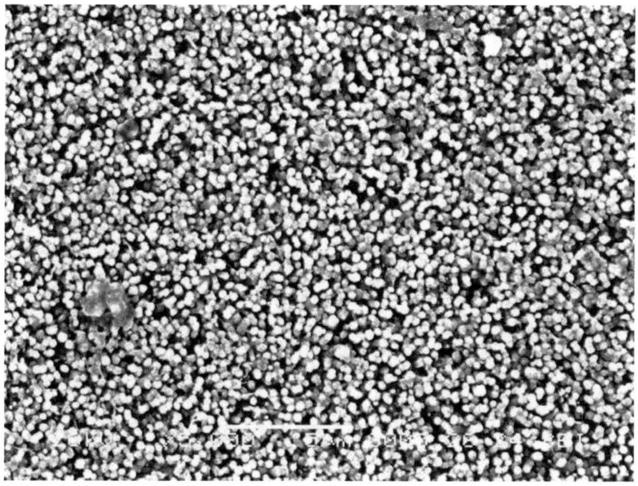

(b)

Fig. 3. Surface morphology on (a) Al-doped (0.1mM), (b) Al-doped (1mM) ZnO thin films.

(a)

(b)

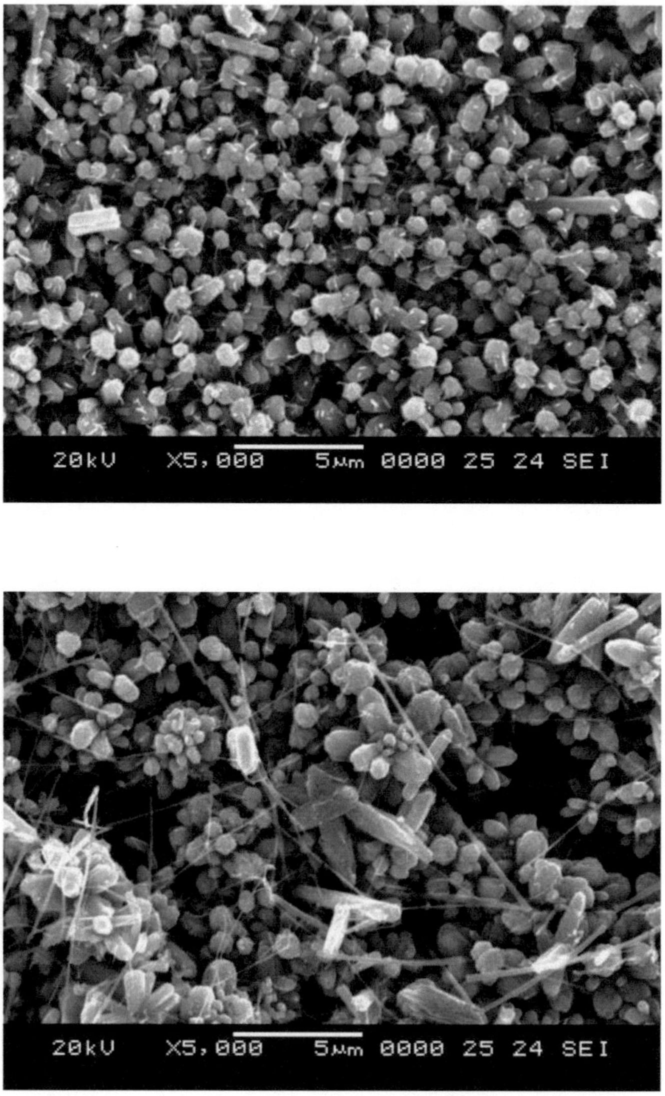

Fig. 4. Surface morphology on (a) Sr-doped (0.1mM), (b) Sr-doped (1mM) ZnO thin films.

(a)

(b)

(c)

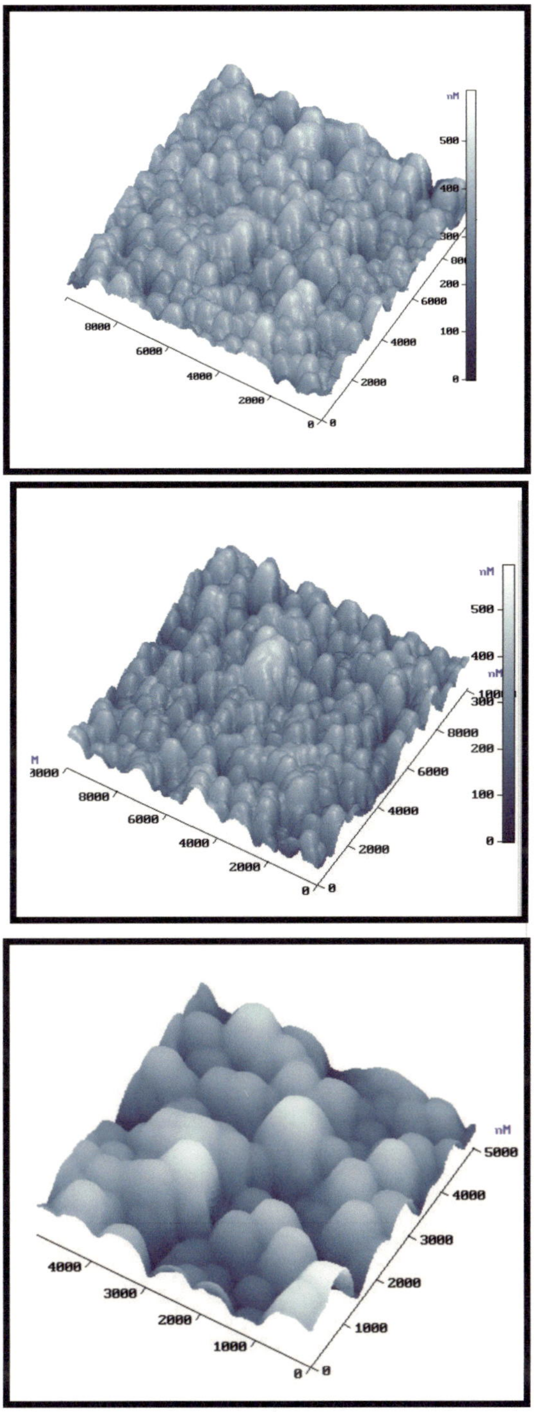

Fig. 5. (a-c) Shows 3D AFM micrographs obtained for ZnO, SZO and AZO samples grown at room temperature under optimized conditions.

5. References

Bunn, C.W. (1935). The lattice-dimensions of zinc oxide. *Proc. Phys. Soc. London.*, 47., 5., 835-842, ISSN 0959-5309.

Liu, C.; Yun, F.; & Morkoç, H, (2005). Ferromagnetism of ZnO and GaN: A Review *J. Mater. Sci: Mat in Electronics.* 16., 9., 555-597, ISSN: 0957-4522.

Tsukazaki, A.; Ortomo, A.; Onuma, T.; Ohtani, M. & Makino, T.; Smiya, M.; Ohtani, K.; Chichibu, S.F.; Fuke, S.; Egawa, Y.S.; Ohno, H.; Kainuma, K. & Kawasaki, M. (2005). Repeated temperature modulation epitaxy for p-type doping and light-emitting diode based on ZnO. *Nat. Mater.*, 4., 1., (December 2004) 42-46, ISSN: 1476-1122.

Özgür, Ü.; Alivov, Ya. I.; Liu, C.; Teke, A.; Reshchikov, M.A.; Doğan, S.; Avrutin, V.; Cho, S.-J. & Morkoç, H. (2005). A comprehensive review of ZnO materials and devices. *J. Appl. Phys.*98., 4., (August 2005) 041301-041404, ISSN 0021-8979.

Wang, Z.L. (2004). Zinc oxide nanostructures: growth, properties and applications. *J.Phys.: Cond. Matter.*, 16., 25., (June 2004) R829-R858, ISSN 0953-8984.

Symposium N, ICMAT IUMRS 2005 3-8 July *2005*, Singapore.

Symposium D, ICMAT 2009, 28 June -3 July, Singapore.

The 6th International Workshop on Zinc Oxide and Related Materials, 5- 7 Aug 2010 in Changchun, China.

Wellings, J.S.; Chaure, N.B.; Heavens, S.N. & Dhamadasa, I.M. (2008). Growth and characterisation of electrodeposited ZnO thin films. *Thin solid films,* 516., 12., (April 2008) 3893-3898, ISSN: 0040-6090.

Yang, Y.; Tay, B.K.; Sun, X.W.; Han, Z.J.; Shen, Z.X.; Lincoln, C.; & Smith, T, (2008). Nanoelectronics Conference, INEC 2008 2 nd IEEE International, Nanyang Technical University, Singapore 24-27 March 2008.

Mitra, P.; Chatterjee, A.P. & Maiti, H.S. (1998). Chemical deposition of ZnO films for gas sensors. *J. Mater. Sci: Mat in Electronics.*, 9., 6., (December & 1998) 441-445, ISSN: 0957-4522.

Ramamoorthy, K.; Arivanandhan, M.; Sankaranarayanan, K. & Sanjeeviraja, C. (2004). *Mater. Chem. Phys.*, 85., 2-3., (June & 2004) 257-262, ISSN: 0254-0584.

Natsume, Y. & Sakata, H. (2000). Zinc oxide films prepared by sol-gel spin-coating. *Thin solid films.*, 372., 1-2., (September & 2000) 30-36, ISSN: 0040-6090.

Chapparro, M.; Maffiotte, C.; Gutierrez, M.T. & Herrero, J. (2003). Study of the spontaneous growth of ZnO thin films from aqueous solutions.*Thin solid films.*, 431., 1., (May & 2003) 373-377, ISSN: 0040-6090.

Water, W. & Yan, Y- S. (2007). Characteristics of strontium-doped ZnO films on love wave filter applications. *Thin solid films.*, 515., 17., (June & 2007) 6992-6996, ISSN: 0040-6090.

Vijayan, T. A.; Chandramohan, R.; Valanarasu, S.; Thirumalai, J.; Venkateswaran, S.; Mahalingam, T. & Srikumar, S.R. (2008). Optimization of growth conditions of ZnO nano thin films by chemical double dip technique. *Sci. Tech. Adv. Mater.*, 9., 035007, ISSN: 1468-6996 .

Vijayan, T. A.; Chandramohan, R.; Valanarasu, S.; Thirumalai, J.; Subramanian, S. P. (2008). Comparative investigation on nanocrystal structure, optical, and electrical properties of ZnO and Sr-doped ZnO thin films using chemical bath deposition method. *J. Mater. Sci.*, 43., 6., (March & 2008) 1776–1782, ISSN: 0022-2461.

Chandramohan, R.; Thirumalai, J.; Vijayan, T. A.; ElhilVizhian, S.; Srikanth, S.; Valanarasu, S. & Swaminathan, V. (2010). Nanocrystalline Mg Doped ZnO Dilute Magnetic Semiconductor Prepared by Chemical Route. *Adv. Sci. Lett.* 3., 3., (September & 2010) 319-322, ISSN: 1936-6612.

Saeed, T. & Brien, P. O'. (1995). Deposition and characterisation of ZnO thin films grown by chemical bath deposition. *Thin solid films.*, 271., 1-2., (December & 1995) 35-38, ISSN: 0040-6090.

Cheng, H- C.; Chen, C- F.; Lee, C- C. *Thin solid films.*, 498., 1-2., (March & 2006) 1, ISSN: 0040-6090.

Sadrnezhaad, S.K. & Vaezi, M.R. (2006). The effect of addition of Tiron as a surfactant on the microstructure of chemically deposited zinc oxide. *Mat. Sci. Engg: B.*, 128., 1-3., (March & 2006) 53-57, ISSN: 0921-5107.

Piticescu, R. R. Piticescu, R. M. & Monty, C. J. (2006). Synthesis of Al-doped ZnO nanomaterials with controlled luminescence. *J. Europ. Cer. Soc.* 26., 14., (March & 2006) 2979-2983, **ISSN:** 0955-2219.

H-M Zhou, D-Q Yi, Z- M Yu, L. Rang, Xiao, Jian Li, (2007). Preparation of aluminum doped zinc oxide films and the study of their microstructure, electrical and optical properties. *Thin solid films.*, 515., 17., (June 2007) 6909-6914, ISSN: 0040-6090.

Joseph, B.; Manoj, P.K. & Vaidyan, V.K. (2006). Studies on the structural, electrical and optical properties of Al-doped ZnO thin films prepared by chemical spray deposition. *Ceramics International.*, 32., 5., (September & 2005) 487-493, ISSN: 0272-8842.

Oral, A.Y.; Batisi, Z.B. & Aslan, M.H. (2007). Microstructure and optical properties of nanocrystalline ZnO and ZnO:(Li or Al) thin films. *Appl. Sur. Sci.* 253., 10., (March & 2007) 4593-4598, ISSN: 0169-4332.

Peiro, A. M.; Ayllon, J. A.; Pearl, J.; Domenech, X. & Domingo C. (2005). Microwave activated chemical bath deposition (MW-CBD) of zinc oxide: Influence of bath composition and substrate characteristics. *J. Crystal. Growth.*, 285., 1-2., (November &2005) 6-16, ISSN: 0022-0248.

Lokhande, B.J. & Uplane, M.D. (2000). Structural, optical and electrical studies on spray deposited highly oriented ZnO films. *Appl. Surf. Sci.*, 167., 3-4., (October & 2000) 243-246, ISSN: 0169-4332.

Srinivasan, G. & Kumar, J. (2006). Optical and structural characterisation of zinc oxide thin films prepared by sol-gel process. *Cryst. Res. Technol.* 41., 9., (September & 2006) 893- 896, ISSN: 0232-1300.

Chou, T- L.; Ting, & J- M. (2005). Deposition and characterization of a novel integrated ZnO nanorods/thin film structure. *Thin solid films.* 494., 1-2., (January & 2006) 291-295, ISSN: 0040-6090.

Tang, Z.K.; Wang, G.K. L.; Yu, P.; Kawaraki, M.; Ohtomo, A.; Koinuma, H. & Segawa, Y. (1998). *Appl. Phys. Lett.* 72., 3270, ISSN: 0003-6951.

Cracium, V.; Elders, J.; Gardeniers, J.G.E. & Boyd, L. W. (1994). *Appl. Phys. Lett.*, 65., 2963, ISSN: 0003-6951.

Look, D.C.; Reynolds, D.C.; W.Litton, C.; Jones, R.L.; Eason, D.B & Cartwell, G. (2002). *Appl. Phy. Lett.*, 81 1830, ISSN: 0003-6951.

Gorla, C.R.; Emanetoglu, N.W.; Liang, S.; Mayo, W.E.; Cu, Y.; Wraback, M. & Shen, H. (1999). *J. Appl. Phys.*, 85., 2602, ISSN 0021-8979.

Kotlyarchuk, B.; Sarchuk, V. & Oszwaldowski, M. (2005). Preparation of undoped and indium doped ZnO thin films by pulsed laser deposition method. *Cryst. Res. Technol.*, 40., 12., (December & 2005) 1118- 1123, ISSN: 0232-1300.

3

Synthesis and morphology control of Eu^{3+} doped M_2O_2S [M=Y, Gd] nanostructures

Rathinam Chandramohan, Jagannathan Thirumalai
and Thirukonda Anandamoorthy Vijayan
Department of Physics, Sree Sevugan Annamalai College, Devakottai - 630 303
India

1. Introduction

Global claim for phosphor materials as efficient sources of energy that can supply sustained competence is growing day by day. The phosphors are facing increased global challenges including high production of rare earth materials, environmental and recycling issues, and necessity to supply devices very quickly that may be outdated rapidly due to new technological developments arising in the industry and market. A number of applications have emerged in recent years that will change the future of the industry and new technologies like nanoscale innovations and specialty phosphors are garnering increased attention. The primary drivers for growth are the expansion of key end-use applications including solid-state lighting and fluorescent lighting. Current research in nanotechnology is focused on new materials, novel phenomena, new characterization technique and fabrication of nano devices.

$Y_2O_2S:Eu^{3+}$ and $Gd_2O_2S:Eu^{3+}$ are excellent materials of current interest (Lei et al., 2010; Liu & Kuang, 2010; Li, et al., 2010; Nakkiran et al., 2007; Thirumalai et al., 2007) owing to their interesting optical and opto-electronic properties. The crystal structure of M_2O_2S (M = Y, Gd and including all lanthanides) are discussed in detail (Delgado da Vila et al., 1997; Sabot & Maestro, 1995; Mikami & Oshiyama, 1998). The crystal symmetry of the above two systems is trigonal, with the space group P3m1 (D^3_{3d}), as determined by X-ray diffraction. These systems are grouped under wide band gap (4.6 - 4.8 eV) semiconductors. $Y_2O_2S:Eu^{3+}$ and $Gd_2O_2S:Eu^{3+}$ as a red phosphor, with its sharp emission line for good calorimetric definition and high luminescence efficiency, is extensively used in the phosphor screen of display devices, fluorescent lamps used for lighting purposes, television sets used for entertainment and information gathering, X-ray imaging instruments used in hospitals and laser instruments used for experimental purposes and, many other electrical and opto-electronic equipments. They employ luminescent materials for (Nakkiran et al., 2007) electronic portal imaging devices (EPID), radioisotope distribution and so on (Yeboah & Pistorius, 2000; Chou et al., 2005). Due to the large size and weight of CRTs, developments of flat-panel displays (FPDs) are of great interest. Among several FPD technologies, liquid-crystal displays (LCDs) dominate the FPD market and plasma display panels (PDPs) are now commercially available in the large area TV market (Yu et al., 2005). New and enhanced properties are expected due to size confinement in nanoscale dimensions that can revolutionize the display devices market in future. Commercially available bulk oxysulfides are quite expensive and are not easily available. So, for the time being, $Y_2O_2S:Eu^{3+}$ and $Gd_2O_2S:Eu^{3+}$ nanostructures are relatively a good choice while compared with the bulk systems. However, for an extensive use in the commercial applications, $Y_2O_2S:Eu^{3+}$ and $Gd_2O_2S:Eu^{3+}$ nanocrystals must be prepared at lower temperatures. Therefore, it is necessary to develop a low-temperature synthesis technology for the growth of oxysulfide nanophosphors. In this background, this chapter has been devoted to the nanophosphors development using these two systems. The realm of novel devices from this wonderful material is yet to be accomplished in full. To give a quantitative report on the state of art of $Y_2O_2S:Eu^{3+}$ and $Gd_2O_2S:Eu^{3+}$ is quite difficult and an attempt has been made to give an account of the synthesis of the nanophosphors in this chapter.

A detailed survey on Y_2O_2S: Eu^{3+} and Gd_2O_2S: Eu^{3+} nanophosphors discusses various synthesis techniques adopted by different research groups as follows. Powder phosphors of $(Y_{1-X}RE_X)_2O_2S$, $(Gd_{1-X}RE_X)_2O_2S$ and $(La_{1-X}RE_X)_2O_2S$ where RE=Eu^{3+}, Tb^{3+}, or Tm^{3+} that were prepared by combustion reactions from mixed metal nitrate reactants and dithiooxamide $(CSNH_2)_2$ with ignition temperatures of 300 – 350 °C (Bang et al., 2004). The $Y_2O_2S:Eu^{3+}$ red phosphor which was prepared by a new method of decomposing the metal complexes $Y(NO_3)_3$•3(DMSO) and $Eu(NO_3)_3$•4(DMSO) in H_2S atmosphere at 900 °C (Guo et al., 2008). Nanocrystalline $Y_2O_2S:Eu^{3+}$ was successfully prepared using a combustion synthesis method employing conventional sulfur flux (Fu et al., 2008). A method of preparing red emitting $Eu^{3+}:Y_2O_2S$ phosphors in which Yttrium sulfite doped Eu is used as starting material and the Eu-activated oxysulfide is obtained from either directly by reducing the sulfite with carbon monoxide, or by first oxidizing sulfite and then reducing the obtained oxysulfate (Koskenlinn et al., 1976). Preparation of spherical Y_2O_2S and $Y_2O_2S:Eu$ particles using a solid–gas reaction of monodispersed precursors with elemental sulfur vapor under an argon atmosphere has been investigated (Delgado da Vila et al., 1997). A one-step solvothermal process developed for the preparation of Eu^{3+} actived yttrium oxysulfide phosphor in ethylenediamine solvent at 280°C through a reaction of yttrium oxide, europium oxide, and sulfur powder (Kuang et al., 2005). Luminescent $Y_2O_2S:Eu^{3+}$ nanoceramics prepared through a gel–polymer thermolysis process employing a urea–formaldehyde resin (Dhanaraj et al., 2003). Nanocrystals of $Y_2O_2S:Eu^{3+}$ synthesized using a two step sol–gel polymer thermolysis method (Dhanaraj et al., 2004). Ensembles of $Y_2O_2S:Eu^{3+}$ widegap semiconductor nano-crystals exhibit ON–OFF fluorescence blinking phenomenon, which mimic II–VI semiconductor quantum-structures synthesized using sol-gel polymer thermolysis method (Thirumalai et al., 2007). Trivalent europium-doped yttrium oxysulfide nanocrystals synthesized using sol–gel thermolysis. A significant blue shift observed in the fundamental absorption edge for the nanocrystals having an average crystallite size (f) in the range 9–15nm indicated a strong quantum confinement with a Bohr exciton radius of 5–13 nm (Thirumalai et al., 2007; Thirumalai et al., 2008). The $Y_2O_2S:Eu^{3+}$ nanocrystallines that were prepared by a new ethanol assisted combustion synthesis method using sulfur contained organic fuel (thioacetamide) in an ethanol-aqueous solution (Xixian et. al., 2006). The luminescence dynamics of optical centers in nanocrystals depending critically on the phonon density of states (PDOS) is quite distinct from that of bulk materials. It is shown that energy transfer (ET) in nanocrystals is confined by discrete PDOS as well as direct size restriction. For applications, the nanoconfinement effects on ET

significantly reduce the efficiency of sensitized or upconversion luminescence (Chen et. al., 2003). The Y_2O_2S:Eu phosphor powders were prepared with a flux fusion method and electrophoretically deposited on an ITO-coated glass substrate to form a thin layer (Tseng et al., 1998). The nanostructured yttrium oxysulfide films prepared via vapor phase growth (V. V. Bakovets et al., 2008). Their first step was the deposition of 50-nm-thick nanostructured yttria films from yttrium dipivaloylmethanate vapor at 525 °C. Next, the films were sulfided in ammonium thiocyanate vapor at temperatures from 800 to 1100 °C. Hexagonal yttrium oxysulfide was obtained at 900 °C and higher temperatures. The investigations of pseudobinary systems Ln_2O_2S---La_2O_2S (*Ln* = Nd, Sm, Eu, Gd, Dy, Yb, Lu, and Y) with complete solid solubility only for the systems like Nd_2O_2S---La_2O_2S and Sm_2O_2S---La_2O_2S; a two-phase region is found for all other systems (M. Leskelä et al., 1976). The solid solubility in the isostructural oxysulfide series is discussed in terms of the differences in ionic radii of the two rare-earth components. The size- (submicrometer-sized) and morphology- (spherical) controlled composite Gd–Eu oxalate particles were prepared in an emulsion liquid membrane (water-in-oil-in-water emulsion) system (Hirai et al., 2002). The oxalate particles thus prepared were calcined in air to obtain $Gd_2O_3:Eu^{3+}$ phosphor particles and in sulfur atmosphere to obtain $Gd_2O_2S : Eu^{3+}$ phosphor particles.
Usually synthesis of any phosphor material would necessitate rigorous conditions such as heavy milling, intermediate and very high temperature heat treatment cycles, etc. Whereas the hydrothermal approach avoids all those difficult and time consuming steps by modifying the reaction parameters to suitable forms so as to react and produce target compound(s) at relatively lesser duration with the capacity to develop desired shapes and dimensions with high crystallanity. The $Y_2O_2S:Eu^{3+}$ and $Gd_2O_2S:Eu^{3+}$ nanostructures show optimum properties (perfect crystalline structure, high stability and good morphological) when they are grown on perfect hydrothermal conditions. In addition to the conventional vapor-phase method, which includes vapor transport and condensation (Kong et al., 2001), metal-organic chemical vapor deposition (Zhang et al, 2004), thermal evaporation (Pan et al, 2001) and solution-phase methods have been developed as alternative ways to synthesize semiconductor nanostructures with different shapes and dimensions. Hydrothermal method is a widely used technique that can control the shape and dimension of nanostructures among all solution-based approaches (Zhang et al, 2002). Unlike conventional vapor-phase methods, the hydrothermal method can produce various nanostructures at a relatively low temperature (below 200 °C) using simple equipments; however, the reaction time required for the growth of nanostructures is too long (usually from a few hours to several days) (Wang & Li, 2003; Wang et al, 2003; Zhang et al, 2002). The various low-dimensional nanostructures, such as nanowires, nanotubes, nanosheets and fullerene like nanoparticles that have been selectively synthesized from rare-earth compounds (hydroxides, fluorides) based on a facile hydrothermal method (Wang & Li, 2003; Wang et al, 2003). The subsequent dehydration, sulfidation and fluorination processes lead to the formation of rare-earth oxide, oxysulfide and oxyhalide nanostructures, which can be functionalized further by doping with other rare-earth ions. An effective method to synthesize $Y_2O_2S:Eu^{3+}$, Mg^{2+}, Ti^{4+} nanoparticles. Tube-like $Y(OH)_3$ were firstly synthesized by hydrothermal method to serve as the precursor. Nanocrystalline long-lasting phosphor $Y_2O_2S:Eu^{3+}$, Mg^{2+}, Ti^{4+} was obtained by calcinating the precursor with co-activators and S powder (Li, et. al., 2010). The afterglow properties of Eu^{3+} activated long lasting Gd_2O_2S phosphor by hydrothermal route. Rod-like $Gd(OH)_3$

were firstly synthesized by hydrothermal method to serve as the precursor. Long lasting $Gd_2O_2S:Eu^{3+},Ti^{4+},Mg^{2+}$ phosphor were obtained by calcinating the precursor with co-activators and S powder (Hang et al., 2008). Hydrothermally prepared $Gd(OH)_3$ nanorod precursor, codoped with Eu, Ti and Mg, was converted into the desired phosphor by calcinating the precursor in CS_2 atmosphere (Mao et al., 2008). Therefore, the development of a simple and fast synthetic route that can control the shape of nanostructures under ambient conditions is the need of the hour and hydrothermal technique has many advantages and technological possibilities. Earlier (Wang & Li, 2003; Wang et al, 2003; Hang et al., 2008; Mao et al., 2008) hydrothermal method was used to synthesize oxysulfide nanotubes and nanorods.

To the best of the author's knowledge, no systematic study has been reported on other nanostructures like nanocrystals, nanosheets, nanobelts, nanotubes, nanorods, nanowires and nanoflowers of this oxysulfide system (Thirumalai et al, 2008; Thirumalai et al, 2009 a, b). Hence maximum efforts were to be put forth in selecting techniques for the synthesis of various nansostructures with good crystallinity, so also any other oxysulfide nanostructures of high research potential activity. Therefore, the present study has been undertaken with a view to synthesizing uncontaminated, highly crystalline $Y_2O_2S:Eu^{3+}$ and $Gd_2O_2S:Eu^{3+}$ adopting two methods of hydrothermal routes owing to the relatively lower temperature with respect to the bulk counterpart (solid-state reaction method): **Template - free method** (two-step synthesis) involving the synthesis of as-prepared $Y(OH)_3$ and $Gd(OH)_3$, followed by subsequent Eu^{3+} doping and sulfurization leading to conversion of $Y_2O_2S:Eu^{3+}$ and $Gd_2O_2S:Eu^{3+}$ nanostructures, respectively, seemed to be a topotactic reaction. **Template - assisted method** (single-step synthesis) involving Anodic Aluminium Oxide (AAO) membranes used for synthesizing of $Y_2O_2S:Eu^{3+}$ and $Gd_2O_2S:Eu^{3+}$ nanostructures. Furthermore, being a low temperature and an easy-to-adopt methodology, this method yields phase pure $Y_2O_2S:Eu^{3+}$ and $Gd_2O_2S:Eu^{3+}$ nanostructures with better reproducibility, definite shape and dimensions. Hydrothermal method is basically a simple, easy and fast synthetic route where the most important synthesis parameters are the precursors, the solution concentration, solution pH value, solvent, temperature and time. Also, this processing route provides the basis for a nearly low cost, low temperature method for the preparation of homogeneous nano-sized ceramics compared to any other existing methods.

2. Formation, structure, and morphology of the $Y_2O_2S:Eu^{3+}$and $Gd_2O_2S:Eu^{3+}$ nanostructures

2.1 Template – free Method

A systematic study has been undertaken on other nanostructures like nanocrystals, nanosheets, nanobelts, nanotubes, nanorods, nanowires and nanoflowers by varying reaction parameters such as solution concentration, pH, growth temperature and reaction time and solvent using template-free method. The effect of these parameters are studied using various characterization studies and the optimized reaction conditions are arrived for $Y_2O_2S:Eu^{3+}$ and $Gd_2O_2S:Eu^{3+}$ nanostructures. The detailed synthesis procedure is discussed already in detail (Thirumalai et al, 2008; Thirumalai et al, 2009 a, b).

Structural studies (XRD) reveal that the products $Y_2O_2S:Eu^{3+}$ and $Gd_2O_2S:Eu^{3+}$ are pure hexagonal phase and they are in good agreement with standard JCPDS [($Y_2O_2S:Eu^{3+}$;

JCPDS # 24-1424) and ($Gd_2O_2S:Eu^{3+}$; JCPDS # 26-1422)] data (Thirumalai et al, 2008; Thirumalai et al, 2009 a, b) and they are highly crystalline in nature. It is already stated that, the conversion from hydroxide to oxysulfide seems to be a topotactic reaction. Firstly, the hydroxide(s) [$Y(OH)_3$ and $Gd(OH)_3$] of nanocrystals / nanoplates, nanosheets, nanobelts, nanotubes, nanorods and nanowires are selectively synthesized were based on the preparation of colloidal hydroxide precipitates at room temperature, and the subsequent hydrothermal treatment at 100 - 180 °C for approximately 12 - 48 hours (Thirumalai et al, 2008; Thirumalai et al, 2009 a, b). The hydrothermal method was shown to be effective in the synthesis of zero- and one-dimensional nanostructures. By the simple tuning of factors such as pH, temperature and concentration, the experimental conditions could be chosen to favor the anisotropic growth of materials. In the present work, nanostructures of hydroxide(s) were successfully obtained through this precipitation-hydrothermal synthetic method by properly tuning the temperature, pH and time, the crystal structures have been found to be responsible for the growth of hydroxide nanostructures with nearly controllable aspect ratios. The conversion of hydroxide to oxysulfide seems to be a topotactic reaction (i.e., the morphology does not change while the phase of the material changes). However in any topotactic reactions, where significant atomic rearrangement due to chemical changes take place, though the morphology remains intact. Yet it is possible to change the resulting morphology. With this important criteria the as-synthesized $Y(OH)_3$ and $Gd(OH)_3$ were converted to $Y_2O_2S:Eu^{3+}$ and $Gd_2O_2S:Eu^{3+}$ nanostructures. To investigate the optimized growth of the $Y_2O_2S:Eu^{3+}$ and $Gd_2O_2S:Eu^{3+}$ nanostructures, SEM and TEM (Thirumalai et al, 2008; Thirumalai et al, 2009 a, b) micrographs of nanocrystals, nanosheets, nanobelts nanotubes, nanorods and nanowires, respectively, were obtained by varying the reaction parameters (pH ~ 6 - 13, 100 - 180 °C, 24 - 48 hrs) and its resulting morphological features and crystallanity are found to be good and were discussed in Table I (Thirumalai et al, 2009 a, b). The studies reveal that by altering the reaction conditions (like solution concentration, pH, temperature and time) the morphology, shape and crystallinity may be perfectly controlled. The SEM and TEM results are discussed and presented in Fig. 1-4. Fig. 1a, 2a and Fig. 3b, 4b show that growth is clearly seen indicating the nanocrystals formed due to combination of spherical and hexagonal-like structure at a pH of 6, 100°C, 48 hrs, respectively. Fig. 1b, 2b and Fig. 3c, 4c show stacked nanosheets of the oxysulfide(s) growth at pH ~ 8, 120°C, 12 hrs, respectively. This may be attributed to the two-dimensional growth tendency of the oxysulfide(s) nanosheets at lower pH. Nevertheless, Fig. 1c, 2c and Fig. 3d, 4d, show the nanobelts of $Y_2O_2S:Eu^{3+}$ and $Gd_2O_2S:Eu^{3+}$ grown at a pH of 8 - 9, 180°C, 24 hrs, respectively. These nanosheets and/or nanobelts curl from the edge, indicating a possible rolling process for the formation of the nanotubes/rods/wires. Fig. 1(d, e, f), 2(d, e, f) and Fig. 3(e, f, g), 4(e, f, g) show that the nanotubes, rods, wires were grown at a pH of 12 - 13, 140 - 180°C, 24 - 48 hrs and has no template to drive the directional growth of nanotubes, rods and wires. The images indicate that the samples are single crystalline in nature and uniformly distributed. Also, based on the above studies it is evident that these nanostructures are stable under thermal treatment, which may be rather useful for their subsequent applications as catalysts. Furthermore, the morphology of these nanostructures is likely to be a near-quantum structure. It is having high surface-to-volume ratio, that plays a major role in the density of singly ionized oxygen vacancies and the charge state of these defects, that may be due to the existence of surface depletion.

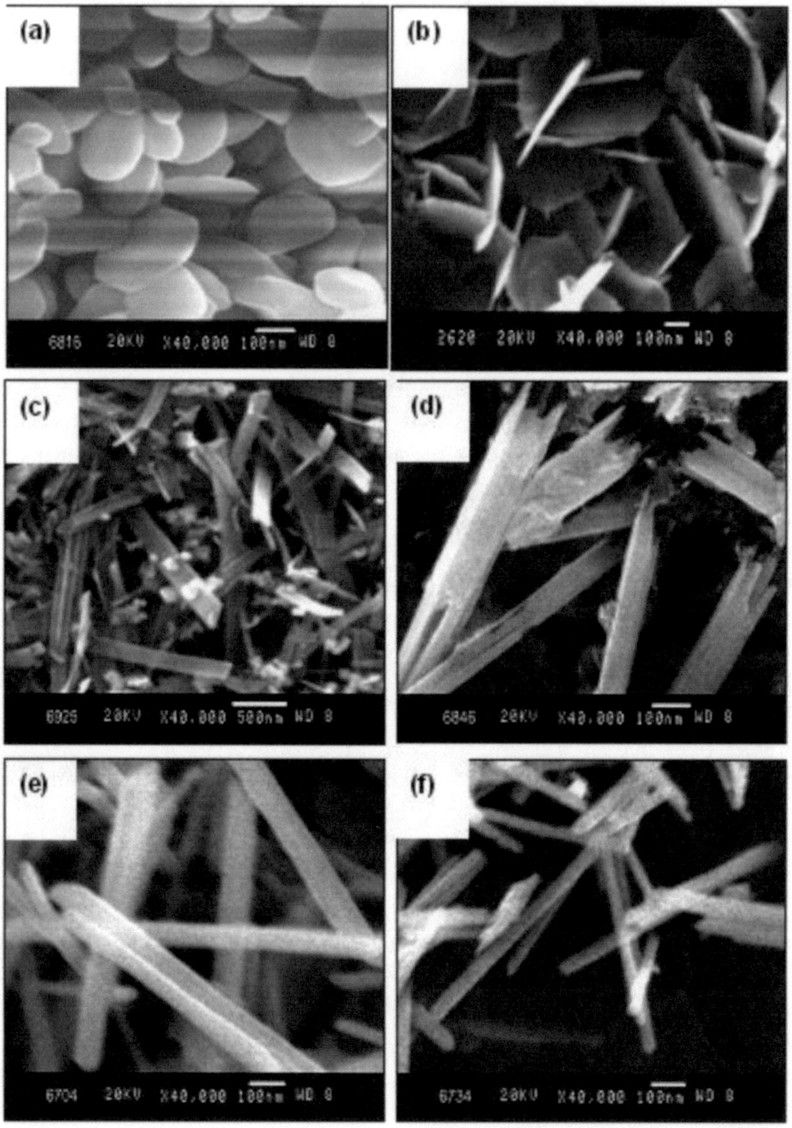

Fig. 1. SEM images show the optimal experimental conditions of $Y_2O_2S:Eu^{3+}$ nanostructures using template – free method.

Fig. 2. SEM images show the optimal experimental conditions of $Gd_2O_2S:Eu^{3+}$ nanostructures using template - free method.

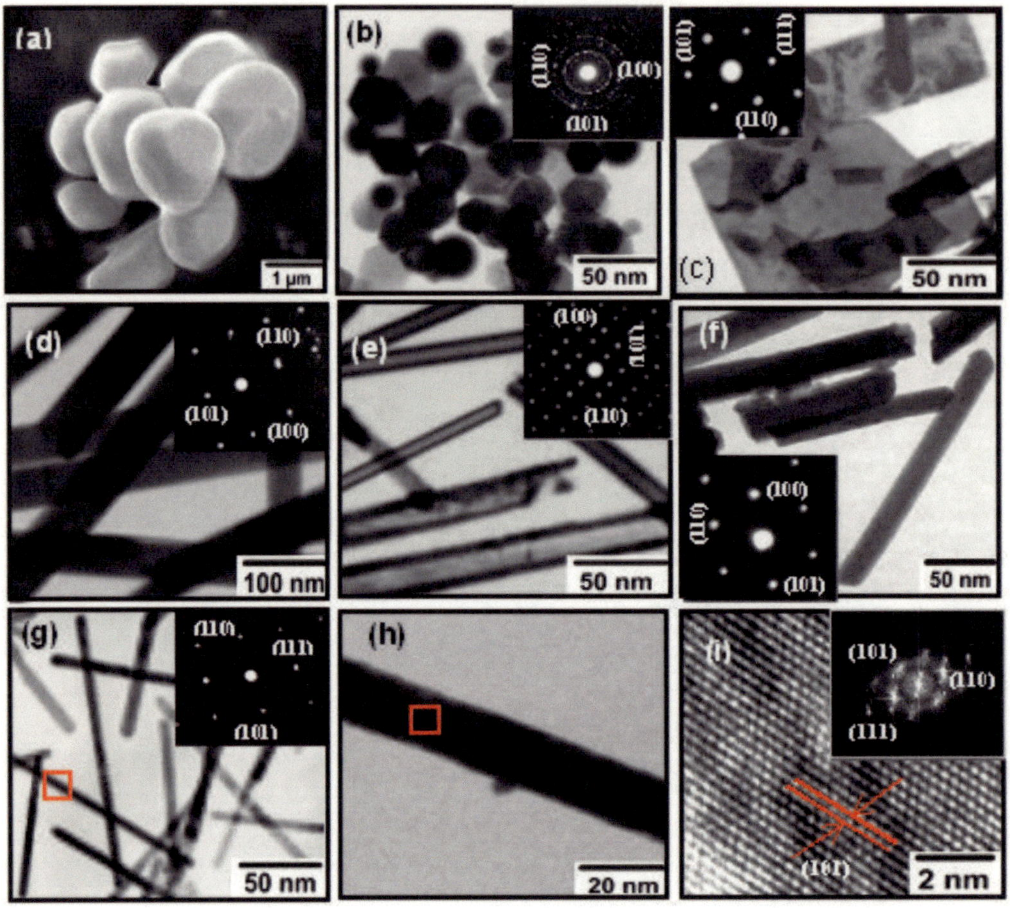

Fig. 3. (a) SEM image of hexagonal-shaped bulk $Y_2O_2S:Eu^{3+}$ (A). The TEM images of $Y_2O_2S:Eu^{3+}$ nanostructures (template - free method) from the sample A_7 to A_{12}: (b) Nanocrystals, (c) Nanosheets, (d) Nanobelts, (e) Nanotubes, (f) Nanorods, (g) Nanowires, (h) a close up of the boxed area in (g) shows a structure of single-nanowire, (i) HRTEM image of $Y_2O_2S:Eu^{3+}$ nanowire. *Courtesy: J. Colloid. Interface. Sci. 336., 2., (2009) 889-897.*

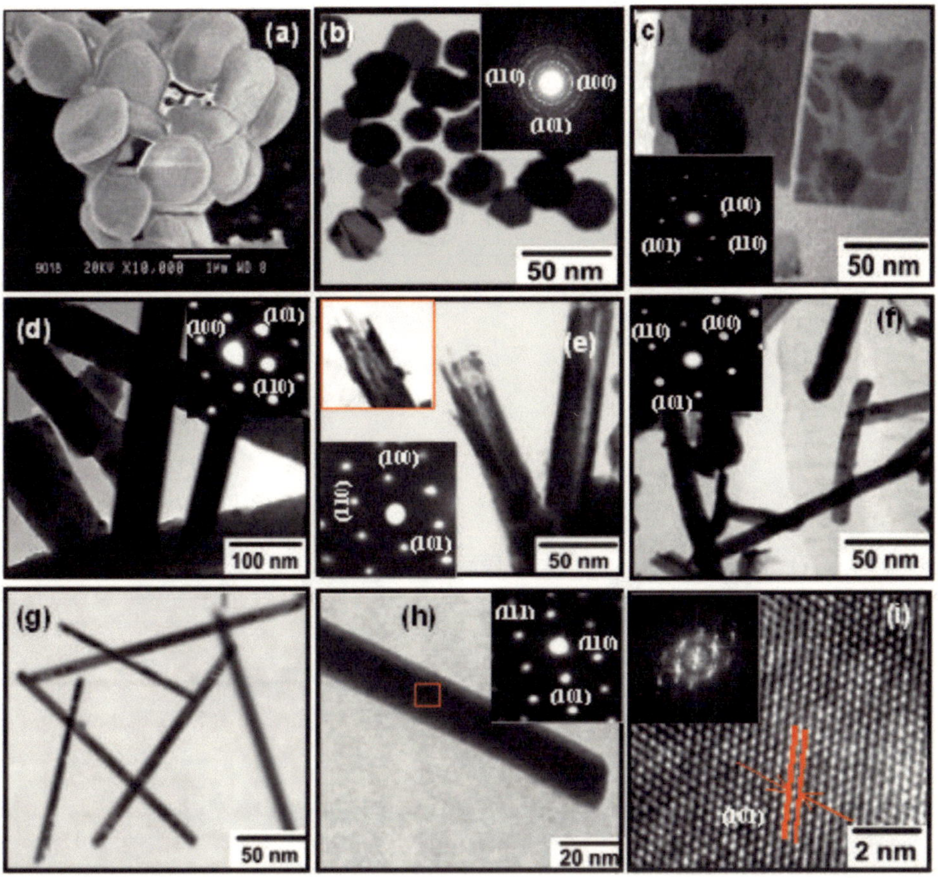

Fig. 4. (a) SEM image of hexagonal-shaped bulk $Gd_2O_2S:Eu^{3+}$ (B). The TEM images of $Gd_2O_2S:Eu^{3+}$ nanostructures (template - free method) from the sample B_7 to B_{12}: (b) Nanocrystals, (c) Nanosheets, (d) Nanobelts, (e) Nanotubes (inset: shows TEM image of the end portion of a single-nanotube), (f) Nanorods, (g) Nanowires, (h) a close up of the boxed area in (g) shows a structure of single-nanowire, (i) HRTEM image of $Gd_2O_2S:Eu^{3+}$ nanowire. The spacing between two adjacent lattice planes is 0.325 nm, which corresponds to the separation of the hexagonal phase lattice planes (110), and the inset shows Fast Fourier transform (FFT) pattern of the corresponding nanowire. *Courtesy: J. Mater. Sci. 44., 14., (2009) 3889-3899.*

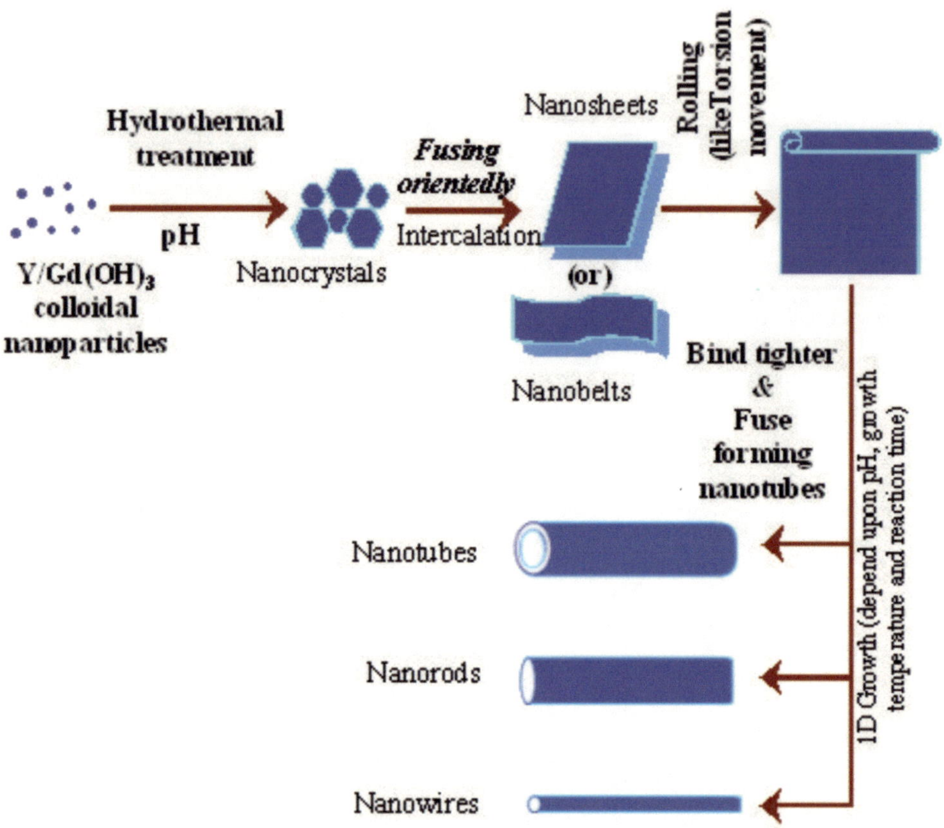

Fig. 5. A schematic diagram of the shape-selective synthesis of doped oxysulfide nanostructures *via* hydrothermal route (template - free method).

2.2 Possible formation mechanism

A schematic diagram of the shape-selective synthesis of doped oxysulfide nanostructures *via* template-free hydrothermal route is shown in Fig. 5. Based on these experimental results and discussions, a possible formation mechanism from nanocrystals to nanowires of hydroxides might be proposed as follows: When the $Y/Gd(NO)_3$ and NaOH solutions are mixed together, the resultant hydroxide nanocrystals can accommodate some Na ions and water molecules between their adjacent surfaces. According to Kerr (Kerr, 1996), these Na ions and water molecules are then exchanged and fused to form nanosheet or nanobelt according to the pH and temperature variation as indicated by the schematic diagram in Fig. 5. As the reaction continues the conformational variation inevitably causes torsional movement within the chains (Colomban, 1999; Mazerolles, 1999), causing the nanosheet / nanobelt to roll up. Because the thin lamellar nanosheet / nanobelt has very good flexibility, they curl and buckle with appropriately 45 degree rotation. As the curling continues, they bind tighter and fuse together at the two edges to form a nanotube with a cylindrical cross section. Furthermore, nanorods and nanowires are obtained according to reaction time, pH and temperature following the same formation kinetics.

3. Template-assisted method

In order to synthesize one-dimensional nanomaterial into a device, a fabrication method that enables well-ordered nanomaterials with uniform diameter and length is important. Template-directed growth is a nanomaterials fabrication method that uses a template, which has arrays of nanopores with uniform diameter and length that is needed for a device. Template-based growth is commonly a solution or colloidal dispersion based process. It is less expensive and readily scalable to mass production. The diameter, density and length of nanotubes, nanorods and nanowires are easily controlled independently. It also offers the advantage of less contamination and is environmentally benign. However, template-based synthesis slightly suffers from the polycrystalline nature of the resultant nanowires and nanorods, in addition to the difficulties to find appropriate templates with pore channels of desired diameter, length and surface chemistry and to remove the template completely without compromising the integrity of grown nanotubes, nanorods and nanowires and is another cumbersome work. In this work the $Y_2O_2S:Eu^{3+}$ and $Gd_2O_2S:Eu^{3+}$ of highly ordered arrays of super-nanostructures like, nanotubes, nanorods and nanoflowers were synthesized by hydrothermal synthesis *via* template assisted synthesis using commercially available AAO templates [Whatman Nuclepore® Inc., (13 mm diameter, 6 μm thickness)]. In comparison with the bulk, this template assisted nanosynthesis route offer several advantages for property study and practical applications. As a whole, to the best of our knowledge, this is the first attempt to prepare such nanostructures from the zero- to three-dimensional scale using both template - free and template - assisted method for these two sytems.

We have extended the method to the synthesis of various nanostructures like nanotubes, nanorods and nanoflowers preparation using template-assisted hydrothermal technique by varying reaction parameters such as solution pH, solvent temperature, time, etc. The porous AAO templates were immersed into a required amount of $[Y/Gd(CH_3COO)_3 \cdot 4H_2O]$ and $[Eu(CH_3COO)_3 \cdot 4H_2O]$ solution, respectively, and evacuated for about 15 min using a vacuum pump to get rid of bubbles within the nanopores. The template was taken out of the solution, rinsed with distilled water and dried in air. This procedure was repeated several times at regular intervals. Subsequently, the template was put into a Teflon-lined stainless steel autoclave, and the solution of $YEu(CH_3COO)_3$ and $GdEu(CH_3COO)_3$, respectively, followed by addition specific amount of Na_2S. The pH of the reaction solution was adjusted in the range from 6 to 9. After that the autoclave was tightly sealed, heated around 150 - 200 °C for 20 hrs and then allowed to cool down to room temperature naturally. For the synthesis of nanoflower-like structures, hexamethylenetetramine (HMT) was used as surfactant. The as-obtained $Y_2O_2S:Eu^{3+}$ and $Gd_2O_2S:Eu^{3+}$ template was subsequently annealed at 300°C for 1 hr under inert Sulphur atmosphere, respectively, to yield the final product.

Structural studies (XRD) reveal that the products $Y_2O_2S:Eu^{3+}$ and $Gd_2O_2S:Eu^{3+}$ are pure hexagonal phase and they are in good agreement with standard JCPDS [($Y_2O_2S:Eu^{3+}$: JCPDS # 24-1424) and ($Gd_2O_2S:Eu^{3+}$: JCPDS # 2 6-1422)] data (Thirumalai et al, 2008; Thirumalai et al, 2009 a,b) and they are highly single crystalline in nature. The patterns show obviously broadened diffraction peaks compared with the bulk $Y_2O_2S:Eu^{3+}$ and $Gd_2O_2S:Eu^{3+}$ systems, signifying the decrease in size of these crystallites. No peaks attributable to other types of oxysulfide(s) are observed in the XRD patterns, indicating the high purity of the phases obtained. The morphology of the resulting samples synthesized by the hydrothermal method was studied using SEM. Using the commercially available AAO templates the morphologies of highly oriented nanoarrays of structures like nanotubes, nanorods and nanoflowers of $Y_2O_2S:Eu^{3+}$ and $Gd_2O_2S:Eu^{3+}$ were synthesized. The optimal experimental conditions and resulting morphologies of $Y_2O_2S:Eu^{3+}$ and $Gd_2O_2S:Eu^{3+}$ nanostructures are given in Table - I. Fig. 6(a1-c2) & Fig. 7(a1-c2) show the SEM micrographs of the synthesized $Y_2O_2S:Eu^{3+}$ and $Gd_2O_2S:Eu^{3+}$ nanotubes, nanorods and nanoflowers, respectively. They were found to be single-crystalline in nature.

To investigate the optimized growth of the $Y_2O_2S:Eu^{3+}$ and $Gd_2O_2S:Eu^{3+}$ nanostructures, SEM micrographs shows highly oriented nanotubes, nanorods and nanoflowers, respectively, that were obtained by varying the reaction parameters (pH ~ 6 to 9, 150 - 200 °C, 20 hrs) and its resulting morphological features and crystallanity are found to be good. The studies reveal that by altering the reaction conditions (like solution concentration, pH, temperature and time) the morphology, shape and crystallinity are perfectly controlled. Fig 7(a1, a2) and Fig 8(a1, a2) show, the AAO templates top-view and cross-sections of the as-synthesized $Y_2O_2S:Eu^{3+}$ and $Gd_2O_2S:Eu^{3+}$ nanostructures, respectively. A cluster of nanotube-like arrays of the oxysulfide(s) growth were observed [Fig 7(b1, b2) and Fig 8(b1, b2)] at a pH of 6, 150 °C, 15 hrs. In Fig 7b1 and Fig 8b1 the tube and/or rod-like growth is clearly seen. The nanotubes and/or rods were grown at a pH of 6 - 9, 160 -180 °C, 20 hrs and the cross-section view reveals that the nanopores are perfectly filled. The nanorods are straight and have a uniform diameter of about 80 nm, which is basically equal to the pore size of the AAO template employed. To investigate the effect of solvent concentration on $Y_2O_2S:Eu^{3+}$ and $Gd_2O_2S:Eu^{3+}$ nanostructures, SEM micrographs were obtained at different solvent concentration (using hexamethyltetramine (HMT) as solvent along with water). The results are discussed and presented in Fig 7, 8(c1 and c2) and Table - I. The sample is composed of a large number of nearly uniform flower-like nanostructures to the surface of the AAO template. They are arranged uniformly in a large area. The average size of the as-obtained nanostructure is about 4-5 μm. A closer inspection reveals that the flower is made up of many thin petals, the thickness of which is ~ 30 nm. Fig. 6c2 and 8c2 show the cross-sectional view of SEM image of the $Y_2O_2S:Eu^{3+}$ and $Gd_2O_2S:Eu^{3+}$ nanorod arrays of length around hundred nanometers for the flower-like structures. Every petal is curled and thin, with a smooth surface and a large surface area. We can see a layer of flowers on the surface of the template. Because of these nanocrystalline sheets like small petals, the flower-like structure on the alumina template surface formed by these nanocrystalline sheets may be called nanoflowers. The nanoflower growth is clearly seen and the cross-section reveals that the nanopores are perfectly filled for concentration of 2 mM. Fig. 6(a2, b2, c2) and Fig. 7(a2, b2, c2), show a slightly discrete, unattached and dislocation free nanotube/rod structure in the direction vertical to the AAO template, respectively, which was different from the shared tube/rod wall between the tubes/rods of the AAO template. The nanotube/rod wall was also constituted by many deposited particles with an average particle size of 20 nm. It can be seen that the product obtained by the hydrothermal reaction with the AAO template had three structural configurations: (i) many fine nanoparticles packed to form tubes/rods, (ii) many parallel (iii) slightly discrete and completely dislocation free tubes/rods constituted nanorod-like arrays vertical to the AAO template, respectively. The formation of 1D structure depends greatly on the reaction kinetics. Such 3D structures with high surface areas can be used relevantly as catalysts, molecular sieves and biosensors (Zhang, 2007).

	Experimental Conditions*	Resulting morphologies
$Y_2O_2S:Eu^{3+}$	(pH~6, 150°C, 20h)	Nanotubes
	(pH~7-8, 200°C, 20h)	Nanorods
	(pH~8-9, 120°C, 12h)	Nanoflowers
$Gd_2O_2S:Eu^{3+}$	(pH~6, 150°C, 20h)	Nanotubes
	(pH~7-8, 200°C, 20h)	Nanorods
	(pH~8-9, 120°C, 12h)	Nanoflowers

Table 1. Optimized growth conditions of $Y_2O_2S:Eu^{3+}$ and $Gd_2O_2S:Eu^{3+}$ (Template - assisted Method) nanostructures

*The as-obtained $Y_2O_2S:Eu^{3+}$ and $Gd_2O_2S:Eu^{3+}$ grown on AAO template was subsequently annealed at 300 °C for 1 hr under inert (A_2 or N_2) / CS_2 / Sulphur atmosphere to yield the final product.

3.1 Possible formation mechanism

Highly oriented single-crystalline $Y_2O_2S:Eu^{3+}$ and $GdO_2S:Eu^{3+}$ nanotubes, nanorods and nanoflowers were synthesized using a template-assisted (AAO and Au coated AAO templates) hydrothermal technique. While Y/Gd $(CH_3COO)_3.4H_2O$: $Eu(CH_3COO)_3.4H_2O$ as precursors act as sources for Y/Gd : Eu ions, during the reaction process, the NaOH not only plays a role as a solvent for lowering the reaction temperature but also acts as a reactant and it is used for achieving basic environment through hydrolysis. Here, Na_2S is acting as a sulfurizing agent. Controlling the pH and reaction time is a key factor in achieving the morphological control. The porous structure of alumina with positive charged walls attracts the ions having negative charge from the solution leading to a preferential deposition at the walls that further grow towards the core. Supersaturation in the growth region is favorable to an anisotropic growth. The shape of a crystal is determined by the relative specific surface energy of each facet of the crystal. Therefore, the initial deposition of nanocrystals is critical for the formation of aligned nanotubes and nanorods. It is inferred that the AAO template may plays an important role in controlling the pH of the reaction mixture and/or local concentration, which leads to an inhomogeneous concentration distribution and may affect the shape development. Immersing the alumina template in the solution containing surfacant (HMT), the solution impregnated through pores to form nanorods immediately. However, the nanorods were not restricted in the pores anymore and extends to the surface of the alumina template, due to forces of adhesion. This in turn may lead to spreading nanorods out of the pore diameter, leading to the possible formation of nanosheets. The nanosheets so formed over the template surfaces comes in intact with the extended nanorods and try to form a stable structure. This situation may lead to the formation of flowery nanostructures. The schematic of the mechanism for nanotube, nanorod and nanoflower generations are shown in Fig. 8.

In summary, this novel template-free and template-based approach has been developed for growing high yield and highly oriented single crystal $Y_2O_2S:Eu^{3+}$ and $Gd_2O_2S:Eu^{3+}$ nanostructures using hydrothermal growth. The pH, growth temperature, reaction time and surfactant are mainly determining the shape of the nanostructures. The morphological studies were performed through SEM and TEM for the optimized conditions, which reveals excellent features and are reported. The morphological studies (SEM and TEM) revealed that the nanostructures (0D, 1D, 2D and 3D), with various structures like nanocrystals, nanosheets, nanobelts, nanotubes, nanorods, nanowires and highly oriented nanoarrays (tubes, rods, flowers) successfully synthesized. The $Y_2O_2S:Eu^{3+}$ and $Gd_2O_2S:Eu^{3+}$ nanostructures was further examined by HRTEM, electron diffraction and FFT images are seen to be single-crystalline in nature. Also, the SEM studies of the nanorods over AAO tempalte (template - assisted method) reveal that they are uniformly distributed throughout the surface exhibiting the superiority of the structures. The studies agree to a great extent with the structural studies results. This technique is found to be highly reproducible and can be extended to large area and large scale fabrication systems. Large-area, non-collapsed and highly-oriented $Y_2O_2S:Eu^{3+}$ and $Gd_2O_2S:Eu^{3+}$ nanostructures are expected as ideal functional components for opto-electronic and nanoscale devices of the next generation.

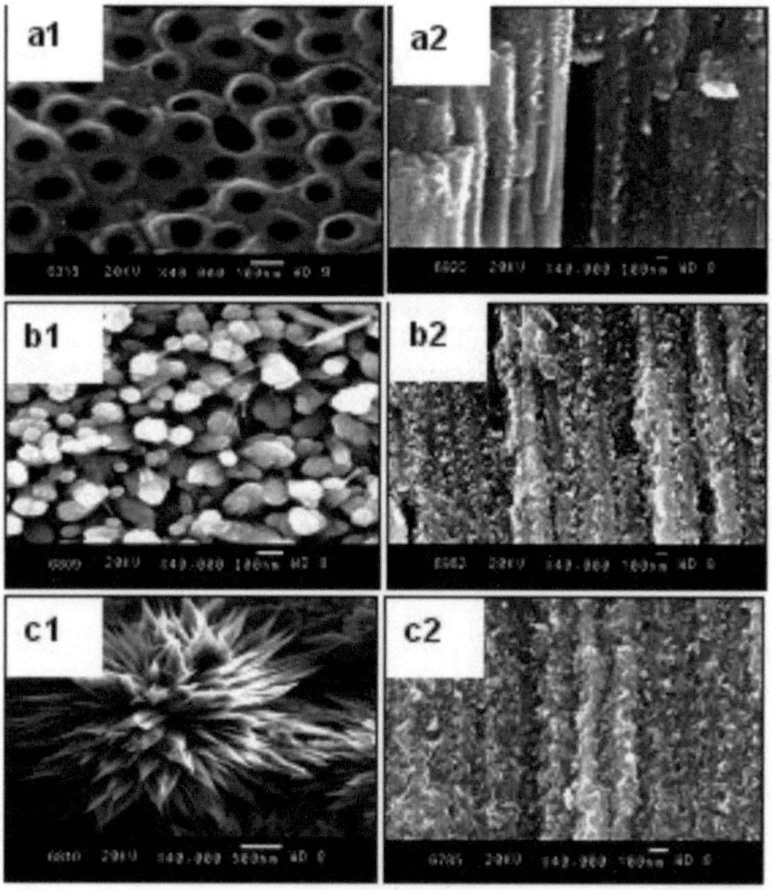

Fig. 6. The SEM images of $Y_2O_2S:Eu^{3+}$ nanostructures (template - assisted method): (a1, a2) Nanotubes, (b1, b2) Nanorods and (c1, c2) Nanoflowers. The SEM micrographs show the cross-sectional views of the samples (a2, b2, c2) corresponding to (a1, b1, c1).

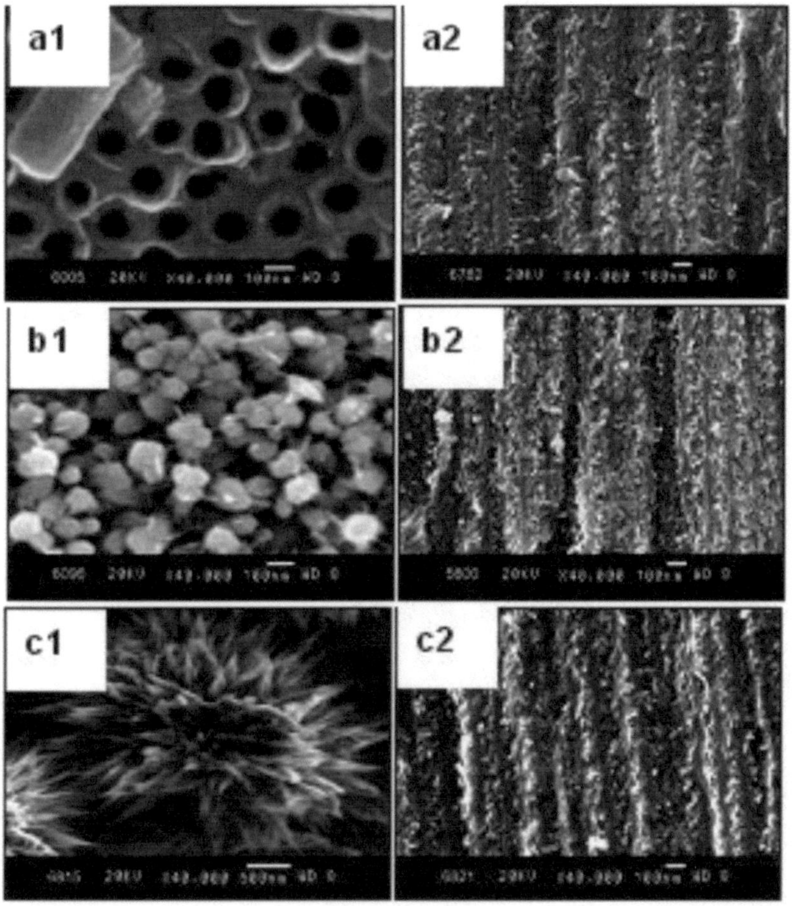

Fig. 7. The SEM images of $Gd_2O_2S:Eu^{3+}$ nanostructures (template – assisted method): (a1, a2) Nanotubes, (b1, b2) Nanorods and (c1, c2) Nanoflowers. The SEM micrographs show the cross-sectional views of the samples (a2, b2, c2) corresponding to (a1, b1, c1).

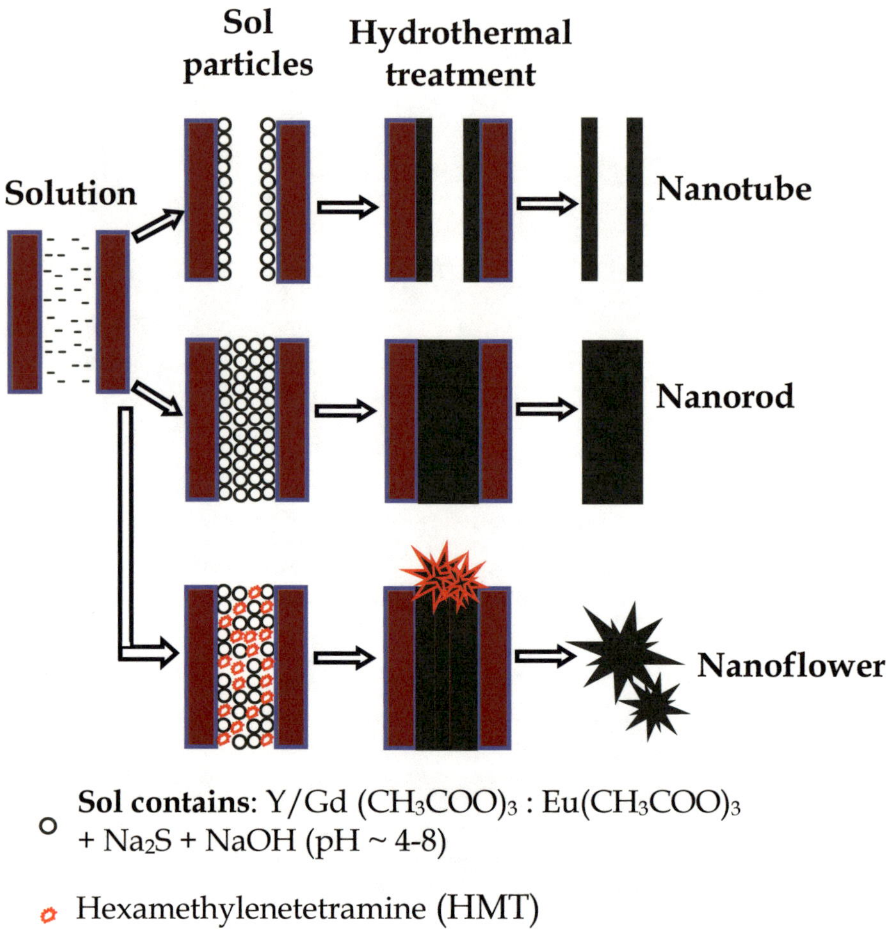

Fig. 8. Schematic illustration of the growth of nanotube, nanorod and nanflower structures by a template-assisted hydrothermal technique.

4. References

Bakovets, V. V.; Levashova, T. M.; Filatova, I. Yu.; Maksimovskii E. A. & Kupcha, A. E. (2008). Vapor phase growth of nanostructured yttrium oxysulfide films. *Inorg. Mater.*, 44., 1., (February & 2008) 67-69, ISSN: 0020-1685.

Bang, J.; Abboudi, M.; Abrams, B. & Holloway, P. H. (2004) Combustion synthesis of Eu-, Tb- and Tm- doped Ln_2O_2S (Ln=Y, La, Gd) phosphors. *J. Lumin.* 106., 3-4., (November & 2003) 177-185, ISSN: 0022-2313.

Chen, X. Y.; Zhuang, H. Z. & Liu G. K.; Li, S. & Niedbala, R. S. (2003). *J. Appl. Phys.*, 94, 1, ISSN 0021-8979.

Chou, T.W. ; Mylswamy, S.; Liu, R.S. & Chuang, S.Z. (2005). Eu substitution and particle size control of Y_2O_2S for the excitation by UV light emitting diodes. *Solid. State.Comm.*, 136., 4., (August 2005) 205-209, ISSN: 0038-1098.

Colomban, P.; Folch, S. & Gruger, A. (1999). Vibrational Study of Short-Range Order and Structure of Polyaniline Bases and Salts. *Macromolecules*. 32., 9., (April & 1999) 3080-3092, ISSN: 0024-9297.

Delgado da Vila, L.; Stucchi, E. B.; Davolos, M. R. (1997). Preparation and characterization of uniform, spherical particles of Y_2O_2S and Y_2O_2S:Eu. *J. Mater. Chem.*, 7., 10., 2113-2116, ISSN: 0959-9428.

Dhanaraj, J.; Jagannathan R. & Trivedi, D. C. (2003). $Y_2O_2S:Eu^{3+}$ nanocrystals—synthesis and luminescent properties. *J. Mater. Chem.*, 13., 7., (May & 2003) 1778-1782, ISSN 0959-9428.

Dhanaraj, J.; Geethalakshmi, M.; Jagannathan, R. and Kutty, T.R.N. (2004). Eu3þ doped yttrium oxysulfide nanocrystals - crystallite size and luminescence transition(s). *Chem. Phys. Lett.*, 387., 1-3., (March & 2004) 23-28, ISSN: 0009-2614.

Fu, Z.; Geng, Y.; Chen, H.; Zhou, S.; Yang, H. K. & J. H. Jeong, (2008). Combustion synthesis and luminescent properties of the Eu^{3+}-doped yttrium oxysulfide nanocrystalline. *Opt.Mater.* 31., 1., (September & 2008) 58-62., ISSN: 0925-3467.

Guo, C.; Luan, L.; Chen, C.; Huang D. & Su, Q. (2008). Preparation of $Y_2O_2S:Eu^{3+}$ *phosphors* by a novel *decomposition* method. *Mater. Lett.* 62., 4-5., (February & 2008) 600-602, ISSN: 0167-577X.

Hang, T.; Liu, Q.; Mao, D. & Chang, C. (2008). Long lasting behavior of $Gd_2O_2S:Eu^{3+}$ phosphor synthesized by hydrothermal routine. *Mater. Chem. Phys.*, 107., 1., (January & 2008) 142-147, ISSN: 0254-0584.

Hirai, T.; Hirano, T. & Komasawa, I. (2002). Preparation of Gd_2O_3 : Eu^{3+} and Gd_2O_2S : Eu^{3+} Phosphor Fine Particles Using an Emulsion Liquid Membrane System. *J. Colloid. Inter. Sci.* 253., 1., (September & 2002) 62-69, ISSN: 0021-9797.

Kerr, T. A.; Wu, H. & Nazar, L. F. Concurrent Polymerization and Insertion of Aniline in Molybdenum Trioxide: Formation and Properties of a $[Poly(aniline)]_{0.24}MoO_3$ Nanocomposite. (1996). *Chem. Mater.* 8., 8., (August 1996) 2005-2015, ISSN: 0897-4756.

Kong, Y. C.; Yu, D. P.; Zhang, B.; Fang, W. & Feng, S. Q. (2001). *Appl. Phys.Lett.*, 78., 407., ISSN 0003-6951.

Koskenlinn, M.; Leskela M. & Niinisto, L. (1976). Synthesis and Luminescence Properties of Europium-Activated Yttrium Oxysulfide Phosphors. *J. Electrochem. Soc.* 123., 1., (January & 1976) 75-78., ISSN: 0013-4651.

Kuang, J.; Liu, Y. & Yuan, D. *Electrochem.* (2005). Preparation and Characterization of $Y_2O_2S:Eu^{3+}$ Phosphor via One-Step Solvothermal Process. *Solid. State. Lett.*, 8., 9., (July & 2005),H72-H74, ISSN: 1099-0062.

Lei, B.; Liu, Y.; Zhang, J.; Meng, J.; Man, S. & Tan, S. (2010). Persistent luminescence in rare earth ion-doped gadolinium oxysulfide phosphors. *J. Alloys. Comp,* 495., 1., (February and 2010) 247-253, ISSN: 0925-8388.

Leskelä M. & Niinistö, L. (1976). Solid solutions in the rare-earth oxysulfide series. *J. Solid. State. Chem.* 19., 3., (November 1976) 245-250, ISSN: 0022-4596.

Li, W; Liu, Y; & Ai, P. (2010). Synthesis and luminescence properties of red long-lasting phosphor $Y_2O_2S:Eu^{3+}$, Mg^{2+}, Ti^{4+} nanoparticles. *Mater. Chem. Phys.*, 119, 1-2., (January & 2010) 52-56, ISSN: 0254-0584.

Liu, Y. & Kuang, J. (2010). Intense visible luminescence from Nd^{3+}-doped yttrium oxysulfide. *J. Lumin.* 130., 3., (March and 2010) 351-354, ISSN: 0022-2313.

Mazerolles, L.; Floch, S. & Colomban, P. (1999). Study of Polyanilines by High-Resolution Electron Microscopy. *Macromolecules*. 32., 25., (November & 1999) 8504-8508, ISSN: 0024-9297.

Mao, S.; Liu, Q.; Gu, M.; Mao D. & Chang, C. (2008). Long lasting phosphorescence of Gd_2O_2S:Eu,Ti,Mg nanorods via a hydrothermal routine. *J. Alloys and Comp.*, 465., 1-2., (October & 2008) 367-374, ISSN: 0925-838.

Mikami, M. & Oshiyama, A. (1998). First-principles band-structure calculation of yttrium oxysulfide.*Phys. Rev. B,* 57., 15., (April &1998) 8939-8944, ISSN: 1098-0121.

Nakkiran, A.; Thirumalai, J.; & Jagannathan, R. (2007). Luminescence blinking in Eu^{3+} doped yttrium oxysulfide ($Y_2O_2S:Eu^{3+}$) quantum-dot ensembles: Photo-assisted relaxation of surface state(s). *Chem. Phys. Lett.*, 436., 1-3., (February and 2007)155-161, ISSN: 0009-2614.

Pan, Z. W.; Dai, Z. R. & Wang, Z. L. (2001). Nanobelts of Semiconducting Oxides. *Science.* 291., 5510., (March & 2001) 1947-1949, ISSN: 0036-8075.

Sabot, J. L. & Maestro, P. (1995). *Kirk-Othmer Encyclopedia of Chemical Technology, Concise, 4th Edition,* John Wiley & Sons, IV-edn, ISBN: 0-471-41961-3.

Tseng, Y.H.; Chiou, B.S.; Peng, C.C. & Ozawa, L. (1998). Spectral properties of Eu^{3+}-activated yttrium oxysulfide red phosphor. *Thin Solid Films.*, 330., 2., (September & 1998) 173-177, ISSN: 0040-6090.

Thirumalai, J.; Jagannathan, R. & Trivedi, D.C. (2007). $Y_2O_2S:Eu^{3+}$ nanocrystals, a strong quantum-confined luminescent system. *J. Lumin.* 126., 2., (October and 2007) 353-358, ISSN: 0022-2313.

Thirumalai, J.; Chandramohan, R.; Sekar M.& Rajachandrasekar, R. (2008). Eu^{3+} doped yttrium oxysulfide quantum structures – structural, optical and electronic properties. *J. Nanopart. Res.* 10., 3., (August & 2007) 455-463, ISSN: 1388-0764.

Thirumalai, J.; Chandramohan, R.; Divakar, R.; Mohandas, E.; Sekar, M. & Parameswaran, P. (2008). Eu^{3+} doped gadolinium oxysulfide (Gd_2O_2S) nanostructures –synthesis and optical and electronic properties. *Nanotechnology.* 19., 39., (October & 2008) 395703-5, ISSN: 0957-4484.

Thirumalai, J.; Chandramohan, R.; Auluck, S.; Mahalingam, T. & Srikumar, S. R. Controlled synthesis, optical and electronic properties of Eu^{3+} doped yttrium oxysulfide (Y_2O_2S) nanostructures. (2009a). *J. Colloid. Interface. Sci.* 336., 2., (August & 2009) 889-897, ISSN: 0021-9797.

Thirumalai, J.; Chandramohan, R.; Valanarasu, S.; Vijayan T. A.; Somasundaram, R. M.; Mahalingam, T. & Srikumar, S. R. (2009b). Shape-selective synthesis and opto-electronic properties of Eu^{3+}-doped gadolinium oxysulfide nanostructures. *J. Mater. Sci.* 44., 14., (July & 2009) 3889-3899, ISSN: 0022-2461.

Wang, X. & Li, Y. (2003). Rare-Earth-Compound Nanowires, Nanotubes, and Fullerene-Like Nanoparticles: Synthesis, Characterization, and Properties.*Chem. Eur. J.* 9., 22., (November & 2003), 5627-5635, ISSN: 0947-6539.

Wang, X.; Sun, X.; Yu, D.; Zou, B. & Li, Y. D. (2003). Rare earth compound Nanotubes. *Adv. Mater.* 15., 17., (September & 2003) 1442-1445, ISSN: 0935-9648.

Xixian, L.; Wanghe, C. & Mingming, X. (2006). Preparation of nano Y_2O_2S:Eu phosphor by ethanol assisted combustion synthesis method. *J. Rare. Earths.*, 24., 1., (February & 2006) 20-24, ISSN: 1002-0721.

Yeboah, C. & Pistorius, S. (2000). Monte Carlo studies of the exit photon spectra and dose to a metal/phosphor portal imaging screen.*Med. Phys.* 27., 2., (February & 2000) 330-339, ISSN: 0094-2405.

Yu, T.; Zhu, Y.; Xu, X.; Shen, Z.; Chen, P.; Lim, C-T.; Thong, J.T-L.& Sow, C-H. (2005). Controlled Growth and Field-Emission Properties of Cobalt Oxide Nanowalls. *Adv. Mater.* 17., 13., (July & 2005) 1595-1599, ISSN: 0935-9648.

Zhang, B. P.; Binh, N. T.; Wakatsuki, K.; Segawa, Y.; Yamada, Y.; Usami, N.; Kawasaki, M. & Koinuma, H. (2004). *Appl. Phys. Lett.*, 84., 4098, ISSN 0003-6951.

Zhang, C.; Tao, F.; Liu, G-Q.; Yao, L- Z. & Cai, W-L. (2008). Hydrothermal synthesis of oriented MnS nanorods on anodized aluminum oxide template. *Mater. Lett.* 62., 2., (January & 2008) 246-248, ISSN: 0167-577X.

Zhang, D.; Fu, H.; Shi, L.; Fang, J. & Li, Q. (2007). Carbon nanotube assisted synthesis of CeO_2 nanotubes. *J. Solid. State. Chem.*, 180., 2., (February 2007) 654-660, ISSN: 0022-4596.

Zhang, J.; Sun, L.; Yin, J.; Su, H.; Liao, C. & Yan, C. (2002). Control of ZnO Morphology via a Simple Solution Route. *Chem. Mater.* 14., 10., (September & 2002) 4172-4177, ISSN: 0897-4756.

4

Emission of semiconductor nanocrystals in photonic crystal environment

Sergei G. Romanov* and Ulf Peschel
Cluster of Excellence "Advanced Engineered Materials"
Institute of Optics, Information and Photonics, University of Erlangen - Nürnberg, Günther-Scharowsky-Str.1, 91058 Erlangen, Germany
**Ioffe Physical Technical Institute, Polytekhnishesckaya ul., 26, 194021 St. Petersburg, Russia*

1. Introduction

Widespread application of colloidal semiconductor nanocrystals (NCs) [1] in light emitting devices is guaranteed by their low production cost, broadly variable emission frequency and relatively long lifetime. In order to exploit the advantages of NCs, the light sources should be equipped with non-dissipative photon management structures with the aim to form the spectrum of the source, to manage the emission diagram and to arrange the radiative recombination of electron excitations in the most purpose-efficient way. Obviously, the electrically biased light sources are the most application relevant [2], but developing the principles of photon management architectures can be performed with systems operating with photoexcitation of the NC emission.

One of the most awaiting realisations of the photon management in light sources is based on the photonic crystals (PhC). The radiation in such sources is controlled through modification of the electromagnetic vacuum [3,4,5]. The pre-condition of such control is the modulation of the photon density of states (DOS) that allows either accelerate or suppress the radiation rate of light emitters in a pre-defined spectral range. The development of PhC-based sources begun with investigation of the spectra and directionality changes of the emission from embedded sources. Nowadays the main aim is to design special defect states that are capable of purposive shaping the emission characteristics.

In PhC-based devices the emitter should be coupled to the outside world only via PhC eigenmodes in order to gain a full control upon emission characteristics. This can be achieved by immersing the emitter in the PhC interior. At the early stage it became clear that the fabrication of the 3-dimensional PhCs for the visible range of the spectrum took a long time due to complexity of required nanofabrication [6,7,8]. Meanwhile the nature offers an easy affordable solution - the opal crystals [9,10], the play of colours of which is based on principles of PhCs. Opals consist of face centred cubic packed lattice of silica spheres [11], i.e., the dielectric permittivity inside the opal is periodically modulated in all three dimensions. The interference of light waves that are scattered in the opal lattice results then in formation of directions, which are forbidden for propagation, if the wavelength is comparable to the lattice constant.

This phenomenon can be formalised in terms of the photonic energy band structure that looks similar to the electron energy band structure of solids. This energy band structure represents the dispersion of propagating modes in the energy-wavevector space. The spectral intervals without propagating modes are called the photonic bandgaps (PBG). Obviously, the complete absence of modes can be achieved only in infinitely large PhC with high refractive index contrast between scatterers and surrounding medium, hence, in reality it is more important to differentiate PhCs with omnidirectional and directional PBG. So far, no omnidirectional PBG was demonstrated in the visible due to the absence of dielectrics with high enough index of refraction [12], but this requirement can be fulfilled in the near-infrared with, e.g., Si or Ge-based inverted opals.

In the visible we ought to deal with directional bandgap crystals. This means that the internal light source, the emission of which is simultaneously coupled to the all available modes of a PhC, will not be blocked completely at any single frequency. Thus, there are two fractions of the energy flow - the one that is modified by the interaction with the structure and the other that leaks unaffected from the crystal. The aim of designing the photon management architectures is to tailor the fraction under control in the purposive way and to maximise the modified fraction of light flow.

The attractiveness of opals for emission manipulation was immediately realised by researches and the very first publications that considered the opal as a PhC were aimed primarily at the emission modification [13,14]. Certainly, the bulk opals in use were structurally very imperfect and the in-void synthesised emitters were randomly distributed over the PhC volume. However, even in such conditions it became possible to establish some links between PBG and emission characteristics. Since that time the tremendous progress was achieved triggered by the invention of artificial opal films of high structural quality [15]. Another crucial development in late 90s was the infiltration of the opals with colloidal NCs as a method that allows to preserve the crystal quality and the refractive index contrast of the opal-based PhC [16]. Nowadays semiconductor NCs are conveniently used in studies of the PhC light sources [17].

The opals discussed in this chapter are the PhC with the directional PBG. We will consider the photoluminescence (PL) from opal-embedded NCs possessing the emission band in the visible under continuous wave (cw) excitation of moderate power. Any time-resolved characteristics and optical non-linearities are outside the scope of our discussion. The aim of this chapter is to demonstrate what kind of emission modification one can achieve with NC that are evenly distributed over the opal-based PhC volume or placed in close vicinity to the PhC surface. We will describe several tests that allow to identify the emission changes and estimate the order of magnitude of these changes.

2. General properties of light sources in photonic crystals

Let us introduce some definitions. In the free space the number of electromagnetic (EM) field modes with frequency $\omega < \omega_0$ is $N(\omega) = \dfrac{\omega^3 V}{3\pi^2 c^3}$. Correspondingly, the mode density is $D(\omega) = \dfrac{dN(\omega)}{d\omega}$ and $D(\omega) \propto \omega^2$ (Fig.1). The dipole emission power is $U = \dfrac{4\pi^2\omega^3}{3V}\left|\vec{d}\right|^2 D(\omega)$ and the emission rate is $\Gamma = U/\hbar\omega \propto \omega D(\omega)$. Using the Golden Fermi rule and following [18] we consider the 2-level system with the transition frequency ω_0 that interacts with EM field of vacuum. The rate of radiative recombination from an excited $|e\rangle$ to a ground $|g\rangle$ state can be written taking into account the field quantization as

$$\Gamma_{e\to g} = \frac{2\pi}{\hbar^2}\sum_{\vec{k},n}\left|\left\langle g,1_{\vec{k}n}\right|\hat{\vec{d}}\bullet\hat{\vec{E}}(\vec{x}_0)\left|e,vac\right\rangle\right|^2\delta(\omega_{\vec{k},n}-\omega_0) \tag{0.1}$$

The summation is taken over all one photon states $\left|1_{\vec{k},n}\right\rangle$ that satisfy the energy conservation law. 2-level atom is located at x_0 and its dipole operator is $\hat{\vec{d}}$.

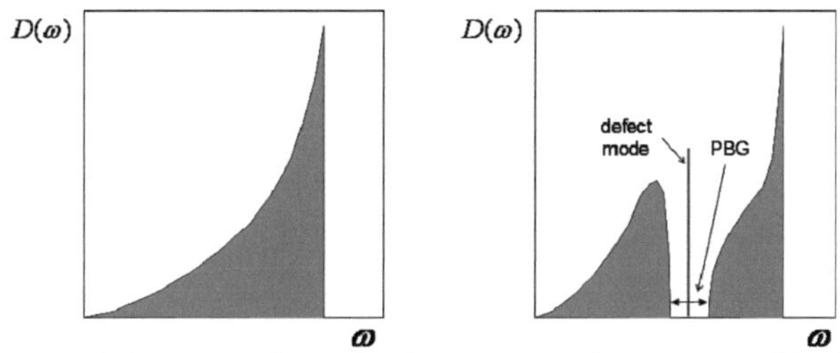

Fig. 1. (a) The spectrum of the optical mode density in a free space. (b) The same spectrum in a PhC with complete omnidirectional PBG. The defect mode is shown at the mid-frequency of the bandgap.

In the PhC this general expression is reduced to the density of modes $D(\omega)$. Let us expand the operator of electric field over the Bloch modes $\vec{E}_{\vec{k},n}(\vec{x})$ with quasimomentum $\vec{k}$ and number n.

$$\hat{\vec{E}}(\vec{x}) = \sum_{\vec{k},n}\sqrt{\frac{\hbar\omega_{\vec{k},n}}{2\varepsilon_0 V}}(\vec{E}_{\vec{k},n}(\vec{x})\hat{a}_{\vec{k},n} + h.c.) \tag{0.2}$$

where $\hat{a}_{\vec{k},n}$ - is the photon annihilation operator in the mode $\vec{k},n$, and the mode field is normalised to crystal volume

$$\frac{1}{V}\int_V dx\varepsilon(\vec{x})\left|\vec{E}_{\vec{k}.n}(\vec{x})\right|^2 = 1 \tag{0.3}$$

Then the emission rate is

$$\Gamma_{e\to g} = \frac{\pi\omega_0\left|\vec{d}\right|^2}{\varepsilon_0\hbar}\rho(\vec{x}_0,\vec{u},\omega_0) \tag{0.4}$$

where $\rho(\vec{x}_0,\vec{u},\omega_0)$ - is the projection of the local density of states (LDOS).

$$\rho(\vec{x}_0,\vec{u},\omega_0) = \frac{1}{V}\sum_{\vec{k},n}\left|\vec{u}\bullet\vec{E}_{\vec{k},n}(\vec{x}_0)\right|^2\delta(\omega_{\vec{k},n}-\omega_0), \tag{0.5}$$

where the matrix element of dipole transition $\vec{d} = \left\langle g \middle| \hat{\vec{d}} \middle| e \right\rangle$ and $\vec{u}$ is the unit vector along $\vec{d}$ direction. It is apparent from (0.4) that the photon emission rate is proportional to the density of photonic modes and the square of the electric field, in particular, the spontaneous emission is suppressed in the interval of low mode density. The full density of modes is the the local density of radiating states that is averaged over the dipole orientation in the unit cell. This is why the PBG in a full DOS $D(\omega) = 2\omega \sum_{\vec{k},n} \delta(\omega^2 - \omega_{\vec{k},n}^2)$ leads to the PBG in the local DOS independently on the emitter orientation.

LDOS differs from DOS by taking into account the distribution of the mode intensity in the primitive lattice cell [19]. In fact, the dipole can add a photon to the emission flux if not only the state is available in the field but also the field magnitude differs from zero in the dipole site. Depending on the dipole positioning in the unit cell of a PhC the LDOS changes and, accordingly, changes the emission rate. For example, the air modes at the upper PBG edge possess the field maximum in the regions of "light" dielectric. Hence, in order to maximise the outcome, the emitter should be positioned in the middle of the PhC voids. If this emitter finds itself in the point of the zero LDOS, only the weak spontaneous emission will be possible even in the non-zero total DOS. At the PBG edges the spontaneous emission can be enhanced, because of higher LDOS. But in the infinitely large PhC obeying the spherical symmetry (the case that cannot be realised experimentally) the Fermi Golden rule fails because of the DOS discontinuity. In the case of finite size PhC such discontinuity is replaced by van Hove singularity and disappears in the case of directional PBG.

In order to consider the stimulated emission one has to introduce the impurity atoms with certain volume density $\sigma(\vec{r})$ that experience the polarization α by the PhC field and possess the inverse population of their energy levels $\operatorname{Im}\alpha < 0$. Polarization of impurity atoms that is excited in point $(\vec{r},t)$ by the mode $E_{\vec{k},n}^{(T)}$ can be expressed as

$$P_{st}^{(1)}(\vec{r},t) = \alpha\sigma(\vec{r})E_{\vec{k},n}^{(T)}(\vec{r})\exp(-i\omega^{(T)}{}_{\vec{k},n} + \delta)t \tag{0.6}$$

In self consisted approach it is necessary to include the polarization wave $\vec{P}_{st}^{(2)}(\vec{r},t)$ that is induced by the field $\vec{E}^{(1)}(\vec{r},t)$:

$$\vec{P}_{st}^{(2)}(\vec{r},t) = \alpha\sigma(\vec{r})\vec{E}^{(1)}(\vec{r},t) \tag{0.7}$$

The polarization $\vec{P}_{st}^{(2)}(\vec{r},t)$ induces the electric wave

$$\vec{E}^{(2)}(\vec{r},t) = \tfrac{1}{2}\beta_{\vec{k},n}^2 l^2 \vec{E}_{\vec{k},n}^{(T)} \exp(-i\omega + \delta)t \quad (0.8) \tag{0.9}$$

the full self consisted stimulated field is expanded in the series

$$E_{\vec{k},n}^{(T)}(\vec{r}) = \sum_{j=0} \vec{E}^{(j)}(\vec{r},t) = \vec{E}_{\vec{k},n}^{(T)}(\vec{r})\exp(\beta_{\vec{k},n} l - (i\omega_{\vec{k},n}^{(T)} + \delta)t) \tag{0.10}$$

On the one hand, the enhancement factor of stimulated emission per unit length $\beta_{\vec{k},n} \propto 1/v_g(\vec{k},n)$ is inversely proportional to the group velocity $v_g(\vec{k},n) = \frac{\partial \omega_{\vec{k},n}^{(T)}}{\partial k_z}$ along, e.g., the z direction. On the other hand, $\beta_{\vec{k},n} = \frac{i\alpha\omega_{\vec{k},n}^{(T)} F_1(\vec{k},n)}{2v_g(\vec{k},n)}$, where the effective density of impurities normalised to the mode intensity $\vec{E}_{\vec{k},n}^{(T)}(\vec{r})$ is

$$F_1(\vec{k},n) = \frac{1}{V}\int_V \sigma(\vec{r})\left|\vec{E}_{\vec{k},n}^{(T)}(\vec{r})\right|^2 d\vec{r} \tag{0.11}$$

Finally, the stimulated emission $Est(\vec{r},t) = E_{\vec{k},n}^{(T)}(\vec{r})\exp(\beta_{\vec{k},n} l - (i\omega_{\vec{k},n}^{(T)} + \delta)t)$ acquires the enhancement in the range of slow waves.

It is worth noting the more elaborated approach to the dynamics of emission in PhC. For example, the strongest modification experiences the emitter whose emission band is located at the edge of the complete PBG due to strong DOS anomaly [20]. The emitter with transition frequency deeply inside PBG will emit photon, but this photon has no chance to propagate and will be absorbed by the same emitter. S. John named this situation as the „dressed" atom [21]. If the transition frequency falls close to the edge, the emission dynamics will be different from the free space emission owing to limited number of modes. This last case bears multiple quantum optics consequences, e.g. laser-like collective emission [22] due to non-Markovian interaction (the interaction that depends on its history) between the atom and the field.

In the case of directional PBG in finite size PhCs the effect of PBG upon the spontaneous emission becomes much less pronounced, first of all, because of the shallow variation of the DOS spectrum in the PBG [23]. Hence, such crystals can be used, primarily, to design PhCs with pre-defined emission indicatrix and for directional enhancement or suppression of the emission in a pre-defined spectral range.

For the inefficient light sources with low quantum efficiency $\eta << 1$, the overall relaxation rate

$$\Gamma_{tot} = \Gamma_{NR}(1 + \eta + O(\eta^2)) \tag{0.12}$$

is defined by the non-radiative transitions Γ_{NR}. Then, the emitter power

$$I = P\Gamma_{rad} / \Gamma_{NR} \tag{0.13}$$

must be proportional the radiative emission rate and correspondingly, to the LDOS as well as the pumping power P. This case is applicable to all experiments described below. It should be emphasised that the power emitted in the narrow homogeneous emission band is proportional to LDOS only if the emission bandwidth is much narrower compared to the spectral interval of the LDOS variation. In the opposite case, namely the emitter with the broad inhomogeneous bandwidth, the LDOS spectrum should be acquired. One way to obtain the LDOS spectrum is to compare the emission of two samples, one PhC-based and one non-PBG reference that possess the same channel of non-radiative recombination [24]:

$$\frac{I_{PhC}}{I_{ref}} = \frac{P_{PhC}}{P_{ref}} \times \frac{\Gamma_{radPhC}}{\Gamma_{rad.ref}} = K \frac{D_{radPhC}(\omega)}{D_{rad.ref}(\omega)} \tag{0.14}$$

This is the true spectrum only in the case of quadratic DOS spectrum in the reference sample and the emitter localisation in the PhC along the constant LDOS surface. In the opposite and more frequent case of the broad distribution of emitters over the unit cell volume, this method produces the correct estimate of specific extrema in LDOS spectrum.

3. Experimental technique

The opal crystals used for impregnation with colloidal NC are the thin film crystals prepared by either sedimentation or by crystallization in the moving meniscus. This technology is well established and documented [15]. The current improvement of crystallization methods aims at eliminating the cracks of the film [25] and better crystallinity [26]. Typical example of the opal film is shown in Fig.2. The symmetry of the opal lattice is very close to the face centred cubic (fcc) symmetry (Fig.3). Opal possess the open porosity that allows embedding the semiconductor NCs in voids between touching spheres (Fig.3a,b). In order to describe the light propagation in the PhC it is convenient to use the lattice representation in the reciprocal space (Fig.3c). If the optical data are collected along the normal to the opal film, this direction corresponds to the ΓL directions in the 1st Brillouin zone. The dashed line on the surface of the Brillouin zone is the line along which the data were obtained along the oblique direction to the film normal.

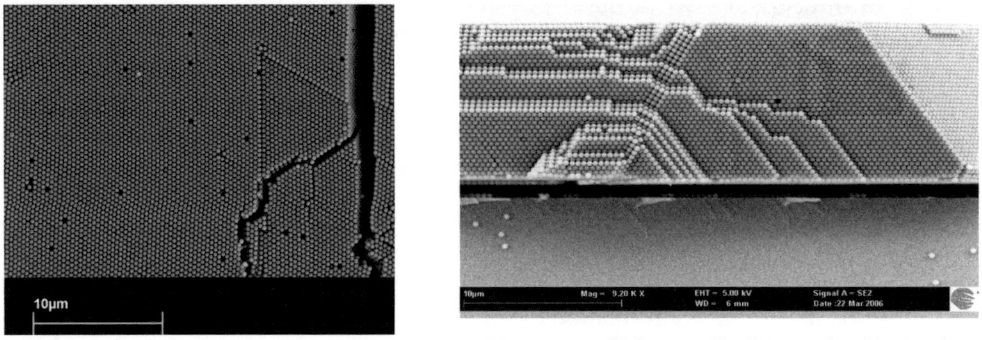

Fig. 2. Scanning electron microscope images. Top view (left panel) shows the (111) plane of fcc lattice at the film surface and the film cracks on the right hand side and side view (right) shows the film cross-section. The opal film is crystallised in the moving meniscus from PMMA spheres of 368 nm in diameter.

Transmission/reflectance spectra of the opal films were measured under white light illumination from a tungsten lamp. The transmitted/reflected light was collected within a solid angle of approximately 2° along different directions with respect to the [111] axis.

The energy band structure of the opal film was calculated under assumption of the fcc lattice symmetry (Fig.3d). This diagram relatively closely corresponds to the PBG structure revealed by the transmission spectra of the opal film measured at different angles of light propagation (Fig.3e). Bragg law approximation allows to associate the most pronounced resonances in these transmission spectra with diffraction at crystal planes.

PL measurements were performed under continuous wave (cw) excitation from an Ar+ gas laser (Fig.4). Typically, the beam was focused in 0.1 mm in diameter spot. Two schemes were used for PL excitation/collection - the back window, in which the PL is collected from the sample side that is opposite with respect to illuminated side, and the front window, in which the PL signal was measured from the same illuminated side of the sample (Fig.5). The PL was typically collected from a solid angle $\Omega = 5^o$. PL spectra were recorded, when the PL intensity becomes stable after each change of the excitation power. In order to trace the anisotropy of the emission, PL spectra were measured at different angles θ with respect to the [111] axis of the opal lattice. In the case of an array of randomly oriented dipoles smaller than the wavelength, the averaging over the sample volume results in an isotropic light source. In this case, the distortion of the spherical wavefront of an isotropic PhC-embedded emitter is a direct consequence of the PBG. In what follows, we will refer to the fraction of the wavefront that is blocked for propagation by the first bandgap as the Bragg cone.

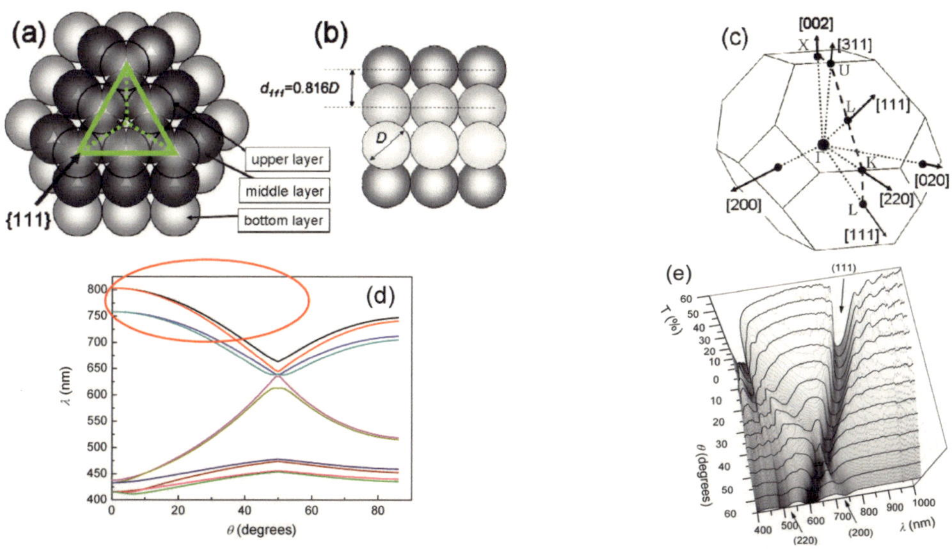

Fig. 3. (a, b) Schematics of the fcc lattice fragment. The {111} family of planes is represented by a tetrahedron. The interstitial voids between touching spheres are clearly seen. (c) Brillouin zone of the fcc lattice. Letters show the main symmetry points. Numbers at arrows name the directions in the reciprocal space in correspondence to the lattice axes in the real space. (d) The energy band structure of the opal crystal assembled from PMMA spheres of 368 nm in diameter (see Fig.2). The ellipse marks the typical range of this diagram, which is relevant to the discussion in this chapter. (e) Example of experimental transmission spectra of the opal film from 368 nm spheres measured under s-polarised light [27]. Numbers are the Miller indices of fcc crystal planes, the diffraction at which corresponds to the transmission minima (compare to panel (d)).

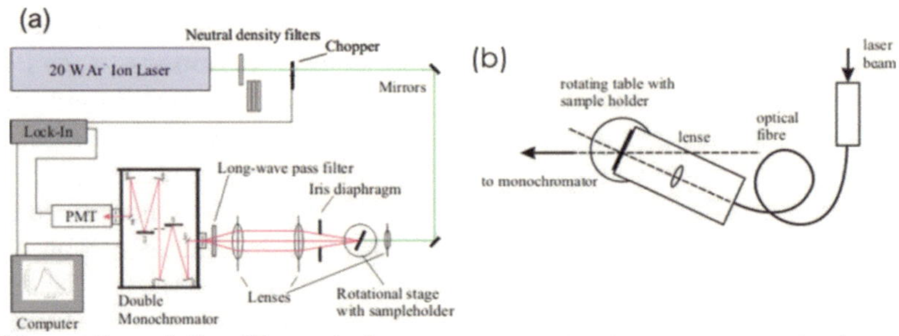

Fig. 4. (a) Schematics of the PL typical measurement set-up - cw excitation and lock-in-based registration detection. (b) Improved excitation conditions to allow constant size of illuminated spot to be preserved while changing the detection angle.

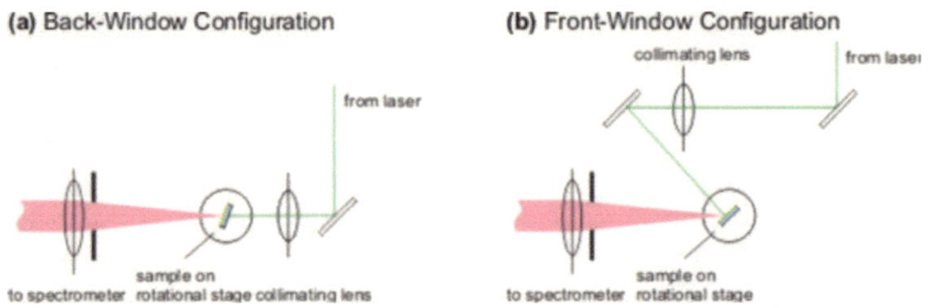

Fig. 5. Schematics of PL measurements in transmission (a) and in reflectance mode (b).

4. Directional suppression of CdTe nanocrystal emission in thin opal film

The opal films for this experiment were prepared from 1% aqueous colloidal solution of latex spheres of D=240 nm in diameter dried on glass slides (1 cm^2). Films crystallize in the randomised fcc lattice, which has the [111] axis as the growth direction. Subsequently, films with a thickness of 20-30 μm were sintered for 2 h at 100°C. CdTe core-shell NCs were synthesized as described elsewhere [28]. A polymer shell was used to prevent the agglomeration of colloidal particles. Infiltration of CdTe colloidal NCs into an opal film was performed by dipping the latter in 0.02 M (referring to Te) CdTe NC aqueous colloidal suspension for 1 min. As a result of electrical charging of these polymer shells, the CdTe NCs are attached to the surface of latex spheres (Fig.6a).

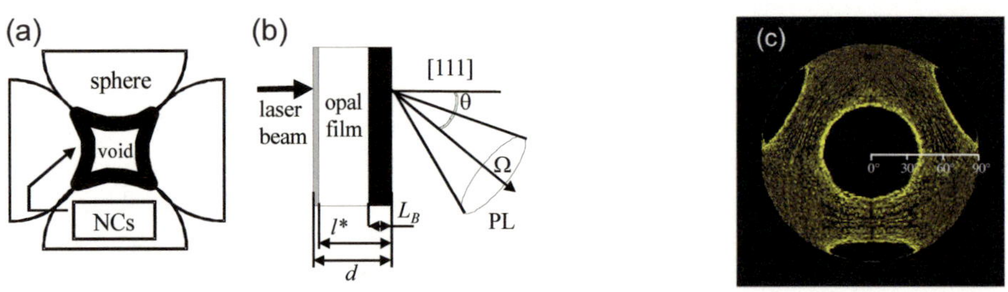

Fig. 6. (a) Schematics of NC CdTe layout in the octahedral opal void. (b) Characteristic length scales that are relevant to the formation of the emission spectrum of opal-embedded NC. (c) The spatial distribution of emission from a point source located inside the opal lattice. View along [111] axis ($\theta = 0^o$). Black holes are the Bragg cones. *Courtesy of D. Chigrin.*

The transmission spectrum of the opal film demonstrates the minimum centred at 2.23 eV (Fig.7a), which manifests the directional (111) bandgap. The position of this minimum corresponds to the Bragg diffraction resonance at the stack of (111) planes in the fcc lattice (Fig.3b) $\hbar\omega_B = 2\pi c/(2\times n_{eff}\times 0.816D)$, where n_{eff} is the effective index of refraction obtained from the effective medium approximation to the opal lattice and c is the light velocity. The relative bandwidth of the transmission minimum $\Delta E/E_B \approx 0.067$ exceeds by 20% the gapwidth calculated for ideally packed opal [29], which is an indication of lattice disorder. Impregnation of the opal with CdTe NCs leads to the "red" shift of this minimum by 0.05 eV due to increase of the refractive index and the transmission reduction at $\hbar\omega > 2.5$ eV, i.e., above the absorption edge of CdTe NCs. The CdTe fraction can be estimated from this shift as 1 to 2 volume % (for different samples) or up to 4% of the void volume. Important, that impregnating the opal with NCs does not destroy the optical quality of the opal-based PhC. The spectral position of the transmission minimum changes rapidly with changing the incidence angle of the light beam with respect to the film normal according to the Bragg law. Due to the destructive influence of opal crystal defects the transmission attenuation in (111) resonance gradually decreases with the angle increase $\theta \geq 70^o$ (Fig.3e) [26].
The relatively narrow linewidth of the NC emission (Fig.7b) compared to the PBG width would not allow for tracing the emission change at different overlaps with the PBG. But the PL bandwidth of NC in the opal appears much broader due to NC interaction with the inner opal surface. Moreover, the PL bandwidth in the CdTe-opal was broadened due to partial degradation of the NC luminescence after intense laser light illumination. On a later stage the excitation power was kept below 6 mW to avoid further degradation of NC emission. The PL spectrum of CdTe-opal collected at $\theta = 70^o$ represents the emission of CdTe NCs in the latex opal without influence of the directional PBG (Fig.7b).

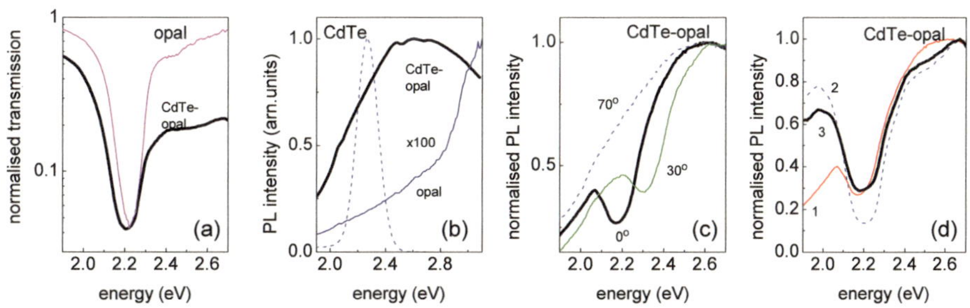

Fig. 7. (a) Transmission spectra at $\theta = 0^o$ of the latex opal assembled from spheres of $D = 240$ нм and the opal impregnated with NC CdTe. (b) PL spectrum of NC in water suspension (dashed line) in comparison to PL spectra of the bare opal (thin line) and CdTe-opal (thick line) obtained at $\theta = 70^o$ under excitation by 351 nm line of Ar+-laser with 1.9 mW power in a spot of 0.1 mm in diameter and collected from a 5° wide solid angle. The PL intensity of bare opal is >100 times weaker in magnitude compared to that of CdTe-opal. (c) PL spectra of CdTe-opal at different angles of collection. (d) Comparison of the CdTe-opal PL spectrum obtained at $\theta = 0^o$ (curve 1) and the reconstructed spectra obtained by multiplication the transmission spectrum and PL spectrum at $\theta = 70^o$ (2) and the same as (2) but with the account taken for the fraction of non-modified emission (3).

If the bandgap is present, the emission flow is PBG-blocked along the Bragg cones (Fig.6c) and the central frequency of the PL minimum coincides with the transmission dip. The angle dependence of PL spectra is clearly seen from comparison of spectra collected at $\theta = 0°$ and 30° (Fig.7c). This angle dispersion of PL minimum follows that of the transmission minimum. In particular, in the Brillouin zone, the $\theta = 30^o$ corresponds to the shift along the LU line from the ΓL towards ΓU direction (Fig.3c).

One can notice that the intensity contrast for the PL dip is reduced by a factor of two compared to the 6-fold reduction in the transmission minimum. To understand this we have to separate the measured emission flux in the ballistic and the diffuse components. The diffuse background is comprised by photons, which experience scattering at lattice defects [30]. In thin film opals the mean free path of photons, l^*, is about 15 μm that is shorter the film thickness (Fig.6b), i.e. scrambling of trajectories of photons emitted at the distance from the film edge $l>l^*$ is expected at any detection angle. Moreover, the scattering results in progressively shallower dip at higher angles of detection due to the longer light path. Another source of unstructured emission is the near-surface emission that comes without attenuation and fills in the PBG minimum. In the bulk opals the latter contribution can be eliminated by bleaching the emitters in the near-surface zone [31], whereas in the thin film opals one can use the photonic hetero-crystal approach (see section 8).

The unstructured contribution to the PL spectrum can be quantitatively estimated using the spectrum of unmodified emission at $\theta = 70^o$ and the transmission spectrum at $\theta = 0^o$ [32]. The result of $I_{PL}(\theta = 0) = I_{PL}(70) \times T(0)$ is shown by curve 2 in Fig.7d, which overestimates the emission suppression. More accurate fit (curve 3) can be obtained taking into account the diffuse light $I_{PL}(\theta = 0) = 3I_{PL}(70) \times T(0) + 0.25I_{PL}(70)$. This consideration proves the substantial diffuse fraction in the light detected along the PBG direction.

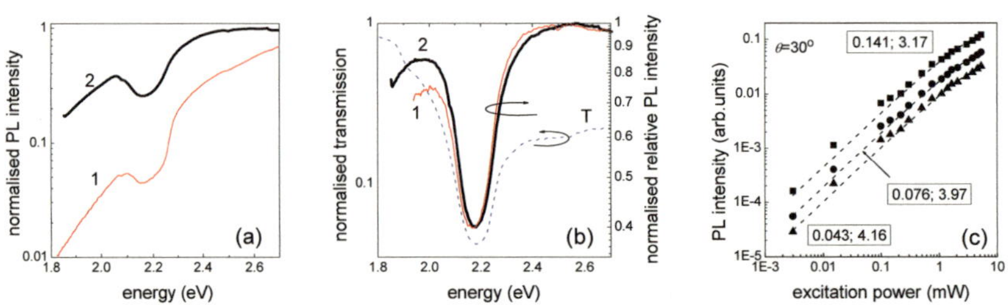

Fig. 8. (a) PL spectra of CdTe-opal with ~1 vol.% (1) and ~2 vol.% (2) CdTe fraction at $\theta = 0^o$. (b) Relative PL intensity spectra of CdTe-opal at low (1) and high (2) CdTe fraction in comparison to transmission spectrum. (c) Input-output characteristics of PL intensity acquired at $\theta = 30^o$ and $\hbar\omega = 2.5, 2.3$ and 2 eV (squares, circles and triangles, respectively). Points –experiment, lines – two-parametric approximation, numbers - fit parameters I_0 and P_0, respectively.

Next test was made to check if the NC-opal spectrum depends on the NC concentration. Obviously, only low volume concentration was explored, because with high fraction of NC the PBG properties of the samples will be altered dramatically due to changing the refractive index contrast and the uneven NC distribution over the opal voids. Fig.8a demonstrates the almost 2-fold increase of the PL intensity followed the doubling of NC concentration. Moreover, the PBG attenuation in the relative PL spectrum remains almost the same (Fig.8b). It is worth noting that obtaining the relative PL spectrum, as it was suggested in early works [33,34], is the very useful method for revealing the PBG effect upon the emission of PhC-coupled light sources especially in the case of complex spectra and weak attenuations.

5. Stimulation of CdTe nanocrystal emission in thin opal film

In the case of an externally pumped emitter inside a PhC of finite size, the field of the photonic mode becomes a superposition of the outgoing waves and the waves reflected from the PhC boundary. For the PBG frequencies the light intensity decays exponentially with distance z into the photonic crystal. Such photonic mode has the form of a standing wave with an envelope function that decays exponentially as $\exp(-\gamma z)$, where γ is the extinction coefficient. Thus, this mode in an opal film is a standing wave formed by interfering evanescent Bloch states inside the crystal and a plane wave in the vacuum.
For a microscopic emitter in the PhC the amplitude of the external field will depend on the distance from the boundary of the PC and will be determined by (i) the strength of the light attenuation in the photonic crystal and (ii) the position of the emitter relative to the nodes and antinodes of the standing evanescent Bloch wave (LDOS). Note that attenuation of light always leads to a decrease of the field and also of the emission rate while the position of the emitter relative to the standing wave of the field can decrease (in the node) or increase (in the antinode) the emission rate. The experimental results show the intensity suppression in the PBG frequency range. As was shown, within the PBG the PL is composed of light emitted at different distances from the PhC boundary. It can be concluded from the transmission spectra that the attenuation length corresponding to the centre of the band gap is about 10 μm (6 μm for an ideal structure) [35]. Thus, only $\sim 1/3$ of the 30 μm thick opal film contributes to the PL intensity observed externally ($\sim 1/5$ for the ideal lattice). These estimates are in good agreement with the measured relative PL intensity, shown in Fig.7b, which exhibits a decrease of PL intensity by a factor of five.
The PL intensity of the NC-opal as a function of the excitation power can be represented by input-output characteristics (Fig.8c). These characteristics exhibit saturation with increasing excitation power for all explored frequencies and can be fitted to the expression

$$I_{PL} = I_0(1 - \exp(-P/P_0)), \qquad (0.15)$$

where P is the excitation power. In this fit the pre-factor I_0 is the power radiated by a saturated emitter and the parameter P_0 in the emission saturation threshold. Since these parameters acquire unique values for a given frequency and angle of detection, they can be represented in a spectral form. The $I_0(\hbar\omega)$ spectrum at 70° is a monotonous function of frequency (Fig.9c). It is measured in PL intensity units and closely resembles the PL spectrum. The parameter $P_0(\hbar\omega)$ is given in excitation power units and corresponds to the projected excitation leading to a complete saturation of the input-output curve.

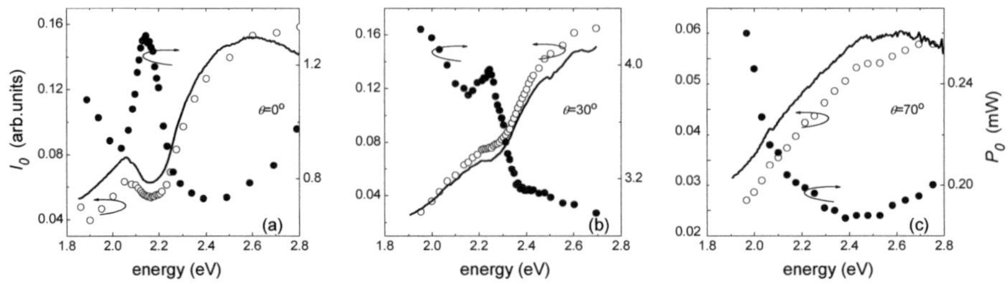

Fig. 9. Spectra of parameters I_0 (open circles) and P_0 (circles) at three detection angles $\theta = 0, 30, 70^o$. PL spectra are shown by lines for comparison.

The $I_0(\hbar\omega)$ spectra at θ= 0° and 30° follow the canvas for that at θ= 70° with the exception of the clearly resolved minimum superimposed on the monotonous background (Fig.8a,b). Moreover, $I_0(\hbar\omega)$ spectra closely resemble the PL spectra. By contrast, the spectra of the saturation threshold $P_0(\hbar\omega)$ at θ= 0° and 30° have their maxima in the bandgap. In particular, P_0 peak is located at the low frequency edge of the bandgap, covers the whole bandgap range and follows the bandgap angular dispersion. The P_0 magnitude is nearly doubled in the gap along the [111] axis, but its resolution becomes worse with increasing detection angle. Such degradation correlates the decrease of ballistic component in the detected emission flow because, neglecting mode re-coupling at the opal-air boundary, the emitter in the ballistic limit radiates in the same mode as detected outside the PhC.

Positioning of emitting NCs along the "heavy" dielectric boundary aligns them with the EM field distribution in the unit cell of the opal that ensures sampling by all NCs the same LDOS. In the case of a saturated emitter and in the presence of an effective non-radiative recombination channel, the radiated power is proportional to the spontaneous emission rate to a given mode (0.13). Therefore, the $I_0(\hbar\omega,\theta)$ is an estimate of the spontaneous emission rate spectrum along a given direction. The good correlation between $I_0(\hbar\omega)$ and the PL spectrum for a given direction (Fig.9) suggests that the spontaneous emission is the dominating process in the radiative relaxation in the opal-embedded NCs.

P_0 measures the range of the emission response to the pump power increase. For a given number of NCs, it depends on several factors. One is the probability of an electron transition between two bands in a NC band structure that is constant in this experiment. Another is the population of these bands. The others are the probability of coupling the emitted photon to the optical mode reservoir of the PhC and the balance between radiative and non-radiative relaxation. At θ= 70° the $P_0(\hbar\omega)$ is dominated by processes in the electronic system of the CdTe NCs (Fig.9c).

The weak variation of the saturation threshold across the emission band corresponds to a uniform density of electrons as a function of energy. Such distribution is the result of non-resonant excitation and fast non-radiative relaxation of the electronic excitations. The CdTe NCs in the opal suffer the surface effects that have an extremely strong influence on the electronic structure. In particular a wide "impurity" band is formed due to the surface states (which can be Tamm-like states, surface defects, and impurity atoms localized at the surfaces). Relaxation of electrons and holes within this large energy band has essentially a hopping character [36,37,38], that provides quite a uniform energy distribution of carriers within the band. Roughly speaking, an electron jumps from one localized state to another with almost the same probability for the states of similar or markedly different energies. Due to Auger processes [39], which are extremely efficient near surfaces, the hopping relaxation is likely to happen. Thus the fact that there is similar character of the dependence of the PL intensity on the pumping for the frequencies below and above the PBG indicate that the energy distribution of carriers is almost pumping-independent in provided experimental conditions.

In turn, the sublinear character of the input-output characteristics suggests that the rate of non-radiative recombination grows super-linearly with the increase of carrier concentration n. This is typical for some processes, e.g., the Auger recombination rate is proportional to n^2 [39]. We can roughly estimate the influence of non-radiative recombination and variation of radiative lifetime on the dependence of PL intensity on pumping. Assuming for simplicity, that PL intensity is proportional to the carrier concentration n and can be characterized by "mean radiative lifetime" τ (the rate of radiative recombination is n/τ), we obtain that at the equilibrium $\partial n / \partial t = 0$ the excitation is equal to the relaxation

$$P = \frac{n}{\tau} + \Gamma_{NR}(n) \tag{0.16}$$

Within the PBG, the PL intensity is influenced by the increase of radiative lifetime τ. We may then ask how does such increase will modify the dependence of I on P in the system with spontaneous recombination and a super-linear dependence of $\Gamma_{NR}(n)$ on carrier concentration? Taking the derivative of the pumping P in eq. (0.16) with respect to the PL intensity I and noting that $I = n/\tau$, we obtain

$$\frac{\partial P}{\partial (n/\tau)} = 1 + \tau \frac{\partial \Gamma_{NR}}{\partial n} \quad \text{or} \quad \frac{\partial I}{\partial P} = \frac{1}{1 + \tau \dfrac{\partial \Gamma_{NR}}{\partial n}} \tag{0.17}$$

Taking into account that $\partial \Gamma_{NR}(n) / \partial n > 0$, we find that the increase of radiative lifetime τ leads to a slower growth of the PL intensity with increased pumping. In other words, the saturation threshold for a frequency within the PBG has to be less than that for a frequency outside the PBG. But this conclusion contradicts the observed PBG-related peak of $P_0(\hbar\omega)$ (Fig.9). To resolve this conflict one has to assume the presence of the amplified spontaneous emission in addition to the spontaneous one. The acceleration of the emission rate is a consequence of applying resonant conditions upon the emitter at a certain frequency. In what follows we will discuss two realizations of resonant modes in opal PhCs, bearing in mind that the modes of the allowed band are propagating ones, whereas the modes of the Bragg cone are the leakage modes of a PBG resonator. Nevertheless, since these resonance conditions are intrinsic to the PhC, we can postulate the omnipresence of the emission enhancement in the incomplete PBG.

5.1. Coupling to slow propagating modes

One source of the emission amplification in PhC can be associated with slowly propagating modes. The complex topology of iso-frequency surfaces in an incomplete PhC gives rise to beam steering effects [40,41,42]. As a result, the actual direction of the energy flow inside a PhC does not necessarily coincide with the mode wavevector $\boldsymbol{k}$, i.e., photons emitted with different wavevectors can propagate along the same direction. The iso-energy surface in k-space of the opal lattice in the PBG range contains eight necks, two per each [111] axis (Fig.6c, 10a) [43]. When this surface crosses the boundary of the 1st Brillouin zone (BZ), the normal component of the group velocity vector $\boldsymbol{v}$ vanishes. This means that all eigenmodes, the wavevectors of which end up at the intersection of the neck with the zone boundary, have the $\boldsymbol{v}$ pointing along the BZ boundary. Due to the topology of the 1st BZ, some eigenmodes with wavevectors along the $[\bar{1}11]$ direction and group velocity pointing in the [111] direction are expected. Because the direction of energy transport in a non-absorbing PhC coincides with the group velocity vector, there should always be some energy flux in the Bragg cone. In what follows, we will refer to modes with wavevectors parallel to the group velocity vector, as type 1 modes and to other modes as type 2 modes [44].

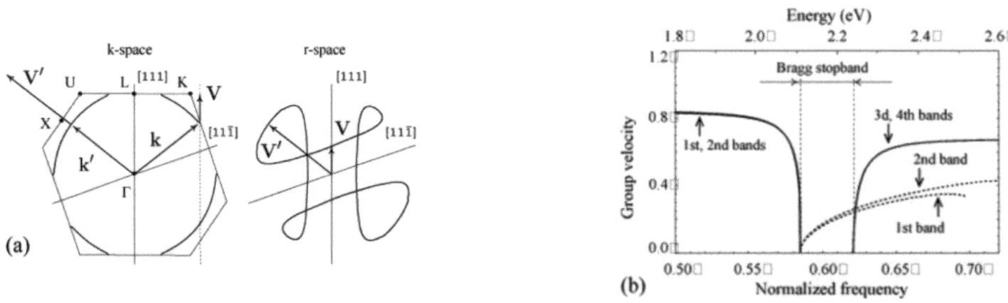

Fig. 10. (a) Iso-frequency (left) and group velocity (right) contours of opal at the frequency within the Bragg bandgap. $\boldsymbol{k}$ and $\boldsymbol{v}$ are vectors of type 2 inhomogeneous waves pointing along the [111] axis. (b) Group velocities of type 1 (solid curves) and type 2 (dash curves) Bloch modes along the [111] direction. The group velocity is given in units of c.

Fig. 10b shows the calculated group velocities along the [111] axis for wavevectors of the XULK section of the first Brillouin zone. Calculations were performed using the plane wave expansion method [45], where the Hellmann-Feynman theorem was used to calculate the group velocity vectors. The best fit to experimental data was obtained for the opal made of 243 nm diameter spheres of $n_{eff} = 1.6$. It is instructive to separate contributions to the energy flux from type 1 and type 2 eigenmodes. For frequencies below the Bragg bandgap, the flux along the [111] direction is solely formed by type 1 modes of the 1st and 2nd photonic bands. Within the PBG, only type 2 modes of the 1st and 2nd bands contribute to the flux. Above the PBG, the flux is composed by the type 1 modes of the 3d and 4th bands as well as by the type 2 modes of the 1st and 2nd bands. For sake of clarity, contributions of the latter are omitted in Fig. 10b, because their frequencies are above the Bragg gap.

The group velocity vanishes at the low frequency bandgap edge in the ballistic limit of an infinite PhC and then grows slowly with increasing frequency, comprising, in average, 1/10 of the group velocity absolute value outside a bandgap. Correspondingly, type 2 modes traverse the opal slowly and interact with the pumped medium for longer time. In agreement with this model, the slowing down of the group velocity towards the low

frequency bandedge is the reason for the "red" shift of the P_0-spectrum maximum with respect to the pseudogap centre (Fig. 9a).
The relative number of type 2 modes is proportional to the ratio of the small solid angle in k-space and the corresponding solid angle in real space:

$$N \sim \sum d\Omega_k / d\Omega \tag{0.18}$$

where the summation is taken over all contributions to the energy flux [43]. Since the type 2 modes originate at the necks of the dispersion surface, the ratio of the solid angles in (0.18) is below 4% over the bandgap. Correspondingly, the contribution of the stimulated emission to the total PL signal is small.

5.2. Coupling to defect modes

The available in the Bragg cone modes are the leakage modes of a PBG resonator. An obvious example of a resonator inside the opal is a lattice defect with its mode in the PBG. The quality factor of defect modes depends on the correlation between the localization length l_{loc} and the opal film thickness t, the distance to the opal boundary, the detuning of the resonant frequency from the PBG centre and the coupling of this resonance to similar defect modes. The radiation coupled to the eigenmode of a single defect is localized in the defect vicinity. In this case, the photon occupation number of this mode and the corresponding strength of the EM field increases under the cw pumping. Consequently, the radiated power acquires a super-linear component due to the backreaction of the emitted radiation upon the radiative transition probability. In the directional PBG, this effect also acquires the related anisotropy. In an opal with relatively high defect concentration, the defect modes are coupled throughout the film and the resulting quality factor is diminished. It is instructive to demonstrate the presence of amplified spontaneous emission simply by plotting the ratio of PL spectra acquired at different excitation power. Such ratio spectra show a peak that corresponds to a dip in transmission spectra indicating that PL intensity grows faster with the increase of pumping inside PBG than outside. There is a shift of the maximum in the ratio spectra to higher photon energy with the detuning of angle of incidence from [111] (Fig.11).

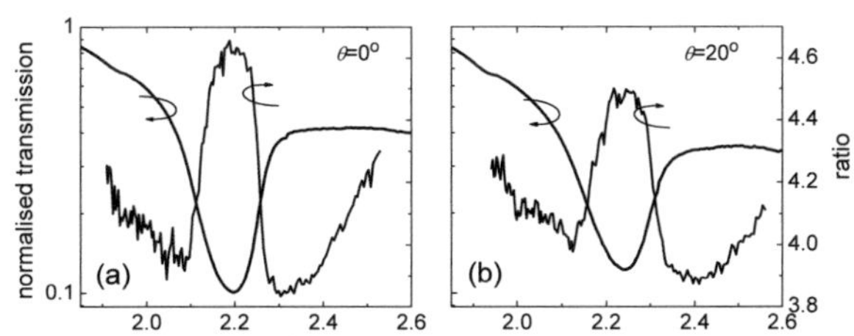

Fig. 11. The ratio of PL spectra acquired at excitation powers 2.54 and 0.49 mW vs. transmission spectrum at $\theta = 0$ and 20° in panels (a) and (b), respectively. The measurements were performed at T=18K in order to reduce the non-radiative relaxation probability.

Faster growth of PL intensity inside the PBG is clearly seen also in Fig.12 (a, b). In Fig.12c one can see that in the PBG spectral range the PL intensity increases much faster with more intensive pumping than outside the PBG. Moreover, this effect gets stronger with the increase of the absolute value of pumping, and this indicates that the functions $I^{in} / I^{out}_{1,2}$ possess more complicated dependence on excitation power than a simple power law [35].

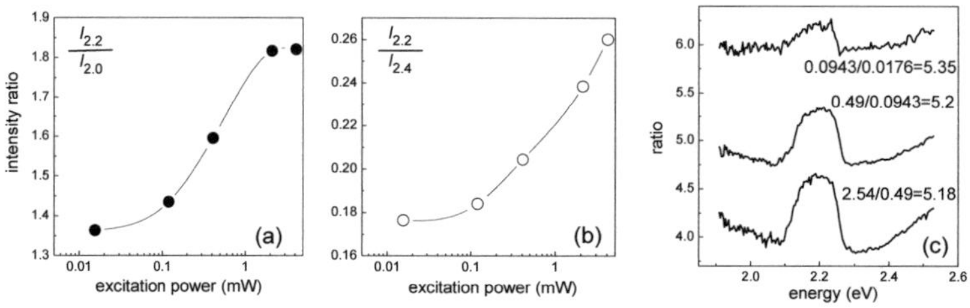

Fig. 12. Ratios of I^{in} / I^{out}_2 and I^{in} / I^{out}_1 at (a) frequencies $\hbar\omega = 2.2$ and 2.0 eV and (b) at 2.2 and 2.4 eV as a function of pumping. (c) Ratios of the PL spectra along [111] direction obtained at different levels of pumping intensities: 0.943 and 0.0176 mW; 0.49 and 0. 0.943mW; 2.54 and 0.49 mW. The peak magnitude increases with increasing the power for the same power increment.

Thus apart from spontaneous emission, which is characterized by an increased radiative lifetime, there is a competing process that is characterized by faster dependence on P within the PBG. Let us check if the emission coupling into localized photon states of lattice defects which are located in the PBG can bring such effect. In disordered media a light wave can be localized due to multiple scattering [46,47], but such localization can be more easily achieved introducing disorder in otherwise periodic media [6]. In the latter case, the localization of light leads to the appearance of a non-zero density of photonic states in the PBG [48,49]. The localized modes, contrary to the evanescent modes, can have large electric field amplitude in the bulk of PhC, and, as a result, the radiative recombination time in such modes is much smaller. Consequently, the recombination is channelled into the localized modes which act as photon reservoirs. According to the Einstein principle, the radiative recombination in a single localized mode can be written as

$$I_{loc} = \frac{n}{\tau_{loc}} + s(P)\frac{n}{\tau_{loc}} \tag{0.19}$$

where τ_{loc} is the spontaneous radiative recombination time and s is the number of photons in the mode. The first and second terms in Eq.(0.19) is the spontaneous and stimulated emission, respectively. In the bulk homogeneous medium at moderate excitations (in case of lasers - below lasing threshold) the relation $s \ll 1$ holds, and therefore the stimulated emission can be neglected. In case of disordered PhC, the photon accumulation in the localized modes can lead to substantial enhancement of stimulated emission. It is obvious that the second term in Eq.(0.19) depends on the pumping power superlinearly. Thus, the experimentally observed increase of parameter P_0 or the peak in ratio of PL intensities at different pumping intensities inside PBG can be explained by stimulated emission through the localized modes.

On the other hand, when a light-amplifying material is inserted into a PC, each localized state can be considered as the analogue of a laser mode. For laser modes the dependence of the PL intensity on the pumping has an abrupt change at threshold. Each laser mode is characterized by its own decay time (or Q-factor) and spatial distribution of the electromagnetic field. In the case of localized defect modes, the spatial distribution of the field can be altered by varying the pumping [50]. Therefore, the threshold intensity is different for different modes. One can conclude that the dependence of the PL intensity upon the pumping averaged over different modes should be some superlinear function, and could explain the faster growth of $I(P)$ within the PBG.

Another disorder-induced modification of the observed PL intensity is related to the scattering of the radiation in the localized mode into a propagating mode. Such a process makes possible the propagation of an emitted photon from the bulk of PhCs to the boundary without attenuation. Consequently, it leads to an increase in the effective thickness of the layer contributing to the detected PL intensity.

6. Emission indicatrix

The emission conditions for dipoles change as soon as the pseudogap overlaps with the emission band (Fig.13a). In the first approximation, the spatial pattern of emission changes according to the shift of the Bragg cone (Fig.6c). In the angle-resolved measurements, the PL intensity for the given direction is proportional to the part of the total radiated power, which is coupled to modes, whose group velocity is within the solid angle selected by the detector aperture. Following procedure presented in [51], one can introduce the radiated power per solid angle in a coordinate space $dP/d\Omega$. This power gives a rate at which the dipole energy is transferred to modes with a group velocity pointing to the observation direction. The fact that due to the beam steering phenomenon several eigenmodes with different wave vectors, $\mathbf{k}_n^{\nu}$, can have parallel group velocity vectors, is taken into account by extra summation, ν, over all wave vectors for which $\hat{\mathbf{x}}\mathbf{V}_{n\mathbf{k}}^{\nu} > 0$ holds. $\hat{\mathbf{x}}$ is a unit vector of the observation direction.

Dipole moments of NC are randomly distributed in space. Consequently, the radiative power should be averaged over the all dipole moment orientations, hence, an ensemble of NC is equivalent to the set of point sources producing an isotropic distribution of wave vectors. An angular distribution of radiative power inside a PhC depends on the topology of the iso-frequency surface of the crystal at the emission frequency. A schematic view of the iso-frequency surface of an opal PhC is presented in Fig.13b for a frequency inside the Bragg gap. To plot the iso-frequency surface, one should calculate a PBG structure for all wave vectors within the irreducible Brillouin zone and then solve the equation $\omega_{n\mathbf{k}} = \omega_0$ for a given frequency ω_0. The iso-frequency surface at the Bragg PBG frequency of the opal deviates from a sphere mostly along the [111] axes, where the Bragg gap openings are located. In the vicinity to openings the iso-frequency surface forms a neck with the alternating negative and positive Gaussian curvature separated by parabolic lines with vanishing curvature. A small Gaussian curvature formally implies bunching of many Bloch eigenwaves with different wave vectors travelling in the same direction due to the crystal anisotropy. Such concentration of radiated power along certain directions is a linear phenomenon and called the photon focusing [52].

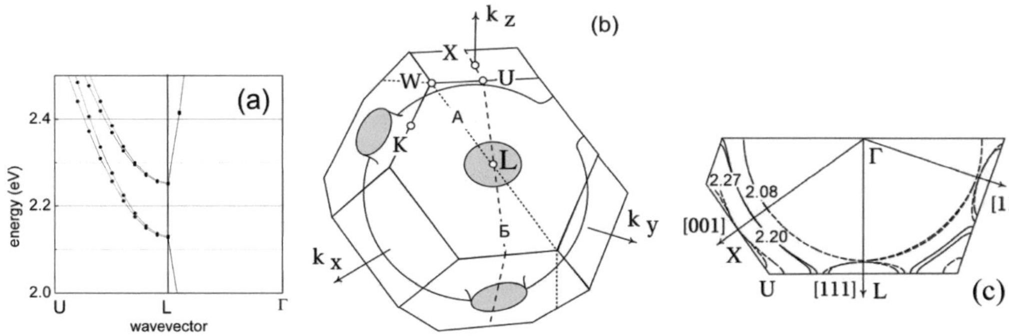

Fig. 13. Fragment of the PBG diagram along ΓLU cross-section. Frequency scale is adjusted to the discussed sample. (b) Illustration of the 3-dimensional iso-frequency surface of the 1st optical mode of an opal PhC for a frequency 2.2 eV inside the Bragg PBG inserted in the 1st BZ of the fcc lattice. (c) Cross-section of the BZ with iso-frequency contours at frequencies 2.08, 2.2 and 2.27 eV (compare to the diagram in panel (a)).

In Fig.13c the iso-frequency contours are presented for three different frequencies. They cut the Brillouin zone through the symmetry points X, U, L and Γ. One can assume that the experimentally measured far-field PL intensity represents the signal, which is averaged over different Brillouin zone cross-sections due to lattice disorder.

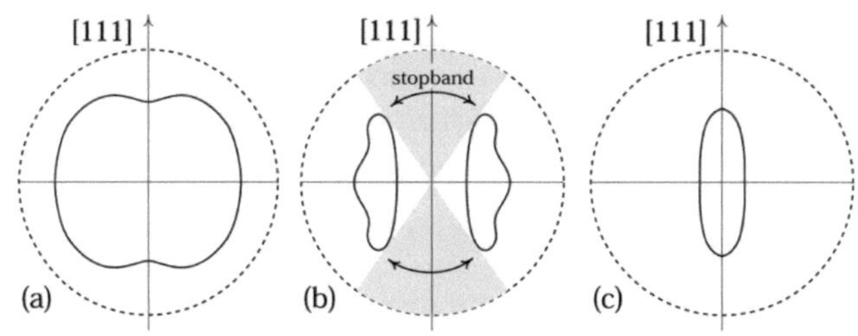

Fig. 14. Wave contours corresponded to frequencies 2.08, 2.2 and 2.27 eV (a, b, c respectively). Only the wave contours corresponded to the first and third bands are shown in the diagrams (a, b and c), respectively. Light grey region in (b) is the Bragg cone projection. A dashed circle is a wave contour in vacuum. Group velocity is plotted in the units of the speed of light in vacuum.

As the frequency remains below the gap, an iso-frequency contour is continuous and almost circular. The Gaussian curvature does not vanish for any wave vector. This implies a small anisotropy in the energy flux inside the crystal. It order to obtain the wave contour in coordinate space, one should plot a ray in the observation direction $\hat{\mathbf{x}}$ starting from the point source position and having the length of the group velocity $\left|\mathbf{V}_{n\mathbf{k}}^{\nu}\right|$. The calculation of the group velocity was discussed earlier in relation to schematics in Fig.10. The wave contour at 2.08 eV is single valued function of observation direction (Fig.14a). Fig.15 shows the angular distribution of the radiated power for the same frequency. The latter is nearly isotropic and resembles, reasonably, the Lambert law (dashed line). To calculate the radiated power, one should sum over all optical modes, which are available at the given frequency.

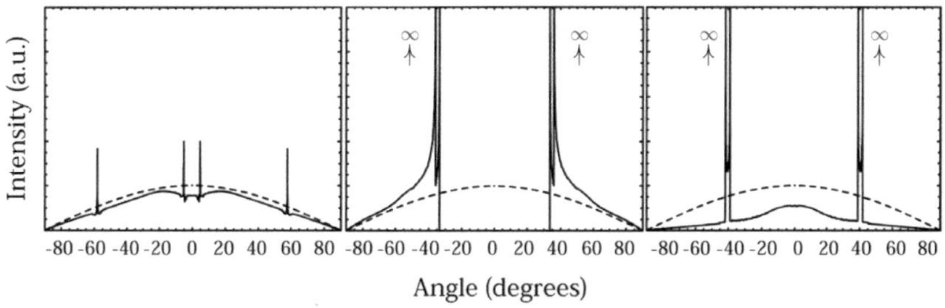

Fig. 15. Angular distribution of the radiated power for the same frequencies as in Fig.14. The radiated power in free space (Lambert law) is shown by dashed line for comparison.

With the frequency increase up to the midpoint of the first PBG, the topology of the iso-frequency contour abruptly changes. The gap developed along the ΓL direction and the iso-frequency contour becomes open (Fig.13c). This topological discontinuity results in a complex contour with alternating regions of different Gaussian curvature. Vanishing curvature leads to the folds of the wave contour (Fig. 14b). The folds in the wave contours yield that two Bloch modes are travelling in any observation direction outside the gap. Contributions from both modes should be taken into account, if the total radiated power is calculated. In the middle panel of Fig.15 the strongly anisotropic angular distribution of the radiative power corresponded to the 2.2 eV is presented, featuring zero intensity in the direction of the gap, and infinitely high intensity spikes in directions of folds of the wave contours occurring at, approximately, $\theta \approx \pm 40^o$ with respect to the [111] axis.

When the frequency crosses the upper boundary of the ΓL Bragg gap, i.e. the gap along ΓL direction closes, the 3rd and the 4th photonic modes come in consideration. In the Fig. 14c a wave contour of the 3rd band at 2.27 eV is depicted. It is single valued function elongated to the [111] axis, which leads to the preferable energy flux in this direction. An angular distribution of the radiative power at this frequency shows infinite intensity spikes at $\theta \approx \pm 40^o$ and a Gaussian-like lobe centred at $\theta = 0^o$ (Fig.15, right). An infinite intensity corresponds to the points of vanishing Gaussian curvature of the 1st and the 2nd photonic bands, while the central lobe is associated with the 3rd and the 4th bands.

Fig.16a shows experimentally obtained directionality diagrams in the vicinity to the Bragg gap. In the first approximation, these patterns satisfy the iso-frequency profiles (Fig. 13c) being the superposition of the isotropically distributed emission from a point source and the Bragg cones of the fcc lattice (Fig.13b). The next iteration takes into account the light focusing occurring due to uneven curvature of iso-frequency contours in a sense of diagrams in Fig.15. In fact, the emission from a thin film opal consists of isotropic unmodified (scattered and near-surface emission) and PBG-moulded ballistic components. The isotropic light source provides the background, the intensity of which varies as $\cos\theta$, and the ballistic component carries the fingerprint of the Bragg gap. Summing the diagrams of ballistic and diffuse components, the details of the experimentally obtained emission indicatrix can be accurately explained [53] (Fig.16b).

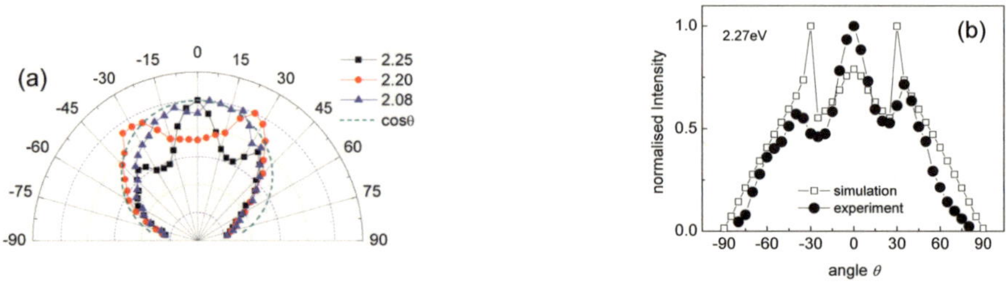

Fig. 16. (a) Directionality diagrams of the emission at specified frequencies across the PBG spectral range. Dashed line shows Lambertian distribution of emitted power. (b) Comparison of calculated and experimental emission directionality diagrams at $\hbar\omega = 2.27$ eV . Simulation includes emission focusing along special directions in the vector diagram. To match the experimental pattern a PBG-unaffected emission component has been added to the simulated pattern as the $\cos\theta$ background.

7. Emission of nanocrystals in inverted opals

Reduction of the volume fraction of "heavy" dielectric compared to that in opals assembled from spheres favours the wider PBG opening [54]. The inverted opals prepared by impregnation of the opal voids with another dielectric and subsequent removing the opal spheres offer such enhancement of PhC performance. Therefore, it is reasonable to examine the luminescent properties of the NC CdTe in the inverted opal. Usually, the semiconductors are used for opal inversion. The close vicinity of NCs to the surface of semiconductor brings another complication to the radiative energy relaxation due to energy transfer between the NCs and the semiconductor. In this section we will consider both aspects of NC to PhC interaction.

TiO_2 inverted opals were prepared by templating in thin film opals. As a first step, opal films consisting of approximately 15-20 monolayers of 300 nm diameter monodisperse polymethyl methacrylate (PMMA) beads were prepared by slow-drying of a sphere suspension on microscope slides. Then these films were dipped into a solution of $TiCl_4$ in HCl, followed by a moisture-promoted hydrolysis and heating at 160°C for 1 h. At this stage, an interconnected TiO_2 framework is formed in the voids between the PMMA beads. After dissolving the polymer beads in tetrahydrofurane, TiO_2 inverted opal films were formed (Fig.17a). The reflectance spectra of the TiO_2 inverted opal display a pronounced (111) diffraction resonance indicating a directional PBG. Its angular dispersion follows the Bragg law (Fig.17b). The strong deviation from a diffraction on (111) planes occurs around $\theta = 40^{o}$ due to the multiple wave diffraction at anticrossing of (111) and (200) resonance dispersions [55]. The relative full width at half maximum (FWHM) of the reflectance peaks is about 10 % (15% in transmission) compared to 5.6 % in opals [56]. Owing to a broader gap, the spectral overlap of pseudogaps along different directions of the Bragg resonance in TiO_2 opal is remarkably larger than in PMMA template. Simultaneously, the Bragg cones are also larger. These factors are particularly important since increasing both, spectral and spatial PBG dimensions is a pre-condition of stronger modification of emission of PhC-embedded light sources.

To realize a PhC light source, a small amount of colloidal CdTe NCs stabilized by thin organic shells was embedded into the replica pores by soaking the films in dilute aqueous colloidal dispersions. CdTe NCs of 3 nm in diameter were used to fit the emission band to the Bragg

PBG of TiO_2 inverted opal. After drying, the CdTe NCs remained attached to the inner surface of the TiO_2 frame through electrostatic interaction with functional groups of thiol stabilizers.

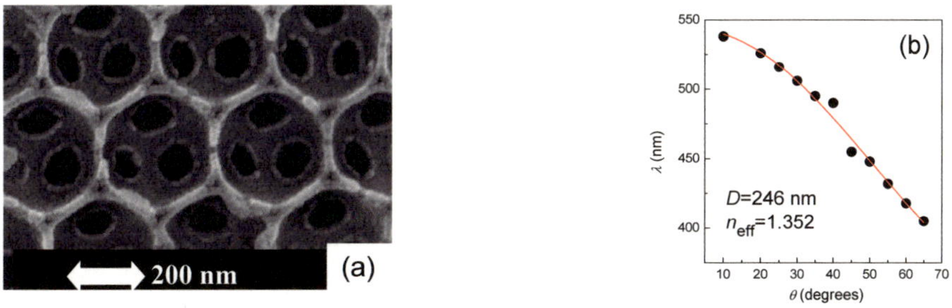

Fig. 17. (a) SEM image of inverted TiO_2-opal. (b) Angle dispersion of the (111) resonance in TiO_2-opal (points) and its approximation by the Bragg law. The diameters of spheres and the effective index of refraction extracted from the Bragg fit are shown.

PL spectra were measured at T = 300 and 18 K under cw excitation by the 457.9 and 351.1 nm lines of an Ar^+ laser. The excitation power was varied between 10^{-5} and 10^{-1} W. PL spectra were recorded in the front window configuration thus allowing the emission to traverse the film. The emission was collected within a 4° solid angle along the direction defined by angle of θ with respect to the [111] axis of the opal fcc lattice.
Under 457.9 nm excitation the PL spectrum at T = 300 K exhibits a band with a maximum at 2.38 eV and a FWHM of 0.26 eV. Under 351.1 nm excitation this band becomes broader and shifts up to 2.44 eV. At T = 18 K this PL band shifts further to 2.47 eV, but its bandwidth is ~1.5 times narrower (Fig.18a). In fact, the low temperature PL band (curve 3) resembles that observed in the NC suspension, where NCs are relatively isolated from the environment. Such evolution of NC emission with changing the excitation conditions has been ascribed to the NC-substrate interaction [57,58]. It is suggested that the TiO_2 template provides a potential relief due to surface defects, which co-ordinates the deposition of NCs.

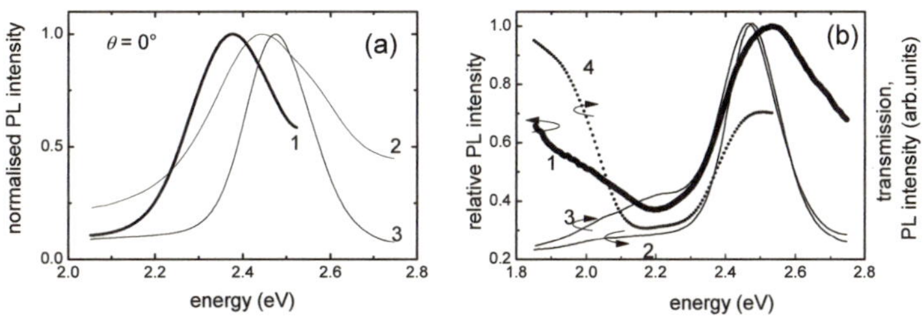

Fig. 18. (a) PL spectra of the CdTe NC-TiO_2 inverted opal at $\theta = 0°$ under 457.9 nm and 351.1 nm line excitation at $T = 300$ K (curves 1, 2, respectively) and at $T = 18$ K (3). (b) Relative PL spectrum (curve 1) of the CdTe NCs-TiO_2 opal obtained as the ratio of PL spectra at $\theta = 0°$ (2) and 70° (3) compared to transmission spectrum (4).

The TiO_2 framework itself has its electronic bandgap at about 3.1 eV thus facilitating the absorption of the 351 nm radiation and enhancing the energy exchange with CdTe NCs. It is further helped by the fact that the band edge emission from TiO_2 is above the spectral range of CdTe PL band. On the other hand, the polymer shells of NCs reduce the energy exchange. Nevertheless, defect states at TiO_2-CdTe interface trap the photogenerated electron-hole pairs in polymer-capped NCs by, e.g., dissociation of excitons [59]. In turn, the radiative recombination of trapped carriers leads to broadening of the observed PL band by an amount comparable to the trap energy range. The narrowing of the NC PL band at low temperature can be understood as the result of reducing the contribution of trapped excitons to the emission flux due to increasing their lifetime. This observation emphasises that the emission of CdTe NCs in TiO_2-opal involves multiple-level relaxation of electronic excitations i.e., this system cannot be treated as a two-level system.

The relative PL spectrum for θ=0° contains the minimum, the position of which is in good agreement with the (111) Bragg gap in the transmission spectrum. As the reference, the PL spectrum of the same sample obtained at $\theta = 70°$ was used (Fig. 18b). The slight dissimilarity of pseudogaps in the transmission and relative PL spectra is due to (i) the difference in coupling of externally (outside the PhC) and internally (inside it) generated light to eigenmodes of the PhC and (ii) the unaffected near-surface emission coupled to free space modes.

The input-output characteristics of CdTe-TiO_2-opal [60] are similar to that of the CdTe-opal (section 5). The P_0 spectrum shows a maximum at the bandgap frequencies (Fig.18). This maximum appears more pronounced under UV-excitation because it is no longer overshadowed by the overlapping minimum, which is blue-shifted following the shift of the PL maximum. A comparison of P_0-spectra obtained at different angles of detection demonstrates that the maximum follows the PBG dispersion and vanishes gradually at larger angles θ of detection, whereas the minimum remains unchanged. The difference with the CdTe-opal sample (section 5) is almost 2-fold increase of the PBG width.

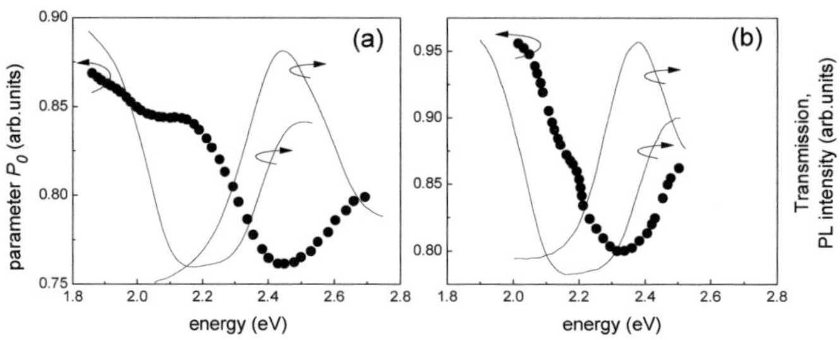

Fig. 19. (a, b) Parameter $P_0(\hbar\omega)$ (dots) in comparison to transmission and PL spectra obtained at $\theta = 0^o$ for excitation by (a) 351.1 and (b) 457,9 nm lines.

The interplay of the minima and maxima in P_0 spectra shows that (i) the minimum corresponds to the faster saturation of spontaneous emission in the PL band maximum that is determined by higher matrix element for electron transitions between excited and ground states in the electronic band structure of the NC, whereas (ii) the maximum results from the emission stimulation due to stronger NC to optical mode coupling provided either by slow propagating modes or localized defect modes.

The drawback of studied inverted opal is higher amount of defects compared to the opal template. These extra defects are added due to inhomogeneous infiltration of the opal voids and uncontrollable lattice contraction due course of annealing. Moreover, the higher refractive index contrast in the TiO_2 opal replica compared to the opal template leads to stronger distortion of the EM field distribution around the defect. This leads to emission losses, e.g., the emission initially radiated to PhC eigenmodes experiences higher chance to be scattered in the Bragg cone [61] and to be transported by defect modes. Since the stimulation takes place only for modes in the PBG direction, this effect becomes buried in the diffuse radiation, leading to suppressing the P_0 peak in the PBG interval. Hence, in spite of higher strength of the PBG effect in studied inverted opals, the emission enhancement becomes counteracted by the influence of disorder.

8. Emission of nanocrystals in hetero-opals

The main idea behind PhC heterostructures is to bring two different PhC in contact (Fig.20a) [62]. The obvious consequence of heterostructuring is the fingerprints of two [63] or more PBG structures in the optical properties. Such a structure contains the interface, which implies a sharp step in the spatial distribution of the EM field. In spite of the fact that the geometrical "thickness" of the interface is less than the light wavelength, the transition region is formed in the vicinity of the interface due to the EM field continuity and a need of light coupling from modes of one PhC to modes of the other. In this region, which extends over several lattice periods, one can expect special conditions for the light propagation [64]. Hetero-opals can be used for achieving a control upon the emission characteristics [65] if one of the opals is infilled with light-emitting NC and used to control the NC-to-mode coupling (opal-source) and another is used to shape the emission spectrum according to its PBG (opal-filter). It appears that the properties of such hetero-opal differ from the linear superposition of properties of the source and filter opals due to the interface interaction [66]. However, the light coupling at the interface remains so far largely unexplored issue [67].

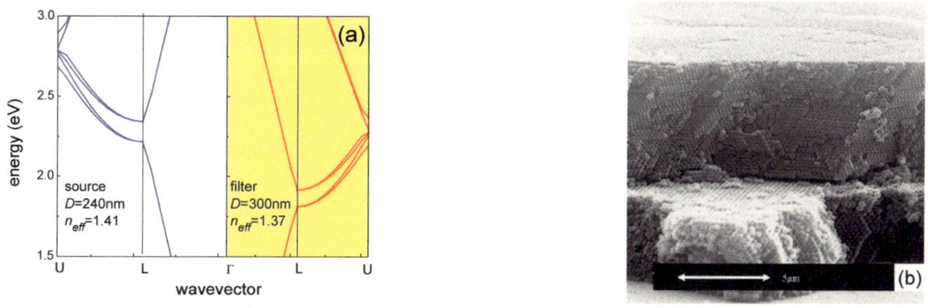

Fig. 20. (a) Band diagrams of the opal-source and the opal-filter in contact. (b) SEM image of the hetero-opal assembled successively from two opal films crystallised from latex spheres of 240 and 300 nm in diameter.

The examined hetero-opal was prepared using the successive assembling of opal films [68]. Opal films studied in this work were grown on glass or quartz slides by crystallization of polystyrene (PS) beads ranging from 240 to 450 nm in diameter. Aqueous 1-5 vol.% colloidal suspensions of PS beads were placed in a Teflon cylindrical cell (7 mm inner diameter) and then the solvent was evaporated under moderate flow of warm air. Typical thickness of opal films was 5-10 μm. The sintering at 90° for 1 h was applied to allow further treatment of films.
Sandwich-type hetero-opals were prepared by self-assembling the top opal film on the surface of the bottom opal film, which was crystallized previously from beads of another diameter. Fig. 20b shows SEM image of a hetero-opal with the bottom film consisting of D=240 nm and the top film - D=300 nm beads. Here and below, we denote this structure as

the 240/300 nm opal. The abrupt interface between films indicates that the formation of the top film proceeds independently on the geometrical profile provided by the bottom one.
The layer-by-layer (LbL) deposition technique, which is based on alternating adsorption of layers of oppositely charged species on the surface, was used for impregnation of opal films with CdTe NCs. LbL technique was originally developed for positively and negatively charged polyelectrolyte pairs [69] and then extended to the assembly of polymer-linked NCs [70]. This method was successfully applied to a variety of substrate materials with flat and highly curved surfaces [71]. Following LbL procedure, the substrate with an opal film was immersed for 30 minutes in a 5 mg/ml aqueous solution of poly(diallyldimethylammonium) chloride (positively charge polyelectrolyte), then thoroughly washed and immersed for 30 minutes in a 10^{-3} M aqueous solution of CdTe NCs capped with thioglycolic acid and, thus, carried a negative charge at appropriate pH [72]. 0.2 M of NaCl was added to both solutions to facilitate a formation of smooth layers. Moderate stirring was applied to solutions to accelerate the mass transfer. Because opal voids are either of 0.41 or 0.23*D* in size and connected via 0.15*D* channels [11], both polyelectrolyte molecules and CdTe NCs can easily access internal pores. We assume that the in-void coating proceeds in the same manner as a deposition on the open surface. The LbL procedure was repeated several times to increase the amount of deposited CdTe NCs. The NC-polyelectrolyte layer formed on the opposite side of the slide was finally removed by washing in acetone and ethanol.
The LbL deposition provides uniformity of the coating thickness and NC environment (Fig.21a), as compared to a direct infiltration of NCs into opals. No leakage of NCs from the bottom opal film back to water takes place during the deposition of the second film judging from the absence of the luminescence of the background solution. Anchoring of NCs is also required to prevent their diffusion to another film of a hetero-opal. The important condition is the need for the emission spectrum of NC to overlap both PBGs in the hetero-opal. Correspondingly, CdTe NC of 2 and 5 nm with respective emission bands centred at 2.29 and 2.08 eV were explored.

Fig. 21. (a) TEM image of 420 nm PS beads, which are LbL-coated by CdTe NCs in one deposition circle. The NCs appear darker due to the higher contrast. (b) Layout of the PL measurements. θ - the angle of incidence and the angle of detection, Ω - the solid angle of the light collection.

A straightforward consequence of the PBG anisotropy in hetero-opals is the spectral and spatial anisotropy of the PL of NCs embedded therein. In Fig. 22 the source and the filter PL spectra are shown in comparison to transmission spectra at different angles of the light detection to demonstrate the correlation between the emission minimum and the corresponding Bragg gap. It is instructive to emphasize the different physical origin of this minimum in the source and the filter PL spectra. The directional minimum of the source PL spectrum is a consequence of the lower mode number available for coupling with emitter within the PBG bandwidth along a given direction. In contrast, the minimum in the filter PL spectrum appears due to the back-reflection of the source radiation from a filter film within the PBG bandwidth of the filter, which does not directly affect the LDOS in the source. In other words, the origin of the filter PL minimum is the same as that of the transmission minimum. The diffuse scattering of emitted photons at the interface and in the filter film can be the reason for the absence of the source PBG minimum in the filter PL spectrum.

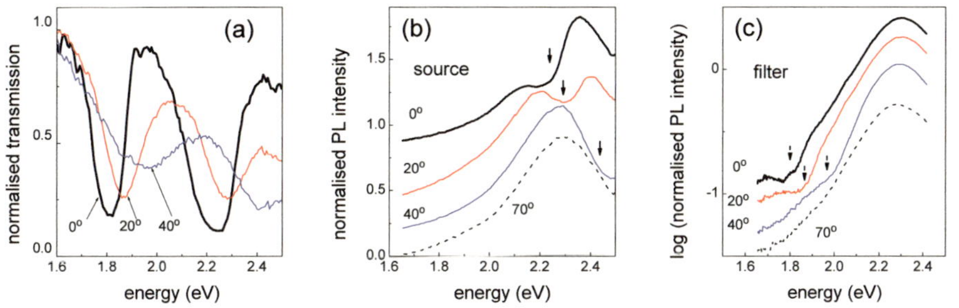

Fig. 22. (a) Spectra of transmission of the 20 μm-thick 240/300 nm source-filter hetero-opal obtained along different directions that are labelled by the angle. (b,c) Source and filter PL spectra of a hetero-opal with 2 nm NC in the source film, respectively. Curves in the panels are shifted vertically for clarity. Dotted line shows The PL spectrum obtained at 70° that is used as a PBG-unaffected reference.

Transmission minima for the light propagating normally to the film planes are centred at 1.81 and 2.24 eV and the FWHM of these resonances is about 9% in transmission (Fig.22a). The transmission minima in such hetero-opals can be traced down to 60° due to the limit applied by enhanced scattering at the interface. The PL spectra of the source and filter films demonstrate minima in agreement with transmission spectra (Fig.22a). The relative PL spectra from the source and filter sides of the hetero-opal demonstrate two minima for the larger NC. These minima can also be noticed in the PL spectra of smaller NC (Fig.23).

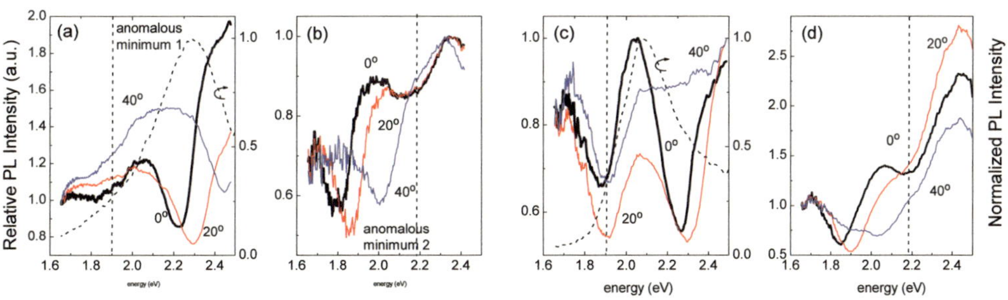

Fig. 23. Relative PL spectra of the source (a, c) and filter (b, d) at different detection angles for NC of 2 (a,b) and 5 nm (c,d) with PL bands centred at 2.29 and 2.08 eV, respectively. Angles are indicated at curves. Dashed PL spectra are obtained at 70°. All spectra are normalised to achieve the same intensity at the "red" edge. Vertical dashed lines indicate the position of "anomalous" minima.

One minimum in relative source and filter spectra moves towards higher energies with increase of the detection angle in accord with the "blue" shift of the (111) Bragg gap. The relative midgap suppression of PL intensity at θ=0° is about 50%. However, the second minimum marked by a dashed line is a stationary one, which shows no shift with changing the angle of detection (Fig. 23). Interestingly to note, that the emission attenuation in the directional and stationary minima are comparable in the case of the 5 nm NCs, the PL intensity of which is evenly distributed between filter and source PBG minima.

Experimental findings so far can be summarized as followed. (a) The emission from the hetero-opal acquires the artificial anisotropy due to the difference of directional Bragg gaps. Important, that this anisotropy is fully under control because the PBG frequencies in a bi-layer opal directly relate to the sphere diameters. (b) The emission spectra acquire the additional non-dispersive minimum, the position of which is correlated but does not coincide with the directional minimum of another opal in a heterostructure. Since such effect was not observed in single light emitting films, its appearance is reasonable to assigned to the light coupling at the interface. The fact that the position of non-dispersive minima is not sensitive to the change of the NC emission band supports this assumption. If so, the spectrum transformation after internally generated light crossing the interface looks like loosing the memory about the PBG directionality.

In order to illustrate this conclusion the angle diagrams of the emission intensity obtained from the source and filter sides of the hetero-opal are compared in Fig.24. The diagrams of the source change their shape like it was discussed with respect to the single light emitting opal film (section 6, Fig.16), moreover, they are broader than the Lambertian diagram. Oppositely, the diagrams of the filter are almost insensitive to changing the frequency and appear much narrower compared to the Lambertian shape. Such phenomenon can be considered as the emission focusing. Such focusing should be controlled by the difference in opal film lattice parameters.

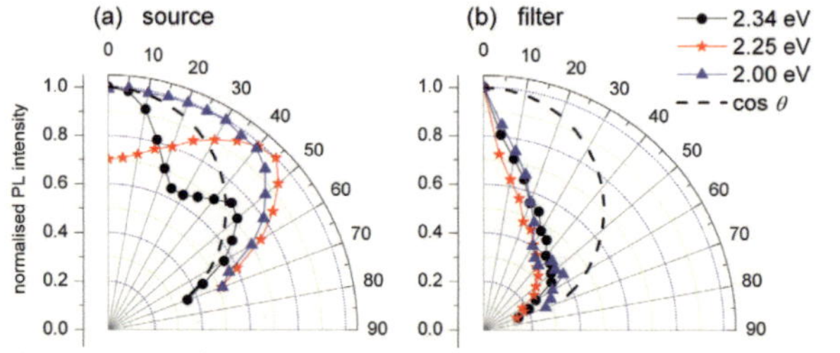

Fig. 24. Emission indicatrices of the normalised to the maximum emission intensity at different frequencies from (a) the opal source and (b) the opal filter in the vicinity to (111) resonances in the 240/300 nm hetero-opal. Dashed line shows the Lambertian diagram.

One can raise a question concerning the mechanism of the emission flux formatting. On the one hand, the absence of memory about the source PBG points to the diffuse character of photon propagation. However, the scattering in the volume of the source (thickness ~14 μm) is not sufficient for complete randomising of the emission flow as demonstrated by the source diagrams. In contrast, the light crossing of the hetero-interface is considered as the sequence of scattering of the incident light, which is transported by the Bloch modes of the source PhC, and following coupling of scattered light to the Bloch modes of the filter PhC [64]. The interface scattering occurs due to the symmetry mismatch between two reservoirs of modes in these PhC lattices. Hence, a fraction of the source radiation propagates to the detector as coupled to the filter eigenmodes and the other fraction - as the uncoupled decaying modes. Therefore, the formatting of outgoing flux proceeds accordingly to the interface coupling conditions, moreover, it is associated with light losses. On the other hand, the light coupling is controlled by the iso-frequencies of two PhC in contact. Fig.25 illustrates this process. In particular, the light propagation from the source with a spherical iso-frequency surface is allowed to all but the directions in the Bragg cones of the filter. With the increase of the emission frequency above the PBG in filter and source crystals, the corresponding iso-frequency surfaces acquire the complex profile leading to rapid variation of the coupling conditions with the frequency.

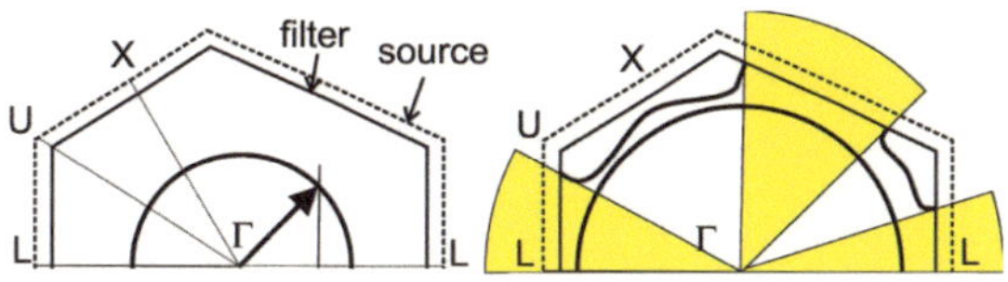

Fig. 25. Schematics of the iso-frequency surfaces in two PhCs with slightly different lattice parameters and the same effective refractive index. (a) Omnidirectional coupling at frequencies below the 1st bandgap in both PhCs. (b) Directional coupling of the emission generated in a source crystal at the frequency below its 1st bandgap to the filter film at the frequency within its 1st bandgap. Shaded cones are the Bragg cones.

The intensity of scattered radiation increases in the source PBG interval with the increase of the angle of light incidence at the interface [73], hence, progressively lower flux couples to the filter. Since scattering equalises the angle distribution of the PL intensity over all directions bringing its spectrum in agreement to the direction-independent DOS spectrum, this is observed in the relative PL spectra of the filter as the anomalous non-dispersive minimum. Thus, effectively, the moving filter PBG minimum squeezes the emission diagram towards the stationary source-related minimum leading to the efficient compressing the filter diagram.

It is instructive to represent the emission anisotropy as the filter-to-source intensity ratio (the anisotropy factor). Such ratio shows a minimum in the filter (111) PBG and the maximum in the source (111) PBG, the spectral separation of which depends on the lattice parameters of the hetero-opal crystals (Fig.26). While this ratio quantifies the emission anisotropy, it is not possible to pin down its absolute value because the substrate introduces the asymmetry in emission outcoupling in a hetero-opal.

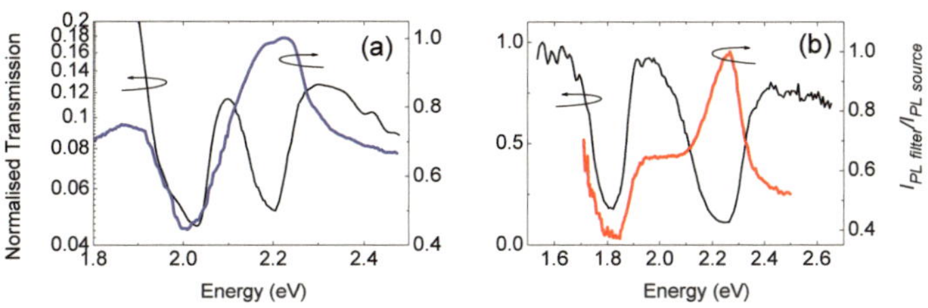

Fig. 26. Anisotropy factor for the source-filter hetero-opals assembled from (a) 240/269 and (b) 240/300 nm opals along the [111] direction.

The shape of the anisotropy factor spectra changes dramatically with the increase of the excitation power and the detection angle. Along the film normal, the minimum of anisotropy factor for the filter remains unchanged along the pumping increase, whereas the maximum for the source PBG monotonously decreases (Fig.27a). The reason of this effect is the stimulated emission in the source (compare to Fig.12c) and the absence of amplification in the filter opal. Along the detection angle increase this picture changes. At $\theta = 20^o$ (i) the extrema of the anisotropy parameter shift to higher frequencies according to PBG dispersions, (ii) the filter minimum becomes also pumping-dependent and (iii) the transition region between resonances becomes distorted. At $\theta = 40^o$ (i) the maximum of anisotropy parameter in the source PBG is replaced with the maximum in between filter and source PBGs and (ii) with the increased pumping the emission anisotropy reduces by three times in its maximum. At $\theta = 70^o$ the pumping dependence of anisotropy disappears and the anisotropy spectrum becomes $\propto 1/I_{PL}(\hbar\omega)$ (Fig.27d).

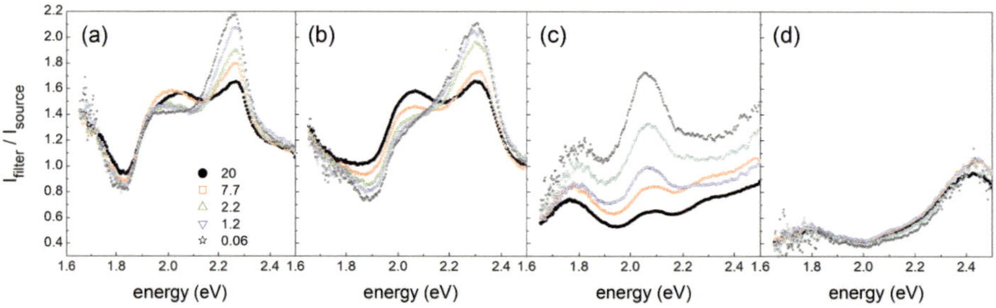

Fig. 27. Evolution of the anisotropy parameter with the increase of the excitation power for different detection angles. The excitation power is specified in the legend in panel (a). Panels (a-d) correspond to the detection angles 0, 20, 40 and 70°.

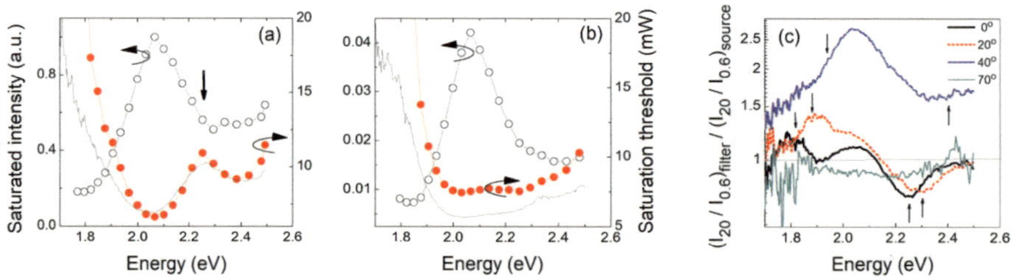

Fig. 28. Spectra of the saturation threshold (circles) and the saturated emission intensity (open circles) for the (a) source and (b) filter opals at $\theta = 0^o$. The ratio of the PL spectra obtained at pumping powers 20 and 0.6 mW are shown by lines for comparison. The arrow in (a) indicates the (111) source PBG. (c) Spectra of enhancement anisotropy parameter for different detection angles. Up- and down-arrows shows PBG of the source and the filter, respectively.

Spectra of the emission anisotropy parameter obtained at 40° show that the emission intensity passed through the interface in the spectral interval between filter and source PBGs (centred at ~2.1 eV) becomes relatively lower with at the increase of the pumping power. This observation reveals an important consequence of the interface coupling – the intensity of the source PL in this range increases much faster compared to that passed through the filter. This non-linear effect can be associated with the increase of the light path in the gain medium of the pumped source opal film. We mentioned that the transmittivity of the interface is much lower for the oblique incident light because of increased interface scattering with possible trapping of scattered light with PBG frequencies at the interface [73]. Taking into account the emission spectrum having the maximum at ~2.05 eV, it is possible to conclude that the scattered radiation is responsible for the source PL increase and that this increase is correlated with the gain spectrum (Fig.23c).

The estimate of the emission rate changes made by using the emission saturation threshold agrees the above picture. The peak in $P_0(\hbar\omega)$ of the source opal coincide the source PBG, whereas no clear features is observed in the spectrum of this parameter for the filter emission (Fig.28a,b). Similarly the spectrum of the ratio of PL intensities measured under pumping of different powers reasonable well corresponds to the $P_0(\hbar\omega)$ spectrum (section 5.2). Then the expression $\left(I_{20}/I_{0.6}\right)_{filter}/\left(I_{20}/I_{0.6}\right)_{source}$ allows to quantify the spectrum of the anisotropy of the emission stimulation (Fig.28c). Since the parameter of enhancement anisotropy is almost 1 for the $\theta = 70^o$, where no PBG influence is expected, it produces the absolute value of enhancement. At $\theta = 0^o$ and 20° one can see the 1.5 time enhancement in the source PBG and the emission suppression in the filter PBG. This once again points to the influencing of the NC emission by filter PBG. At higher angles the influence of PBG becomes indistinguishable compared to the 2.5 times enhancement of the interface-trapped emission at 2.05 eV. It is worth noting that the development of the latter band is detectable at lower angles as well, but its magnitude is comparable to the PBG enhancement. Looking back to Fig. 27, this enhancement is the reason for the corresponding peak of the anisotropy parameter.

9. Modification of nanocrystal emission by the local field of colloidal crystals

The control over the emission spectrum and directionality of light sources can be achieved by tailoring their EM environment, as discussed earlier [3,4,7,6]. So far we considered the emission of sources embedded in opals. Such configuration leads, in the first instance, to the PBG-related suppression of spontaneous emission due to reduction of the mode density and pushing emission flow away from the PBG direction. Simultaneously, the emission coupled to the resonance modes in the same PBG interval experiences stimulation. At the PhC surface the strong local fields are developed at PBG frequencies. The local field pattern of colloidal crystals has been visualized using the near-field optical microscopy [74,75]. Although such field exponentially decays with increasing the distance from the PhC surface, it can promote the emission from NC within the wavelength-scale distance [76]. If so, the fraction of the enhanced emission will be added on top to the spontaneous emission of NC in contrast to the case of the PhC-embedded NCs, when the stimulated emission is added on top of the suppressed spontaneous emission (Fig.29a).

The idea of the spatially separated NC-PhC light source is shown in Fig.29b. In such configuration the dipole transition probability of an oscillator placed in between a 2D PhC and a dielectric substrate depends on the NC location with respect to nodes and antinodes of the EM field, hence, the emission of a light source radiated in the near-field zone of the PhC can be either enhanced or suppressed, but the enhancement effect prevails [77,78].

The PBG emission enhancement should not be confused with traditional ways of increasing the brightness of light emitting devices based on grating couplers and randomly textured surfaces [79,80]. These techniques increase the external quantum efficiency by extracting that fraction of emission, which is normally trapped in the structure due to the total internal reflectance. With the increase of the refractive index contrast, such gratings can be considered in terms of a slab 2D PhCs, in which the wave optics phenomena co-exist with geometrical optics [81]. To date, the efficient out-coupling of radiation has been realized by coupling the radiation to leaky guided modes with frequencies above the light cone in emitters nanopatterned as 2D PhCs [82].

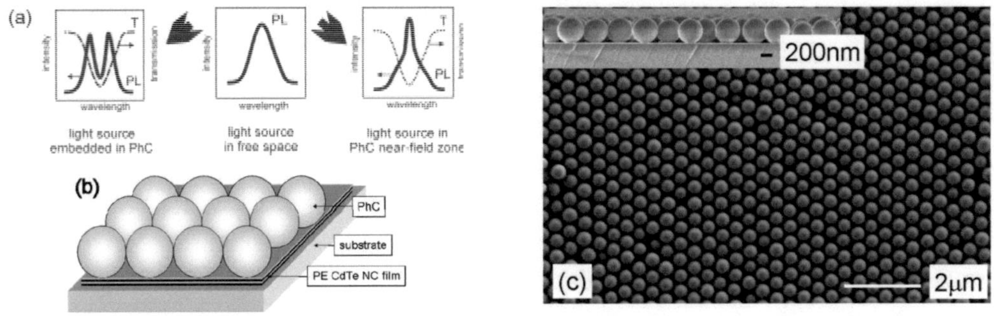

Fig. 29. Schematics of the emission spectrum modification for the light source inside a 3D PhC (left) and at the PhC slab surface (right). Note the different origins of the transmission minima. On the left it occurs due to diffraction at the stack of crystal planes. On the right it occurs due to excitation of leaky guided modes at the PhC surface. (b) Schematics of the structure consisted of a glass substrate with LbL-deposited CdTe NC and LB deposited monolayer of silica spheres. (c) SEM image of a LB monolayer of 519 nm silica spheres on a glass substrate. Inset: the SEM side view of LB monolayer.

From a technology point of view sandwiched NC-opal structures look simpler compared to NC-impregnated opals. However, the light source should be thin enough to be accommodated in the near-field zone of the PhC (Fig.20c) [83].

9.1. Transmission, diffraction and photoluminescence spectra of sandwich sources

Uniform thin films containing 2 nm CdTe NCs [84] were prepared by LbL assembly [85]. The following cyclic procedure was used: (i) dipping of the glass substrate into a solution (5 mg/ml in 0.2 M NaCl) of poly(diallyldimethylammonium chloride) (PDDA) for 10 min, (ii) rinsing with water for 1 min; (iii) dipping into aqueous suspensions of the negatively charged NCs (ca. 10^{-6} M) for 10 min; (iv) rinsing with water again for 1 min. Each cycle of this procedure results in a 'bilayer' consisting of a polymer/NC composite. 20 bi-layers were deposited in order to increase the brightness of luminescence. The thickness of a 20 bi-layer film is about 60 nm. Since the volume fraction of NCs is about 27%, the average RI of the PE-CdTe film is about 1.9. In what follows these films are referred as NC films (NCFs).
Silica spheres of the nominal diameter D=519 nm were hydrophobised with 3-trimethoxysilylpropyl methacrylate. The Langmuir-Blodgett (LB) technique, with a low barrier speed of 6 cm^2 min^{-1}, was used to compress sphere arrays in hexagonally packed monolayers (1L) on the surface of doubly distilled deionised water and to transfer them to

the substrate. Using this technique, the monolayer of spheres was deposited at a surface pressure of 4mN/m [86] on to the LbL NCF substrate (inset, Fig.2). Subsequent monolayers were deposited after drying the deposited ones. Both mono- and multiple-layers (up to five layers (5L)) of the SiO_2 spheres were prepared. Monolayers of spheres form the LB colloidal PhC possessing the PBG of (2+1)-dimensionality. Although no 3D lattice is formed due to the lack of lateral alignment between monolayers, the LB crystal remains ordered within each monolayer of spheres and as a periodic stack of monolayers [87,88].
The PL spectra of NCFs were excited by the 457.9 nm line of an Ar-ion laser. A laser spot size of 5 mm in diameter was selected in order to mimic the emission from a large-area display. The excitation conditions - the laser power and the spot size, were maintained constant at all angles of the detection in order to eliminate excitation-related variation of the PL spectra (Fig.4b). The light collection cone was restricted to 6° by an aperture between the sample and collecting lens. Emission was collected in the forward direction by exciting the NCF through the substrate and detecting after passing the PhC film and in the backward direction - similarly excited but not traversed the PhC film.

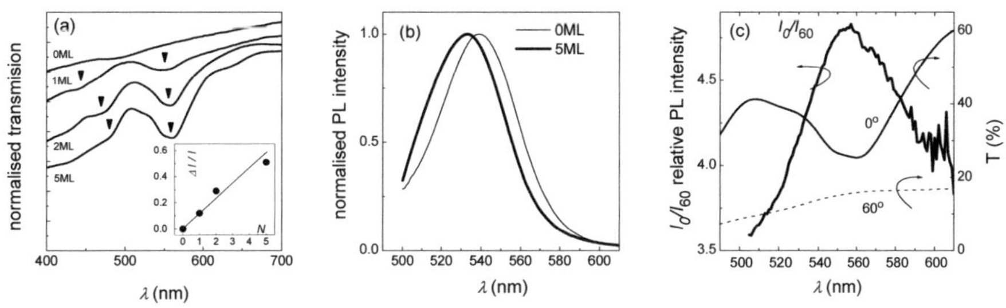

Fig. 30. (a) Transmission spectra of samples with different number, N, of sphere monolayers along the LB crystal normal. Inset - relative light attenuation at the transmission minimum as a function of N. The short wavelength minimum is the result of the resonance splitting in symmetric and antisymmetric combinations. This splitting decreases along the increase of the number of monolayers. (b) Normalized PL spectra of bare NCF (thin line) and 5L LB-coated NCF (thick line) at $\theta = 0^o$. (c) Relative PL spectrum in comparison to far-field transmission spectra obtained at angles $\theta = 0^o$ and 60°.

The *s*-polarized far-field transmission spectrum of the 1ML sample at $\theta = 0^o$ on a glass substrate (Fig.30a) shows the well-defined minimum of 0.08 relative FWHM centred at 548 nm, which provides 12% attenuation of the incident light. The central wavelength of this minimum increases slightly with the increasing number of layers, N. For multiple layer coatings, the transmission significantly decreases towards shorter wavelengths due to light scattering at package irregularities. Transmission spectra of multilayers demonstrate the increase of the relative attenuation $\Delta I / I_0$ with increasing N as $\Delta I / I_0 \sim N^{0.9}$ (inset Fig.30a), where I_0 is the extrapolated transmission in the absence of the minimum. This, practically, linear dependence shows that the attenuation in the LB crystals is an additive function per number of layers. Additional shallow minima were detected at, approximately, 444, 470 and 481 nm for the 1L, 2L and 5L samples, respectively (Fig.30a). The 2L and 5L samples also possess the transmission minimum centred at 1093 nm due to the Bragg diffraction on a stack of monolayers [87], which is not shown.

Fig.31a,b shows the angular dispersion of the transmission minima in a 2L LB-NCF sample. There are I, II, III, IV branches of minima, which are degenerate at $\theta = 0^o$. Calculated dispersion of leaky guided modes in a 2D grating is superimposed on the transmission pattern. The correspondence is observed only between I and (1,0) in s-polarised light and between III and $(0,\bar{1})$ branches in p-polarised light.

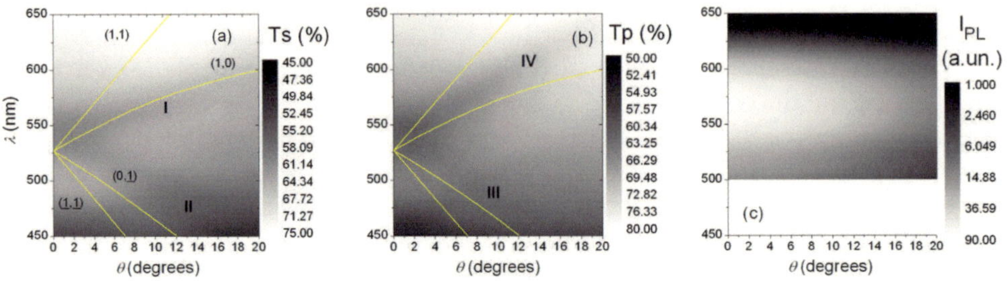

Fig. 31. (a, b) Transmission spectra of CdTe-1L-LB sample under s- and p-polarised light. Rome numbers denote the transmission minima. Lines are calculated eigenmodes of 2D slab PhC possessing the hexagonal lattice of scatterers and the effective refractive index of the 1ML of silica spheres. (c) The PL band of this sample.

The emission band of the NCF assembled from 2 nm NCs is centred at 540 nm (Fig.30b) and possesses the 0.085 relative FWHM. At $\theta = 0^o$ the emission band of the 0.09 relative FWHM of the 5L-LB NCF is centred at 533 nm. The blue shift of the emission band of the LB-NCF sample with respect to that of the bare NCF is the result of the LB coating as was proved by the multiplication of the PL spectrum of the bare NCF by the transmission spectrum of the 5L LB that reproduces such bandshift. The broadening of the PL band of LB-NCF sample could be induced by the chemical modification of the NC surface during the course of the LB film deposition. PL spectra of the LB-NCFs look similar for the different number of LB monolayers. The relative PL spectrum $I(\lambda)_{\theta=0} / I(\lambda)_{\theta=60}$ shows the maximum centred at 555 nm, which coincides with the minimum of the transmission spectrum (Fig.30c). The angle-resolved PL spectrum at $\theta = 60^o$ was used as the reference (Fig.30c).

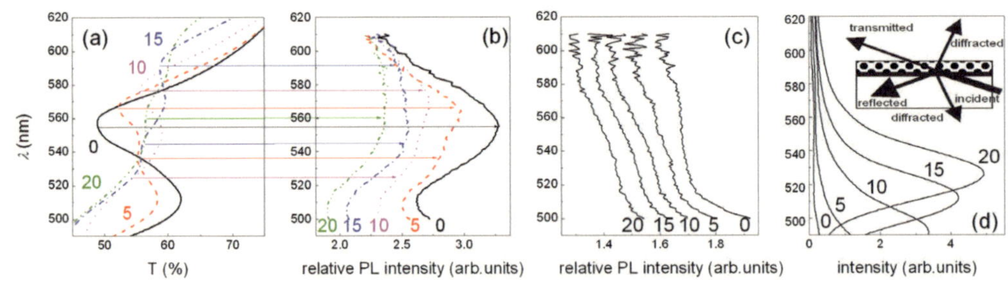

Fig. 32. (a) Angle-resolved transmission spectra of 2L LB-NCF under s-polarized light. Numbers show the angles of light incidence. (b) Angle-resolved relative PL spectra. Numbers show the detection angles. Arrows indicate the correspondence between transmission minima and PL bands. (c) Angle-resolved PL spectra of the uncoated NCF. Numbers show the detection angles. (d). Diffraction spectra at angles α shown at curves. Inset: Schematics of the light transmission and the first order diffraction in LB coating.

The overlap of the PBG with PL band takes place at $0^o \le \theta \le 20^o$ (Fig.31). It is immediately seen that the PBG effect upon the emission intensity is small. Over this range the maxima in the relative PL spectra correlate the minima of the transmission spectra (Fig.32 a, b). It is worth noting that the PBG-related modulation of the PL intensity is proportional to the attenuation in the transmission minimum (Fig.32a,b). In the case of the uncoated NCF no angular dependence of the PL spectrum was detected (Fig.32c)

9.2. Emission indicatrix

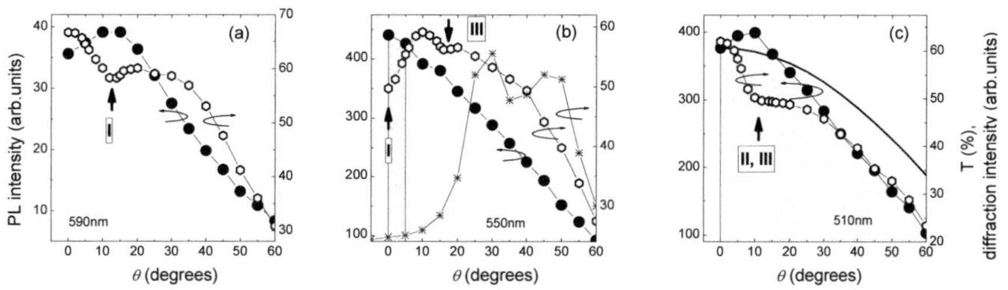

Fig. 33. Angle diagrams of emission intensity (solid circles) at $\lambda = 590, 550$ and 510 nm in comparison to transmission diagrams (open hexagons). These diagrams are compared to the angle diagram of diffraction intensity at $\lambda = 550$ nm (stars) in panel (b) and the Lambert diagram (line) in panel (c). Shaded areas show the emission power between 0 and 5º - S_{550}^{5} (b) and 0 and 60º - S_{510}^{60} (c). Arrows mark the angle position of transmission minima. Rome numbers at arrows indicate the dispersion branch as indicated in Fig.31 a,b.

The PL angle diagrams, $I_\lambda(\theta)$, of a 2L-LB NCF obtained at wavelengths of 510, 550 and 590 nm are shown in Fig.6 together with corresponding transmission diagrams $T_\lambda(\theta)$. The transmission diagrams show the minima superimposed on a smooth background. The angular width of these minimum is less than $\pm 10^o$ with respect to the mid-gap direction. Comparison of emission and transmission diagrams reveals that the PL intensity peaks along the direction of the transmission minimum [89]. This observation correlates with the observation of the relative PL maxima at transmission minima (Fig.32 a,b). It is worth noting that $I_\lambda(\theta)$ does not follow the Lambert diagram $I_\lambda \propto \cos\theta$ (Fig.33c, 34a). To quantify the radiation power emitted by the NCF, we calculated the total area under the angle diagram $S_\lambda^{60} = \int_{\Delta\theta} I_\lambda(\theta) d\theta$, where the angle interval $\Delta\theta = 60^o$ (Fig.33c, 34a). Assuming the even azimuth distribution of the emission intensity and taking into account that the emission intensity becomes sufficiently low at $\theta \ge 60^o$, the S_λ^{60} is proportional to the power radiated at a given wavelength. Integration of the S_λ^{60} over the studied spectral range estimates the total emitted flux.

In order to characterize the effect of PhC coating upon the emission directionality, the S_λ^{60} area was normalized to that under the Lambert-like diagram, $S_\lambda^L = \int_{\Delta\theta} I(\lambda, \theta = 0^o) \cos\theta d\theta$, where the intensity $I(\lambda, \theta = 0^o)$ is taken from the experimental data and $\Delta\theta = 60^o$. Deviation of the ratio $S_\lambda^{60} / S_\lambda^L$ spectrum of NCF-LB from that of bare NCF occurs in the PBG range: the width of the PL indicatrix is reduced by 14% at PBG wavelengths compared to that outside the PBG. Similar observations apply to samples with the 1L, 2L and 5L-thick crystals.

The quantity, S_λ^5, which is proportional to the emission flux propagating within the angular cone $\Delta\theta$ from 0° to 5° at a given wavelength along the film normal was calculated to map the emission diagram distortion in the angle-wavelength plane. S_λ^5 was then normalized to S_λ^{60} to obtain the fraction $S_\lambda^5 / S_\lambda^{60}$. The spectrum of $S_\lambda^5 / S_\lambda^{60}$ shows that the spectral band of enhanced emission is centred at 555 nm as a counterpart of the transmission minimum (Fig.34b). In contrast, the $S_\lambda^5 / S_\lambda^{60}$ fraction for the bare NCF increases monotonically across the same wavelength range.

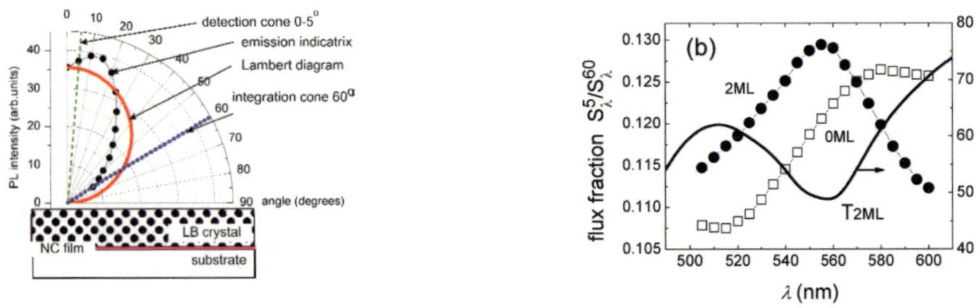

Fig. 34. (a) The definition of the 5°- and 60°-wide flux fractions - dashed and dotted lines, correspondingly. Dots show the PL indicatrix at $\lambda = 590$ nm. Line is the Lambertian indicatrix. (b) Spectra of the emission flux fraction propagating along the film normal within a five degree wide cone for bare NCF (squares) and 2L-LB-NCF (circles) in comparison to the transmission spectrum of the 2L-LB-NCF (line).

By increasing the principal angle of the 5°-wide probe section the enhancement band follows the dispersion of the minimum of the far-field transmission spectra. This means that the radiation enhancement occurs in the direction of the transmission minimum, as it follows from comparison of Figs.35a and 32a. At higher angles, the enhancement becomes less pronounced in response to the decreasing attenuation in the corresponding transmission minimum. It is worth noting the resolution of other transmission minima is too poor to generate a corresponding flux maximum.

The diffraction in the 2D grating that is conveniently used in lightning devices for extraction of guided light [90,91]. For example, 1ML of SiO_2 spheres was able to diffract the beam of 4% bandwidth from the emission trapped in a glass plate between the light source and the monolayer of spheres [92]. The diffraction spectra were measured in order to exclude the diffraction outcoupling as the reason for the PL intensity increase. They were obtained keeping the sum of the incidence and diffraction angle constant $\theta + \alpha = 76^o$. The dispersion of these resonances satisfies the expression

$$m\lambda_0 = \Lambda(\pm \sin\theta \pm \sin\alpha) \qquad (0.20)$$

or, in particular, $\lambda_0 = \Lambda(\sin\theta + \sin(76^o - \theta))$, where the grating period is $\Lambda = 456$ nm. This is the half of the period $a = \sqrt{3}D$ of the trigonal lattice for the wave vector of incident light propagating along the ΓK direction in the Brillouin zone of the 2D hexagonal lattice with the distance between centres of spheres of 526 nm. Such diffraction produces resonances at shorter wavelengths at $\theta \leq 20^o$ compared to the eigenmodes of the 2D PhC (Fig.32d). Moreover, the PhC-induced band in the relative PL spectra (squares in Fig.35b) follows the dispersion branch I of transmission minima and does not satisfy the dispersion (0.20). Therefore, the enhanced PL band cannot be explained by the diffraction at the grating. This conclusion is also supported by comparison of the angle diagram of the intensity of the diffracted beam at $\lambda_0 = 550$ nm (crosses in Fig.33b) and the emission intensity diagram at the same wavelength. Hence, we can abandon the mechanism [93,94,95] assigning the increase of the PL intensity to diffractive outcoupling of the emission.

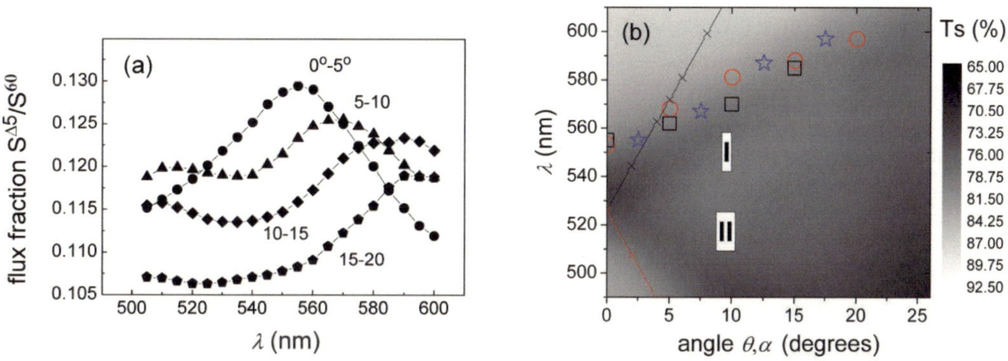

Fig. 35. (a) Spectra of the 5°-wide emission flux fractions propagating along different directions. (b) The transmission pattern of a 2L LB coating under s-polarized light. Squares and circles represent the dispersion of the enhancement band in the forward- and backward-measured relative PL spectra, respectively. Stars show the dispersion of peaks in radiation directionality spectra of panel (a). Dispersion of diffraction (0.20) is shown by crosses.

Another mechanism, the diffraction of the second order, assumes interaction of counter-propagating modes with the grating period that gives rise to the out-coupled beam with $\lambda_0 = 2n_{eff}\Lambda / m$, where $m > 1$. This beam propagates within a very narrow radiation cone along the grating normal [96,97]. However, this diffraction was also not observed.

More adequate explanation takes into account the PBG structure of a 2D lattice of dielectric spheres [98,99]. This model explains reasonably well the observed far-field transmission spectra and their angular dependence. Moreover, this theory predicts considerable broadening of the minima bandwidth due to light leakage to the substrate [100]. The spectral position of the first bandgap in such PhC corresponds to the sphere diameter and the next gap takes place at a wavelength, which is lower by a factor of 1.21. The observed minima at 548 and 444 nm for 1L LB film are in good agreement with this model. As was experimentally demonstrated, a 2D model remains valid for the description of the PBG structure at wavelengths $\lambda \approx D$ for LB colloidal multilayers [87].

Dips in transmission spectra of LB films appear due to light losses for the excitation of eigenmodes, which propagate in the LB film plane. At the PBG resonance, e.g. for $\theta \approx 0^o$ at $\Lambda / \lambda \approx 1$, the local field is enhanced by one (in the case of a PhC on a substrate [99,101]) or two orders of magnitude (for a PhC membrane [77]) due to large evanescent components of the EM field. The strong local field on a scale of the PhC lattice constant speeds up the radiative recombination rate as compared to that in bare NCF. The apparent independence of the magnitude of the emission enhancement upon the number of monolayers in multiple-layer LB films correlates with the surface-related nature of the observed effect.

With this picture in mind, the PL spectrum transformation can be explained as follows. In the case of the bare NCF, the emission is coupled to the substrate and free space. In the presence of an LB coating, the formation of "hot spots" of the local field at PBG wavelengths at PhC surface leads to a higher emission rate of NCs at these spots. Since the relative PL spectrum demonstrates a peak at PBG wavelengths, one can assume that the acceleration of the emission rate in "hot" spots prevails over the rate decrease at "cold" spots.

The PBG origin of the enhancement leads to the emission directionality in agreement with the angular dispersion of the PBG minimum. If NCs emit into eigenmodes of the PhC, then the outcoupling of this radiation is the reciprocal process with respect to the coupling of external radiation to PhC modes in a transmission experiment.

Interestingly, the emission indicatrix (Fig.33) in the studied case appears more sensitive to the PBG directionality compared to the PL spectra (Fig.30b), but the maximum in the directionality spectra agree the enhancement bands of the relative PL spectra (Fig.35b)

The strong argument in favour of the PBG-related enhancement mechanism is the correlation between the relative LB-NCF PL enhancement bands measured in the backward direction, for which the emission does not pass the LB crystal, and the minima in LB crystal transmission spectra (Fig.36). The structure of the forward and backward relative PL spectra and the dispersions of these bands are similar, i.e., enhancement reveals itself in opposite directions simultaneously, although less efficient in the backward direction. This property correlates the reciprocity of the LB-NCF sample transmission and the leakage character of guided modes in the LB-NCF sample. These observations point to the fact that the NCF forms an interacting unit with the LB PhC.

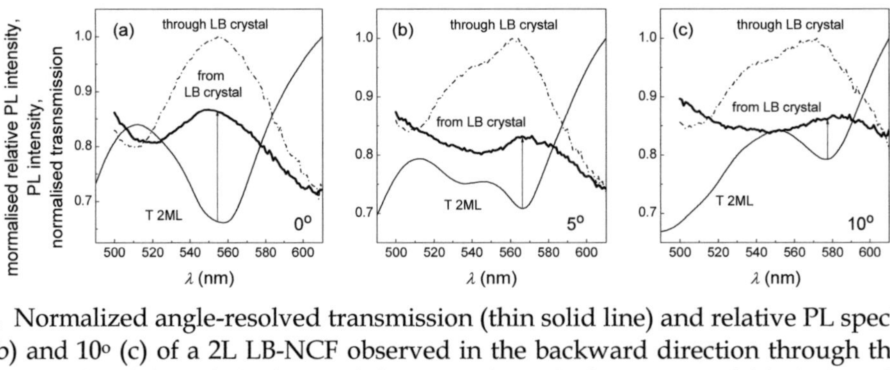

Fig. 36. Normalized angle-resolved transmission (thin solid line) and relative PL spectra at 0 (a), 5 (b) and 10° (c) of a 2L LB-NCF observed in the backward direction through the glass substrate (thick solid) and the forward direction through the LB crystal (dash-dotted lines). Arrows indicate the centres of transmission minima.

10. Summary

In this review we demonstrated different aspects of the PBG control on the emission of semiconductor nanocrystals that is coupled to optical modes of colloidal crystal-based thin film photonic crystals. We discussed (i) changes of the emission spectra and the emission directionality, (ii) methods used to recognise and estimate the emission modification, (iii) physical mechanisms behind the emission control and (iv) energy transfer between the nanocrystals and the carcass of photonic crystals. This basic information allows to estimate pros and contras of photonic crystal-integrated light sources and their prospects in the design of future lightning devices.

The described experiments were limited to the emission control in the spectral range of the lowest frequency directional bandgap, namely, the (111) bandgap in the opal crystals. The directional nature of this bandgap reduces dramatically the strength of the PBG effect because the light flow can escape the photonic crystal using remaining unblocked propagation directions. Preparing inverted opals from high refractive index dielectrics can at least partially lift up this problem due to smaller escape angle range.

Exploring the range of high order bandgaps is also in progress, but due to the nature of the photonic bandgap diagram, the (111) gap in opals is the only true one in terms of the absence of photon modes. All other high order bandgaps are, in fact, merely diffraction resonances with contribution of slow propagating modes. This uncertainty blurs up the physical mechanisms of emission modification. Obviously, in the case of inverted opals with an omnidirectional bandgap the physical nature of the emission control can reveal itself in a full extent, but such crystals were not yet reported for the visible for very fundamental reasons.

We did not also address any works related to optimisation of the NC-to-PhC interaction for following reasons. First, the PBG effect on emission of embedded sources strongly depends on the ordering of the crystal lattice. In this sense the photonic crystals prepared by nanolithography provide the precise control upon the light source positioning and overall lattice regularity, but at the cost of time-consuming and expensive technologies and small crystal volumes. Second, the design of light sources with specific functionality requires corresponding structuring of the photonic crystal, e.g., formation of resonators. Third, the optimisation usually exploits already established operation principles, but does not lead to new phenomena. Instead, we concentrated the attention on photonic crystals prepared by colloidal assembly because their combination with colloidal nanocrystals can benefit from inexpensive technology of both components, while preserves all possibilities for emission control offered by other realisations of photonic crystal-enhanced light sources. Overall, along the progress in improving the colloidal crystal ordering and methods of their structuring, this approach looks prospective for up-scaled production.

Acknowledgments

Authors gratefully acknowledge valuable contributions of their collaborators and colleagues N. Gaponik, A. Eychmueller, A.L. Rogach, M. Bardosova, C.M. Sotomayor Torres, D.N. Chigrin, V.G. Solovyev, R. Kian and R. Zentel.

11. References

1 Yin Y., A. P. Alivisatos, Colloidal nanocrystal synthesis and the organic–inorganic interface, *Nature*, **437** 664-670 (2005)

2 Rogach A. L., N. Gaponik, J. M. Lupton, C. Bertoni, D. E. Gallardo, S. Dunn, N. Li Pira, M. Paderi, P. Repetto, S.G. Romanov, C. O'Dwyer, C. M. Sotomayor Torres, A. Eychmüller, Light-Emitting Diodes with Semiconductor Nanocrystals, *Angewandte Chemie Int. Edition*, **47**, iss.35, 6538 - 6549 (2008)

3 Parcell E.M., Spontaneous emission probabilities at radio frequencies, *Phys.Rev.*, **69**, 681 (1946)

4 Bykov V. P., Spontaneous emission in a periodic structure, *Sov. Phys. JETP* **35**, 269-273 (1972).

5 Bykov V. P., Spontaneous emission from a medium with a band spectrum, *Sov. J. Quant. Electron.* **4**, 861-871 (1975).

6 John S., Strong localization of photons in certain disordered dielectric superlattices, *Phys.Rev. Lett.* **58**(23), 2486-2489 (1987).

7 Yablonovitch E., Inhibited Spontaneous Emission in Solid-State Physics and Electronics, *Phys. Rev. Lett.* **58**, 2059-2062 (1987).

8 Joannopoulos J. D., S. G. Johnson, J. N. Winn, R. D. Meade, Photonic Crystals: Molding the Flow of Light, 2nd Edition, Princeton University Press, Princeton and Oxford, 2008, 284p.

9 Sanders J. V. , Colour of Precious Opal, *Nature*, **2049**, 1151-1154 (1964).

10 Darragh P. J. , A. J. Gaskin, B. C. Terrell & J. V. Sanders, Origin of Precious Opal, *Nature*, **209**, 13-16 (1966).

11 Balakirev V.G., V.N.Bogomolov, V.V.Zhuravlev, Y.A.Kumzerov, V.P.Petranovskii, S.G.Romanov, LA.Samoilovich, Three-dimensional superlattices in the opals, *Crystallography Reports*, **38**, 348-353 (1993).

12 Lopez C., Material aspects of photonic crystals, *Advanced Mater.*, **15**, N20, 1679-1704 (2003).

13 Astratov V.N., V.N.Bogomolov, A.A. Kaplyanskii, A.V. Prokofiev, L.A. Samoilovich, S.M. Samoilovich, Yu.A. Vlasov, Optical Spectroscopy of Opal Matrices with CdS Embedded in its Pores: Quantum Confinement and Photonic Band Gap Effects, *Il Nuovo Cimento*, **17D**, 1349-1354 (1995).

14 Romanov S.G., A.V.Fokin, V.V.Tretiakov, V.Y.Butko, V.I.Alperovich, N.P.Johnson, C.M.Sotomayor Torres. Optical properties of ordered 3-dimensional arrays of structurally confined semiconductors. *J.Cryst.Growth*, **159**, 857-860 (1996).

15 Xia Y., B. Gates, Y. Yin, and Y. Lu, Monodispersed Colloidal Spheres: Old Materials with New Applications, *Adv. Mater.*, **12**, No. 10 693-713 (2000).

16 Gaponenko S.V., Bogomolov V.N., Kapitonov A.M., Prokofiev A.V., Eychmueller A., Rogach A.L., Electrons and photons in mesoscopic structures: Quantum dots in a photonic crystal, *JETP Letters*, **68**, №2, 142-147 (1998).

17 Lodahl P., A. F. van Driel, I. S. Nikolaev, A. Irman, K. Overgaag, D. Vanmaekelbergh, W. L. Vos, Controlling the dynamics of spontaneous emission from quantum dots by photonic crystals, *Nature*, **430** , 654-657 (2004).

18 Boedecker G., C. Henkel, Ch. Hermann, O. Hess, Spontaneous emission in photonic structures: Theory and simulation, pp.23-42 in "Photonic crystals - Advances in Design, Fabrication and Characterization", eds. K.Busch, S.Loelkes, R.B. Wehrspohn,. H.Foell, 354 p Wiley-VCH, Weinheim, 2004.
19 Sprik R., B. A. van Tiggelen, and A. Lagendijk, Optical emission in periodic dielectrics, *Europhys. Lett.* **35**, 265 (1996).
20 Vats N., S. John, K. Busch, Theory of fluorescence in photonic crystals, *Phys.Rev. A*, **65**, 043808-1-13 (2002).
21 John S.and J. Wang, Quantum optics of localized light in a photonic band gap, *Phys. Rev. B* **43**, 12 772-12789 (1991).
22 Vats N. and S. John, Non-Markovian quantum fluctuations and superradiance near a photonic band edge, *Phys. Rev. A* **58**, 4168-4185 (1998).
23 Busch K. and S. John, Photonic band gap formation in certain self-organizing systems, *Phys. Rev. E* **58**, 3896-3908 (1998).
24 Koenderink A.F.,L.Bechger,A.Lagendijk,W.L.Vos, An experimental study of strongly modified emission in inverse opal photonic crystals, *phys.stat.sol.*(a) **197** №.3,648–661 (2003).
25 Chabanov A. A., Y. Jun, and D. J. Norris, Avoiding Cracks in Self-Assembled Photonic Band Gap Crystals, *Appl. Phys. Lett*. **84**, 3573 (2004).
26 Khunsin W., G. Kocher, S. G. Romanov, C. M. Sotomayor Torres, Quantitative analysis of lattice ordering in thin film opal-based photonic crystals, *Adv. Funct. Mater.*, **18**, 2471-2479 (2008).
27 Romanov S. G., Specific features of polarization anisotropy in optical reflection and transmission of colloidal photonic crystals, *Phys. Solid State*, **52**, N 4, 844-854 (2010)
28 Rogach A. L., L. Katsikas, A. Kornowski, D. Su, A. Eychmueller, H. Weller, Synthesis and Characterization of Thiol-Stabilized CdTe Nanocrystals, *Ber. Bunsenges. Phys. Chem.* **100**, 1772-1778 (1996).
29 Romanov S. G., T. Maka, , C. M. Sotomayor Torres M. Müller, R. Zentel, D. Cassagne, J. Manzanares-Martinez and C. Jouanin, Diffraction of Light from Thin Film PMMA Opaline Photonic Crystals, *Phys. Rev. E*, **63**, 056603-1-5 (2001).
30 Nikolaev I. S., P. Lodahl, and W. L. Vos, Quantitative analysis of directional spontaneous emission spectra from light sources, in photonic crystals, *Phys.Rev. A* **71**, 053813-1-10 (2005).
31 Koenderink A. F. and W. L. Vos, Optical properties of real photonic crystals: anomalous diffuse transmission, *J. Opt. Soc. Am. B*, **22**, No. 5 1075-1084 (2005).
32 Romanov S. G., T. Maka, C. M. Sotomayor Torres, M. Müller, R. Zentel, Emission in an SnS_2 Inverted Opaline Photonic Crystal, *Appl. Phys. Lett.*, **79**, 731-733 (2001).
33 Romanov S.G., A.V. Fokin, V.I. Alperovich, N.P. Johnson, R.M. De La Rue, The Effect of the Photonic Stop-Band upon the Photoluminescence of CdS in Opal, *Physica Status Solidi a*, **164**, 169-173 (1997).
34 Bogomolov V. N., S. V. Gaponenko, I. N. Germanenko, A. M. Kapitonov, E. P. Petrov, N. V. Gaponenko, A. V. Prokofiev, A. N. Ponyavina, N. I. Silvanovich, S. M. Samoilovich, Photonic band gap phenomenon and optical properties of artificial opals, *Phys.Rev. E*, **55**, 7619-7625 (1997).

35 Romanov S.G., M. A. Kaliteevski, C.M.Sotomayor Torres, J. Manzanares Martinez, D. Cassagne, J.P. Albert, A.V. Kavokin, V.V. Nikolaev, S.Brand, R.A. Abram, N. Gaponik, A. Eychmueller, A.L. Rogach, Stimulated emission due to light localization in the bandgap of disordered opals, *Phisica Status Solidi (c)*, **1**, 1522-1530 (2004).
36 Golub J. E., R. Eichmann, and P. Thomas, Dimensional crossover of exciton hopping relaxation in mesoscopic structures, *Europhys. Lett.* **47**(5), 628-631 (1999).
37 A.A. Kiselev, Hopping-induced energy relaxation with allowance for all possible versions of intercenter transitions, *Semiconductors* **32**(5), 504-508 (1998).
38 Baranovskii S. D., T. Faber, F. Hensel, and P. Thomas, On the Description of Hopping - Energy Relaxation and - Transport in Disordered Systems, *J. Non-Cryst. Solids* 198-**200**, 222-225 (1996).
39 Zegria G. G. and V. A. Kharchenko, New mechanism of Auger recombination of nonequilibrium current carriers in semiconductor heterostructures, *Sov.Phys.JETP*, **74**(1), 173-181 (1992).
40 Yeh P., Optical Waves in Layered Media, John Wiley&Sons, New York, 1998; ISBN 0-471-82866-1.
41 J Russell P.St., Optics of Floquet-Bloch waves in dielectric gratings, *Appl. Phys. B*, **39,** 231-246 (1986).
42 Kosaka H., T. Kawashima, A. Tomita, M. Notomi, T. Tamamura, T. Sato, S. Kawakami, Superprism phenomena in photonic crystals, *Phys. Rev. B*, 58 R10096-99 (1998).
43 Romanov S.G., D. N. Chigrin, T. Maka, M. Müller, R. Zentel, C. M. Sotomayor Torres, Directionality of light emission in three-dimensional opal-based photonic crystals, *Proc. SPIE Int. Soc. Opt. Eng.* **5825**, 160-172 (2005).
44 Romanov S. G., D.N. Chigrin, C. M. Sotomayor Torres, N. Gaponik, A. Eychmüller, A. L. Rogach, Emission Stimulation in a directional bandgap of a CdTe-loaded opal photonic crystal, *Phys.Rev.E.* **69**, 046606-1-4 (2004).
45 Johnson S.G., J.D. Joannopoulos, Block-iterative frequency-domain methods for Maxwell's equations in a plane-wave basis,*Optics Express*, **8** (2001) 173-190
46 Wiersma D.S. and A. Lagendijk, Light diffusion with gain and random lasers.*Phys. Rev.E.*, **54**(4), 4256-4265 (1996).
47 John S. and G. Pang, Theory of lasing in a multiple-scattering medium, *Phys. Rev.A.*, **54**(4), 3642-3652 (1996).
48 Vlasov Yu. A., M.A. Kaliteevski, and V.V. Nikolaev, Different regimes of light localization in a disordered photonic crystal, *Phys. Rev. B* , **60**(3), 1555-1562 (1999)
49 Kaliteevski M. A., J. Manzanares Martinez, D. Cassagne, and J. P. Albert, Disorder-induced modification of the transmission of light in a two-dimensional photonic crystal, *Phys.Rev.B.*, **66**, 113101-1-4 (2002).
50 Soest G. van, F. J. Poelwijk, R. Spirk, and A. Lagendijk, Dynamics of a random laser above threshold, *Phys. Rev. Lett.* **86**, 1522-1525 (2001).
51 D.N. Chigrin, Radiation pattern of a classical dipole in a photonic crystal: Photon focusing, *Phys. Rev.* E **70**, 056611-1-12 (2004).
52 Etchegoin P. and R. T. Phillips, Photon focusing, internal diffraction, and surface states in periodic dielectric structures, *Phys. Rev. B* **53**, 12674-12683 (1996).

53 Romanov S. G., D. N. Chigrin, V. G. Solovyev, T. Maka, N. Gaponik, A. Eychmüller, A. L. Rogach, and C. M. Sotomayor Torres, Light emission in a directional photonic bandgap, *Phys.stat.sol. (a)*, **197**, 662-672 (2003)
54 Ho K. M., C. T. Chan, and C. M. Soukoulis, Existence of a photonic gap in periodic dielectric structures, *Phys. Rev. Lett.* **65**, 3152-3155 (1990)
55 S. G. Romanov, U. Peschel, M. Bardosova, S. Essig, K. Busch, Suppression of the critical angle of diffraction in thin film colloidal photonic crystals, *Phys. Rev. B,* **82** (2010) in press.
56 Reynolds A., F. López-Tejeira, D. Cassagne, F. J. García-Vidal, C. Jouanin, and J. Sánchez-Dehesa, Spectral properties of opal-based photonic crystals having a SiO2 matrix*Phys. Rev. B.*, **60**, 11422-11426 (1999).
57 Zhukov E. A., Y. Masumoto, H. M. Yates, M. E. Pemble, E.A. Muljarov, C. M. Sotomayor Torres and S. G. Romanov, Interface Interactions and the Photoluminescence from Asbestos-Templated InP Quantum Wires, *Microstructures and Superlattices*, **27**, No. 5/6, pp. 571-576 (2000).
58 Bardosova M., S.G. Romanov, C.M. Sotomayor Torres, N. Gaponik, A. Eychmueller, Y.A. Kumzerov, J. Bendall, Effect of template defects in radiative energy relaxation of CdTe nanocrystals in nanotubes of chrysotile asbestos, *Micro- and Meso-porous materials,* **107**, 212-218 (2007).
59 Mikhailovsky A.A., A.V. Malko, J.A. Hollingsworth, M.G. Bawendi, V.I. Klimov, Multiparticle interactions and stimulated emission in chemically synthesized quantum dots, *Appl.Phys.Lett.,* **80**, 2380 (2002).
60 Solovyev V. G., S. G. Romanov, C. M. Sotomayor Torres, M. Müller, R. Zentel, N. Gaponik, A. Eychmüller and A. L. Rogach. Modification of the spontaneous emission of CdTe nanocrystals in TiO_2 inverted opals, *J. Appl.Phys.*, **94**, 1205-1210 (2003).
61 Astratov V.N., A.M. Adawi, S. Fricker, M.S. Skolnick, D.M. Whittaker, P.N. Pusey, Interplay of order and disorder in the optical properties of opal photonic crystals, *Phys. Rev. B*, **66**, 165215-1-13 (2002)
62 Stefanou N., V. Yannopapas, A. Modino, Heterostructures of photonic crystals: frequency bands and transmission coefficients, *Computer Physics Commun.* **113**, 49-77 (1998)
63 Romanov S. G., H. M. Yates, M. E. Pemble, R. M De La Rue, Opal-Based Photonic Crystal with Double Photonic Bandgap Structure, *J. Phys.: Cond. Matter* **12**, 8221-8229 (2000).
64 Istrate E., E. H. Sargent, Photonic crystal heterostructures and interfaces, *Rev. Modern Physics*, **78**, 455-481 (2006).
65 Gaponik N., A. Eychmüller, A.L. Rogach, V.G. Solovyev, C.M. Sotomayor Torres, S. G. Romanov, Structure-related optical properties of luminescent hetero-opals, *J. App. Phys.*, **95**, 1029-1035 (2004).
66 Romanov S. G., N. Gaponik, A. Eychmüller, A.L. Rogach V. G. Solovyev, D.N. Chigrin and C. M. Sotomayor Torres, Light emitting opal-based photonic crystal heterojunctions, in "Photonic Crystals" eds. K. Busch, S.Lölkes, R. Wehrspohn, H. Föll, Wiley-VCH, Weinheim, ISBN-13: 978-3-527-40432-2, 132-152 (2004).
67 Romanov S.G., Optical characterization of opal photonic hetero-crystals, in "Frontiers of Multifunctional Integrated Nanosystems" eds. E. Buzaneva and P. Scharff, Kluwer Acad. Publ., ISBN978-1-4020-2171-8, 309-330 (2004).

68 Jiang P., G. N. Ostojic, R.Narat, D. M. Mittleman, V.L. Colvin, The Fabrication and Bandgap Engineering of Photonic Multilayers, *Adv. Mater.* **13**, No. 6, 389-393 (2001)
69 Decher G., Fuzzy Nanoassemblies: Toward Layered Polymeric Multicomposites. *Science,* **277** 1232-1237 (1997).
70 Rogach A. L., D. S. Koktysh, M. Harrison, N. A. Kotov, Layer-by-Layer Assembled Films of HgTe nanocrystals with Strong Infrared Emission, *Chem. Mater.*, **12**, 1526-1528 (2000).
71 Caruso F., Nanoengineering of Particle Surfaces. *Adv. Mater.*, **13**, 11-22 (2001).
72 Gaponik N., D. V. Talapin, A. L. Rogach, K. Hoppe, E. V. Shevchenko, A. Kornowski, A. Eychmüller, and H. Weller, Thiol-Capping of CdTe Nanocrystals: An Alternative to Organometallic Synthetic Routes, *J. Phys. Chem.* B **106**, 7177-7185 (2002).
73 Ding B., M. Bardosova, I. Povey, M. E. Pemble, and S.G. Romanov, Engineered Light Scattering in Colloidal Photonic Heterocrystals, *Adv. Funct. Mater.*, **20**, 853-860 (2010).
74 Ohtaka K., H. Miyazaki, T. Ueta, Near-field effect involving photonic bands, *Mat. Science and Engineering B*, **48** 153- 161 (I 997).
75 Fluck E., N. F. van Hulst, W. L. Vos, and L. Kuipers, Near-field optical investigation of three-dimensional photonic crystals, *Phys.Rev. E* **68**, 015601-1-4 (2003).
76 Koenderink A. F., M. Kafesaki, C. M. Soukoulis, V. Sandoghdar, Spontaneous emission rates of dipoles in photonic crystal membranes, *J.Opt.Soc.Am.*B, **23**, 1196-1206 (2006)
77 Inoue M., Enhancement of local field by a two-dimensional array of dielectric spheres placed on a substrate, *Phys.Rev.B*, **36**, 2852-2862 (1987).
78 Kurokawa Y., H. Miyazaki, Y. Jimba, Optical band structure and near-field intensity of a periodically arrayed monolayer of dielectric spheres on dielectric substrate of finite thickness, *Phys.Rev B* **69**, 155117-1-9 (2004).
79 Rigneault H., F.Lemarchand, A.Sentenac, H.Giovannini, Extraction of light from sources located inside waveguide grating structures, *Opt.Lett.*, **24**, 148-150 (1999).
80 Joyce W.B., R.Z. Bachrach, R.W.Dixon, D.A.Sealer, Geometrical properties of random particles and the extraction of photons from electroluminescent diodes, *J.Appl.Phys.*, **45**, 2229-2254 (1974).
81 Bermel P., C. Luo, L.Zeng, L. C. Kimerling, J. D. Joannopoulos, Improving thin-film crystalline silicon solar cell efficiencies with photonic crystals, *Optics Express*, **15**, №25, 16986-17000 (2007)
82 Boroditsky M., T. F. Krauss, R. Coccioli, R. Vrijen, R. Bhat, E. Yablonovitch, Light extraction from optically pumped light-emitting diode by thin-slab photonic crystals, *Appl. Phys.Lett.*, **75**, 1036-1039 (1999).
83 Romanov S.G., M. Bardosova, M. Pemble, C.M. Sotomayor Torres, N. Gaponik A. Eychmüller, Emission pattern of planar CdTe nanocrystal light source coated by two-dimensional Langmuir-Blodgett photonic crystal, *Materials Science and Engineering C,* **27**, 968-971 (2007) .
84 Gaponik N., A.L. Rogach, Aqueous synthesis of semiconductor nanocrystals in "Semiconductor Nanocrystal Quantum Dots. Synthesis, Assembly, Spectroscopy and Applications", Rogach, A. (Ed.), Springer, Wien, N ew York, 2008, 372 p., ISBN: 978-3-211-75235-7
85 Shavel V., N. Gaponik, A. Eychmüller. The Assembling of Semiconductor Nanocrystals, *Eur. J. Inorg. Chem.*, **18** 3613-3263, (2005).

86 Bardosova M., P. Hodge, V.Smatko R.H. Tredgold and D. Whitehead, A new method of forming synthetic opals, *Acta Phys. Slovaca*, **54**, 409-415 (2004).
87 Romanov S.G., M. Bardosova, M. Pemble, C.M. Sotomayor Torres, (2+1)-dimensional photonic crystals from Langmuir-Blodgett colloidal multilayers, *Applied Physics Letters*, **89**, 43105 (2006).
88 Romanov S.G., M. Bardosova, D.E. Whitehead, I. Povey, M. Pemble, C.M. Sotomayor Torres, Erasing diffraction orders - opal versus Langmuir-Blodgett colloidal crystals, *Applied Phys. Lett.*,**90**,133101 (2007).
89 Romanov S.G., M. Bardosova, I.M. Povey, M. Pemble, C.M. Sotomayor Torres, N. Gaponik and A. Eychmüller, Modification of emission of CdTe nanocrystals by the local field of Langmuir-Blodgett colloidal photonic crystals , *J. Appl. Phys.* **104**, 103118 1-8 (2008).
90 Alferov Zh.I., V.M. Andreev, S.A.Gurevich, R.F.Kazarinov, V.R.Larionov, M.N. Mizerov, E.L.Portnoi, Semiconductor lasers with the light output through the diffraction grating on the surface of the waveguide layer, *IEEE J. Quant. Electronics*, **11**, 449-451 (1975).
91 Turnbull G. A., P. Andrew, M. J. Jory, W. L. Barnes, and I. D. W. Samuel, Relationship between photonic band structure and emission characteristics of a polymer distributed feedback laser, *Phys.Rev. B*, **64**, 125122-1-6 (2001).
92 Yamasaki T., K. Sumioka, and T. Tsutsui, *Appl.Phys.Lett*, Organic light-emitting device with an ordered monolayer of silica microspheres as a scattering medium, **76**, 1243-1245 (2000).
93 Dakks M.L., L.Khun, P.F.Heidrich, B.A. Scott, Grating Coupler For Efficient Excitation Of Optical Guided Waves In Thin Films, *Appl.Phys.Lett.*, **16**, 523-525 (1970).
94 Matterson B. J., J. M. Lupton, A.F. Safonov, M. G. Salt, Wi. L. Barnes, I. D. W. Samuel, Increased Efficiency and Controlled Light Output from a Microstructured Light-Emitting Diode, *Adv. Mater.* **13**, 123-127 (2001).
95 Ziebarth J. M., A. K. Saafir, S. Fan, M. D. McGehee, Extracting Light from Polymer Light-Emitting Diodes Using Stamped Bragg Gratings, *Adv. Funct. Materials*, **14**, 451 (2004).
96 Streifer W., D.R. Scifres, R.D. Burnham, Coupled wave analysis of DFB and DBR lasers *IEEE J. Quant. Electronics*, **13**, 134-141 (1977).
97 Kanskar M., P.Paddon, V.Pacradouni, R. Morin, A. Busch, J.F,Young, S.R. Johnson, J.MacKenzie, T. Tiedje, Observation of leaky slab modes in an air-bridged semiconductor waveguide with a two-dimensional photonic lattice, *Appl.Phys.Lett.*, **70**, 1438-1440 (1997).
98 Miyazaki H.T., H. Miyazaki, K.Ohtaka, T.Sato, Photonic band in two-dimensional lattices of micrometer-sized spheres mechanically arranged under a scanning electron microscope, *J. Appl. Phys.*, **87**, 7152-7159 (2000).
99 Miyazaki H.T., H. Miyazaki, K.Ohtaka, T.Sato, Photonic band in two-dimensional lattices of micrometer-sized spheres mechanically arranged under a scanning electron microscope, *J. Appl.Phys.*, **87**, 7152-7159 (2000).
100 Kurokawa Y., H.Miyazaki, Y.Jimba, Light scattering from a monolayer of periodically arrayed dielectric spheres on dielectric substrates, *Phys.Rev.B*, **65**, 201102-1-4 (2002).
101 Kurokawa Y., H. Miyazaki, Y. Jimba, Optical band structure and near-field intensity of a periodically arrayed monolayer of dielectric spheres on dielectric substrate of finite thickness, *Phys.Rev B* **69**, 155117-1-9 (2004).

5

A review of high nanoparticles concentration composites: semiconductor and high refractive index materials

Igor Yu. Denisyuk
Doctor of Science, professor, head of the chair of Quantum Sized Optical Systems of State University of Information Technologies, Mechanics and Optics
Russia, 197101, Saint-Petersburg, Str. Kronverskii 49

Mari Iv. Fokina
Assistant professor of State University of Information Technologies, Mechanics and Optics
Russia, 197101, Saint-Petersburg, Str. Kronverskii 49

1. Introduction

At the present time homogeneous optical and optoelectronic media such as glasses, monocrystals and polymer materials are widely used. Each of these media has a specific set of properties. For example, the polymer materials allow producing flexible and transparent films, for example film OLED. They are cheap. The technology of polymers treatment is very simple and suitable for some applications. However the holes and electrons mobility in polymer materials are many time less in comparison with inorganic well known semiconductors. Another problem is a relative high exciton decay energy in polymer with value of 100 meV that result on temperature dependence of photogeneration.
The possibility of combining the different properties into a single material should be rather useful. It is impossible to solve this problem by traditional ways because the properties reflect the internal structures of these different materials.
The method of nanostructuring provides the possibility of combining the properties of polymers and crystals. The resulting nanocomposite is the mechanical mixture of inorganic semiconductor distributed uniformly in the polymer matrix. Under the condition of uniform distribution of nanoparticles and if the size of such nanocrystals is small (2-5 nm), they don't distort an incident light wave and the light scattering is low. If to use high refractive index nanoparticles such as ZnS, CdS, ZnO, TiO2 incorporation of nanoparticles into polymer will result on significant increasing of refractive index of material. Same time these material has a proprieties of homogenous semiconductor material because of small nanometer size distance between semiconductor nanoparticles and easy tunnel transportation of charge carrier. Nanocomposite with high concentration of small size nanocrystals becomes effectively a homogeneous medium, having semiconductor proprieties of inorganic material with low scattering and good flexibility and processability of polymer. The set of properties of this mixture is determined by both components, namely polymer and nanocrystals, and

by the ratio of concentrations of them. The main efforts of research directed to develop quasi- homogenous nanocomposite material with nanoparticles and polymer matrix comparable content for photonics application areas: photoresist for nanolithography, microoptics, organic solar cell and OLED.

2. Crystalline lattice of small nanoparticles

Physics of semiconductor theory use macroscopic charge carrier statistic parameters of bulk crystals for all structures types including microns and sub- microns elements in microchips. It is correct approximation now as the sizes of microchips elements more than 100 nm are more time larger in comparison to interatomic distances. In contrary, typical nanoparticles, for example CdS or ZnS with the size of 1 - 3 nm include a few atomic layers and its interatomic distance can differ essentially from same of bulk crystals. Now are a few of works where these effects are investigated.
Indeed, if to shrink elements down to the nanometer scale, creating nanodots, nanoparticles, nanorods and nanotubes a few tens of atoms across, they've found weird and puzzling behaviors unexpected for bulk and micron sized material.

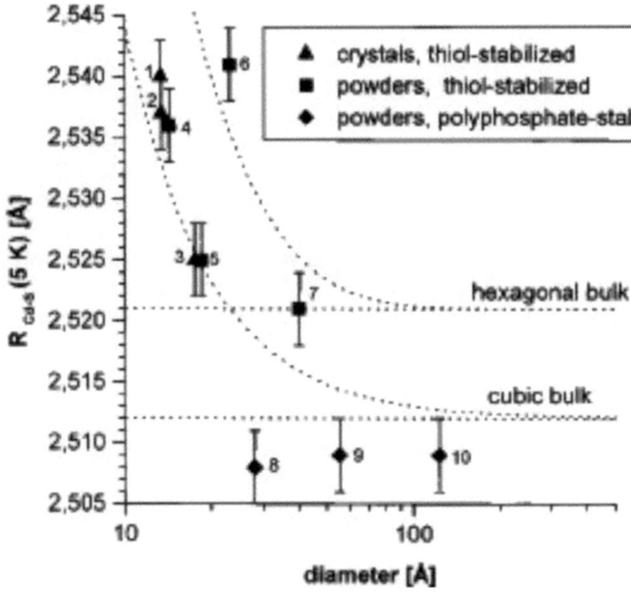

Fig. 1. Mean Cd-S distances RCd-S as a function of the size of CdS nanocrystals deduced from EXAFS experiments. Bulk values of the cubic and the hexagonal phase of CdS are also indicated (dotted).

Certainly at decreasing of nanoparticles sized up to few nanometers that correspond to some tens atoms its crystalline lattice change essentially from same of bulk crystal and it is appear instability of crystalline lattice. Large surface of small particle will result on augmentation of influence of surface states on crystalline lattice of nanoparticles.
There are a lot of work that show size and surface effects of crystalline lattice change in nanometers size nanoparticles.

In the work [1] it is shown nanoparticles crystalline lattice parameter dependence from organic substance settled on its surface and its size. Micron size particles, for example CdS have crystalline lattice similar to bulk crystals at decreasing of nanoparticles sizes down to nanometers range its lattice change essentially.

In Fig. 1 show Cd-S distance dependence from nanoparticles diameter and organic shell nature.
A few CdS samples have been investigated, the particles differing in size and crystal structure:
- three samples of polyphosphate-stabilized nanoparticles with cubic crystal structure (diameters approximately 3.0-12.0 nm),
- four samples of thioglycerol-stabilized particles (hexagonal and cubic (see below), diameters from 1.4 to 4.0 nm), and
- three samples of crystallized nanoparticles, the structures and superstructures of which are known from SC-XRD (diameters from 1.3 to 1.7 nm).
Microcrystalline CdS of hexagonal crystal structure was used as a reference substance. The smallest "particles" investigated consist of a three-dimensional network of $Cd_8(SR)_{16}$ units (R) thioglycerol), which may serve as a model for the surface of thiol-stabilized CdS nanocrystals. The other two crystallized clusters ($Cd_{17}S_4(SR)_{26}$ and $Cd_{32}S_{14}(SR)_{36}$, R) mercaptoethanol and 1-mercapto-2-propanol, respectively) may be regarded as fragments of the cubic (zinc blende) phase of CdS and appear tetrahedrally. Thus, at least for the latter three samples, "diameter" is to be taken only as a reference point for the "size" of the particles. EXAFS spectra have been taken in transmission mode at the Cd K-edge in the energy range from 26.4 to 29.0 kV at temperatures between 5 and 296 K.
All of the samples clearly showed the Cd-S coordination shell, whereas the Cd-Cd coordination (second shell) was visible in particles larger than 3.0 nm only. Figure 1 shows the dependence of the Cd-S bond length as a function of the particle diameter deduced from the EXAFS analysis. The quality of the data allows us to divide the plot into three regions. Samples 8-10 exhibit a slight contraction of the Cd-S bond with decreasing particle size, which is due to the minimization of the surface energy, obviously unhindered by the ionically bound stabilizers.
In contrast, the covalently bound stabilizers of samples 1-5 expand the Cd-S bond. This expansion is larger for the smaller particles, and it becomes larger as the steric interaction among the stabilizers comes into effect. For samples 1-3, the bond lengths determined by EXAFS match very well those from the SC-XRD analysis. From P-XRD, sample 7 is assigned to the hexagonal crystal phase, by means of which the larger mean bond length, compared to samples 8 and 9 (similar size but cubic phase), is explainable. Like sample 7, sample 6 is prepared at elevated temperatures, which makes it likely that this sample also belongs to the thermodynamically stable hexagonal crystal phase (P-XRD does not allow an unambiguous assignment). This guess is corroborated by the "out of order" bond length (Fig. 1) and by the analysis of the third moments of the pair distribution function (Fig. 1). According to the surface-to-volume ratio of the nanocrystals, this quantity increases with decreasing particle size but is divided into two groups: on one hand, all particles clearly belonging to the cubic structure (four equivalent Cd-S bonds lead to C3) for the largest particles) and on the other hand, samples 6 and 7, with distinctly elevated anharmonicity (in hexagonal CdS, three Cd-S bonds are equivalent, and one differs from those). Thus, for sample 6, we are in a position

to state the assingnment to the hexagonal phase by means of EXAFS spectroscopy, which has not previously been possible, by applying P-XRD and HRTEM. The Debye temperatures and static disorders are extractable from the EXAFS data together with the bond lengths and anharmonicities, as mentioned above. The Debye temperatures that were determined increased slightly with decreasing particle size, which points to a stiffening of the Cd-S bonds. When compared to the bulk value, all of the nanoparticles displayed an elevated static disorder, which, in the first instance, increases with decreasing size according to the surface-to-volume ratio (samples 10-6). For the very small clusters of samples 1-5, again slightly reduced static disorders are observed (in good agreement with the SC-XRD of samples 1-3). Possibly, this finding is a hint toward different regimes of particle growth: thermodynamically controlled growth leads to a crystallizable species, whereas subsequent Ostwald ripening yields larger and less specific colloids.
At decreasing of nanoparticles sizes up to 1 nanometers distances in crystalline lattice between metal atoms will increase that result of moving from crystalline to amorphous state of material in result. Transformation from crystalline to amorphous form of small particles well known for Fe_2O_3, Se, inorganic materials.

3. Methods of preparation and stabilization of semiconductor nanoparticles

3.1 Inorganic nanoparticles based nanocomposit

From 90 years was develop the main methods of nanoparticles synthesis. Few nanoparticles types now is commercial available from Aldrich and other commercial supplier. Methods of the nanoparticles synthesis can be divided into three main groups.
1. Synthesis of semiconductor nanoparticles in solutions of the corresponding salts by controlled addition of anions (or cations) or by hydrolysis [2];
2. Preparation of nanoparticles as a result of phase transformations [3];
3. The synthesis of nanoparticles in aerosols [4].
Preparation of nanocomposite material having both high nanoparticles concentration, absence of its coagulation and homogeneous optical and semiconductor proprieties is a mostly difficult problem. Usually it accomplished by preparation of suitable nanoparticles with modified surface and then to its incorporation in polymer, having surface active proprieties often.
For consideration of nanoparticles state in nanocomposite, we should assume that nanoparticles can interact with polymer matrix with formation of ordered polymer layers in its surface. In the work [5] was made numerical study of state of nanoparticles (fullerene) in polymer matrix and received interesting results on this bi- phase system elastic proprieties.
The purpose of this work is to investigate the effect of nanoparticle size on elastic properties of polymeric nanocomposites using MD simulations. For this, molecular models of a nanocomposite were constructed by reinforcing amorphous polyethylene (PE) matrix with nano sized buckminister fullerene bucky-ball (or simply bucky-ball). Bucky-balls of three different diameters (0.7, 1.2 and 1.7 nm, respectively) were utilized to incorporate size effect in the nanocomposites. To represent them as a generic nanoparticle system, all bucky-balls were configured as rigid body. This is necessary because a bucky-ball embedded inside the polymer matrix may deform excessively depending on its size and may overshadow the composite mechanical properties attributed to filler size. The assumption of rigid bucky-ball will ensure that the shape of filler does not contribute to variation in elastic properties. The

assumption may be unrealistic for large diameter buckyballs, it is a reasonable assumption for small bucky-balls and solid nanoparticles. In addition to this shape constraint, the volume fraction of the filler, matrix characteristics (density, molecular weight, molecular weight distribution, branch content, degree of crystallinity, etc.) and their force interaction with the nanoparticle were kept constant in all nanocomposites. Molecular models of the neat PE matrix were also developed for comparison. Elastic properties of the neat and nanocomposite systems were then evaluated using four different modes of deformation, namely, unidirectional tension and compression, and hydrostatic tension and compression, respectively.

Molecular models of nanocomposites were developed by symmetrically placing a spherical fullerene bucky-ball in the PE matrix, as shown schematically in Fig. 2. The dashed box in Figs. 2 a and b indicates the periodic cell or unit cell that was simulated by MD. Three types of bucky-balls, C60, C180 and C320 (subscripts denote number of carbon atoms), were used to incorporate the size effect. All bucky-balls were infused in matrix by approximately 4.5 vol%. Periodic boundary conditions were employed to replicate the unit cells in three dimensions. In nanocomposites, the PE matrix was represented by united atom (UA) -CH2- units. The initial structure of the matrix was constructed by positioning the bucky-ball at the center of the unit cell and by randomly generating PE chain(s) on a tetrahedron lattice surrounding the bucky-ball.

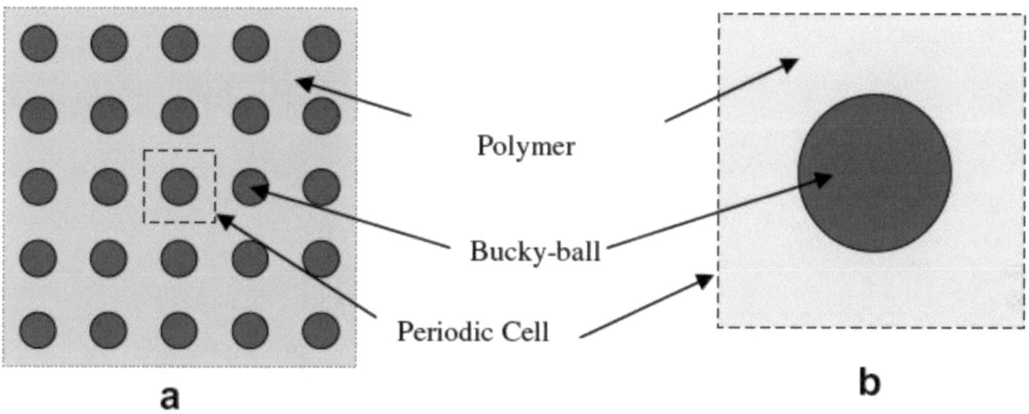

Fig. 2. (a) Schematic diagram of polymer nanocomposites, (b) periodic cells used for MD simulations.

Numerical simulation shown that elastic properties of nanocomposites are improved appreciably with the infusion of bucky-balls in PE matrix. The trend shows that with the increase in filler size, the extent of enhancement in elastic properties is gradually reduced. The result is somewhat surprising because in all cases the volume fraction was maintained constant (4.5%).

It can be concluded from this observation that size of the filler has considerable influence on polymer density even with non-bonded inter-molecular interactions between polymer and nanoparticle. The effect can be well understood from the radial density distribution of PE for both neat and nanocomposites, as shown in Fig. 3 in which the distribution is constructed by measuring local densities of PE at various radial distances starting from the center to the half-length of the periodic box. It is interesting to find that local densities are not constant along the radial distance. A 200–250% increase in polymer density exists for all nanocomposites at a distance close to the nanoparticle. At further distances, the distribution fluctuates in a similar manner as in the neat polymer system. The fluctuating character is inherent because mass needs to be conserved [6]. The collective contributions of these factors yielded a decreasing trend in polymer bulk density with the increment of filler size. It appears from the analysis that polymer density distribution plays the foremost role in size effect. However, it is not elucidated why size difference influence polymer density. The discernible contribution from filler size can be realized from radial distribution plot as shown in Fig. 4.

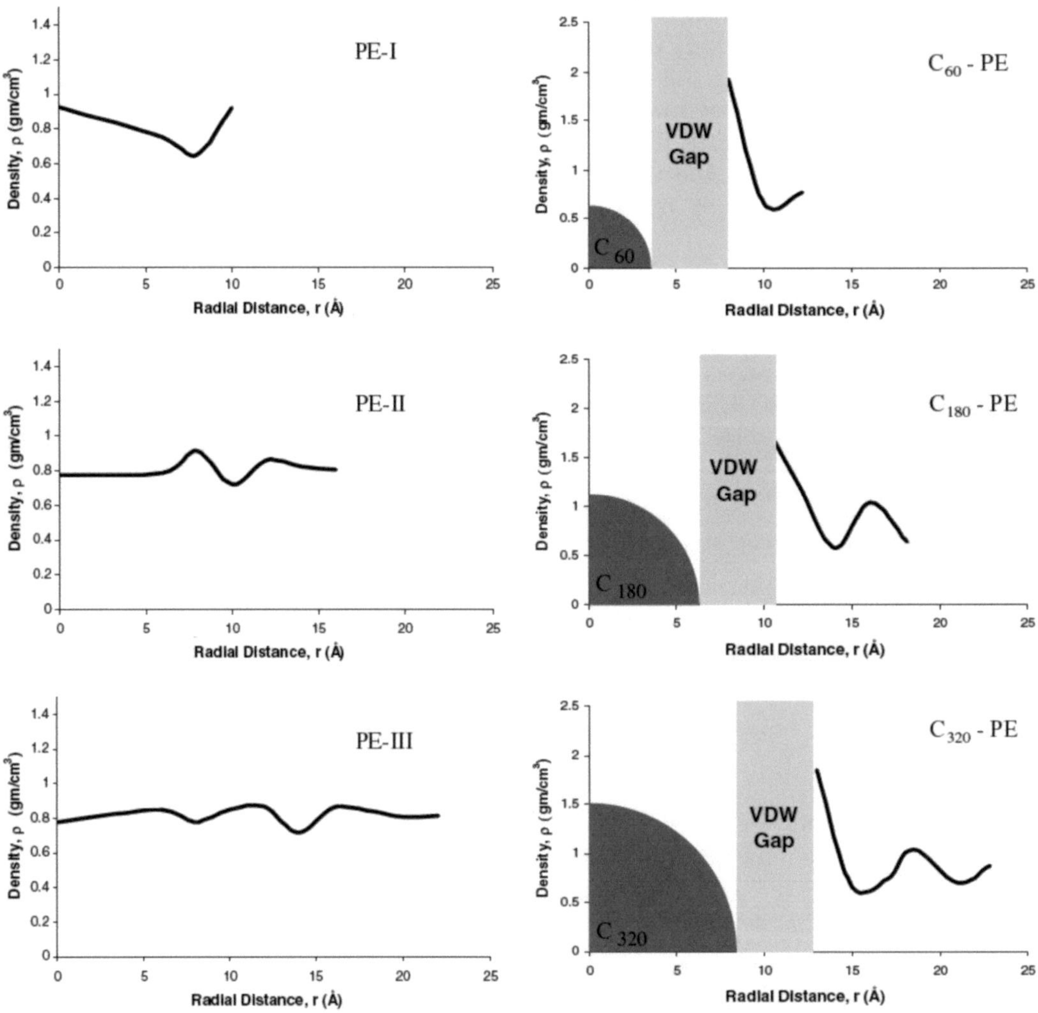

Fig. 3. Radial density distribution of various: (a) neat PE and (b) nanocomposite models. Space occupied by nanoparticles is schematically shown by the quarter circles.

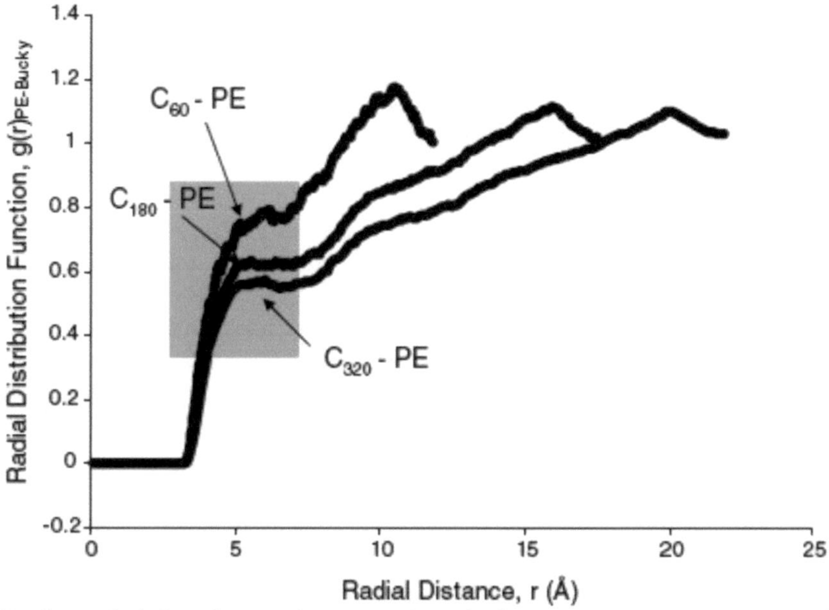

Fig. 4. PE-Bucky radial distribution functions (RDF) of various nanocomposite models.

It is known that the radial distribution function for any atom pairs gives a measure on how corresponding atoms are distributed in three-dimensional space due to VDW interactions. Hence, g(r)PE-Bucky refers to radial distribution of PE atoms with respect to Bucky-ball atoms. As atomic position of all bucky-balls were fixed, a plot of g(r)PE-Bucky would thus provide information about the polymer distribution due to interaction with a nanoparticle. Fig. 4 reveals that the size of Bucky-ball has strong influence on the g(r) plot. It is observed that the value of g(r) assumed zero from 0 to 3.4 A ° for all nanocomposites, then increases with radial distance. The zero value refers to the VDW thickness h. It is also evident that *h* does not depend on filler size. It is quite expected because parameters describing LJ potentials are identical for all nanocomposites and the nature of the h is known to be governed by such interactions between nanoparticle and polymer [7]. However, the relative distribution of polymer atoms towards the nanoparticle, as indicated by the variation in g(r) at a particular radial distance, is quite different with the change in filler size. It is obvious from Fig. 4 that more atoms are tending to disseminate across the polymer–nanoparticle interface as the size of buckyball decreases.

It appears from the above discussion that with the reduction in filler size, the bulk density of polymer and the attractive interaction energy between polymer and nanoparticle at the interface increase substantially. Enhancements of these parameters are then translated to improved elastic moduli.

Preparation of nanocomposite with high nanoparticles content.

According to above numerical simulation and discussion there are strong interaction between polymer and nanoparticles. At rapprochement of nanoparticles up to distance of 3 – 5 nm nanocomposite behavior change dramatically. In real nanocomposite with nanoparticles size of 2-3 nm these condition of nanoparticles – nanoparticle distance about 3 nm begin with about 10 vol. % nanoparticles concentration. So, 10 % is a border between usual nanocomposite with no nanoparticles interaction and high nanoparticles concentration composite where nanoparticle - nanoparticle interaction play the main role in behavior formation of whole composition.

At preparation of nanocomposite with high nanoparticles concentration a contradiction is appear: to avoid nanoparticles coagulation we need to increase interaction between them and the polymer matrix, but at the same time those interactions with polymer will result in hardening of the composites in result of polymer cross-linking over nanoparticles.
In most part research this contradiction has been avoided by use of very fast drying of material [8]. At fast drying nanoparticles have not enough time for coagulation and solid material keep good distribution take place in solution. In that works, the dangerous stage of particles interactions passed fast and the composites kept their transparency. For example, if a layer is prepared by spin coating, the process of solid coating preparation occupies only some part of a second. Certainly, this method is not suitable for preparing thick nanocomposite layers and bulk nanocomposites.
Next two works show possibility to obtain high nanoparticles concentration in material that suitable to obtain high refractive index homogenous nanocomposite material based on high concentration of ZnS or CdS semiconductor nanocrystals in polymeric matrix. Same homogenous nanocomposite with high semiconductor nanoparticles concentration around 20 - 30 vol % are suitable as a homogenous semiconductor materials, so synthesis of these materials were described in detail here.
One example of high concentration ZnS nanocomposit thin film preparation give the work [8]. In the work was used previously prepared ZnS-monomers prepolymer that was UV-cured at once after spin coating in thin film. UV curing technology was used to rapidly prepolymerize the ZnS–macromer system, and then a radical polymerization process was carried out to complete the polymerization reaction. Another advantage of using this macromer is that the macromer in solution has a viscosity, which is favorable for spin-coating to form films. Furthermore, a polymerizable moiety as capping agent has also been utilized to modify the surface of ZnS nanoparticles in order to immobilize the ZnS particles into the polymer. Research on the preparation of nanoparticles–polymer composites using polymerizable surfactants, ligands or capping agents has previously been reported [9, 10]. This approach can effectively avoid the phase separation and results in transparent composites because the functionalized inorganic particles with polymerizable vinyl groups can be copolymerized with the monomers to form integrated polymeric materials.
In this work firstly was being synthesized the thiophenol (PhSH)–4-thiomethyl styrene (TMSt)-capped ZnS nanoparticles with high concentration in DMF. Then a UV curable urethane-methacrylate macromer (UMM) was introduced into the nano-ZnS containing DMF solution. The ZnS nanoparticles were immobilized into the polymer matrix via copolymerization of the macromer (UMM) with 4-thiomethyl styrene (TMSt) bound on the surface of ZnS particles to synthesize a series of nano-ZnS–poly(urethane-methacrylate macromer) (PUMM) transparent composite films with high refractive indices. The structure and composition of the thiolcapped ZnS nanoparticles were characterized via TEM, X-ray, FTIR and chemical analyses. The thermal properties, optical properties and microstructure of the nanocomposite films were investigated in detail.
Detailed description of high concentration nanocomposite synthesis method according to work [8].

3.2 Materials

Anhydrous zinc acetate, N,N-dimethylformamide (DMF), thiophenol (PhSH), thiourea, 2,4-tolylene diisocyanate (TDI), dibutyltin dilaurate (DBTL), 2-hydroxyethyl methacrylate (HEMA) and other chemical reagents were of analytical grade and were used without further purification. 4-Vinylbenzyl chloride (w 95% GC grade, Fluka) and 2,2-dimethyl-2-hydroxyacetophenone (Darocur 1173 from Ciba Special Chemicals) were used as received. 2,2′-Dimercaptoethyl sulfide (MES) was synthesized as reported previously [11]. Synthesis of 4-thiomethyl styrene (TMSt) TMSt was synthesized from 4-vinylbenzyl chloride in a manner similar to that reported in the literature [12]. 15.3 g of 4-vinylbenzyl chloride (0.1 mol), 9.12 g of thiourea (0.12 mol), 200 ml of ethanol and 0.08 g of p-methoxyphenol as inhibitor were put into a four-necked flask fitted with a reflux condenser. The reaction mixture was stirred at reflux temperature for 4 h under N_2 flow and then was cooled. 75 ml of 20% solution of sodium hydroxide were added to the above mixture, and the resulting solution was immediately heated to 80 C and continuously stirred at 80 C for 0.5 h. Finally, the resulting solution was cooled to room temperature and 100 ml of $CHCl_3$ were added. The organic phase was separated out and washed with distilled water until neutral. Then the organic phase was dried over anhydrous Na_2SO_4 and the product (TMSt) was obtained by removing the organic solvent under reduced pressure. Yield 75%, n_{20} d ~ 1.625.
The method for preparing thiol-capped ZnS nanoparticles (TCZnS) was similar to that reported in ref. [8].

3.3 Preparation of ZnS–PUMM nanocomposite films

UV curable macromers (UMM)-DMF solution of desired weight ratio, containing 2 wt% of Darocur1173 and 0.5 wt% of AIBN as initiator were mixed with 3 ml of thiol-capped colloidal ZnS-DMF solution according to the required doping content of ZnS particles in the films. The mixture solutions were concentrated to a suitable viscosity at room temperature under vacuum, and the resulting viscous solution was spincoated on silicon wafers or quartz plates at 1000–3000 rpm.
The coated films were dried under vacuum for 10 min at 45 uC, and then were exposed to the UV radiation of a medium pressure mercury lamp of 2 kW for 3 min. After UV curing reaction, the films were cured at 70 uC for 2 h, 100 uC for 1 h, 120 uC for 1 h and treated at 160 uC for 0.5 h.

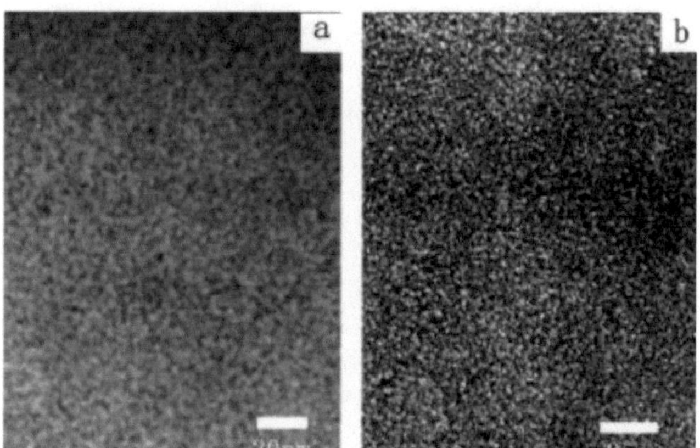

Fig. 5. TEM micrographs of the nanocomposite films of (a) TCZnS32 and (b) TCZnS79.

Resulting TEM micrographs of the nanocomposite films TCZnS32 and TCZnS79 are show in Fig. 5. The ZnS nanoparticles ranging from 2 to 5 nm are uniformly dispersed inside the polymer matrix and the ZnS nanoparticles remain their original size without aggregation after immobilization into the polymer matrix, indicating that the thiol capping agents and polymer play an important role in stabilizing and dispersing nanoparticles.

The main proprieties of nanocomposites with different content of ZnS nanoparticles show in Table 1.

The WXRD pattern of ZnS nanoparticles synthesized in the work shows broad peaks typical of samples in the nanosize regime (Fig. 6). The peaks in the diffraction pattern appearing at 2h values of 28.5, 47.5 and 56.3u correspond to (111), (220) and (311) planes of the cubic structure of sphalerite ZnS. The resulting ZnS crystallite structure is in accordance with that reported previously.

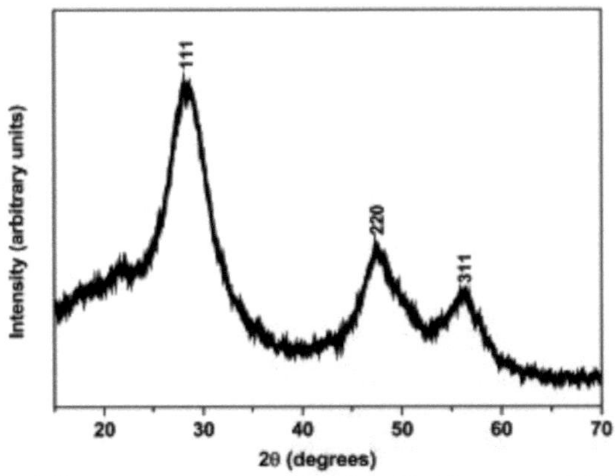

Fig. 6. Wide-angle X-ray diffraction pattern of PhSH–TMSt-capped ZnS nanoparticles.

The FTIR spectrum of the PhSH–TMSt-capped ZnS nanoparticles is shown in Fig. 7.

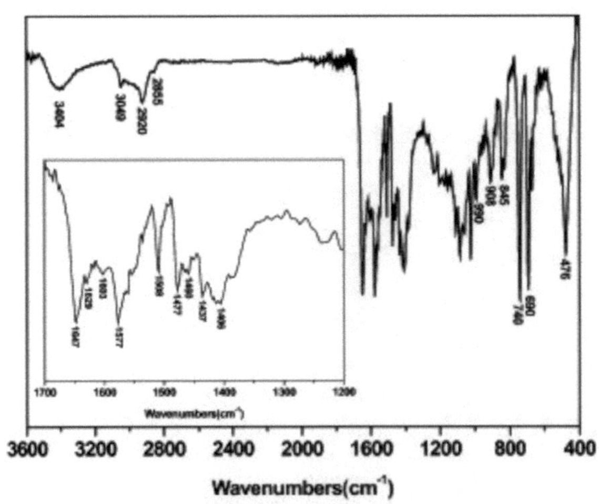

Fig. 7. FTIR spectrum of PhSH–TMSt-capped ZnS nanoparticle powder, dried from colloidal ZnS–DMF solution in vacuum.

The peaks at 2920, 2855 and 1406–1477 cm^{-1} are assigned to the characteristic vibration of the methylene groups in TMSt. The peaks assigned to C-C vibrations of benzene rings are observed at 1603, 1577, 1508 and 690–990 cm^{-1}. The stretching vibration band of vinyl groups on TMSt located at 1629 cm^{-1} also appears in Fig. 7 (insert), although its intensity is very weak. The absorption peak of the S-H vibration at 2550–2565 cm^{-1} is not observed in the IR spectrum, indicating that the mercapto groups of TMSt and PhSH molecules were bound to the ZnS nanoparticle surface. The band of the CLO stretching vibration on residual DMF molecules is also observed at 1647 cm^{-1} which is lower than that of free DMF molecules at 1667 cm^{-1}. This shift reflects that there is a relatively strong interaction between DMF molecules and the surface of colloidal ZnS nanoparticles.28 In addition, the broad peak near 3404 cm^{-1} in the IR spectrum may be from the absorption of traces of moisture or adsorbed water associated with DMF.

The chemical composition of the thiol-capped ZnS nanoparticles was determined by EA and ICP-AES analyses. Anal. Found: C, 30.5; H, 2.66; S, 25.30; Zn, 42.90. The relative molar ratio of S to Zn was calculated to be 1.2 on the basis of the above quantitative analyses. This result is in good accordance with the EDAX result for ZnS particles, which showed that the ratio of the number of S to that of Zn is 1.18 within an accuracy of 2%. If all of the capping agents (RSH) are capped on the ZnS particles, the molar ratio of Zn^{2+} : S^{2-} : RS^- was calculated to be 1 : 0.6 : 0.6, based on the feed ratio (Zn^{2+} : RSH ~ 1 : 0.6), and the contents of carbon and hydrogen on the ZnS particles were also calculated to be 31.8 and 2.78, respectively. The contents of these two elements agree well with the results of chemical analyses (EA and ICP-AES) within experimental error. Therefore, it can be deduced that almost all of the capping agents, PhSH and TMSt, were capped to the ZnS particles and the molar ratio of Zn^{2+} : S^{2-} : RS^{-2} for the PhSH–TMSt-capped ZnS particles was determined to be 1 : 0.6 : 0.6.

3.4 Nanocomposite films

FTIR spectra of the ZnS–PUMM nanocomposite films for the TCZnS16–TCZnS79 are shown in Fig. 8 (X is weight percent of PhSH–TMSt-capped ZnS particles in the films). The absorption peaks at 3300 and 1657 cm^{-1} are attributed to N-H and CLO bonds of urethane linkages and the latter covers the characteristic peaks of CLO of methacrylate groups.

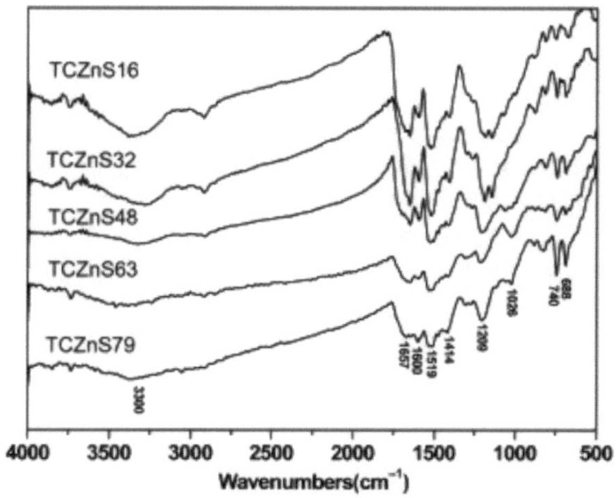

Fig. 8. FTIR spectra of nanocomposite films of TCZnS16–TCZnS79.

The CLO band gradually decreases in intensity with increasing TCZnS nanoparticles, compared with the intensity of the characteristic absorption band of phenyl (688–740 and 1600 cm^{-1}). This result indicates that an increasing amount of TCZnS with high phenyl content is immobilized into the PUMM matrix. The IR absorbances of the CLC double bonds at 1629–1639 cm^{-1} for the methacrylate groups and the capping agent (TMSt) disappear, indicating that they have completely polymerized. Fig. 9 illustrates TGA curves of pure PUMM, TCZnS16, TCZnS48 and TCZnS86 films at a heating rate of 10 ^{o}C min^{-1} under nitrogen atmosphere from 50 to 750 ^{o}C.

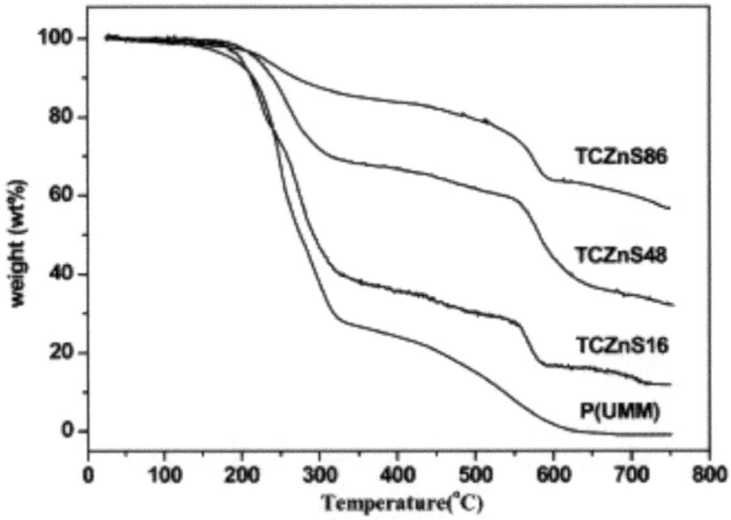

Fig. 9. TGA curves of nanocomposite films of TCZnS16, TCZnS48 and TCZnS86 at a heating rate of 10 oC min-1 under nitrogen flow.

The nanocomposite films have the initial decomposition temperatures of 201, 204 and 203 ^{o}C for TCZnS16, TCZnS48 and TCZnS86 respectively, and these values relate to the decomposition temperature of the polymer matrix (PUMM). There are two obvious weight loss regions: between 200 and 330 ^{o}C, and from 550 to 600 ^{o}C. The weight loss between 200 and 330 ^{o}C is predominantly attributed to the weight loss of the polymer matrix. The secondary weight loss at 550–600 ^{o}C is considered to be the weight loss of another part of the polymer and the thermal decay of the partial thiol-capped agents on the surface of ZnS nanoparticles. As shown in Table 1, the residues of the nanocomposite films TCZnS16–86 at 750 ^{o}C are in the range of 11.5–56.5% and they increase with increasing TCZnS content in the films. By and large, these char yields are in agreement with the theoretical weight fraction of inorganic ZnS contained in the films, indicating that the ZnS particles were successfully incorporated into the polymer matrices. This result also effectively supports the chemical analysis results for the TCZnS particles. Fig. 10 shows DSC curves of PUMM, TCZnS16, TCZnS48 and TCZnS79 films at a heating rate of 10 ^{o}C min^{-1} under nitrogen flow.

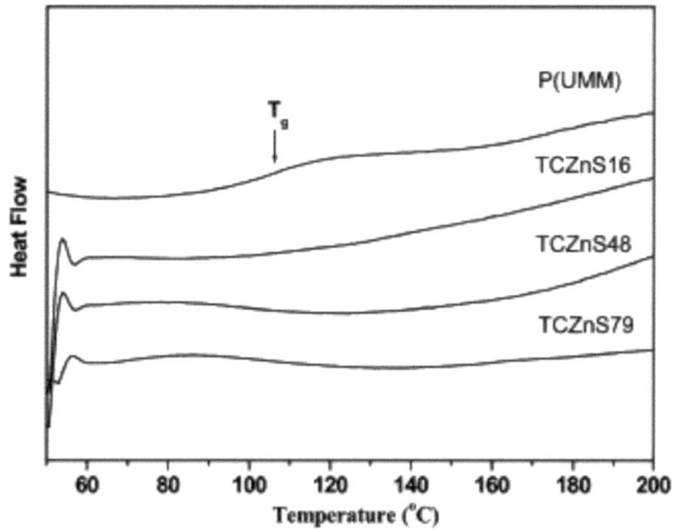

Fig. 10. DSC curves of PUMM, TCZnS16, TCZnS48 and TCZnS79 films at a heating rate of 10 oC min-1 under nitrogen flow.

The pure PUMM polymer exhibits a glass transition temperature (Tg) of about 106 °C. However, no significant thermal transition peaks are observed for the ZnS–PUMM nanocomposites below 200 °C. Also was used the torsion braid analysis (TBA) to measure the thermal transition behavior of the nanocomposite samples. The Tg of the polymers still are not observed. This suggests that nanocomposites have higher rigidity and crosslinking density due to the incorporation of the thiol-capped ZnS nanoparticles, which restricted the motion of the polymer chain segments. Thus, it may be that the glass transition temperature of the polymer is close to the decomposition temperature of it. Thermal analyses indicate that the ZnS nanoparticles were successfully immobilized into the polymer matrix and the nanocomposite films exhibit a good thermal stability.

Maximal weight concentration of thiol capped ZnS is around 86% as at higher concentration nanocomposite films with good mechanical properties cannot be obtained.

Another example of ZnS and CdS high concentration nanocomposite give the work [13, 14].

In the work has involved nanoparticles stabilization primarily by steric barriers. So, each nanoparticle should be covered by quite a thick shell linked with the surface of nanoparticle. This was accomplished by having a multi-atom chain connecting the acid group and aromatic group of the shell molecule.

Was used UV curable monomers with an acid group at one end and a vinyl group at the other and low viscosity at room temperature to accomplish both bonding to nanoparticle surface and, same time, possibility to UV-curing. Based on these requirements one suitable substance has been chosen: 2-carboxyethyl acrylate (CEA). This substance has an acrylic group for curing and can be used as shell material because of the acid group. This molecule has only a short distance between the groups, and only a relatively thin shell would be expected before polymerization. And we know that a thin shell formed at synthesis of nanoparticles is not enough at high nanoparticles concentration in a thermoplastic matrix. Certainly, we would expect additional monomer units to add to the shell monomers during the UV curing reaction.

The method to introduce the ZnS nanoparticles into CEA found is the following: ZnS nanoparticles with the shell of 5-Phenylvaleric acid have been put into toluene and heated for 10 hours at 80 °C. This operation is needed to remove residue water from the nanoparticles surface. After the nanoparticles powder was dried in air at 80 °C during 10 minutes, it was put into CEA. Ultrasonic dispersion for about 30 – 40 minutes in apparatus was done.

After dissolution of ZnS into CEA, a shell of CEA is formed at the surface of each nanoparticle. As the result ZnS - CEA nanocomposite dispersion (or solution) is stable for a long time.

UV curing of nanocomposite was made by usual way by addition of photoinitiator and curing a film with UV light. Experimental conditions: photoinitiator Dimethoxy phenyl acetophenone 0,1 w%, film thickness 100 um, UV light 365 nm, 5 mW/cm^2, room temperature, time of curing 10 minutes. Maximal ZnS volumetric concentration in the compositions was 25%. The resulting RI of the UV cured film was 1,65, compared with an RI of 1,45 for the pure CEA film. Thus, the RI increase is 0,20. Dependence of RI on nanoparticles concentration is shown in Fig. 11 Maximal nanoparticles concentration has been limited by viscosity increase up to the point of a non-flowing composition.

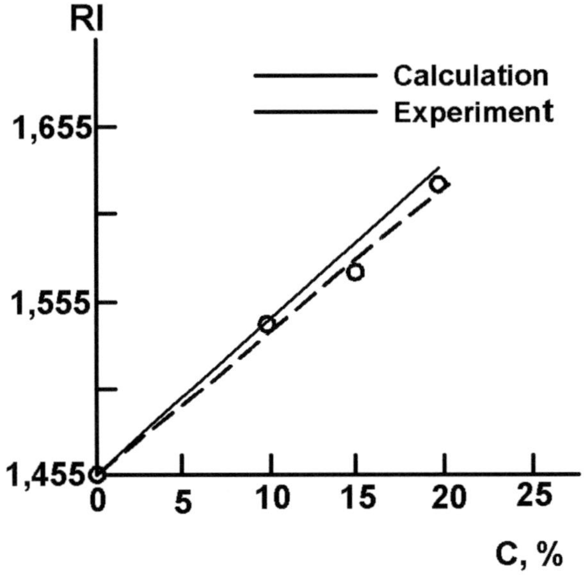

Fig. 11. Dependence of nanocomposite RI on nanoparticles concentration.

TEM photo (Fig 12) shows inside structure of nanocomposite. The photo was made of a nanocomposite cured immediately after preparation. The nanoparticles concentration is 20 vol.%. Note that the distribution of particles is almost uniform, which explains the transparency.

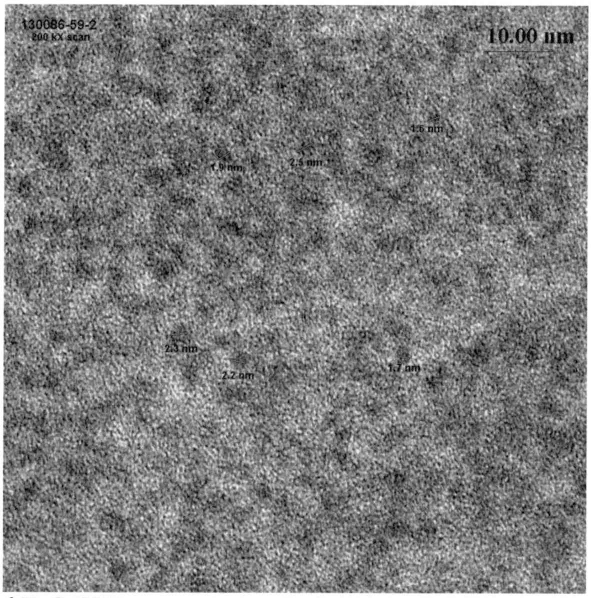

Fig. 12. TEM photo of ZnS-CEA nanocomposite (nanoparticles concentration is 20 vol. %)

4. Electron and hole transport over disperse nanocomposite systems and nanolayers

At present time charge transport over disperse semiconductor were investigated mostly for organic conjugated materials. For high concentration nanocomposite materials are used same understanding. In this chapter will be considered charge transport over disperse organic materials as the nearest analog of nanocomposite.

At organic materials when atoms are bonded together to form a molecule, the upper atomic orbitals interact with each other to form delocalized molecular orbitals while the deep atomic orbitals are still localized in the atomic potential well (Fig. 13 a) [15].

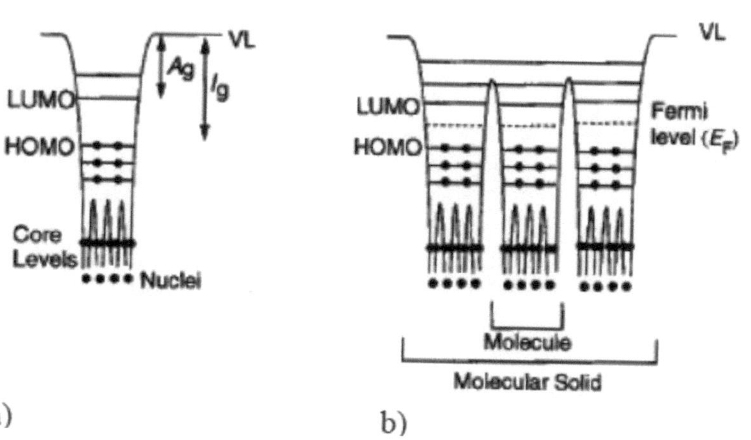

Fig. 13. Electronic structures of (a) a polyatomic organic molecule or a single chain polymer and (b) an organic solid.

When the orbital overlap occurs directly between the nuclei of the atoms, the orbitals form σ-bond and the sideway overlapping of the orbitals form п-bonds. Materials having п-bonds orbitals are named as conjugated materials same time. It is the п electrons which mainly determine the electronic and optical properties of the molecule. In ground state, the п-electrons form the п-band and the highest energy п-electron level is known as the highest occupied molecular orbital (HOMO). In excited state, the п-electrons form the п-band and the lowest energy п-electron level is known as the lowest unoccupied molecular orbital (LUMO). The HOMO resembles the valence band and the LUMO resembles the conduction band in the inorganic semiconductor concepts. The energy separation between the HOMO and the vacuum level corresponds to the gas phase ionization energy (I_g) and that between the LUMO and the vacuum level corresponds to the gas phase electron affinity (A_g) [2.1]. In an organic solid, the molecules or polymer chains are packed closely together and result in an electronic structure as shown in Figure 13b. It can be observed that the electronic states are localized to individual molecules with narrow intermolecular band widths.

4.1 Excitons in organic

When an electron has been excited from the ground state orbital to a higher orbital, it leaves a hole in the ground state orbital. The resulting bound state of an electron and a hole due to the Coulombic interaction is called an exciton. In general, excitons can be divided into two classes. If it is delocalized with radius much larger than the interatomic spacing, it is a Mott-Wannier type of exciton. On the other hand, if it is localized and tightly bound, it is called a Frenkel exciton. In organic materials, since the excitations are often localized on either individual molecules or a few monomeric units of a polymer chain, the excitons are highly localized and are considered to be Frenkel excitons which usually have large binding energy of some tenths of an eV or even higher. For example, in the case of Alq_3, a commonly used OLED material, the exciton binding energy is ~ 1.4 eV.

4.2 Transport

According statement in beginning of this paragraph, properties of disperse systems will be considered in example of conjugated polymers charge transport theory and abbreviation of polymer semiconductor is suitable for other disperse systems included nanoparticles in polymer matrix. According to this in mostly part of recent works nanoparticles like fullerene C_{60} are investigated with large organic molecules like phthalocyanine and perylen together as component of polymeric disperse compositions.[16]

Electronic properties of polymers can be described in terms of semiconductor physics [17]. The particular framework of one dimensional periodic media is well suited to the basic understanding of an isolated polymer chain [18]. Polymers are bonded by strong covalent bonds. As π-orbitals overlap is weaker than s-orbitals overlap, the energy spacing (band gap) between bounding and antibounding molecular orbitals is larger for the π-π* difference than for the σ -σ* one. One can thus, in a first approach, limit the band study to the π-π* molecular orbitals. Those are respectively the HOMO (for Highest Occupied Molecular Orbital) and LUMO (for Lowest Unoccupied Molecular Orbital), in terms of molecular physics. They are also the usual valance (VB) and conduction bands (CB) of

semiconductor physics, respectively σ -bonds then only contribute to the stability of the molecular structure.
In a real material also, 3-dimensional interactions play a major role in transport properties, even dominating the transport which becomes an interchain hopping process. Small molecules are bounded by weak interactions in the condensed state: Van der Waals forces. There results a weak coupling between them, the resonance integral t_1 is thus small (tenth of eV at most) [19], resulting in narrow flat bands. Mobility is thus a priori smaller in small molecules, owing to a large effective mass

$$m^* = \frac{n^2}{2}\left(\frac{\partial^2 E}{\partial k^2}\right)^{-1}$$

There can of course be exceptions to such a rule, the interdistance spacing can be small and molecular materials can in fact possess a rather large mobility. The first electrically pumped injection organic laser was indeed made from small molecules (a tetracene single crystal) [20].
Transport and mobility in organic materials require a knowledge of the charged species. A review of transport properties is given by Schott [21]. Energy levels of the charges are usually determined by cyclic voltametry for materials in solution. They can be characterized by XPS or UPS (X-ray and UV photoelectron spectroscopies) for solid materials. In small molecules, charged species are localized spatially, they are simply the cation (positive) and anion (negative) radicals. In polymers, the electron–phonon coupling leads to the so-called polarons which are charges dressed by a reorganization of the lattice [22]. Polarons may be regarded as defects in conjugated polymer chains. Such defect stabilises the charge which is thus self-trapped as a consequence of lattice deformation. So in the vast majority of organic semiconductors, transport bears all characteristics of a hopping process in which the charge (cation or anion) propagates via side to side oxidation–reduction reactions (Fig. 14.).

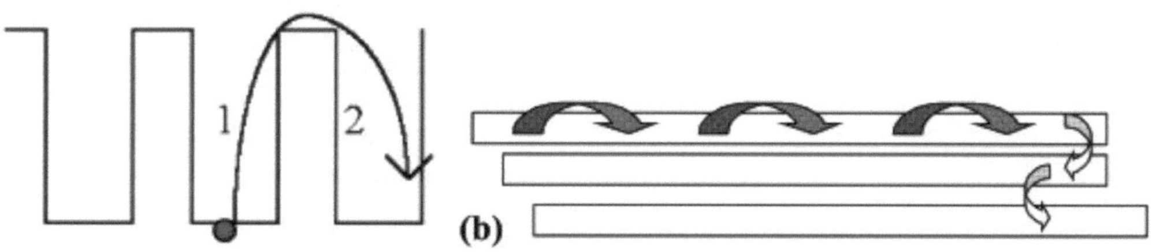

Fig. 14. Hopping process between molecules 1 and 2. b- Intra- (full arrows) and intermolecular (broken arrows) charge-transport.

One must distinguish between intramolecular charge transport along a conjugated polymer chain and intermolecular charge transport between adjacent molecules or polymer chains (Fig. 14 b.).
The former which is specific to conjugated polymers is the most efficient. Charge mobility in organics is field dependent, especially in the law mobility materials in which it usually follows phenomenologically a Poole–Frenkel law: $\mu \alpha \exp(\sqrt{E})$ [23]. Mobility can be

experimentally determined by photo-current transients (time of flight) [24], field effect transistor saturation currents [25], space charge limited currents [26] or impedance spectroscopy [27]. Mobilities in organic semiconductors are usually rather small: from 10^{-2} in well ordered conjugated polymers (liquid crystalline polyfluorene), down to 10^{-8} $cm^2/(V \cdot s)$ in guest–host polymer systems (dye doped poly-vinylcarbazole - PVK. Electron and hole mobilities differ by orders of magnitude in a single material; in small molecules such as the widely studied tris (8-hydroxyquinolinolato) aluminium - Alq3 - as well as in conjugated polymers such as the famous poly-paraphenylvinylene - PPV. The lowest mobilities are usually dispersive, which is the result of a distribution of mobilities [28]. Mobility can increase by up to two decades upon applying a voltage, being eventually very large above 1 MV/cm in conjugated polymers [29]. Mobility is increased by orders of magnitude when the molecular packing is improved. This is achieved by molecular ordering. Single crystals have the best performances, electron mobility in fullerene C60 single crystals is 2.1 $cm^2/(V \cdot s)$ [30], but it is reduced by at least 3 orders of magnitude by imperfect purification and uncontrolled crystallization [31], as well as by oxygen traps.
Charge transport is also improved by purification or deposition conditions; for instance, mobility becomes non-dispersive in Alq3 upon purification (oxygen induces traps) [32] and it becomes non-dispersive in soluble PPV derivatives upon selection of the solvent used for deposition [33]. Mobility is usually low and dispersive in randomly distributed polar molecules, but it is increased significantly when the dipoles are organized [34]. A record non dispersive electron mobility of up to $2*10^{-4}$ $cm^2/(V \cdot s)$ was recently achieved in an air stable amorphous glassy molecular material [35]. It is important that the mobility always drops by at least 2 orders of magnitude with impurities or defects (traps).
Light absorption and photogeneration process depend from nature of nanoparticles. If nanoparticles is inorganic semiconductor like CdS or CdSe, processes of photogeneration and absorption are similar to inorganic bulk crystals, is nanoparticles are pigment nanoparticles like phthalocyanine for example, photogeneration and absorption processes are similar to same in organic.
Charge transport over disperse system depend from concentration of charge traps in material (Fig. 11.), if concentration is low and distance between it is long, mobility become low. In nanocomposite (composition of inert polymer material and semiconductor nanoparticles) condition of charge transport over it take part at high nanoparticles concentration only. Lowest concentration limit is about 5 volumetric % of nanoparticles in material, but usually about 30 - 40 vol. % concentration is optimal for obtain charge transport over disperse nanocomposite material as over homogeneous media. Same time achievement of 30 vol. % of nanoparticles in material is a difficult task for technology.

5. Application of semiconductor nanostructures: solar cells, OLED

Important area of semiconductor nanoparticles application is n-type semiconductor layers in organic multilayer thin film structures as addition to usual p-type organic semiconductor.
Typical structure of organic solar cell as well as structure of OLED (organic light emitting device) consist organic thin layers: p-type for holes injection (OLED) or extraction (solar cell), n-type - for electrons injection or extraction in corresponding application.

5.1 Principle of operation of solar cell (PV - photo Voltaic) and OLED structures

Before discussing the development of organic PVs the basic principles are outlined. Almost all organic solar cells have a planar-layered structure, where the organic light-absorbing layer is sandwiched between two different electrodes. One of the electrodes must be (semi-) transparent, often Indium–tin-oxide (ITO), but a thin metal layer can also be used. The other electrode is very often aluminum (calcium, magnesium, gold and others are also used). Basically, the underlying principle of a light-harvesting organic PV cell (sometimes referred to as photodetecting diodes) is the reverse of the principle in light emitting diodes (LEDs) (see Fig. 15) and the development of the two are somewhat related [36].

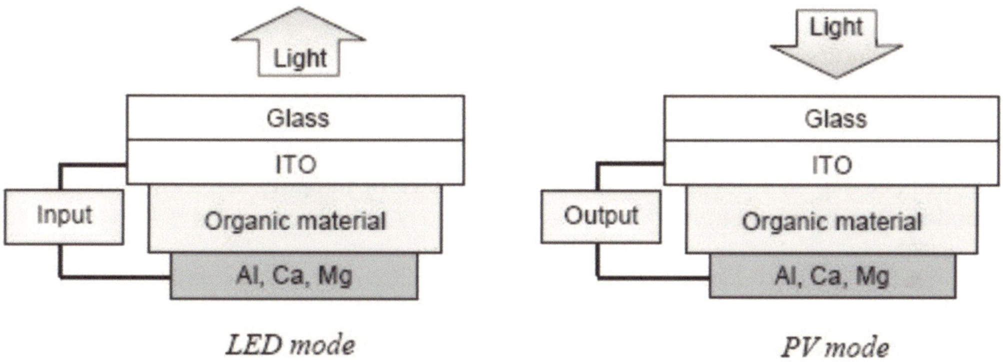

Fig. 15. A PV device (right) is the reverse of a LED (left). In both cases an organic material is sandwiched between two electrodes. Typical electrode materials are shown in the figure. In PVs electrons are collected at the metal electrode and holes are collected at the ITO electrode. The reverse happens in a LED: electrons are introduced at the metal electrode (cathode), which recombine with holes introduced at the ITO electrode (anode).

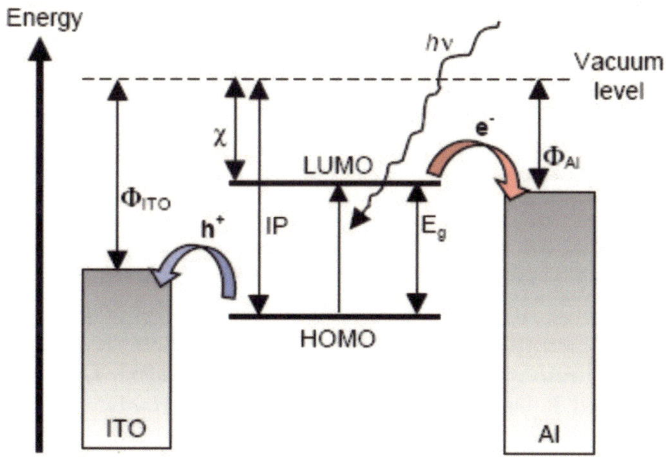

Fig. 16. Energy levels and light harvesting. Upon irradiation an electron is promoted to the LUMO leaving a hole behind in the HOMO. Electrons are collected at the Al electrode and holes at the ITO electrode. F: workfunction, c: electron affinity, IP: ionisation potential, Eg: optical bandgap.

In LEDs an electron is introduced at the low-workfunction electrode (cathode) with the balanced introduction of a hole at the high-workfunction electrode (anode). At some point the electron and the hole meets, and upon recombination light is emitted [37]. The reverse happens in a PV device. When light is absorbed an electron is promoted from the highest occupied molecular orbital (HOMO) to the lowest unoccupied molecular orbital (LUMO) forming an exciton (see Fig. 16).

In a PV device this process must be followed by exciton dissociation. The electron must then reach one electrode while the hole must reach the other electrode. In order to achieve charge separation an electrical field is needed, which is provided by the asymmetrical ionisation energy/workfunctions of the electrodes. This asymmetry is the reason why electron-flow is more favoured from the low-workfunction electrode to the high workfunction electrode (forward bias), a phenomenon referred to as rectification. The light harvesting process along with the positioning of energy levels is depicted in Fig. 16.

In the solid phase, the HOMOs and LUMOs of adjacent molecules may interact and form a conduction band (CB) and a valance band (VB) respectively (this will be described below). The shape of the CB and VB changes when the organic material is put into contact with electrodes (see Fig. 17), depending on the conductance of the polymer and on whether the electrodes are connected or not.

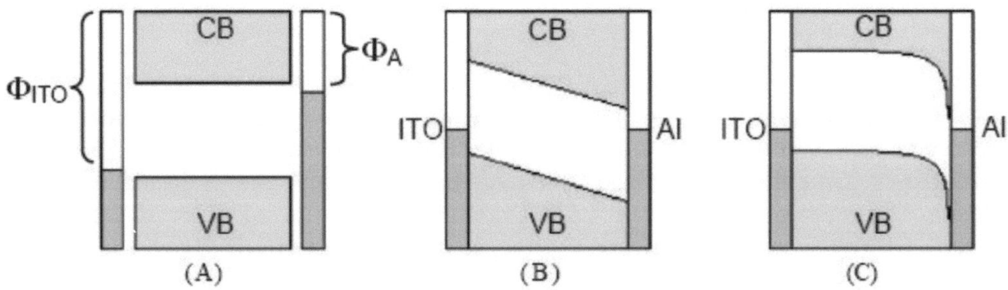

Fig. 17. The relative energy levels of the electrodes, CB and VB are shown in three situations, with no external bias. (A) CB and VB are shown along with the low-workfunction electrode (Al) and the high workfunction electrode (ITO) when isolated from each other. (B and C) The cell is assembled and short circuited, causing alignment of the electrode potentials. In (B), an insulating organic material is used. In C a hole-conducting polymer is used forming a Schottky junction at the high-workfunction electrode.

If the cell is short circuited the Fermi levels of the electrodes align (B and C), and in doing so the CB and VB are pulled skew. In B the polymer material is an insulator. This gives a field profile that changes linearly through the cell. In C a hole-conducting (p-type) semiconductor is used (most polymers are much better hole conductors than electron conductors). If the material is doped or illuminated charge carriers are generated. Due to the p-conduction properties, the generated holes are allowed to redistribute freely and they will flatten the bands approaching the high-workfunction electrode (a Schottky junction). The distance over which the CB and VB exhibit curvature is called the depletion width. In B the depletion width extends throughout the material. In C the depletion width is less than half the material thickness. Under external bias the relative electrode potentials can be changed, depending on the size and direction (forward or reverse) of the bias.

5.2 Comparison of inorganic and organic PV

In a crystalline inorganic semiconductor with a 3D crystal lattice the individual LUMOs and HOMOs form a CB and a VB throughout the material. This is fundamentally different from most organic dye semiconductors where the intermolecular forces are too weak to form 3D crystal lattices. Consequently the molecular LUMOs and HOMOs do not interact strongly enough to form a CB and VB. Thus charge transport proceeds by hopping between localised states, rather than transport within a band. This means that charge carrier mobility in organic and polymeric semiconductors are generally low compared to inorganic semiconductors. Also, charge separation is more difficult in organic semiconductors due to the low dielectric constant. In many inorganic semiconductors photon absorption produces a free electron and a hole (sometimes called charge carriers), whereas the excited electron is bound to the hole (at room temperature) in organic semiconductors. Conjugated polymers lie somewhere between the inorganic semiconductors and organic dyes. In general, excitons are considered to be localised on specific chain segments. However, there are cases where excitons seem to be delocalised. In these cases, the excitons are referred to as polarons [38, 39].

In simple PV devices and diodes based on organic semiconductors the primary exciton dissociation site is at the electrode interface (other sites include defects in the crystal lattice, absorbed oxygen or impurities) [40]. This limits the effective light harvesting thickness of the device, since excitons formed in the middle of the organic layer never reaches the electrode interface if the layer is too thick. Rather they recombine as described above. Typical exciton diffusion distances are on the order of 10 nm.

In fact inorganic PV have the best proprieties of charge carrier photogeneration and transport, polymer materials have the worst same proprieties. Organic crystalline dyes is lie between inorganic semiconductors and polymers. However flexible solar cell can be made in the basis of polymer films mostly, organic crystalline dyes can be introduced to structure as a thin layer, but inorganic semiconductors layers can not to be introduced in this structure absolutely.

From this point of view use of polymer based nanocomposite included semiconductor nanocrystals of different types (inorganic semiconductors - TiO_2, CdS; fullerenes; carbon nanotubes; organic crystalline dyes in nanocrystalline state) is the single way to combine high semiconductor proprieties of inorganic and crystalline materials with good flexibility of polymer film.

There are another advantage of inorganic semiconductor nanocrystals for PV and OLED application:

All semiconductors have a proprieties of production of single atom oxygen at presence of oxygen molecules and its illumination by light, especially in UV region. Presence of water vapor made this process more effective. Both PV and OLED device use at presence of light certainly and oxygen and water from air. Single atom oxygen produced by action of light is very reaction active chemical. Its action to thin nanometers sized polymeric structure will result on fast its destruction and degradation.

In the work [41] show results of degradation of PV structure at action of oxygen and the light. In fact presence of oxygen and the light illumination are standard conditions of PV and OLED structure application as oxygen and water vapors go easily over polymer film of flexible structure.

Fig. 18 shows the degradation of the photocurrent of a plastic solar cell under $1000 W/cm^2$ illumination inside the sealed container. Already within 1 h the effect is very pronounced. It also shows how a temporary exposure of the cell to air leads to an accelerated decrease of photoefficiency. The photocurrent seems to become inhomogeneous, suggesting selective degradation. Higher degrees of degradation around the rim of the cell could be an effect of a mechanism of lateral degradion through diffusion of oxygen or water vapor.

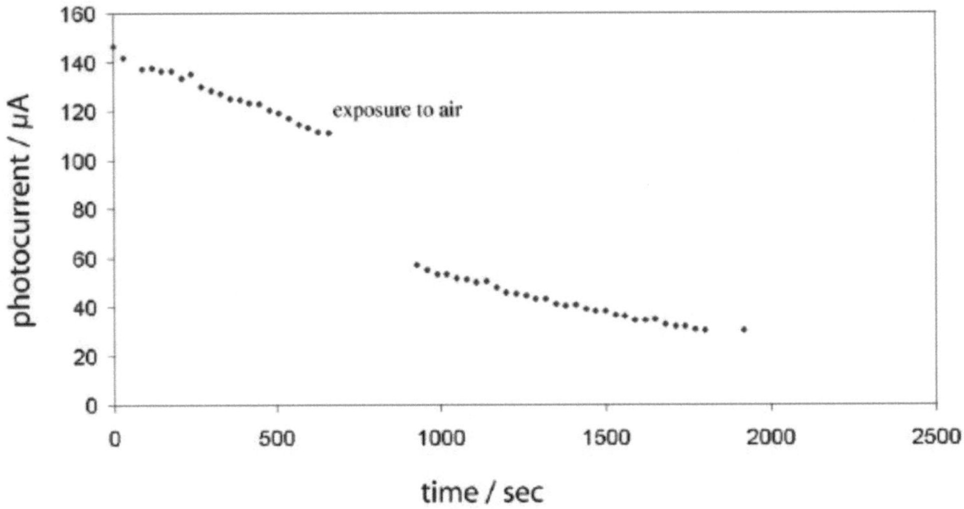

Fig. 18. Decay of photocurrent (Isc) in Cell as a result of degradation under 1000 W/cm2 illumination in an inert atmosphere, followed by exposure to the atmosphere.

From this point of view use of inorganic nanocrystals, more stable to oxygen action is preferable too.

In despite of evident advantage of inorganic and organic nanocrystals application in PV and OLED devices, practical examples are not numerous because of difficulties of thin layer preparation from new and unknown materials nanocomposites as well as because of nanocomposites preparation.

Now mostly applicable are: fullerene C_{60} nanoparticles, carbon nanowires, TiO_2 and CdS nanocrystals: all n-type semiconductors.

5.3 Example of PV structure based on fullerene

Excitons do not dissociate readily in most organic semiconductors. The idea to overcome this obstacle is use a heterojunction: to use two semiconductor materials with different electron affinities and ionisation potentials. This will favour exciton dissociation: the electron will be accepted by the material with the larger electron affinity and the hole by the material with the lower ionisation potential.

One of the most used acceptors in heterojunction cells is the fullerene C_{60} nanoparticles (or large organic molecules in some publication) [42]. Besides having a high electron affinity, C_{60} is fairly transparent and also has fair electron conductance (10^{-4} Scm^{-1}). This makes fullerenes a good component in PV cells. The first report of a conducting polymer/C_{60} cell came in 1993 by Sariciftci et al. [43]. In one study, fullerene was vacuum sublimed onto a MEH-PPV layer that was spin coated on ITO-covered glass . Au was used as the electron-collecting electrode (see Fig. 19).

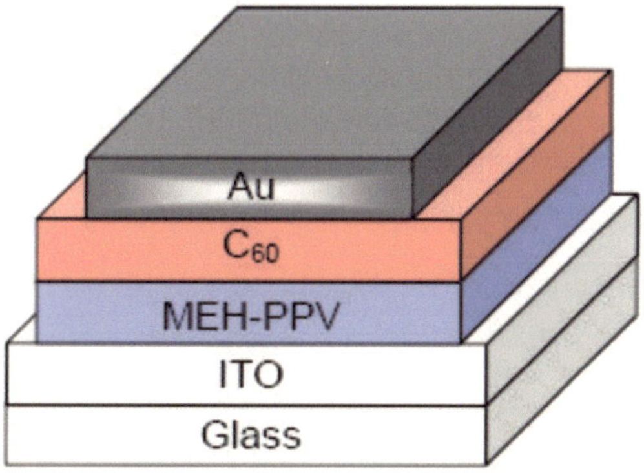

Fig. 19. A two-layer heterojunction photovoltaic cell with C60 fullerene nanoparticles. The electron accepting C_{60}-layer contacts the Au electrode, while the electron donating MEH-PPV layer contacts the ITO electrode.

In this example a defect of structure is a low light absorbing area close to border between C_{60} and MEH-PPV semiconductors. It is clear that exciton dissociation is most effective at the interface in heterojunction cells, thus the exciton should be formed within the diffusion length of the interface. Since typically diffusion lengths are in the range of 10 nm, this limits the effective light-harvesting layer. However, for most organic semiconductors the film thickness should be more than 100nm in order to absorb most of the light. It follows that thicker film layers increase light absorption but only a small fraction of the excitons will reach the interface and dissociate. This problem can be overcome by blending donor and acceptor, a concept called dispersed (or bulk) heterojunction (see Fig. 20) [41].

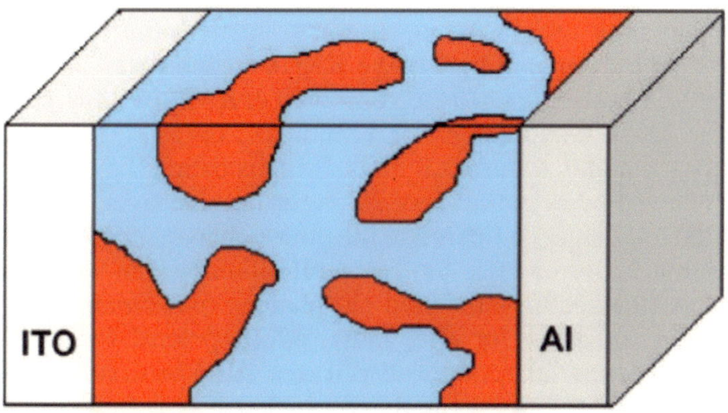

Fig. 20. Dispersed heterojunction between a transparent ITO electrode and an Al electrode.

Certainly, this structure can be prepared by mixing of polymer or polymer like materials only and use of nanocomposites here is without alternative. In 1994 Yu [44] made the first dispersed polymer heterojunction PV cell by spincoating on ITO covered glass from a solution of MEH-PPV and C_{60} in a 10:1 wt-ratio. Finally, Ca was evaporatedonto the organic layer. The cell showed a photosensitivity of 5.5 mA/W, an order of magnitude larger than the photosensitivity of the pure polymer.

One limitation of this approach is the relative low solubility of fullerenes in normal solvents. This problem was solved when Hummelen et al. [45] synthesiseda number C_{60}-derivatives with increased solubility in 1995, which allowed the fullerene content to be as high as 80% in the prepared films. Using a methano-functionalised fullerene derivative Yu et al. [46] repeated the fabrication procedure with a polymer/fullerene ratio of 20/80, the contacts were made of ITO and Ca, and the cell had a QE of 29% and a PCE of 2.9% (under monochromatic light, intensity at 20 mW/cm2). Thus a substantial increase compared to earlier polymer/fullerene mixtures. In 2000 Shaheen et al. [47] reportedhigh QE values of 85% in a PPV-derivative and fullerene heterojunction cell, with a PCE of 2.5%. All approaches described above are to use high concentrated nanocomposites based on fullerene C_{60} nanoparticles stabilized by linking to polymer chain. Different structures of fullerenes based nanocomposites materials are show in Fig. 21.

It is clear that the control of morphology in dispersed heterojunction devices is a critical point. The degree of phase separation and domain size depend on solvent choice, speed of evaporation, solubility, miscibility of the donor and acceptor etc. One strategy towards increasing control is to covalently link donor and acceptor. In 2000 Stalmach et al. [48] synthesized PPV-C_{60} diblock copolymers through controlled living radical polymerization. The same year Peeters et al. [49] synthesized a number of p-phenylene vinylene oligomers (OPV) attachedto C_{60} and investigated their use in PV devices. Peeters found that charge separation lifetimes was dependent on the number of repeating oligomer units. Thus, charge separation lifetimes were much longer for 3–4 units compared to 1–2 repeating units. A cell consisting of the longest oligomer (4 repeating units) between aluminium and a PEDOT-PSS covered ITO electrode had an ISC of 235 mA/cm^2 and a VOC of 650 mV, but a relatively low FF of only 0.25. Van Hal et al. [50] made a similar study on fullerene oligio (thiophene)-fullerene triads varying the number of monomer units. In agreement with Peeters they found that a certain length is needed in order to observe charge transfer upon excitation. Thus photoinduced charge transfer was much more pronounced for 6 monomer units, compared to an oligomer with 3 monomer units. Such model studies are important in understanding charge transfer and light harvesting in greater detail. In 2003 Krebs et al. [51] have made an interesting study on a dyad consisting of a poly(terphenylene cyanovinylene) terminated with an ADOTA dye. The dye is a cation and the assembly thus resembles a soap molecule and have the ability to form LB films. By spincoating the dyad on an ITO covered glass substrate, followed by evaporation of Al on the organic layer, the short circuit current of the dyad was 100 fold larger compared to the pure polymer.

While some control is introduced by covalently linking the donor and acceptor in polymer/oligomer-C60 assemblies the final morphology may suffer from phase separation and clustering of the fullerene (or dye) units, which potentially limits efficient charge separation due to low donor/acceptor interfacial area. Also, increased phase separation may disrupt the continuity of the phases and reduce the charge transport properties of the material, due to inefficient hopping between different domains, reducing overall performance. This may be a critical point, since intramolecular charge recombination might occur at a fast rate. One way to control a bicontinuous phase separation andinsure a large interfacial area between donor and acceptor is to covalently graft fullerene moieties onto the donor-polymer backbone (Fig. 21), so-called double-cable polymers (due to their p/n type conduction properties).

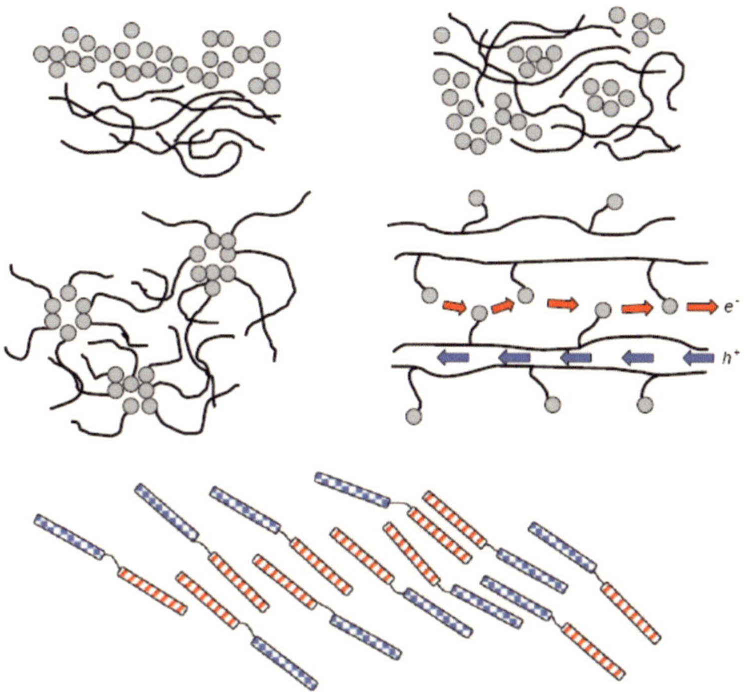

Fig. 21. Different morphologies of heterojunction cells based on fullerene nanocomposites. Top, left: Two-layered structur e of fullerenes and polymer chains. Top, right: dispersed heterojunction. Middle, left: fullerenes with polymer chains attached. Middle, right: self-assembled layered structure of double-cable polymers. Bottom: self-assembled layered structure of diblock copolymers. The layered structure of double-cable polymers and diblock copolymers are expected to facilitate efficient electron and hole transport.

These assemblies have been intensively investigated in recent years as promising components in PV devices [52], but they are also interesting as components in molecular electronics. The first reports of polymers bearing fullerene on the side chains came in 1996 by Benincori et al. [53] The polymer was a polythiophene. Cravino et al. [54] synthesiseda fullerene-thiophene double-cable and used it in a solar cell. However the PV device had limited cell efficiency due to low levels of fullerene content (C_{60} was attached to 7% of the repeating units). It is important that the fullerene content reaches the percolation threshold to insure efficient electron transport. Thus as the complexity of the designed systems increase the more critical it becomes to optimize design parameters. Even though the synthesized double cable polymers have shown some promising results, we have yet to see the fully optimized double-cable polymers. An alternative to polymer-fullerene double-cables is block copolymers consisting of a donor and acceptor block. In general block copolymers are known to phase separate and form ordered domains similar to the double-cable polymers. In 2003 Krebs et al. [55] synthesiseda block copolymer consisting of an electron acceptor block and an electron donor block. The backbone was polyacetylene, and by using phenyl and pentafluorophenyl as side groups the HOMO and LUMO of the individual blocks could be tuned so that hole or electron conductance is favored.

5.4 D well ordered bulk heterojunction based on carbon nanotubes

Dispersed heterojunction described above ensure augmentation of effective surface of junction between n and p- types of semiconductors, but same time irregular structure of n and p- types areas will result on augmentation of length of charge transportation from n and p semiconductors border to the electrodes (Fig. 20). Certainly if to make regular 3-D structure of different phases of heterojunction, it is possible to diminish charge transportation length up to thickness of the layer. This structure can to resolve problem of relative high exciton energy dissociation of organic material. Indeed, high exciton energy dissociation in organic in comparison with inorganic will result on diminish of organic solar cell efficiency (problem was described in details above of this paragraph). Same time use of polymers for flexible solar cell thin film have no alternative in classical technique as polymers materials only allow to form thin flexible film. However, if to use technique of self- assembly, it is possible to prepare polymer free thin film consist from inorganic nanoparticles and to obtain inorganic high efficiency and stable solar cell based on nanoparticles assembly structure.

However, for practical making of these structure are need to prepare 3-D elements with tens nanometers sizes with artificial distribution of n- and p- types of semiconductor nano-sized areas that is too difficult at the present time. Now only way to make so is to use self-assembly technique for preparation of nanometers sized vertical columns from carbon nanotubes well ordered. The spaces between columns should be filled by another semiconductor. This technique will allow to produce well ordered heterojunction with charges transport distances minimized in comparison to structure in Fig. 20. At the present time investigations of self- assembly carbon nanotubes processes is in beginning only. There are a lot of both scientific and technological problems. The first practical realization of this way give a work [56]. In the fallowing text will be considered semiconductor solar cell strictures based on single carbon nanotubes and its combination with other types of nano-semiconductors: CdS nanoparticles and porphyrines organic pigments. The principle of self-assembly structure shown in Fig. 22.

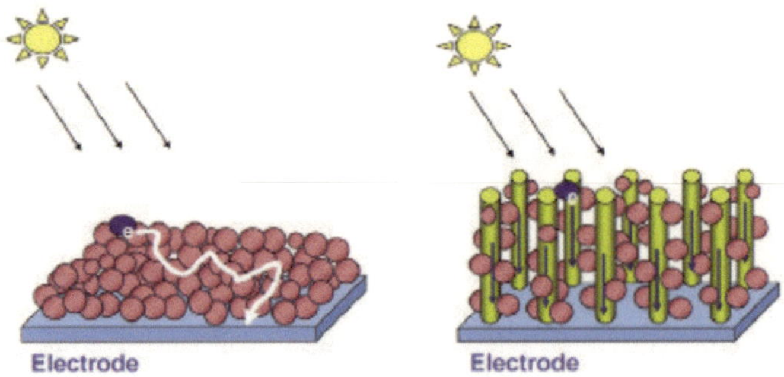

Fig. 22. Illustration of random charge carrier transport in bulk heterojunction (in the left) - charge transportation length is long and well oriented 3-D nanostructure carbon-nanotube-directed charge transport in an organized hybrid assembly.

Certainly should be take into account that each nanotubes have diameter of 1,5 nm and it is very difficult to made really this structure.

Commercially available semiconductor carbon nanotubes (SWNTs) contain both metallic and semiconducting nanotubes with different chirality. The work function of SWNT bundles is known to be about -4.8 eV versus absolute vacuum scale (AVS). Carbon nanotubes possess a bandgap in the range of 0-1.1 eV, depending upon their chirality and diameter.

Semiconducting carbon nanotubes undergo charge separation when subjected to bandgap excitation. Two different approaches can be considered for the use of carbon nanotubes in solar cells (Fig. 23):

(i) direct bandgap excitation of semiconducting nanotubes; or (ii) the use of conducting tubes as conduits to improve the transport of charge carriers from light-harvesting nanoassemblies.

The methods employed to deposit carbon nanotubes as thin films on a conducting surface for use as photoresponsive electrodes in solar cells are discussed in the following sections. Examples of the two strategies presented in Fig. 23 are also illustrated.

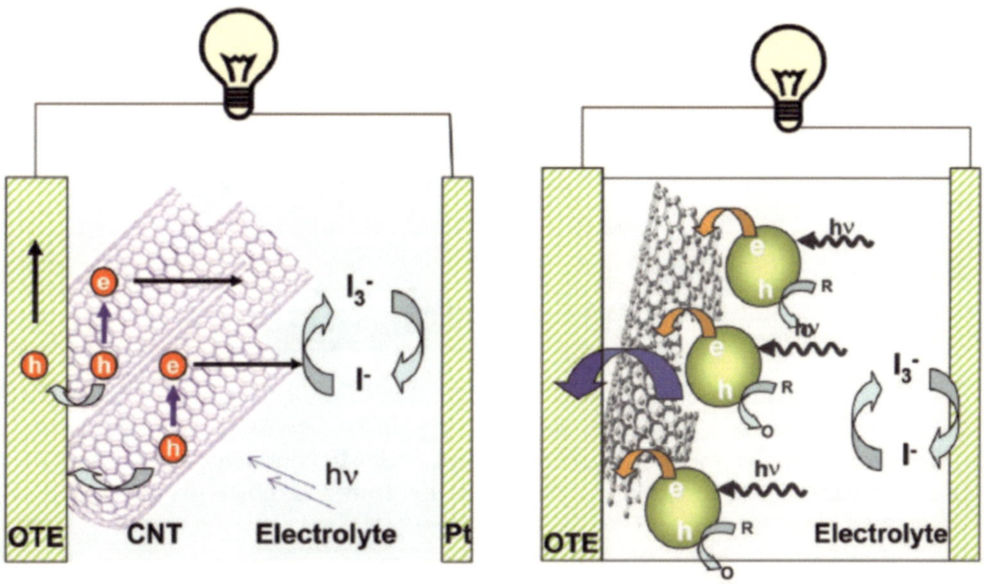

Fig. 23. Strategies to employ carbon nanotubes in photochemical solar cells: (left) by direct excitation of carbon nanotubes; (right) by excitation of light-harvesting assemblies anchored on carbon nanotubes. The electrons and holes generated by photoexcitation are referred to as e and h, respectively. One of these charge carriers is collected at the electrode surface and the other one is scavenged by the oxidized (O) or reduced (R) form of the redox couple in the electrolyte.

Photoinduced charge separation in SWNT films An interesting semiconducting property of SWNTs is their ability to respond to light. For example, the photoresponse of carbon nanotubes filaments was realized in early years from the elastic response of the aligned bundles between two metal electrodes [57]. Avouris and coworkers [58] have monitored hot carrier luminescence from ambipolar carbon nanotube field-effect transistors (FETs). The holes and electrons injected via an external circuit produce emission resulting from electron-hole recombination in the system. The recent report of bandgap fluorescence from a semiconducting SWNT sample rich in individual nanotubes has made it possible to correlate optical properties with individual tube species as a result of their well-defined optical transitions. Spectroscopic studies have demonstrated that the relaxation of electrons and holes to the fundamental band edge occurs within 100 fs after photoexcitation of the second van Hove singularity of a specific tube structure [59]. These early studies confirm the ability of carbon nanotubes to possess a band structure that can undergo electron-hole charge separation with visible light excitation (Fig. 24).

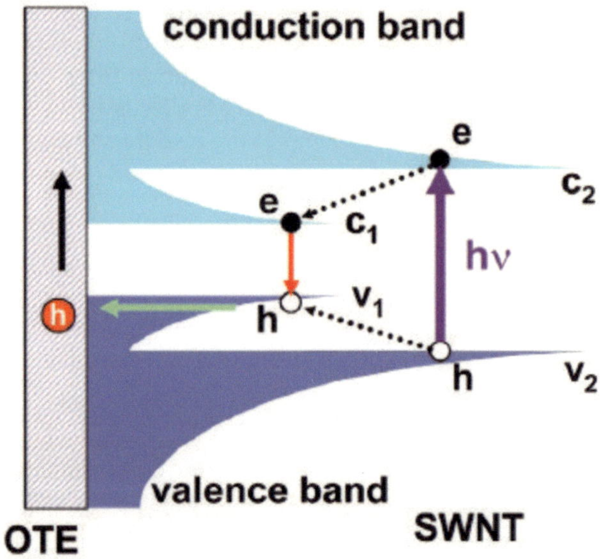

Fig. 24. Schematic illustrating the density of states of a single carbon nanotube. Photogenerated holes are captured at the collecting electrode surface, resulting in current generation in a photoelectrochemical cell. C1 and C2 refer to conduction bands and V1 and V2 refer to valence bands. e and h refer to the electron and holes generated following photoexcitation of the SWNTs.

In order to use photogenerated charge carriers for generating electricity, it is important that they are separated before undergoing recombination. However, spatially confined charge carriers in the nanotube are bound by Coulombic interactions with the bound pair referred to as an exciton [60]. Most of these excitons from higher C2 and V2 levels relax via interband transitions to the low-lying C1 and V1 levels of the fundamental gap to produce a second sub-bandgap exciton. A small fraction of the excitons are able to dissociate and form unbound electron-hole (e-h) pairs. The dissociation of excitons to create the charge-separated state thus becomes an important process to tap them into photocurrent generation. The charge separation in carbon nanotubes can be probed using femtosecond laser pump-probe spectroscopy. This technique is useful to investigate the ultrafast processes that occur following the excitation of carbon nanotubes or semiconductor materials. In a typical experiment, the absorption changes in the sample are recorded at different delay times following excitation with a short laser pulse. Difference absorption spectra at various delay times have been recorded by exciting SWNT suspensions in THF with a 387 nm laser pulse (pulse width 130 fs). (A Clark MXR-2010 laser system and Ultrafast Systems detection setup was used for these measurements.) Representative transient absorption spectra and the decay of absorption at 700 nm are shown in Fig. 25.

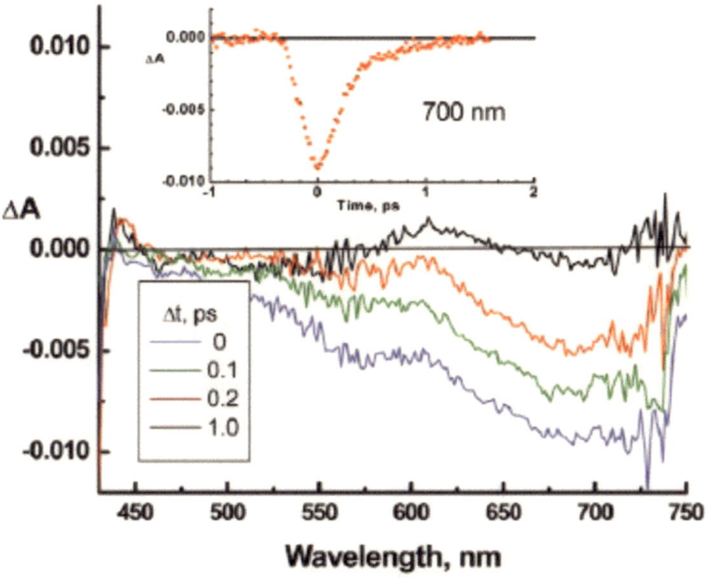

Fig. 25. Time-resolved transient absorption spectra of an SWNT suspension in THF (flow cell) recorded using a 387 nm laser pulse (pulse width 150 fs; Δt = 0 corresponds to the end of the pulse). The inset shows the bleaching recovery at 700 nm.

The photoexcitation causes the bleaching of SWNT absorption in the red region. The broadness of the bleaching band essentially arises from the diversity of tube diameters, chiral angles, and the aggregation of nanotubes.

The bleaching in the visible region, which corresponds to the C2-V2 transition, recovers in ~1 ps as the bound electron-hole pairs or excitons relax to the low-lying C1-V1 state. The dynamics of the transient bleaching recovery and the decay of the emission in the infrared arising from charge recombination in the fundamental gap have been studied recently by Ma et al. [59]. They observed that the electron-hole pairs accumulate in the fundamental gap (C1-V1) and their lifetime (10-100 ps) is dependent on the excitation intensity. Based on the difference between the emission decay and transient absorption recovery, these researchers highlighted the involvement of charge trap states as the additional contributing factors responsible for electronic transitions. The presence of such surface states are likely to stabilize the photogenerated charge carriers and contribute to the overall photocurrent generation. Such enhanced charge separation is crucial for increasing the probability of charge collection at the electrode surface. The transient bleaching observed following laser pulse excitation shows that there is a significant number of charge carriers produced in the SWNTs. The obvious question is whether one can collect the photoinduced charge carriers generated in SWNTs suitably for photocurrent generation, similar to the photovoltaic application of other semiconductors.

5.5 SWNT-semiconductor hybrids

In photoelectrochemical cells based on nanostructured or mesoscopic semiconductor films, the electron transport across particles is susceptible to recombination loss at the particle grain boundaries. The use of a nanotube support to anchor light-harvesting assemblies (e.g. semiconductor particles) provides a convenient way to capture photogenerated charges and transport them to the electrode surface. An illustration of these two scenarios can be seen in Fig. 23.

SWNTs are an ideal candidate as conduit for collecting and transporting charges across light-harvesting assemblies. Of particular interest is a CdS-SWNT composite that is capable of generating a photocurrent from visible light with unusually high efficiency [61]. The luminescence of CdS is quenched by SWNTs. Transient absorption experiments have confirmed the quick deactivation of excited CdS on a SWNT surface, as the transient bleaching recovers in about 200 ps.

In order to test the hypothesis of electron transfer between excited CdS and SWNT in the composite film, CdS particles are deposited on SWNT electrodes (referred to as OTE/SWNT/CdS) [62]. The SWNT film was first deposited on the OTE using the electrophoretic deposition method described earlier. The electrode was then sequentially immersed in solutions containing Cd^{2+} and S^{2-} to form CdS nanocrystallites. The electrode was thoroughly washed with deionized water between the two immersions so that only adsorbed Cd^{2+} ions react with S^{2-}. It may be noted that such an ion adsorption precipitation method is similar to methods employed for casting nanostructured films of metal chalcogenides on oxide films [63]. It is interesting to note that Cd^{2+} ions readily adsorb on SWNTs and react with S^{2-} to form CdS nanocrystallites with a characteristic onset absorption around 500 nm.

It is employed the OTE/SWNT/CdS electrode in a photoelectrochemical cell containing acetonitrile solution with 0.1% triethanolamine as a sacrificial electron donor. Triethanolamine undergoes irreversible oxidation as it scavenges photogenerated holes from the electrode surface. Photocurrent generation is seen when the CdS-modified SWNT film is irradiated with visible light ($\lambda > 380$ nm). An open circuit voltage of ~200 mV and a short circuit current of 6.2 μA were recorded. The dependence of the IPCE on the excitation wavelength is shown in Fig. 26.

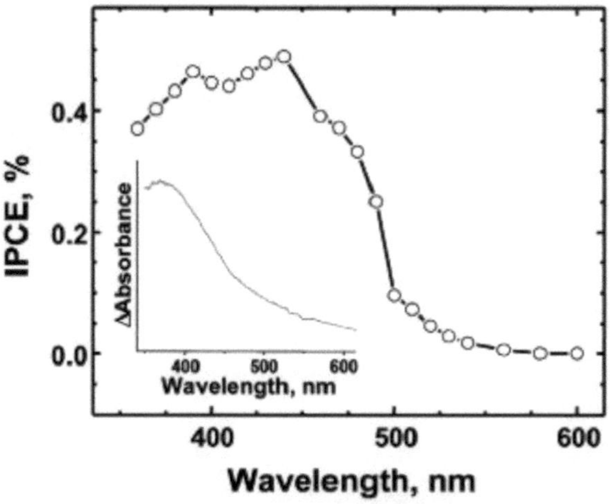

Fig. 26. IPCE of OTE/SWNT/CdS. Inset: absorbance difference between OTE/SWNT/CdS and blank SWNT film.

The onset of IPCE is seen at ~500 nm and closely follows the absorbance characteristics of CdS (see inset of Fig. 26). The observed photocurrent is dominated by the initial excitation of CdS as is evident from the photocurrent action spectrum. Furthermore, the anodic current observed with SWNT/CdS films confirms the direction of the electron flow from CdS to the collecting electrode mediated by the SWNT network. The ability of the CdS-SWNT nanocomposite system to undergo photoinduced charge separation opens up new ways to design light-harvesting assemblies.

5.6 TiO_2 nanoparticles based PV

In the work [64] was used TiO2 nanoporous film in structure of dye sensitized solar cell. Nanoporous film is real alternative to high concentration nanocomposite. Preparation of high concentrated homogenous nanocomposite is a difficult task as nanoparticles tend to coagulated at high concentration. Same time nanoporous film have uniform distribution of pores with its high volumetric concentration - up to 60 %. If to fill nanopores by polymer we obtain stable nanocomposite material (porous TiO_2 film - see Fig. 27).

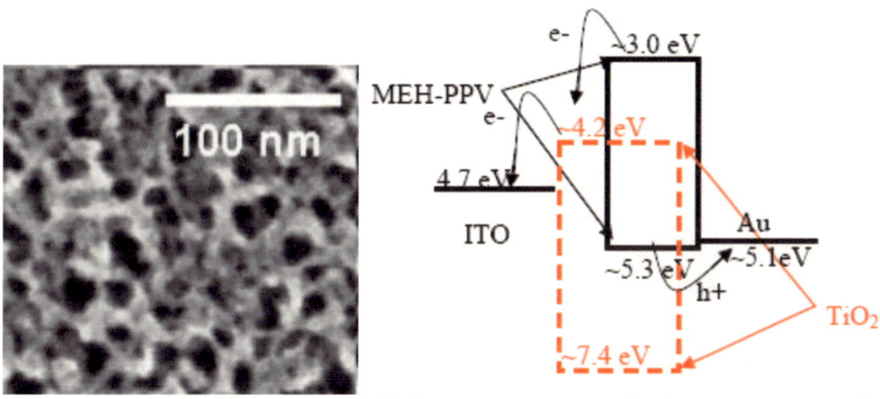

Fig. 27. SEM image of highly porous TiO2 layers consisting of a 3D interconnected network of anatase crystallites used for device fabrication. Right - Energy diagram of ITO/TiO2/MEH-PPV/Au.

With the control of the nanostructured morphology, metal oxides are believed to act as promising alternatives as the electron acceptor and transporter in bulk-heterojunction solar cells. Among the metal oxides, TiO_2 is a very good candidate for this purpose because the use of nanocrystalline TiO_2 as electron accepting electrode for dye-sensitized solar cells has shown an overall power conversion efficiency as high as 10%.
Mesoporous TiO_2 films were deposited on conducting glass ITO or SnO_2. By varying the temperature in the nanoparticle synthesis, the average particle diameter was altered between 20.5 and 41.5 nm. XRD measurements indicate that the TiO_2 particles were pure anatase within the detection limit of 3%–5%. Also, transmission electron microscopy indicates that the particles were crystalline. The film porosity was 57.5%±1.5% and was independent of the average particle size [64]. If to use MEH-PPV polymer as a p-type semiconductor, reported maximum EQE is 6%, short circuit current density - 0.4 mA/cm2 and power conversion efficiency - 0.17% under 100 mW/cm2 white light illumination [65].

Takahashi et al. [66] reported short circuit current density of 0.35 mA/cm^2 and power conversion efficiency of 0.13% under the irradiation of AM 1.5 illumination (100 mW/cm^2). The short circuit current could be improved by about 3 times and power conversion efficiency could reach 0.47% by blending MEH-PPV with [2- [2- [4- (dimethylamino) phenyl]-ethenyl] -6- methyl -4H- pyran –4– ylidene] propanedinitrile (DCM). All results are comparable to same received with carbon based nanocomposites described above.

6. Conclusion

At present time application of semiconductor nanoparticles in different devices like solar sell and light emitting devices is a reality. However there are some problems connected with low value of charge carrier mobility and high exciton dissipation energy in disperse nanocomposite material. This problem can be solved by following way:
- augmentation of nanoparticles concentration in nanocomposite for preparation of hybrid material in which charge carrier transportation go over nanoparticles and not polymer matrix. Limitation: loss of flow proprieties and hence loss of good processability of material. Maximal level of nanoparticles concentration is limited in 20 - 25 vol % of nanoparticles concentration. This value is enough for charge carrier transport.
- formation of nanostructure in material layer to provide directed charge carrier transport between electrodes with minimal path length. This can be made by nanoimprint or self assembly methods. At present time there are the first examples of these nano- ordered structures.

Both problems can be solved and nanocomposite materials go to factory scale use application. For example, in conference (7-th International Conference ELECTRONIC PROCESSES IN ORGANIC MATERIALS (ICEPOM-7) Ukraine, Lviv, May 26 – 30, 2008 was being announced plastic nanostructured solar sell as a new product for commercial production since 2009: T. Yoshida, M. Matsui, K. Funabiki, H. Miura, Y. Fujishita Plastic solar cells employing electrodeposited nanostructured zno and organic photosensitizer, developed by Center of Innovative Photovoltaic Systems (CIPS), Gifu University, Yanagido, Japan. These results were published in the work [67].

7. References

1. A. Eychmu, J. Phys. Chem. B 104, 6514 (2000)
2. R. Rossetti, J. L. Ellison, J. M. Gibson and L. E. Brus, J. Chem. Phys. 80, 4464 (1984)
3. M. W. Peterson, M. T. Nenadovic, T. Rajh, R. Herak, O. Micic, J. P. Goral and A. J. Nozik, J. Phys. Chem. 92, 1400 (1988)
4. M. S. El-Shall, W. Slack, W. Vann, D. Kane and D. Hanley, J. Phys. Chem. 98, 3067 (1994), I. Yu. Denisyuk, T. R. Williams and J. Ed. Burunkova, Mol. Cryst. Liq. Cryst., 497, (2008.)
5. Ashfaq Adnan, C.T. Sun, Hassan Mahfuz A molecular dynamics simulation study to investigate the effect of filler size on elastic properties of polymer nanocomposites // Composites Science and Technology 67 (2007) 348–356
6. Odegard GM, Clancy TC, Gates TS. Modeling of the mechanical properties of nanoparticle/polymer composites. Polymer 2005; 46(2) p. 553–62.

7. Binder K. Monte carlo and molecular dynamics simulations in polymer science. New York (USA): Oxford University Press; 1995.
8. C. Lu , Z. Cui, Z. Li, B. Yang and J. Shen, J. Mater. Chem., 13, 526 (2003)
9.J. H. Golden, H. Deng, F. J. DiSalvo, J. M. J. Fre´chet and P. M. Thompson, Science, 268, 1463 (1995)
10. J. H. Golden, F. J. DiSalvo, J. M. J. Fre´chet, J. Silcox, M. Thomas and J. Elman, Science, 273, 782 (1996)
11. C. Gao, B. Yang and J. Shen, J. Appl. Polym. Sci., 75, 1474 (2000)
12. P. Barbaro, C. Bianchini, G. Scapacci, D. Masi and P. Zanello, Inorg. Chem., 33, 3180 (1994)
13. Igor Yu. Denisyuk, Todd R. Williams, Julia E. Burunkova Hybrid optical material based on high nanoparticles concentration in UV-curable polymers - technology and proprieties // Mol. Cryst. Liq. Cryst., Vol. 497, pp. 142–153, 2008
14. Todd R. Williams, Igor Yu. Denisyuk, Julia E. Burunkova Filled polymers with high nanoparticles concentration - synthesis, optical and rheological proprieties // Journal of Applied Polymer Science, Volume 116 Issue 4, P. 1857 - 1866, 2010
15. H. Ishii, K. Sugiyama, E. Ito, and K. Seki, Adv. Mater. 11, 605 (1999)
16. J. M. Nunzi, C. R. Physique, 3, 523 (2002)
17. C. Kittel, Introduction à la physique de l'état solide, Bordas, Paris, (1972)
18. C. Cojan, G.P. Agrawal and C. Flytzanis, Phys. Rev. B, 15, 909, (1977)
19. J. Lange and H. Bässler, Phys. Stat. Sol. B, 114, 561 (1982)
20. J.H. Schön, C. Kloc, A. Dodabalapur and B. Batlogg, Science, 289, 599 (2000)
21. M. Schott and C. R. Acad. Sci. Paris Sér., 4, 381 (2000)
22. D. Emin, in Handbook of Conducting Polymers, edited T.A. Skotheim, M. Dekker, (1996), Vol. 2. p.
23. W.D. Gill, in Photoconductivity and Related Phenomena edited J. Mort, D.M. Pai, Elsevier, (1976) p. 63.
24. R.G. Kepler, P.M. Beeson, S.J. Jacobs, R.A. Anderson, M.B. Sinclair, V.S. Valencia and P.A. Cahill, Appl. Phys. Lett., 66, 3618 (1995)
25. G. Horowitz, Adv. Mater., 10, 365 (1998).
26. P.W.M. Blom, M.J.M. De Jong and J.J.M. Vleggaar, Appl. Phys. Lett., 68, 3308 (1996)
27. H.C.F. Martens, J.N. Huiberts and P.W.M. Blom, Appl. Phys. Lett., 77, 1852 (2000)
28. H. Scher, in Photoconductivity and Related Phenomena edited J. Mort, D.M. Pai, Elsevier, (1976) p. 63
29. M.N. Bussac and L. Zuppiroli, Phys. Rev. B, 55, 15587 (1997)
30. J.H. Schön, C. Kloc, R.C. Haddon and B. Batlogg, Science, 288, 656 (2000)
31. J.H. Schön, S. Berg, C. Kloc and B. Batlogg, Science, 287, 1022 (2000)
32. G.G. Malliaras, Y. Shen, D.H. Dunlap, H. Murata and Z.H. Kafafi, Appl. Phys. Lett., 79, 2582 (2001)
33. A.R. Inigo, C.H. Tan, W. Fann, Y.-S. Huang, G.-Y. Perng and S.-A. Chen, Adv. Mater., 13, 504 (2001)
34. C. Sentein, C. Fiorini, A. Lorin and J.M. Nunzi, Adv. Mater., 9, 809 (1997)
35. H. Murata, G.G. Malliaras, M. Uchida, Y. Shen and Z.H. Kafafi, Chem. Phys. Lett., 339, 161 (2001)]
36. H. Spanggaard and F. Krebs, Solar Energy Materials & Solar Cells, 83, 125 (2004)
37. J.H. Burroughes, D.D.C. Bradley, A.R. Brown, R.N. Marks, K. Mackay, R.H. Friend, P.L Burns and A.B. Holmes, Nature, 34, 539 (1990)

38. U. Rauscher, H. B.assler, D.D.C. Bradley and M. Hennecke, Phys. Rev. B, 42, 9830 (1990)
39. E.L. Frankevich, A.A. Lymarev, I. Sokolik, F.E. Karasz, S. Blumstengel, R.H. Baughman and H.H. Horhold, Phys. Rev. B, 46, 9320 (1992)
40. L.J. Rothberg, M. Yan, F. Papadimitrakopolous, M.E. Galvin, E.W. Kwock and T.M. Miller, Synth. Met., 80, 41 (1996)
41. T. Jerankoa, H. Tributscha, N.S. Sariciftcib and J.C. Hummelen, Solar Energy Materials & Solar Cells, 83, 247 (2004)
42. Md.K.H. Bhuiyan and T. Mieno, Thin Solid Films, 441, 187 (2003)
43. N.S. Sariciftci, D. Braun, C. Zhang, V.I. Srdanov, A.J. Heeger, G. Stucky and F. Wudl, Appl. Phys. Lett., 62, 585 (1993)
44. G. Yu, K. Pakbaz and A.J. Heeger, Appl. Phys. Lett., 64, 3422 (1994)
45. J.C. Hummelen, B.W. Knight, F. LePeq, F. Wudl, J. Yao and C.L. Wilkins, J. Org. Chem., 60, 532 (1995)
46. G. Yu, J. Gao, J.C. Hummelen, F. Wudl and A.J. Heeger, Science, 270, 1789 (1995)
47. S.E. Shaheen, C.J. Brabec, N.S. Sariciftci, F. Padinger, T. Fromherz and J.C. Hummelen, Appl. Phys. Lett., 78, 841 (2001)
48. U. Stalmach, B.d. Boer, C. Videlot, P.F.v. Hutten and G. Hadziioannou, J. Am. Chem. Soc., 112, 5464 (2000)
49. E. Peeters, P.A.v. Hal, J. Knol, C.J. Brabec, N.S. Sarciftci, J.C. Hummelen and R.A.J. Janssen, Phys. Chem. B., 104, 10174 (2000)
50. P.A. van Hal, J. Knol, B.M.W. Langeveld-Voss, S.C.J. Meskers, J.C. Hummelen and R.A.J. Janssen, J. Phys. Chem. A., 104, 5974 (2000)
51. F.C. Krebs, M. Jørgensen, Macromolecules, 35, 7200 (2002)
52. A. Cravino and N.S. Sariciftci, J. Mater. Chem., 12, 1931 (2002)
53. T. Benincori, E. Brenna, F. Sannicol, L. Trimarco and G. Zotti, Angew. Chem., 108, 718 (1996)
54. A. Cravino, G. Zerza, M. Maggini, S. Bucella, M. Svensson, M.R. Andersson, H. Neugebauer, C.J. Brabec and N.S. Sariciftci, Monatsh. Chem., 134, 519 (2003)
55. F.C. Krebs and M. Jørgensen, Polym. Bull., 50, 359 (2003)
56. P. V. Kamat, Nanotoday, 4, 315 (2006)
57. Y. Zhang, , and S. Iijima , Phys. Rev. Lett., 82, 3472(1999)
58. M. Freitag, Nano Lett., 4, 1063(2004)
59. Y.-Z. Ma, J. Chem. Phys., 120, 3368 (2004)
60. C. L.Kane, and E. Mele, J., Phys. Rev. Lett., 90, 207401(2003)
61. L. Sheeney-Haj-Ichia, Angew. Chem. Int. Ed. 44, 78 (2004)
62. I. Robel, Adv. Mater., 17, 2458(2005)
63. S. Hotchandani, and P. Kamat, J. Phys. Chem., 96, 6834 (1992)
64. N. Kopidakis, N. R. Neale, K. Zhu, J. van de Lagemaat, and A. J. Frank Appl. Phys. Lett., 87, 202106 (2005)
65. P. M. Sirimanne, T. Shirata, L. Damodare, Y. Hayashi, T. Soga, T. Jimbo, Solar Energy Materials and Solar Cells, 77, 15 (2003)
66. K. Takahashi, K. Seto, T. Yamaguchi, J.-I. Nakamura, C. Yokoe, and K. Murata, Chem. Lett. 33, 1042 (2004)
67. T. Dentania, K. Nagasakaa, K. Funabikia, J. Jinb, T. Yoshidac, H. Minourac and M. Matsui, Dyes and Pigments 77, Issue 1, (2008), 59-69

6

Diluted magnetic semiconductor nanocrystals in glass matrix

N. O. Dantas[1], E. S. Freitas Neto [1] and R. S. Silva [1,2]
[1]*Laboratório de Novos Materiais Isolantes e Semicondutores (LNMIS), Instituto de Física, Universidade Federal de Uberlândia, 38402-902, Uberlândia, Minas Gerais, Brazil*
[2]*Instituto de Ciências Exatas e Naturais e Educação (ICENE), Licenciatura em Física, Universidade Federal do Triângulo Mineiro, 38025-180, Uberaba, Minas Gerais, Brazil*

1. Introduction

Diluted magnetic semiconductor (DMS), are semiconductors in which a magnetic impurity is intentionally introduced; a small fraction of the native atoms in the hosting non-magnetic semiconductor material is replaced by magnetic atoms. The main characteristic of this new class of compounds is the possibility of the onset of an exchange interaction between the hosting electronic subsystem and electrons originating from the partially-filled *d* or *f* levels of the introduced magnetic atom (Erwin et al., 2005; Norris et al., 2008). Once that onset is reached in the above-mentioned exchange interaction, it enables the control of both the electronic, and the optical properties of the end material, using external fields in regimes hardly achieved with other classes of materials. Slightly transitions in metal-doped II-VI and IV-VI semiconductor, as for instance, $Cd_{1-x}Mn_xS$, $Pb_{1-x}Mn_xS$, and $Pb_{1-x}Mn_xSe$, is a typical diluted magnetic semiconductor, in which a small amount of Mn^{2+} is substitutionally incorporated into the hosting CdS, PbS and PbSe semiconductor crystal structure (Ji et al., 2003; Silva et al., 2007; Dantas et al., 2008; Dantas et al., 2009).
Quantum confinement effects can be considered with the incorporating of magnetic ions in semiconductors NCs, modifying the optical, magnetic, and electronic properties in relationship to semiconductor bulk. The transition metal ion (Mn^{2+}) d-electrons, usually located in the band gap region of the hosting semiconductor, are available to promote exchange interactions to the sp-band electrons of the hosting semiconductor (Fudyna, 1988). The sp-d exchange interaction taking place in II-VI, and IV-VI DMS (as for instance in $Cd_{1-x}Mn_xS$, $Pb_{1-x}Mn_xS$ and $Pb_{1-x}Mn_xSe$) provides a unique interplay between optical properties and magnetism, which could be strongly-dependent upon the doping mole fraction (x) (Silva et al., 2007; Dantas et al., 2008; Dantas et al., 2009). By varying the material's doping profile (x) a fine tuning of the semiconductor band gap energy can be achieved. Furthermore, quantum size effects caused by the shrinking in DMS bulk II-VI and IV-VI, as in nanosized particles, enhance the optical and the magnetic properties, even further. In addition, in the presence of applied magnetic fields the sp-d interaction involving electrons, holes, and the hosted magnetic ions is affected, as a result modifying the DMS properties and providing the material basis for new applications in magneto-optical,

magneto-transport, spintronics, lasers, and infrared devices. Theoretical models to explain the incorporation of magnetic impurities in nanocrystals are reported in the literature (Erwin et al., 2005; Dalpian & Chelikowsky, 2006; Norris et al., 2008), such as the "self-purification" mechanisms which are explained through energetic arguments. These mechanisms show that the formation energy of magnectic impurities increases when the NCs size decreases. Moreover, the binding energy of the impurities in the crystalline faces is highly dependent on the semiconductor material, such as the crystal structure, and NCs shape.
Nanocrysttaline structures doped with a small amount of magnetic impurities are obtained from a controlled process known as thermal diffusion of precursor ions, for NC formation in conditions of thermodynamic equilibrium. In this context, $Cd_{1-x}Mn_xS$, $Pb_{1-x}Mn_xS$, and $Pb_{1-x}Mn_xSe$ NCs have been synthesized by fusion method in glass matrixes. In this chapter we report the synthesis process of nanocrystals in a glass matrix, the synthesis routes of diluted magnetic semiconductor $Cd_{1-x}Mn_xS$, $Pb_{1-x}Mn_xS$, and $Pb_{1-x}Mn_xSe$ nanocrystals grown on a borosilicate glass matrix and their investigation by experimental techniques of optical absorption (OA), photoluminescence (PL), electron paramagnetic resonance (EPR) spectra, atomic force microscope (AFM), and x-ray diffraction (XRD).

2. Synthesis of nanocrystals in glass

Nowadays, one of the biggest interests is the nanostructured systems production which present desired physical properties to technological applications, and that are of low cost. Among the materials, which satisfy these necessities, there are nanocrystal-doped glasses (Woggon, 1997; Gaponenko, 1998). These are interesting in the physical property studies of low-dimensionally structures, and its optical transitions of electrons in quantum confinement regime (Bányai & Koch, 1993; Woggon, 1997; Gaponenko, 1998). Since the origin of nanocrystal-doped glasses, the optical fiber-based communication systems, which before had the entire amplifying process and optical signal processing performed in an electronically way, begin to amplify and process the optical signals through the use of fully optical devices, in which the use of those devices considerably enhance the quality in signal transmission. The first evidences of nanocrystals existence in glasses undergone through thermal annealing were given by Rocksby at about 1930's (Woggon, 1997). Since the second half of 20th century, companies like Corning Glass Industries, Schott Optical Glass, Hoya, and Toshiba, have been using quantum dot-doped glasses (Gaponenko, 1998). Although, the semiconductor-doped glasses potential to application in optical devices still be a well discussed subject in diverse papers, the research continues focused in the direction of a more primary stage, where the main objective is the understanding of physics involved in this kind of material.

2.1. Nanocrystals' growth kinetics

The semiconductor nanocrystals' growth kinetics in doped glass matrixes is the resulted precipitation in a supersaturated solid solution by dopers, controlled by the diffusion process of the solved semiconductor materials in the glass matrix (Woggon, 1997; Gaponenko, 1998). The solid solution is defined as been constituted by a unique phase in which more than one atomic specie is inserted, and for which the atom identity, that occupy one or more sites, in the solution is variable (Zarzycki, 1991). In this solid solution, the

precursor elements in the glass matrix, which can move, themselves by diffusion, are considered the solutes the glass matrixes are the solvents while the quantum dots are the solid phase or the precipitated from the process. For an occurrence of precipitation, the solution must be supersaturated, i. e., the solute concentration must exceed the saturation value at a given temperature and pressure. The appearing of a new phase happens by the discreet particles formation with well-defined, well-arranged interfaces, in an aleatory way, within the original phase.
These growth kinetic processes, in general, can be divided in three different stages: the nucleation, the normal growth, and the coalescence or competitive growth (Zarzycki, 1991; Gaponenko, 1998).

2.1.1. Nucleation

In the temperature, where there is an appreciable atomic mobility, there is also a continuum rearrangement of atoms in thermal disturbance. If the phase is thermodynamically unstable, these rearrangement domains have a temporary existence, so then they are destroyed, and replaced by others. When the phase is metastable, such fluctuations are potential sources of a stable phase, and don't become permanent. The fluctuation effects can produce dots that are different in size, shape, structure, or composition.
In the simplest classical model, which was proposed by Volmer and Weber (Volmer & Weber, 1926), and Becker and Döring (Becker & Döring, 2006), it is assumed that embryos of the processes have uniform structure, composition, and identical properties to those of future phase, and only differ in shape and size. The shape, in question, is the one that results from the minimum free energy formation, which will be connected to the interface nature. If it is assumed that, in first approximation, the surface energy is independent of crystallography orientation, and the elastic deformation energy is negligible, the embryos will have a spherical shape. The embryos' size is dependent of the thermodynamic stability condition. When two phases coexist in different homogeneous regions, a phase transition or formation of a different phase within another can occur. "Nucleation", is the process of forming a new phase within an existing phase, separated by a well-defined surface.
In a supersaturated solid solution there is an excess of solute in the solvent (the solute concentration exceeds the saturation value at a given temperature and pressure). This excess can be turned into a precipitate, if the nucleation process happens. The quantum dots-doped glasses are examples of materials created by nucleation in a supersaturated solid solution, where there is the coexistence of phases: the solvent (glass matrix), the solute (doper), and the precipitate (quantum dots).
Assuming that within a determined volume (matrix) a coexistence of disperse atoms (solvent) with particles forming atoms (doper) occur, and defining g^m as a free energy per disperse atom, and g^c as the free energy per crystal atom, it is obtained that the free energy of the compound of particles varies from an amount ΔG, when the quantum dot nuclei are formed. This variation can be given by:

$$\Delta G = \Delta G' = \left(\frac{4\pi R^3}{3}\right)\left(\frac{g^c - g^m}{V}\right) + 4\pi R^2 \gamma \tag{1}$$

The term (g^c-g^m/V) of the Eq. (1), represents the free energy variation per unit of volume, R is the radius of quantum dot nucleus, V is the volume per particle in the quantum dots, γ is the surface energy per area unit.
When the matrix is supersaturated by dopers which will form the semiconductor crystalline phase, the first term of the Eq. (1) is negative, while, the second one is positive. As the terms are proportional to R^3 and R^2, respectively, it can be concluded that the second term influence will be lower when R increases and the curve ΔG versus R will increase until it reaches a maximum value and after it will decrease, as it is represented in Fig. 1.

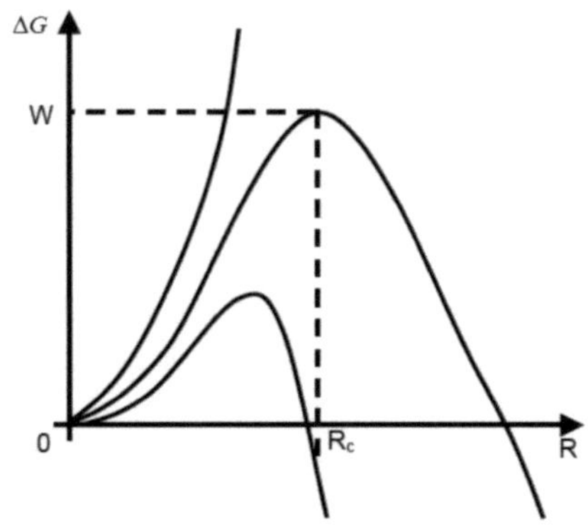

Fig. 1. Free energy variation ΔG as a function of the particle radius (R) (Christian, 1965).

The position of this maximum is given by:

$$\frac{\partial \Delta G^{`}}{\partial R} = 0 \tag{2}$$

what leads to a critical radius R_c of the quantum dot nucleus, given by:

$$R_c = \frac{2\gamma V}{(g^c - g^m)} \tag{3}$$

A particle of radius R_c will be in a instable equilibrium situation. If the radius is lower than R_c, the particle tends to be re-dissolved, once an increase in radius leads to an increase of ΔG. If the radius is greater than R_c, the particle tends to grow, once an increase in radius leads to decrease in ΔG. The particles with $R < R_c$ are called "embryos", while the ones with $R > R_c$ are called "nuclei".
The free energy variation, in a transformation, also depends on quantum dot size, which are formed in the semiconductor phase. The quantum dot radius depends on the number of particles that are dispersed in the glass matrix, and also on equilibrium concentration for the semiconductor phase. Therefore, from Gibss-Thomson's equation, the free energies can be related to semiconductor concentration in the glass matrix, as follows:

$$\left(g^m - g^c\right) = KT \ln\left[N(R)/N(\infty)\right] \tag{4}$$

Here, *N(R)* is the equilibrium concentration for the semiconductor species in quantum dots with radius *R*, *N(∞)* is the equilibrium concentration for the semiconductor species that are dispersed in the glass matrix, *K* is the Boltzmann constant, and *T* is the temperature.
The critical radius (R_c) for any volume, in terms of this equation, is expressed by:

$$R_c = \frac{2\gamma V}{KT\ln\left[N(R)/N(\infty)\right]} \tag{5}$$

which results in:

$$N(R) = N(\infty)\exp\left(\frac{2\gamma V}{KTR_c}\right) \tag{6}$$

According to Eq. (6), it is possible to determine the equilibrium concentration for quantum dots of radius *R*. In the equilibrium, the quantum dots should not increase or decrease in size, i. e., the absorbed species rate must be equal to released species rate. In Fig. 2 is presented a typical curve of these concentrations.
It is observed, in Fig. 2, that the point, where the curve intercepts the line of doping concentration existing in the matrix, and this point defines the critical radius, from which the quantum dot nuclei will grow. It is also observed that the quantum dots are completely re-dissolved when the temperature is increased from T_2 to T_3. In this case, the dissolution rate is proportional to the difference between the equilibrium concentration and the existing concentration in matrix.

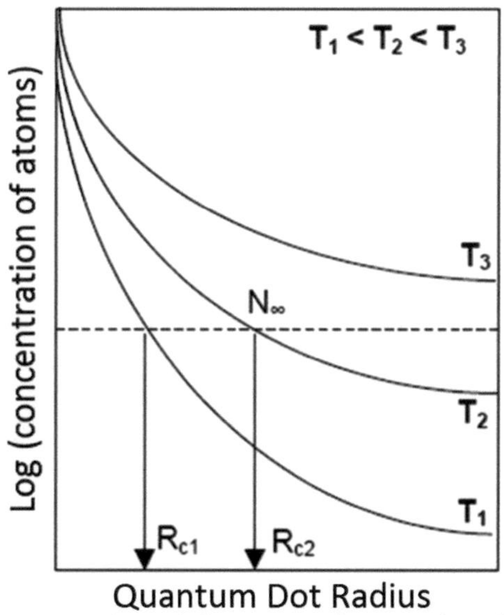

Fig. 2. The equilibrium concentration of the atoms dispersed in the matrix as a function of the quantum dot radius for three different temperatures (Barbosa et al., 1997).

Thus, the quantum dots with smaller radii will be re-dissolved much faster than those with larger radii, which would lead to their size dispersion. It is clear that, when *N(R)* is below of *N(∞)*, there will be no growth of any quantum dot nucleus. The ratio between *N(R)* and *N(∞)* is used as a supersaturation measure, given by:

$$\Delta = \frac{N(R)}{N(\infty)} \tag{7}$$

A matrix always will be supersaturated when $\Delta > 1$. In terms of supersaturation, the critical radius can be written as:

$$R_c = \frac{2\gamma V}{kT\ln(\Delta)} \tag{8}$$

The supersaturation degree is also defined as follows:

$$\Delta_m = \frac{[N(R) - N(\infty)]}{N(\infty)} = \Delta - 1 \tag{9}$$

2.1.2. Growth

The theoretical considerations, on crystals growth description, are based on three general models, on the type of liquid-crystal interface and the nature of active sites for crystallization (Zarzycki, 1991): normal growth (or continuum growth); growth determined by processes of bi-dimensional nuclei formation and subsequent increase; and coalescence or competitive growth. For briefly, it will be considered just the first, and last mechanism.

2.1.2.1. Normal growth

With the decrease in solution supersaturation during the initial stages of nucleation, the so called normal growth process starts. During this process, the nuclei that reached a critical radius increases in size, while the others are re-dissolved in the matrix (Zarzycki, 1991).

2.1.2.2. Coalescence or competitive growth

When the supersaturation degree of the matrix decreases a lot, i. e., almost all semiconductor material is already incorporated in a nucleus, the stage denominated as coalescence or competitive growth takes place. There is a competition in which the larger nanocrystals grow from the smaller ones. The study of this process is known as Coarsening Theory of Lifshitz-Slyozov, and leads to a size distribution with same name (Zarzycki, 1991 ; Gaponenko, 1998). This distribution has the peculiarity of being asymmetric around its average values, with an sudden cut to the particles with larger size, and a huge dispersion to the smaller ones. In practice, these different stages occur simultaneously in the real growth process, however, it is possible to consider each stage separately for theoretical purposes.

2.2. Diluted magnetic semiconductor nanocrystals formation: magnetic impurities incorporation

A study of Mn incorporation in semiconductors II-VI and IV-VI was performed, assuming that the adsorption of this magnetic ion occurs on surface of the three crystallography faces ((1 1 1), (1 1 0), and (0 0 1)) of these semiconductors. The obtained results, from the density functional theory, show that the binding energy of Mn, on surface of the crystallography faces, is dependent of the crystalline structure of the semiconductor materials. For crystalline structures of *zinc-blend* type (ZnS, CdS, ZnSe, and CdSe), the binding energy is in the range of 2 to 7 eV. While, for the crystalline structures of *rock-salt* type (PbS, and PbSe), this energy is around of 2 eV (Erwin et al., 2005).

The magnetic impurities incorporation in nanocrystals, produces changes in optical, magnetic, and structural properties of these materials (Furdyna, 1988; Bacher et al., 2005). The exchange interactions, between the levels sp of atoms in semiconductor nanocrystals and the level d of Mn^{2+} ions, completely modify these nanocrystals properties. Most of the semiconductor nanocrystals have a diamagnetic phase. However, with the magnetic impurities incorporation, forming a diluted magnetic semiconductor, this material starts to present paramagnetic, ferromagnetic, anti-ferromagnetic, or spin glass phases, even more, they modify the lattice parameter of semiconductor nanocrystals, and Mn-Mn exchange interactions in closer Mn ions also occur.

The electronic configuration on Mn ions introduced in diluted magnetic semiconductor is A-($3d^5$) or A_o ($3d^5$ + h (holes)) (A- is the negatively charged c, and A_o denotes the neutral center). Studies show that there are three types of Mn centers, when it is incorporated to semiconductor materials. The first is formed to the Manganese in Mn^{3+} state, which is found in $3d^4$ electronic configuration with the spin in ground state S = 2, considered as a neutral accepter center A_o ($3d^4$). The second type of center occurs when the Manganese, in Mn^{3+} state, imprisons an electron and strongly bind it to the d layer, where it starts to show an electronic configuration $3d^5$ with S = 5/2, denoted by A-($3d^5$). This second type of Mn center become negatively charged, been able to attract and weakly bind a hole, forming a third center , denoted by A_0 ($3d^5$ + h) (Sapega et al., 2002).

2.3. Synthesis of DMS NCs in a glass matrix

In this section, we describe the main synthesis of DMS NCs in glasses, methods which have been being developed in recent years, by our research group. From the adequate composition, the masses of powder compounds, that will form the glass matrix as well as the DMS NCs, are measured, mixed, and homogenized.

The first step of sample preparation consisted of melting powder mixtures in an alumina crucible at high temperature for a determined time. In the sequence, a quick cooling to room temperature was undergone to the crucible containing the melted mixture. In the second step, was carried out a thermal annealing of the previously melted glass matrix at specific temperature for several hours aiming to enhance the diffusion of precursor ions into the host matrix.

The DMS NCs, which were synthesized by this methodology, growth kinetics can be explained based on the described models in sections 2.1 and 2.2. The optical, magnetic, and structural properties of these DMS NCs will be presented in the following sections.

2.3.1. Synthesis of $Pb_{1-x}Mn_xS$ NCs

$Pb_{1-x}Mn_xS$ NCs embedded in an oxide glass matrix were synthesized by the fusion method. The synthesis process proceeds as follows. First, the Mn-doped SiO_2–Na_2CO_3–Al_2O_3-PbO_2–B_2O_3+S (wt %) powder was melted in an alumina crucible at 1200°C for 30 min. Then, it was cooled down to room temperature. After that, thermal annealing treatment proceeded at 500°C. Finally, spherically shaped $Pb_{1-x}Mn_xS$ NCs were formed in the glass matrix. In order to study the effects of the synthetic process on the magnetic properties of DMS NCs, four $Pb_{1-x}Mn_xS$ samples with x-concentration varying from 0% until 40% denominated as SNABP: $Pb_{1-x}Mn_xS$, have been synthesized under different thermal treatments, with annealing times of 2, 4, 8, and 10 h, respectively (Silva et al., 2007).

2.3.2. Synthesis of $Pb_{1-x}Mn_xSe$ NCs

The semimagnetic $Pb_{1-x}Mn_xSe$ samples were synthesized in glass matrix SNABP with nominal compositions $40SiO_2 \cdot 30Na_2CO_3 \cdot 1Al_2O_3 \cdot 25B_2O_3 \cdot 4PbO$ (mol %) adding 2Se (wt %), in which the incorporation of Mn^{2+} ions varies between $0 < x < 5\%$. The preparation process consisted of melting the powder mixtures in an alumina crucible at 1200°C for 30 min. In the sequence, the crucible containing the melted mixture underwent quick cooling to room temperature. In a second step, thermal annealing of this melted and cooled glass matrix was carried out at 500°C for several hours, in order to enhance the diffusion of Pb^{2+}, Mn^{2+}, and Se^{2-} species into the host matrix, or produce a rearrangement of ions entering the formation of the NCs. This annealing process produces good quality $Pb_{1-x}Mn_xSe$ NCs showing small size distribution of dots. The same process holds for undoped PbSe quantum dot formation. Two types of Mn-doped samples were grown: (i) SNABP matrix only doped with x-content of Mn, and labeled SNABP: xMn; (ii) SNABP templates, labeled SNABP: $Pb_{1-x}Mn_xSe$ containing $Pb_{1-x}Mn_xSe$ NCs with the percentage of manganese-to-lead in the range $0 < x < 5\%$ (Dantas et al., 2009).

2.3.3. Synthesis of $Cd_{1-x}Mn_xS$ NCs

$Cd_{1-x}Mn_xS$ NCs were synthesized in a glass matrix SNAB with a nominal composition of $40SiO_2 \cdot 30Na_2CO_3 \cdot 1Al_2O_3 \cdot 29B_2O_3$ (mol %) adding 2(CdO+S) (wt %), and Mn-doping concentration (x) varying with respect to Cd-content from 0% to 10%. The first step of sample preparation consisted of melting powder mixtures in an alumina crucible at 1200°C for 30 min. Then, the crucible containing the melted mixture underwent quick cooling to room temperature. In the second step, thermal annealing of the previously melted glass matrix was carried out at 560°C for 10 hours in order to enhance the diffusion of Cd^{2+}, Mn^{2+}, and S^{2-} species into the host matrix. As a result of the thermal annealing, $Cd_{1-x}Mn_xS$ NCs were formed in the glass template. There were two classes of samples for different Mn concentrations: (i) doped with Mn and denominated SNAB: xMn and (ii) NC samples $Cd_{1-x}Mn_xS$ NCs were denominated SNAB: $Cd_{1-x}Mn_xS$ (Dantas et al., 2008).

3. Optical Properties of DMS NCs

The optical properties with the Mn^{2+} ions incorporation in $Pb_{1-x}Mn_xS$, $Pb_{1-x}Mn_xSe$, and $Cd_{1-x}Mn_xS$ NCs, which were grown in glass matrixes through the methodologies described in section 2.3, were investigated by Optical Absorption (OA), and/or Photoluminescence spectroscopy techniques. The obtained results will be presented and discussed as follows.

3.1. Optical Absorption (OA) of $Pb_{1-x}Mn_xS$ NCs

The optical absorption spectra of samples SNABP: $Pb_{1-x}Mn_xS$ (x = 0, 0.003, 0.005, and 0.010) are shown in Fig. 3. Strong blue shift with respect to the optical absorption of bulk PbS (band gap at 0.28 eV) is clearly observed (see Fig. 3) in all SNABP: $Pb_{1-x}Mn_xS$ samples, indicating the quantum confinement effect of carriers within the as-produced PbS NCs.

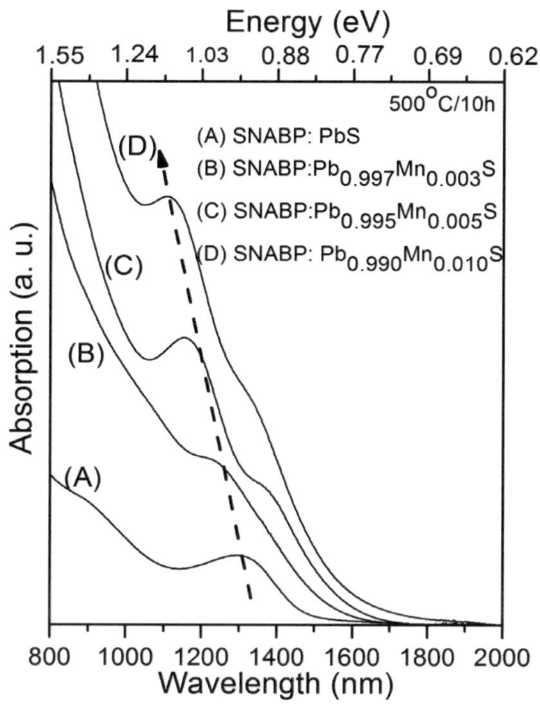

Fig. 3. Room temperature optical absorption spectra of SNABP: $Pb_{1-x}Mn_xS$ samples. (Silva, 2008)

Furthermore, with the introduction of the magnetic impurity (Mn^{2+}) in the PbS NC lattice the optical properties are completely modified due the exchange interaction (sp-d) between the electronic subsystem of the PbS NC and electrons originated from the partially-filled Mn^{2+} ions (Lee et al., 2005). This exchange interaction, scaling with the x-content, responds for the relative blue shift of the effective band gap observed in the OA spectra of samples SNABP: $Pb_{1-x}Mn_xS$, as shown in Fig. 3. More specifically, effective band gaps of 0.95 eV (1307 nm), 1.00 eV (1238 nm), 1.07 eV (1158 nm) and 1.12 eV (1111 nm) were observed for x-contents of 0, 0.003, 0.005 and 0.010, respectively.

3.2. Optical Absorption (OA) of $Pb_{1-x}Mn_xSe$ NCs

Optical absorption spectra recorded for different samples have provide strong evidences of Mn^{2+} ion incorporation into the PbSe NCs samples, labeled as SNABP: $Pb_{1-x}Mn_xSe$ ($x>0$). Due to the exchange interaction (**sp-d** hybridization) between electronic subsystems, the incorporation of magnetic ion in NCs modify the confined electronic states and thus, the optical properties of the quantum dots (Furdyna, 1988; Ohno, 1998). This exchange interactions causes the blue shift of optical resonance, proportional to the Mn-concentration x, between $Pb_{1-x}Mn_xSe$ and PbSe NCs.

Figure 4 shows this effect for SNABP: $Pb_{1-x}Mn_xSe$ samples embedded with $Pb_{1-x}Mn_xSe$ with $x = 0, 0.005, 0.01$ and 0.05%. The observed NC blue shift changes from 0.84 eV (1476 nm) for $x=0$ to 0.89 eV (1398 nm) for $x = 0.05\%$. Effects associated to the spatial confinement can be estimate since bulk semiconductor lead-salt $Pb_{1-x}Mn_xSe$ samples have rock salt crystal structure with a direct band-gap, at the **L**-point of the Brillouin zone, with a value ranging between 0.28 eV, for bulk PbSe, and 3.4 eV, for the bulk MnSe which displays hexagonal structure. The appearance of well defined subband peaks in the absorption spectra demonstrates a relatively small size distribution and good quality of these SNABP: $Pb_{1-x}Mn_xSe$ samples synthesized by fusion method. The absorption peak observed near 570 nm, for SNABP: xMn sample, is attributed to the presence of 0.05% of Mn^{2+} ion.

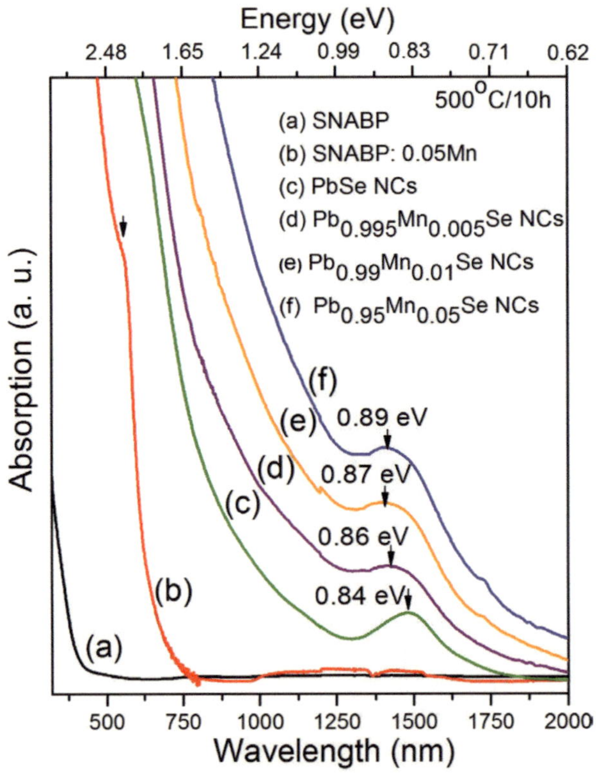

Fig. 4. Room temperature optical absorption spectrum as a function of the wavelength: for a (a) glass matrix; (b) for glass matrix doped with $x = 0.05\%$ Mn (SNABP: xMn) and $Pb_{1-x}Mn_xSe$ NCs (SNABP: $Pb_{1-x}Mn_xSe$) for Mn-concentration: (c) $x = 0$; (d) $x = 0.005\%$; (e) $x = 0.01\%$; and (f) $x = 0.05\%$. (Dantas et al., 2009)

Atomic force microscopy (AFM) images of these samples (shown in section 5.2) confirm the same average size for both the PbSe NCs and $Pb_{1-x}Mn_xSe$ NCs. Then, this blue shift in the OA spectra (Fig. 4), between PbSe and $Pb_{0.95}Mn_{0.05}Se$ NCs, was associated to the incorporation of Mn^{2+} ions in the PbSe dot structure.

3.3. Optical Absorption (OA) of $Cd_{1-x}Mn_xS$ NCs

The OA spectra of the $Cd_{1-x}Mn_xS$ NC samples, that were synthesized as described in section 2.3.3, were obtained using a spectrophotometer Varian-500 operating between 175–3300 nm. OA spectra provided other evidence of Mn^{2+} ion incorporation in the SNAB: $Cd_{1-x}Mn_xS$ and SNAB: xMn samples.(Wang et al., 2004; Levyayb et al., 1998) The introduction of magnetic impurities in semiconductors modified NCs' optical properties as a result of *sp-d* exchange interactions between electrons confined in dot states and located in the partially filled Mn^{2+} states, causing a blue shift with increasing x in the band gap of $Cd_{1-x}Mn_xS$ samples compared to undoped CdS NCs.

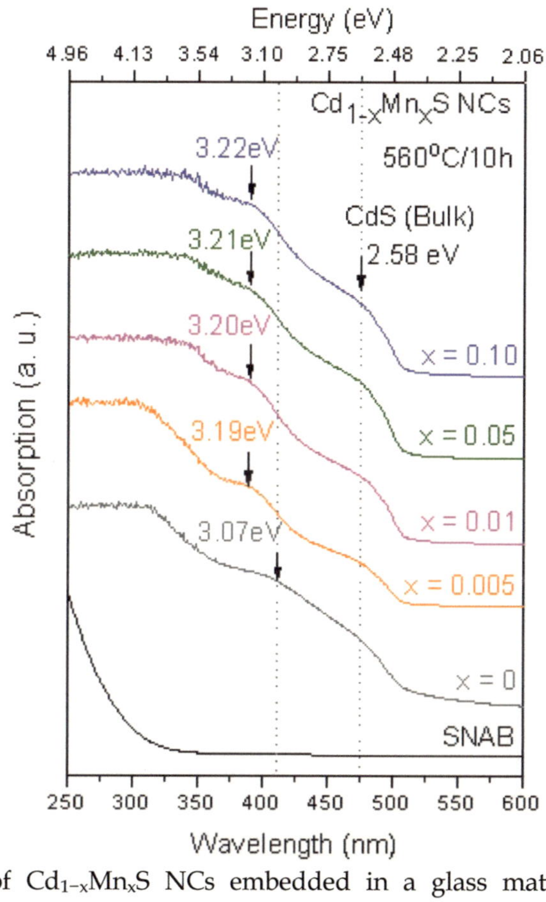

Fig. 5. OA spectra of $Cd_{1-x}Mn_xS$ NCs embedded in a glass matrix for samples with concentrations: x = 0, 0.005, 0.01, 0.05, and 0.10. The blue shift is marked by lines and arrows. (Dantas et al., 2008)

Figure 5 shows this effect in embedded $Cd_{1-x}Mn_xS$ NCs samples, for x=0, 0.005, 0.01, 0.05, and 0.10. Note the blue shift of band gap varying from 3.07 eV (403 nm) to 3.22 eV (385 nm). The band gap of $Cd_{1-x}Mn_xS$ semiconductor varies between 2.58 eV (CdS bulk) and 3.5 eV (MnS bulk). The appearance of well defined subband peaks in the absorption spectrum demonstrates the high quality of the synthesized samples and the relatively small distribution of the NCs.

Using a simple confinement model based on effective-mass approximation (Brus, 1984), the energy of the lowest exciton state in the microcrystallites of radius R smaller than the exciton Bohr radius a_B can be estimated by the following expression: $E_{conf} = E_g + \left(h^2\pi^2/2\mu R^2\right) - 1.8\left(e^2/\varepsilon R\right)$, where E_g is the energy gap of material (bulk), μ the reduced effective mass; e the elementary charge, ε the dielectric constant, and the estimated average radius for CdS NCs was R ~ 2.2 nm.
As all samples were subjected to the same thermal annealing at 560°C for 10 hours, is expected that the $Cd_{1-x}Mn_xS$ NCs have the same size as the corresponding CdS NCs, in agreement with AFM data which will be shown in section 5.3. Therefore, the OA resonance blue shift that was observed in Fig. 5, which increases with increasing Mn^{2+} concentration, occurred due to the incorporation of Mn^{2+} ions into CdS NCs.

3.4. Photoluminescence (PL) of $Cd_{1-x}Mn_xS$ NCs

Through the OA spectra, shown in Fig. 5 (section 3.3), we have confirmed that there was the formation of $Cd_{1-x}Mn_xS$ NCs with quantum confinement properties as well as bulk-like properties. The formation of these NCs can also occur, even in the quick cooling, for unannealed samples, which we shall call of $Cd_{1-x}MnS$ NCs as-grown samples. The room temperature PL spectra of the $Cd_{1-x}Mn_xS$ NCs as-grown samples were acquired with a spectrofluorometer (NanoLog - Horiba JY).
Figure 6(a) shows the PL spectra at room temperature of $Cd_{1-x}Mn_xS$ NCs as-grown samples with different nominal x-concentrations: 0, 0.005, 0.050, and 0.100. The samples were excited at the absorption band edge of the NCs with bulk-like properties (510 nm), which implies that the electrons are occupying the bottom of conduction band (CB).
Our results clearly indicate that nonradiative decay paths are present, alongside the radiative recombination of the electron-hole pairs (E_{e-h}), which implies that the energy levels Mn^{2+} ($^4T_1$), trap (1), and trap (2) are occupied by electrons. The recombination aspects are well described in the diagram depicted in Fig. 6(b) which shows the emission I between the levels $^4T_1$ and $^6A_1$, characteristic of the d orbital of Mn^{2+} ion when it is substitutionally incorporated in semiconductors II-VI (Zhou et al., 2006; Beaulac et al., 2008; Archer et al., 2007). In these materials the Mn^{2+} ions can be substitutionally incorporated in two distinct sites: one in the NC-core (labeled as S_I), and other near the NC-surface (labeled as S_{II}).
The traps (1) and (2) are deep defect levels attributed to the CdS (Smyntyna et al., 2007), whose origins are not quite clear until the moment. Despite of its not so clear origin, we assume that, these two defect levels are possibility related to the V_{Cd}-V_S divacancy centers with different orientations, in analogy to the studies performed for CdSe NCs with hexagonal wurtzite structures (Babentsov et al., 2005), a material that presents great similarities with ours. It was shown that there are two energetically different divacancies: an oriented along the hexagonal c-axis (assigned to trap (1)), and other oriented along the basal Cd-Se bond directions (assigned to trap (2)). This divacancy model could also be used to explain the origins of the two deep traps (1) and (2) in CdS NCs with hexagonal wurtzite structures as well as DMS $Cd_{1-x}Mn_xS$ NCs. These considerations are quite reasonable since the wurtzite structure is a common phase for CdS NCs grown in a glass matrix (Cheng et al., 2006; Xue et al., 2009) as well as for $Cd_{1-x}Mn_xS$ NCs ($0 < x \leq 0.500$) (Jain, 1991).

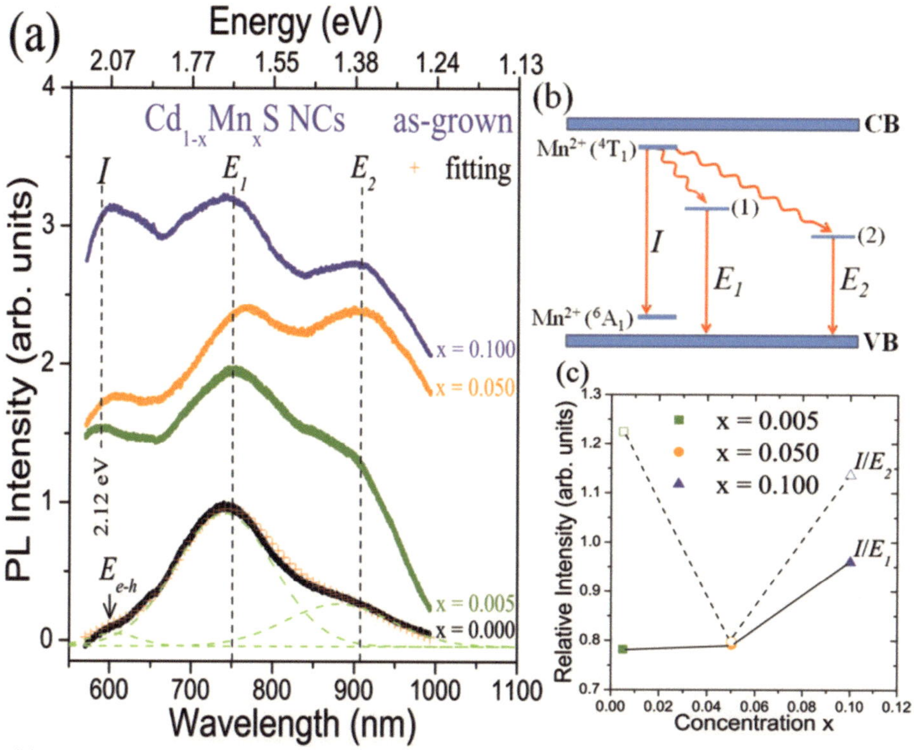

Fig. 6. (a) Room temperature PL spectra of as-grown $Cd_{1-x}Mn_xS$ NCs embedded in the glass matrix SNAB for samples with x = 0, 0.005, 0.050, and 0.100. (b) Schematic diagram showing the radiative recombinations *I*, E_1, and E_2 by the straightened arrows. The nonradiative transitions from the level $^4T_1$ to trap-levels (1) and (2) are represented by the wavy arrows. (c) Relative intensities between the emissions *I* and E_1 (solid line), and the emissions *I* and E_2 (dashed line).

From these two traps, there are the emissions E_1 and E_2 which can be observed in Fig. 6(a) for all samples. The PL spectra of CdS NCs (x = 0.000) were fitted using three like-gaussian components associated to these emissions (E_1 and E_2) as well as the radiative recombination of the electron-hole pairs ($E_{e\text{-}h}$). It is clear that the emissions E_1 and E_2 are more intense than the emission $E_{e\text{-}h}$, showing that the nonradiative processes are dominant from the CB bottom to trap-levels (1) and (2). Figure 6(a) also shows that an overlap, between the emissions $E_{e\text{-}h}$ and *I* for the $Cd_{1-x}Mn_xS$ NCs samples with x = 0.005, 0.050, and 0.100, takes place. Undoubtedly, this overlapped band is more intense than the band of the emission $E_{e\text{-}h}$ observed for CdS NCs samples, which indicates that majority contribution of the overlapped band can be attributed to emission of M^{2+} ion ($^4T_1$ - $^6A_1$).
Related to *I* emission was observed that for $Cd_{1-x}Mn_xS$ NCs grown in a colloidal solution, it is suppressed due to nonradiative processes when the Mn^{2+} ions are incorporated in the site S_{II} (Zhou et al., 2006). Thus, the strong emission *I*, observed in the PL spectra of the $Cd_{1-x}Mn_xS$ NCs samples (see Fig. 6(a)), gives evidence that the Mn^{2+} ions are substitutionally located in the core of nanoparticles, i.e., in the site S_I. Figure 6(c) shows the behavior of relative intensity between the emissions *I* and E_1 (I/E_1), and the emissions *I* and E_2 (I/E_2), for different x-concentration. The enhancement in relative intensity I/E_1 shows that the

increasing in x-concentration favors the incorporation of Mn^{2+} ions in the NC-core (S_I). On the other hand, the I/E_2 relative intensity curve presents a completely different shape, where a decrease in the I/E_2 is followed by an increase with the augment of x-concentration. Such kind of behavior can be understood taking into account nonradiative processes, in order that electron transfers occur from the level $^4T_1$ to the trap-levels (1) and (2), as schematically depicted in the Fig. 6(b) by the wavy arrows. The values I/E_2 are larger than I/E_1 for x = 0.005 and 0.100, indicating that, there is a smaller energy transfers (via electrons) to trap (2) than to trap (1) for these two x-concentrations. While for x = 0.050, the energy transfers from the level $^4T_1$ to the trap-levels (1) and (2) are practically equals. These results indicate that possibly there is a competition between the electron transfers for the trap-levels dependent of the x-concentration.

4. Magnetic properties of DMS NCs

The physical properties of semiconductor nanocrystals change with the incorporation of magnetic impurities, making them diluted magnetic semiconductors. This phenomenon occurs due to the sp-d exchange interaction, which involves Mn ions and electrons in conduction band or holes in the valence band, when there are Mn-doped semiconductors II-VI or IV-VI. This exchange interaction causes changes in the magnetic properties of the nanocrystals, influencing the sp-d exchange interactions of these materials.

4.1. Electron Paramagnetic Resonance (EPR)

A study of the Manganese's magnetic properties in host matrixes can be performed from the EPR technique. The studies by this technique, to the obtained results, confirm that the oxidation state of Mn is 2+, in which the Mn^{2+} ions belong to the group 3d. The free ions of this group present, in their ground configuration, the 3d incomplete layer, which is responsible for the paramagnetism. In the presence of a crystalline lattice, the Mn^{2+} ions start to have the splitted by the crystalline field. This splitting produces a reduction in the orbital movement contribution of the magnetic moment, been the magnetism from these ions fundamentally attributed to the electronic spin (Silva, 2008).
The crystal structure of the hosting semiconductor has a strong influence into the incorporated Mn^{2+}-ions, as observed in the EPR spectrum recorded from the glass sample embedded with $Pb_{1-x}Mn_xS$, $Pb_{1-x}Mn_xSe$, and $Cd_{1-x}Mn_xS$ NCs, favouring a strong electron spin-nucleus spin interaction. The experimental spectra of Mn-doped in NC samples can be modelled by the spin Hamiltonian (Dantas et al., 2008):

$$\hat{H} = \mu_e \hat{S} \cdot g_e \cdot \vec{B} + D\left[S_z^2 - S(S+1)/3\right] + E\left(S_x^2 - S_y^2\right) + A\hat{S} \cdot \hat{I}\,, \tag{10}$$

where the first term is represent the Zeeman interaction with μ_e, g_e, and $\vec{B}$ being the Bohr magneton, the Lande factor and the applied magnetic field, respectively., in which the second and third two terms describe the zero-magnetic field fine-structure splitting due to spin-spin interaction of electrons, which is nonzero only in environments with symmetries lower than cubic, and the fourth term ($A\hat{S} \cdot \hat{I}$) is stemmed from the hyperfine interaction between electron and nuclear spins. In a magnetic field the spin degeneracy of Mn^{2+}-ions will be lifted by the Zeeman splitting, resulting in six energy levels classified by magnetic electron spin quantum number M_S. Due to hyperfine splitting, each of these transitions will be split into six hyperfine levels characterized by the magnetic nuclear spin quantum number M_I. The main hyperfine lines in the spectra are due to allowed $M_S = \pm 1$ transitions with $\Delta M_I = 0$, whereas other lines from forbidden transitions (due to breakdown of selection rule) with nonzero ΔM_I may also be observed. Hence, the typical EPR of Mn^{2+} with electron ($S = 5/2$) and nuclear ($I = 5/2$) spins is composed of 30 lines - five fine structure transitions, each splitting into six hyperfine lines. The incorporation of metal-transition in NC sites cause changes in coordination states modifies the crystal field. These changes are analyses by EPR spectra. Typical EPR of Mn^{2+} with electron ($S = 5/2$) and nuclear ($I = 5/2$) spins is composed of 30 lines - five fine structure transitions, each splitting into six hyperfine lines. For the Mn-doped in PbS, PbSe and CdS NCs the hyperfine lines were only observed for the central $M_S = 1/2 \leftrightarrow -1/2$ transition. The location of Mn^{2+} is presented in two different sites (surface and core) in NCs can be determinate by EPR measurements. In addition, the interaction constants A, D, and E depend strongly on the characteristics of the crystal field. For instance, when a Mn^{2+} ion is located close to or on the NC surface, a large structural difference between the NC and the glass matrix results in a larger hyperfine constant A and larger D and E values. Hence the EPR spectrum varies when the local structure of Mn^{2+} ion in the NC changes (Silva et al., 2007; Dantas et al., 2008; Dantas et al, 2009).

4.1.1. EPR of $Pb_{1-x}Mn_xS$ NCs

Figure 7 shows the EPR spectra for different Mn-concentrations in $Pb_{1-x}Mn_xS$ NCs, to the samples that were annealed at 500°C for 10 hours. The six hyperfine transitions were observed due to the electron spin-nucleus spin interaction. The lines width comes from the sum of two contributories, $\Delta H = \Delta H_I + \Delta H_D$, where ΔH_I is the width which appears in the spectra cause intermolecular processes, and ΔH_D is the width related to "spin-spin" interaction between the first neighbours of Mn (Hinckley & Morgan, 1965). These "spin-spin" interactions are proportional to r^{-3}, where r is the average distance between the Manganese atoms (Mn-Mn). To bigger distances than 55 Å ($r > 55$ Å) the lines are narrow and exclusively determined by ΔH_I. In this case, the EPR spectra of Mn^{2+} ion are sixtet degenerated spin what assures the presence of six hyperfine interaction lines. On the other hand, to smaller distances than 9 Å ($r < 9$ Å), the six hyperfine interaction lines become wider and due to the hyperfine structure collapse, the spectra appear as a single line as observed for samples with Mn-concentration of $x = 0.40$. The arrow-indicated structures which appear around 345 mT, are caused by different Mn^{2+} ions localizations in PbS quantum dots (core or surface).

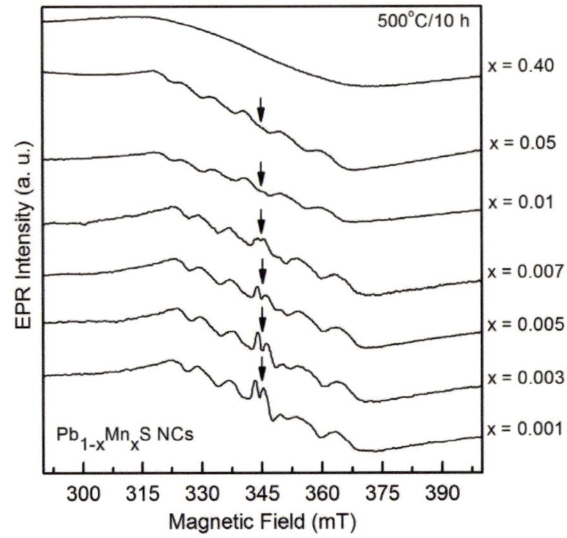

Fig. 7. EPR spectra of $Pb_{1-x}Mn_xS$ NCs for x = 0.001, 0.003, 0.005, 0.007, 0.01, 0.05, and 0.40. (Silva, 2008)

Figure 8(a) shows the EPR spectra for the SNABP: PbS, SNABP: 0.005Mn, and SNABP: $Pb_{0.995}Mn_{0.005}S$ samples, that were annealed at 500°C for 10 hours. It can be perceived that the hyperfine interactions are more clearly observed to SNABP: $Pb_{0.995}Mn_{0.005}S$ samples. These interactions, caused by the Mn^{2+} ion presence in the crystalline field from PbS quantum dots, produce a splitting in this fine structure (electronic transition $+1/2 \leftrightarrow -1/2$) into six hyperfine transitions due to electron spin-nucleus spin interaction.

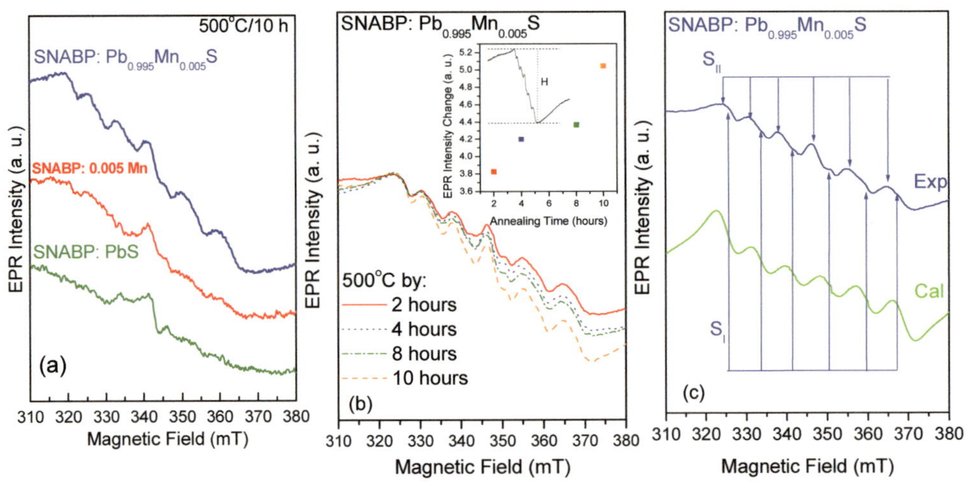

Fig. 8. (a) EPR spectra for the SNABP: PbS, SNABP: 0.005Mn e SNABP: $Pb_{0.995}Mn_{0.005}S$ samples (Silva, 2008). (b) Changing in the EPR intensity signal of the SNABP: $Pb_{0.995}Mn_{0.005}S$ sample that were annealed at 500°C by increasing times [modified from (Silva et al., 2007)]. (c) Experimental and calculated EPR spectra of $Pb_{0.995}Mn_{0.005}S$ NCs [modified from (Silva et al., 2007)].

The EPR spectra intensity modifies with the increasing in the annealing time at 500°C, as shown in Fig. 8(b). This increasing in intensity of EPR signal is caused by a greater incorporation of Mn^{2+} ions, in PbS NCs which increase their density in the glass matrix, as observed by Atomic Force Microscopy images in Fig. 12.
For $Pb_{1-x}Mn_xS$ NCs that were grown in the glass matrix SNABP, the EPR spectra are caused by the contributions of Mn^{2+} ions incorporated in the core (Signal S_I), and/or on the surface (Signal S_{II}) of the PbS NCs. The signals S_I and S_{II} simulations was performed using the WINEPR and SINFONIA Brucker's softwares, being the resulting simulated signal compared with the obtained EPR spectrum to the SNABP: $Pb_{0,995}Mn_{0,005}S$ sample and shown in Fig. 8(c). It was observed the signal S_{II} predominance over the signal S_I, giving strong evidences of a higher Mn^{2+}-concentration near to the NC-surface. The average values obtained to hyperfine constants (A) are of A_{SI} = 8.20 mT and A_{SII} = 9.37 mT, with the electron g-factor equal to 2.005 (Silva et al., 2007). The different hyperfine constant values between the $Pb_{1-x}Mn_xS$ quantum dots and the respective bulk material, are attributed to quantum confinement of the electrons, promoting a higher interaction between the electron spin and the Mn^{2+} ion nucleus (Ji et al., 2003).

4.1.2. EPR of $Pb_{1-x}Mn_xSe$ NCs

A high sensitivity Bruker ESP-300 spectrometer, operating in the *X*-band microwave frequency (9.5 GHz), with swept static field and the usual modulation and phase sensitive detection techniques, was used to record the EPR spectra of the $Pb_{1-x}Mn_xSe$ NCs samples that were synthesized by methods described in section 2.3.

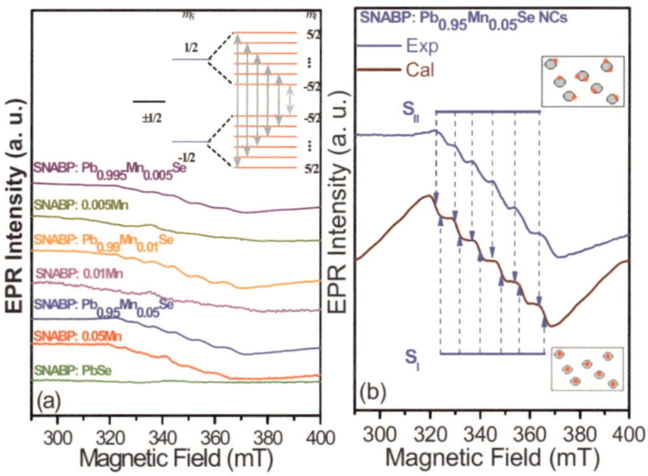

Fig. 9. Left panel: EPR spectra for pure PbSe and for semimagnetic $Pb_{1-x}Mn_xSe$ NCs, with $x = 0.005$, $x = 0.01$, and $x = 0.05$. In each panel, lines show the difference between undoped and doped PbSe samples. Right panel: EPR spectra of $Pb_{0.95}Mn_{0.05}Se$ NCs measured in the *X*-band and at room temperature and the computed EPR spectra obtained by a summation of two spectra with A = 7.3 mT and 7.9 mT, corresponding to Mn^{2+} sites inside labeled as S_I and on the surfaces labeled as S_{II} of NCs, for a system with $S = 5/2$, $I = 5/2$, D = 30 mT, E = 4 mT, and g = 2.005. (Dantas et al., 2009)

Figure 9(a) shows EPR spectra of a set of $Pb_{1-x}Mn_xSe$ NCs. The broad background line, in all spectra, indicates the magnetic interaction between Mn^{2+} ions inside the $Pb_{1-x}Mn_xSe$ structure.

The presence of the six hyperfine lines, as shown in the inset, confirms the uniform incorporation of Mn^{2+} ions into the dot. Figure 9(b) shows fairly good agreement between experimental (theoretical) spectrum of $Pb_{0.95}Mn_{0.05}Se$ NCs, measured in the *X*-band and at room temperature and shown as solid blue (dotted green) line. This EPR simulation is composed of two spectra: i) A broad line that uses a fine structure constant, $A = 7.3$ mT corresponding to Mn^{2+} ions inside the NC, ii) a stronger fine structure constant $A = 7.9$ mT corresponding to ions located close to the surface. The hyperfine parameters are $D = 30$ mT, $E = 4$ mT, and the Mn parameters are $S_{Mn} = 5/2$, $I = 5/2$, and $g_{Mn} = 2.005$.

Analysis of the EPR spectra (shown in Fig. 9) provided further evidence for the presence or absence of doped PbSe dots. The incorporation of Mn^{2+} ions in $Pb_{1-x}Mn_xSe$ NCs is confirmed by the presence of the six central hyperfine lines in the EPR spectrum. These hyperfine lines are not present in glass samples only doped with Mn^{2+} ions.

4.1.3. EPR of $Cd_{1-x}Mn_xS$ NCs

The room-temperature modifications in the electronic states induced by Mn^{2+} ion incorporation into CdS NCs (forming $Cd_{1-x}Mn_xS$ NCs) were examined by EPR using a high sensitivity Bruker ESP-300 spectrometer operating in the *X*-band microwave frequency (9.5 GHz).

Figure 10(a) shows EPR spectra for the synthesized sample set. The broader background signal observed in the EPR spectra is due to the spin-spin interaction between Mn^{2+} electrons when they are incorporated into the $Cd_{1-x}Mn_xS$ NCs.

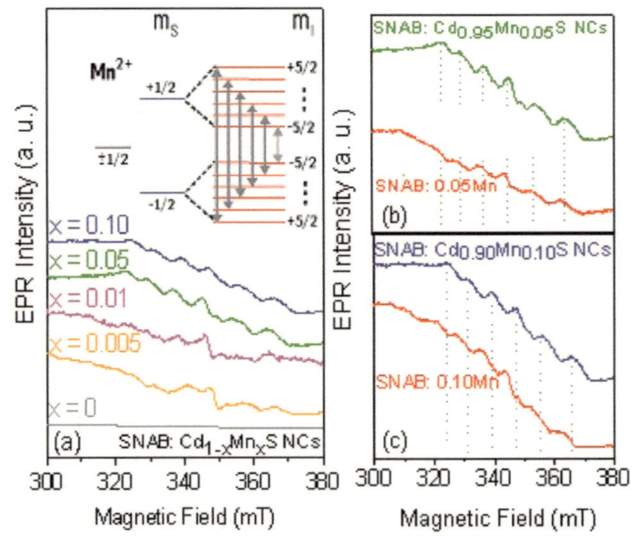

Fig. 10. Panel (a) room temperature EPR spectra of $Cd_{1-x}Mn_xS$ NCs for samples with concentration: $x = 0, 0.005, 0.01, 0.05$, and 0.10. Panels (b) and (c) EPR spectra of glass matrix doped with *x*Mn and $Cd_{1-x}Mn_xS$ NCs for x=0.05 and 0.10, respectively. (Dantas et al., 2008)

The six line structure confirms the uniform incorporation of Mn^{2+} ions into the host CdS NC structure, whereas the broader EPR line indicates the presence of magnetic exchange interaction when two Mn^{2+} ions get close enough and are found in small agglomerate islands. The difference between the glass samples (SNAB: xMn and SNAB: $Cd_{1-x}Mn_xS$ NCs) is observed in the EPR spectra [Figs. 10(b) and 10(c)]. Note that the hyperfine interaction due to Mn^{2+} ions results from the presence of a crystalline field in CdS NCs. The hyperfine interaction observed in the spectrum of glass samples doped with Mn^{2+} is not as evident as those observed for $Cd_{1-x}Mn_xS$ NCs. This difference could possibly be attributed to the formation of small islands of crystalline phase MnO. (Mukherjee et al., 2006)

It has already been reported that Mn^{2+} ions are incorporated into two distinct sites of NCs; when found at the dot core, it produces the EPR signal S_I, when located on or near the NCs surface it produces the EPR signal S_{II}. (Silva et al., 2007; Zhou et al., 2006) This analysis is strongly supported by good agreement between the experimental EPR spectrum and the simulated one. In Fig. 11, the EPR spectrum of $Cd_{0.95}Mn_{0.05}S$ NCs, measured in the X-band and at room temperature, is shown as a solid blue line and the calculated one as a solid green line.

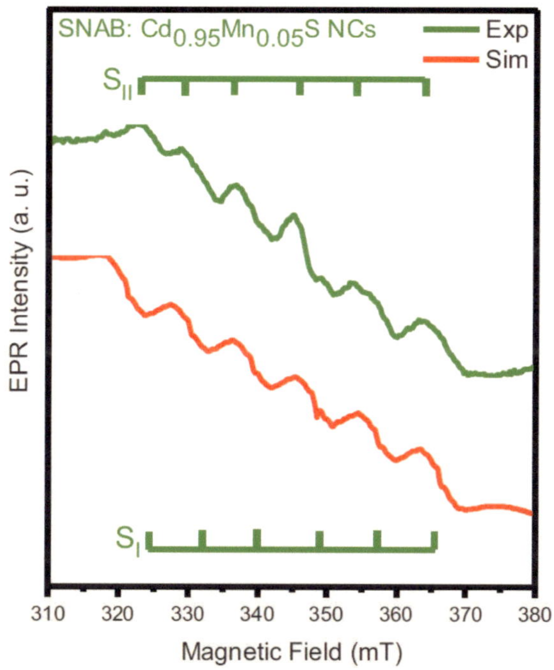

Fig. 11. Experimental and simulated EPR spectra of $Cd_{0.95}Mn_{0.05}S$ NCs. The simulation uses superposition of two spectra with A = 7.6 mT and 8.2 mT, corresponding to Mn^{2+} ions located inside (labeled as S_I) and on the surfaces (labeled as S_{II}) of NCs for a system with parameters: $S = 5/2$, $I = 5/2$, D = 40 mT, E = 5 mT, and g_e = 2.005. (Dantas et al., 2008)

The EPR simulation was obtained by a sum of two spectra with A = 7.6 and 8.2 mT, corresponding to Mn^{2+} ions located inside the NC (labeled as S_I) and near the NC surface (labeled as S_{II}) for a dot system with parameters: $S = 5/2$, $I = 5/2$, D = 40 mT, E = 5 mT, and g_e = 2.005. On the other hand, as the annealing time increases, the probability of finding

magnetic ions inside NCs occupying neighboring lattice sites as well as the number of antiferromagnetic spin correlated clusters increases. This combination of effects enhances the dipolar interaction and increases the lattice distortions on the Mn^{2+} sites. Furthermore, the accumulation of Mn^{2+} ions on the NCs surfaces also strengthens Mn–Mn interactions. (Silva et al., 2007; Zhou et al., 2006; Jian et al., 2003) As a consequence, the intensity of the broader background peak is increased.
EPR spectra provided evidence that Mn^{2+} ions are incorporated at two distinct sites: at the core and/or at the surface of CdS NCs. Influences of CdS NCs crystalline field and the presence of MnO cluster phases could be confirmed by the presence of six hyperfine lines assigned to the Mn^{2+} ions in the samples.

5. Structural properties of DMS NCs

The structural properties of the $Pb_{1-x}Mn_xS$, $Pb_{1-x}Mn_xSe$, and $Cd_{1-x}Mn_xS$ nanocrystals that were grown in glass matrixes, through the methodologies described in section 2.3, their spatial distribution, size homogeneity, and the nanocrystal shapes were analyzed by Atomic Force Microscopy using the contact mode. It was also employed the X-Ray Diffraction technique to investigate the crystalline structure of the $Pb_{1-x}Mn_xSe$ NCs samples.

5.1. Atomic Force Microscopy (AFM) of $Pb_{1-x}Mn_xS$ NCs

In the Fig. 12 are shown atomic force microscopy (AFM) images that were obtained for the SNABP: $Pb_{0.995}Mn_{0.005}S$ NCs samples, which were annealed at 500°C for 2, 4, 8, and 10 hours. It can be observed that with the increase in annealing time, the $Pb_{0.995}Mn_{0.005}S$ NCs density increases as well as their average size. The obtained values were 4.2 nm (2 hours), 4.3 nm (4 hours), 4.4 nm (8 hours), and 4.8 mn (10 hours).

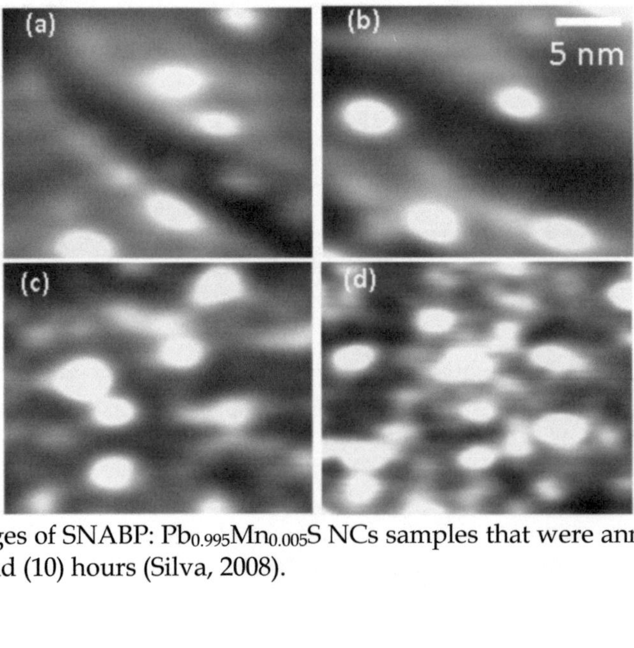

Fig. 12. AFM images of SNABP: $Pb_{0.995}Mn_{0.005}S$ NCs samples that were annealed at 500°C for (a) 2, (b) 4, (c) 8 and (10) hours (Silva, 2008).

The increase in the $Pb_{0.995}Mn_{0.005}S$ NCs density within host glass matrix results in the intensity increase of optical absorption spectra (see Fig. 3), as well as, the increase in EPR signal intensity (Fig. 8(b)).

5.2. Atomic Force Microscopy (AFM) of $Pb_{1-x}Mn_xSe$ NCs

To confirm the NCs formation on glass templates we have taken AFM images, as shown in Fig. 13, for (a) PbSe NCs and (b) $Pb_{0.95}Mn_{0.05}Se$ NCs samples. The 3D and the 2D (inset) morphologies of $Pb_{1-x}Mn_xSe$ NCs can be noted. As shown in Figures 13(a) and 13(b), is possible to estimate the average size for PbSe and $Pb_{1-x}Mn_xSe$ NCs as 5.2 nm, and with 6% the size distribution, from these AFM images.

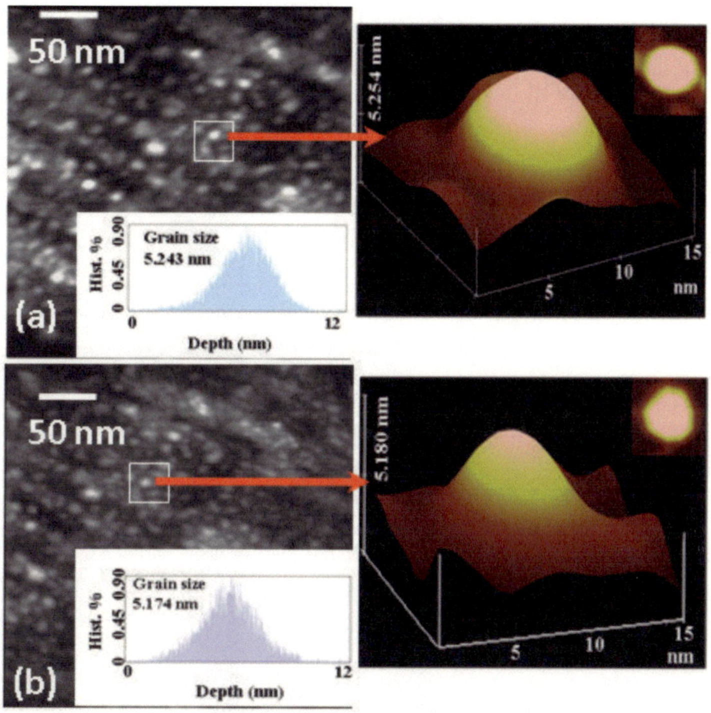

Fig. 13. AFM images of NCs showing size distribution below 6% for: (a) PbSe (b) $Pb_{0.95}Mn_{0.05}Se$. AFM image in 3D e 2D (inset) illustrates the morphology of $Pb_{1-x}Mn_xSe$ NCs, for x = 0 and 0.05. (Dantas et al., 2009)

These AFM data have confirmed that both PbSe NCs and $Pb_{1-x}Mn_xSe$ NCs have the same average size (5.2 nm). Thus, undoubtedly, the observed blue shift in OA spectra of these samples, shown in Fig. 4 (section 3.2), is related to incorporation of Mn^{2+} ions into PbSe NCs.

5.3. Atomic Force Microscopy (AFM) of $Cd_{1-x}Mn_xS$ NCs

Figure 14 shows the AFM images that were recorded for glass samples embedded with CdS NCs [Fig. 14(a)] and $Cd_{0.95}Mn_{0.05}S$ NCs [Fig. 14(b)] in order to confirm both the NCs formation and OA data for size distribution, shown in Fig. 5 (section 3.3).

The average NCs size estimated from these AFM images is at around R ~ 2.3 nm with corresponding size distributions of 5% and 9%, respectively. The AFM morphology of isolated NCs is shown in two-dimension (2D) and three-dimensions (3D) images. Thus, is observed that all samples are single-phase materials with hexagonal wurtzite structure, which is a very common phase for CdS NCs grown in glass matrix (Cheng et al., 2006; Xue et al., 2009), as well as for $Cd_{1-x}Mn_xS$ NCs ($0 < x \leq 0.500$) (Jain, 1991).

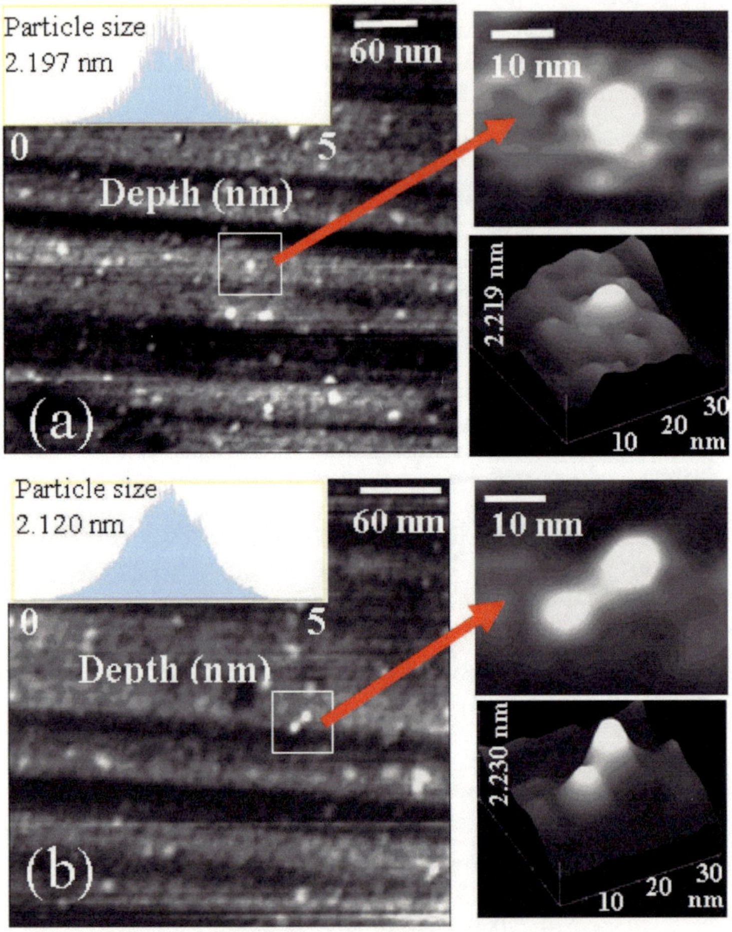

Fig. 14. AFM images showing nearly 5% size distribution and dot size R ~ 2.3 nm. The 2D and 3D images illustrate the morphology of (a) CdS NCs and (b) $Cd_{0.95}Mn_{0.05}S$ NCs. (Dantas et al., 2008)

As it was expected, the AFM data confirm that both CdS NCs and $Cd_{1-x}Mn_xS$ NCs have the same average size. Thus, it was proved that the incorporation of Mn^{2+} ions into CdS NCs provokes a blue shift in OA spectra of nanoparticles, according to the results shown in Fig. 5.

6. X-Ray Diffraction (XRD) of $Pb_{1-x}Mn_xSe$ NCs

The XRD patterns of the SNABP: $Pb_{1-x}Mn_xSe$ samples ($x \geq 0$) for $x = 0$ and $x = 0.05\%$ are shown in Fig. 15(a). It is noted that the typical bulk PbSe rock salt crystal structure is preserved for the $Pb_{1-x}Mn_xSe$ dot samples having Mn-concentration $x \leq 0.05\%$. Nevertheless, the characteristic XRD peaks is shifted towards lower diffraction angle values as the Mn^{2+} incorporation in the hosting PbSe structure increases, as shown in Fig. 15(b), and this is a clear indication that a decrease in the lattice constant is occurring. We have estimate the lattice constant (Cohen Method) of the structure using the (111), (200), and (220) peaks of the XRD spectrum. Here, the average lattice crystal constant found for $Pb_{1-x}Mn_xSe$ samples is 6.130 Å for x = 0% (PbSe) and 6.127 Å for 0.05% incorporation of Mn^{2+} into the NCs. This monotonic decrease observed in the lattice constant can be attributed to the replacement of Pb^{2+}-ions, having larger ionic radius (119 pm) in the rock salt PbSe crystal structure, by Mn^{2+}-ions with smaller ionic radius (83 pm).

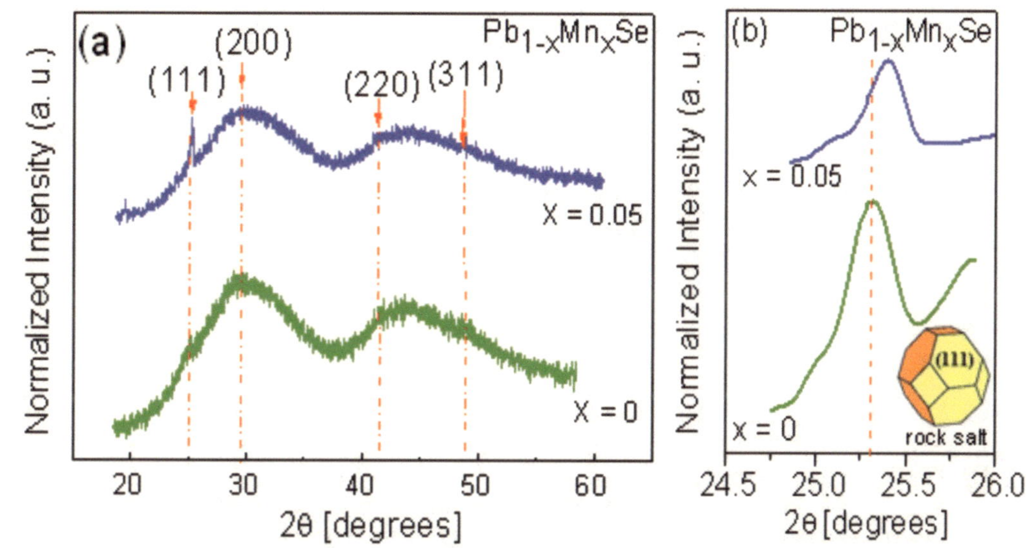

Fig. 15. (a) XRD measurements of PbSe (lower line) and $Pb_{0.95}Mn_{0.05}Se$ (top line) NCs embedded in glass matrix. (b) The effects associated to the incorporation of Mn^{2+} ions into PbSe NCs are seen as an intensity increase (left panel) and a shift to higher 2θ diffraction angle (right panel) of the (111) peak. Probably, the diffusion is enhanced along the (111) direction. (Dantas et al., 2009)

Once the grown samples have very small Mn-concentration, the crystalline structures for undoped and doped NCs remained rock salt and showed similar values for the lattice constants. However, it is expect a decrease in the crystalline quality of samples containing higher concentration of Mn, with the possible occurrence of MnSe clusters inside the $Pb_{1-x}Mn_xSe$ NCs.

7. Conclusions

In this chapter, it was presented the main advances, which have been obtained by us in last few years, related to study of Diluted Magnetic Semiconductor Nanocrystals (DMS NCs) in glass matrixes.

Probably the first time, $Pb_{1-x}Mn_xS$, $Pb_{1-x}Mn_xSe$, and $Cd_{1-x}Mn_xS$ DMS NCs were successfully grown in a glass matrix by the fusion method, when subjected to an adequate thermal annealing.

Our results, obtained by the experimental techniques, showed that it was possible to control the optical, magnetic, and structural properties of these DMS NCs, confirming the high quality of the synthesized samples.

Therefore, we have proved that the use of glass matrixes, as host material for DMS NCs, is an excellent alternative, since they provide great stability to the nanoparticles, with a relatively low cost of synthesis.

We believe that this chapter may be useful for further investigations on the optical, magnetic, and structural properties of Mn-doped nanocrystals.

8. Acknowledgements

The authors gratefully acknowledge the financial support from the Brazilian agencies: MCT/CNPq, Capes, Fapemig and FUNEPU. We are also grateful to our collaborators: Denis Rezende de Jesus (D. R. Jesus), Fernando Pelegrini (F. Pelegrini), Gilmar Eugênio Marques (G. E. Marques), Henry Socrates Lavalle Sullasi (H. S. L. Sullasi), Ilde Guedes da Silva (I. Guedes), Newton Martins Barbosa Neto (N. M Barbosa Neto), Paulo César de Morais (P. C. Morais), Qu Fanyao (Qu Fanyao), and Walter Elias Feria Ayta (W. E. F. Ayta).

9. References

Archer, P. I.; Santangelo, S. A. & Gamelin, D. R. (2007). Direct Observation of sp-d Exchange Interactions in Colloidal Mn^{2+}- and Co^{2+}-Doped CdSe Quantum Dots. *Nano Lett.*, Vol. 7, Issue 4 (March 2007), pag. 1037, 7 pages, ISSN: 1530-6984

Babentsov, V.; Riegler, J. Scheneider, J.; Ehlert, O.; Nann, T. & Fiederle, M. (2005). Deep level defect luminescence in cadmium selenide nano-crystals films. *J. Cryst. Growth*, Vol. 280, Issue 3-4 (July 2005), pag. 502, 7 pages, ISSN: 0022-0248

Bányai, L. & Koch, S. W. (1993). *Semiconductor Quantum Dots*, World Scientific Publishing Co. Pte. Ltd., ISBN: 9810213905, Singapore

Barbosa, L. C.; Reynoso, V. C. S; de Paula, A. M.; de Oliveira, C. R. M.; Alves, O. L.; Craievich, A. F.; Marotti, R. E.; Brito Cruz, C. H. & Cesar, C. L. (1997). CdTe quantum dots by melting heat treatement in borosilacate glasses. *J. Non-Cryst. Sol.*, Vol. 219, Issue 1 (October 1997), pag. 205, 7 pages, ISSN: 0022-3093

Beaulac, R.; Archer, P. I.; van Rijssel, J.; Meijerink, A. & Gamelin, D. R. (2008). Exciton Storage by Mn^{2+} in Colloidal Mn^{2+}-Doped CdSe Quantum Dots. *Nano Lett.*, Vol. 8, Issue 9 (August 2008), pag. 2949, 5 pages, ISSN: 1530-6984

Bacher, G.; Schömig, H.; Scheibner, M.; Forchel, A.; Maksimov, A. A.; Chernenko, A. V.; Dorozhkin, P. S.; Kulakovskii, V. D.; Kennedy, T. & Keinecke T. L. (2005). Spin-spin interactions in magnetic semiconductor quantum dots. *Physica E*, Vol. 26, Issue 1-4 (February 2005), pag. 37, pages, ISSN: 1386-9477

Becker, R. & Döring, W. (2006). Kinetische Behandlung der Keimbildung in übersättigten Dämpfen. *Ann. Phys.*, Vol. 416 (reprinted from Vol. 24 (1935)), Issue 8 (March 2006), pag. 719, 34 pages, ISSN: 0003-4916

Brus, L. E. (1984). Electron-electron and electron-hole interactions in small semiconductor crystallites: The size dependence of the lowest excited electronic state. *J. Chem. Phys.*, Vol. 80, Issue 9 (May 1984), pag. 4403, 7 pages, ISSN: 0021-9606

Cheng, Y; Wang, Y.; Bao, F. & Chen, D. (2006). Shape Control of Monodisperse CdS Nanocrystals: Hexagon and Pyramid. *J. Phys. Chem. B*, Vol. 110, Issue 19 (April 2006), pag. 9448, 4 pages, ISSN: 1089-5647

Christian, J. W. (1965). *The theory of Transformation in Metal and Alloy*, Pergamon Press, ISBN: 0080440193, Oxford

Dalpian, G. M. & Chelikowsky, J. R. (2006). Self-Purification in Semiconductor Nanocrystals. *Phys. Rev. Lett.*, Vol 96, Issue 22 (June 2006), pag. 226802, 4 pages, ISSN: 0031-9007

Dantas, N. O.; Neto, E. S. F.; Silva, R. S.; Jesus, D. R. & Pelegrini, F. (2008). Evidence of $Cd_{1-x}Mn_xS$ nanocrystal growth in a glass matrix by the fusion method. *Appl. Phys. Lett.*, Vol. 93, Issue 19 (November 2008), pag. 193115, 3 pages, ISSN: 0003-6951

Dantas, N. O.; Silva, R. S.; Pelegrini, F. & Marques, G. E. (2009). Morphology in semimagnetic $Pb_{1-x}Mn_xSe$ nanocrystals: Thermal annealing effects. *Appl. Phys. Lett.*, Vol. 94, Issue 26 (June 2009), pag. 263103, 3 pages, ISSN: 0003-6951

Erwin, S. C.; Zu, L.; Haftel, M. I.; Efros, A. L., Kennedy, T. A. & Norris, D. J. (2005). Doping semiconductor nanocrystals. *Nature*, Vol. 436, Issue 7047 (July 2005), pag. 91, 4 pages, ISSN: 0028-0836

Furdyna, J. K. (1988). Diluted magnetic semiconductors. *J. Appl. Phys.*, Vol. 64, Issue 4 (August 1988), pag. R29, 36 pages, ISSN: 0021-8979

Gaponenko, S. V. (1998). *Optical Properties of Semiconductor Nanocrystals*, Cambridge University Press, ISBN: 0521582415, Cambridge

Hinckley, C. C. & Morgan, L. O. (1965). Electron spin resonance linewidths of mangane (II) ions in concentrated aqueous solutions. *J. Chem. Phys.*, Vol. 44, Issue 3 (1965), pag. 898, 8 pages, ISSN: 0021-9606

Jain, M. K. (1991). *Diluted Magnetic Semiconductors*, World Scientific Publishing Co. Pte. Ltd., ISBN: 9810201761, Singapore

Ji, T.; Jian, W.B.; Fang, J. (2003). The first synthesis of $Pb_{1-x}Mn_xSe$ nanocrystals. *J. Am. Chem. Soc.*, Vol. 125, Issue 28 (June 2003), pag. 8448, 2 pages, ISSN: 0002-7863

Jian, W. B.; Fang, J.; Tianhao, J. & He, J. (2003). Quantum-size-effect-enhanced dynamic interactions among doped spins in $Cd_{1-x}Mn_xSe$ nanocrystals. *Appl. Phys. Lett.*, Vol. 83, Issue 16 (October 2003), pag. 3377, 3 pages, ISSN: 0003-6951

Lee, S.; Dobrowolska, M. & Furdyna, J. K. (2005). Effect of spin-dependent Mn^{2+} internal transitions in $CdSe/Zn_{1-x}Mn_xSe$ magnetic semiconductor quantum dot systems *Phys. Rev. B*, Vol. 72, Issue 7 (2005), pag. 075320, 5 pages, ISSN: 1098-0121

Mukherjee, S.; Pal, A. K.; Bhattacharya, S. & Raittila, J. (2006). Magnetism of Mn_2O_3 nanocrystals dispersed in a silica matrix: Size effects and phase transformations. *Phys. Rev. B*, Vol. 74, Issue 10 (September 2006), pag. 104413 , 10 pages, ISSN: 1098-0121

Norris, D. J.; Efros, A. L. & Erwin, S. C. (2008). Doped Nanocrystals. *Science*, Vol. 319, Issue 5871 (March 2008), pag. 1776, 4 pages, ISSN: 0036-8075

Ohno, H. (1998). Making Nonmagnetic Semiconductors Ferromagnetic. *Science*, Vol. 281, Issue 5379 (1998), pag. 951, 6 pages, ISSN: 0036-8075

Sapega, V. F.; Moreno, M.; Ramsteiner, M.; Däweritz, L. & Ploog, K. (2002). Electronic structure of Mn ions in (Ga,Mn)As diluted magnetic semiconductor. *Phys. Rev. B*, Vol. 66, Issue 7 (2002), pag. 075217, 6 pages, ISSN: 1098-0121

Silva, R. S.; Morais, P. C.; Fanyao, Q.; Alcalde, A. M.; Dantas, N. O. & Sullasi, H. S. L. (2007). Synthesis process controlled magnetic properties of $Pb_{1-x}Mn_xS$ nanocrystals. *Appl. Phys. Lett.*, Vol. 90, Issue 25 (June 2007), pag. 253114, 3 pages, ISSN: 0003-6951

Silva, R. S. (2008). *Síntese e Estudo das Propriedades Ópticas e Magnéticas de Pontos Quânticos de* $Pb_{1-x}Mn_xS$ *Crescidos em Matrizes Vítreas*, Tese (Doutorado em Física)-Universidade de Brasília, 98, [21]f., Brasília

Smyntyna, V.; Skobeeva, V. & Malushin, N. (2007). The nature of emission centers in CdS nanocrystals *Radiation Measurements*, Vol. 42, Issue 4-5 (2007), pag. 693, 4 pages, ISSN: 1350-4487

Volmer, M. & Weber, A. (1926). Nucleus formation in Supersaturated Systems. *Z. Phys. Chem.*, Vol. 119, Issue (1926), pag. 277, 25 pages, ISSN: 0942-9352

Woggon, U. (1997). *Optical Properties of Semiconductor Quantum Dots*, Springer - Verlag, ISBN: 3540609067, Berlin

Xue, H. T. & Zhao, P. Q. (2009). Synthesis and magnetic properties from Mn-doped CdS/SiO_2 core-shell nanocrystals. *J. Phys. D: Appl. Phys.*, Vol. 42, Issue 1 (January 2009), pag. 015402, 5 pages, ISSN: 0022-3727

Zarzycki, J. (1991). *Glasses and the vitreous state*, Cambridge University Press, ISBN: 0521355826, Cambridge

Zhou, H.; Hofmann, D. M.; Alves, H. R. & Meyer, B. K. (2006). Correlation of Mn local structure and photoluminescence from CdS:Mn nanoparticles. *J. Appl. Phys.*, Vol. 99, Issue 10 (May 2006), pag. 103502, 4 pages, ISSN: 0021-8979

7

$KTiOPO_4$ single nanocrystal for second-harmonic generation microscopy

Loc Le Xuan, Dominique Chauvat, Abdallah Slablab and Jean-Francois Roch
Ecole Normale Superieure de Cachan
France

Pawel Wnuk and Czeslaw Radzewicz
University of Warsaw
Poland

1. Introduction

Nonlinear second-harmonic generation (SHG) microscopy has become a commonly used technique for investigating interfacial phenomena(Kemnitz et al., 1986; Shen, 1989) and imaging biological samples.(Moreaux et al., 2000) Different non-centrosymmetric nanometric light sources have been recently studied in this context, e.g. organic nanocrystals.(Shen et al., 2001; Treussart et al., 2003) For those systems, resonant optical interaction leads to an enhancement of the nonlinear response but also to parasitic effect that is detrimental for practical applications, namely photobleaching due to two-photon residual absorption.(Patterson et al., 2000) Conversely, inorganic non-centrosymmetric materials with far-off resonance interaction avoid this limitation.(Johnson et al., 2002; Long et al., 2007) Recent achievements have been obtained using $KNbO_3$ nanowires as a tunable source for sub-wavelength optical microscopy(Nakayama et al., 2007) and $Fe(IO_3)_3$ nanocrystallites as promising new SHG-active particles with potential application in biology.(Bonacina et al., 2007) However, either the dimensions of the used crystals are still of the order of the micrometer along one axis,(Nakayama et al., 2007) or the corresponding bulk material is not easily grown,(Bonacina et al., 2007) so that the crystal characteristics are not directly available. A complementary approach consists in considering a well-known SHG-active bulk material and investigating its properties in nanoparticle form. Different materials have been considered, e.g. BaTiO3 (Hsieh et al., 2009) and ZnO (Kachynski et al., 2008). In this view, we were among the pioneers in this domain, considering the well-known KTP material. Potassium titanyl phosphate ($KTiOPO_4$, KTP) is a widely used nonlinear crystal.(Zumsteg et al., 1976) Studies on this material have focused on the optimized growth of large-size single crystals, which have found numerous applications in laser technology for efficient frequency conversion.(Driscoll et al., 1986)
Here we show that KTP particles of nanometric size (nano-KTPs) are an attractive material for SHG microscopy. Under femtosecond excitation and in ambient conditions, a single nano-KTP with a size around 60 nm independently determined with atomic force microscopy (AFM), generates a perfectly stable blinking-free second-harmonic signal which can be easily detected in the photon-counting regime. Furthermore, we demonstrate that this single nanocrystal can

be characterized in situ by retrieving the orientation of its 3-axis with respect to the optical observation axis. This analysis uses both nonlinear polarimetry(Brasselet et al., 2004) and defocused imaging with a model that has been adapted from similar techniques developed for single-molecule fluorescence.(Bohmer & Enderlein, 2003; Brokmann et al., 2005; Sandeau et al., 2007)
Such characterized nanocrystal can have potentially many applications in nano-optics or biology, and downscaling their size is a crucial step forward. From the point of view of detection techniques, we have developed two methods for investigating SHG nanocrystals and decrease the minimum detectable size for a SHG nanocrystal. In particular both techniques exploit the coherent characteristics of the second harmonic emission. First, a coherent balanced homodyne detection is associated to SHG microscopy, allowing us to detect nano-objects with high signal-to-noise ratio and phase sensitivity.(Le Xuan et al., 2006) Second, broadband femtosecond pulses are used to improve the second harmonic signal, with a perspective of spectral manipulation for coherent control of the nonlinear process.(Wnuk et al., 2009)

2. Photostable second harmonic generation from single KTP nanocrystals

The KTP nanoparticles are extracted from the raw powder that remains in the trough at the end of the flux-growth process which leads to the synthesis of large sized KTP single crystals.(Rejmankova et al., 1997) We then apply a size-selection procedure by centrifugation which was previously worked out for optically active nanodiamonds (Treussart et al., 2006) and which leads to a colloidal solution of non-aggregated particles.
Analysis with dynamic light scattering (DLS) of this solution reveals a strong peak in the size distribution centered around 150 nm (Fig.1). Centrifugation at a higher speed or for a longer duration acts as a low-pass filter on the size distribution, reducing the average peak value from 150 nm to 80 nm (30 min. at 5000 rpm), and 60 nm (10 min. at 11000 rpm).

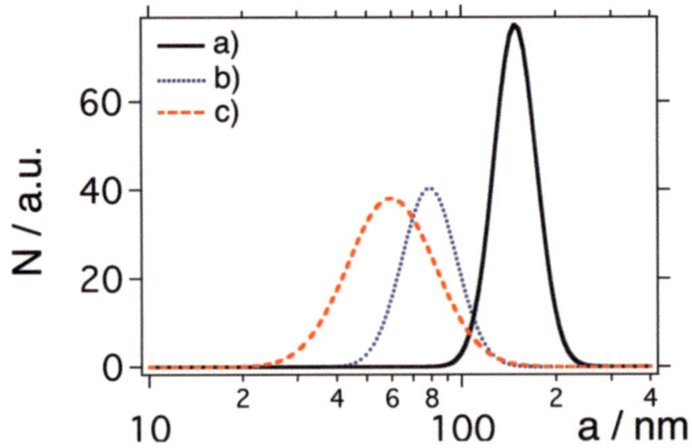

Fig. 1. Size selection of the nanoparticles in the polymer solution. DLS spectra for successive centrifugations (a) 5 min. at 5000 rpm, (b) 30 min. at 5000 rpm, (c) 10 min. at 11000 rpm, and corresponding to an average particle size of (a) 150 nm, (b) 80 nm, (c) 60 nm.

Since we are interested in the study of a well-isolated single nanocrystal, the sample is prepared so as to enable identification of the same single nano-object under AFM, a classical white-light optical microscope, and a home-made scanning SHG microscope (Figure 2). We designed structured samples by first spin-coating the colloidal solution on a glass cover-slip, then etching it through a mask to leave square areas with nano-KTPs directly on glass. Figure 3a shows one such area labeled with photo-written numbers by the focus laser itself at high mean power (100mW at a repetition rate of 86MHz) and imaged using transmission microscopy in white-light illumination. Figure 3b corresponds to a 10×10 μm^2 zoom inside this domain. Well-contrasted spots are clearly visible and are attributed to the strong elastic scattering from nano-KTPs resulting from the mismatch between the index of refraction of glass ($n \approx 1.5$ in the visible spectrum) and the average one of KTP ($n \approx 1.8$). Figure 3c is an AFM image of exactly the same domain taken in non-contact operation mode. It indeed reveals a large number of objects with nanometric size.

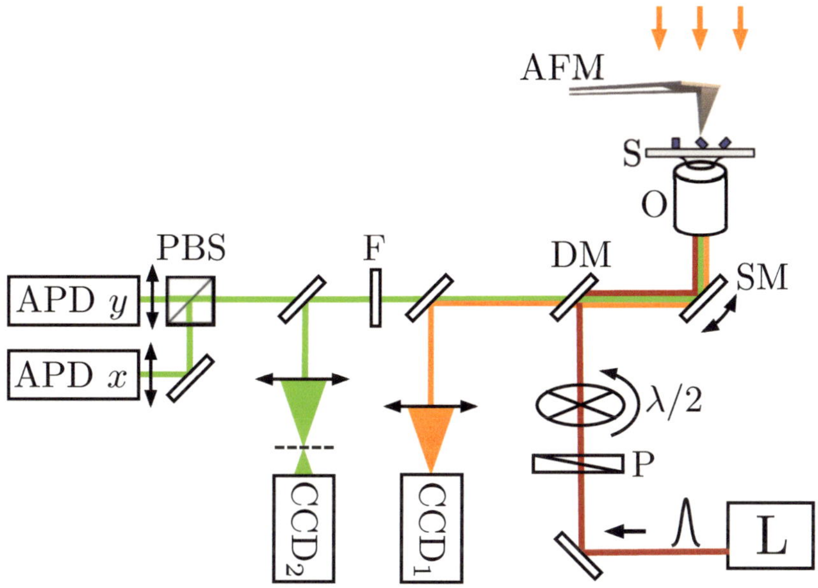

Fig. 2. Experimental setup. AFM: atomic force microscope mount on top of the optical set-up. L: femtosecond laser (86 MHz repetition rate, ≈ 100 fs pulse duration, $\lambda = 986$ nm), P: polarizer, $\lambda/2$ half-wave plate, DM: dichroic mirror, SM: scanning mirror, O: microscope objective (N.A. = 1.40, $\times$ 100), S: sample, CCD_1 white-light camera, CCD_2 sesitive CCD camera, PBS: polarization beam splitter, APD: avalanche photodiode functioning in photo-counting regime

KTP is a non-centrosymmetric orthorhombic crystal with large nonlinear coefficients, of the order of $\chi^{(2)}_{zzz} = 2d_{33} \approx 34$ pm/V along its 3-axis, when excited in the near infrared.(Sutherland, 1996) We use a SHG microscope, as described in Figure 2, to identify if the scattering objects identified in Figures. 3a an 3b correspond to optically-active particles with nonlinear response. Femtosecond infrared light pulses (100 fs) at 86 MHz repetition rate are tightly focused on the sample with a high numerical aperture microscope objective (NA=1.4, $\times$100). Since the nanocrystal size is much smaller than the wavelength, we can neglect any phase shift between the fundamental and second-harmonic fields. Compared to propagation in a bulk nonlinear crystal, there is no phase-matching requirement, thus allowing us to excite the nanocrystal at any wavelength. Here we choose the excitation wavelength

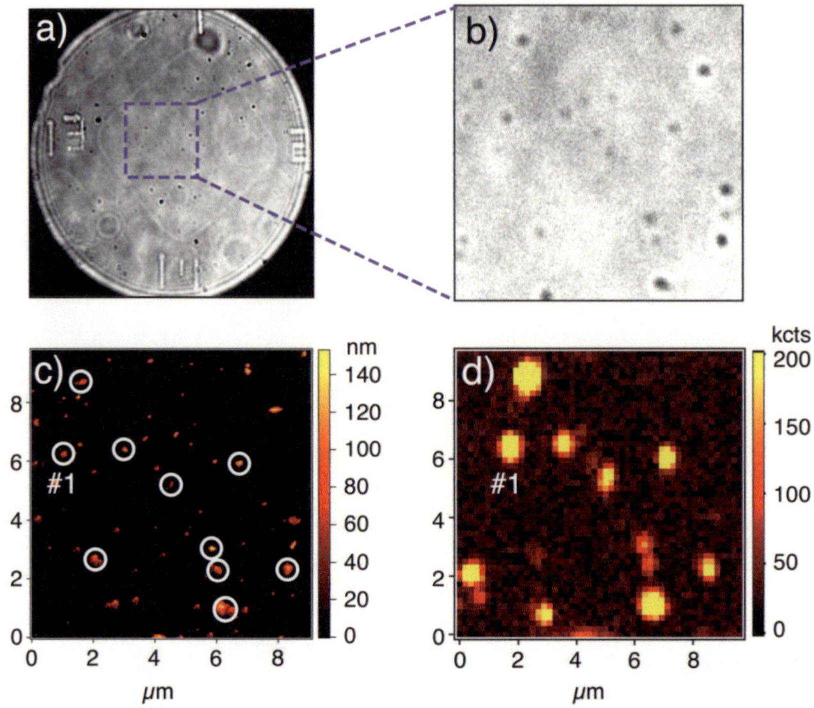

Fig. 3. Correspondence of nanoparticles in images taken with different observational techniques. (a) Overview in white-light microscopy of the structured sample. The image displays a square area with nano-KTPs deposited on the glass surface and surrounded by a grid of thin polymer film. One square is unambiguously located by photoetched numbers on this grid. (b) $10\times10\ \mu m^2$ zoom of the white-light image. (c) Corresponding $10\times10\ \mu m^2$ AFM scan. (d) Corresponding raster-scan SHG image. Circles in (c) pinpoint the nanoparticles in the AFM image which are SHG-active. Note that a small triangular area in the bottom-left corner is not scanned by the AFM in (c), thus missing one SHG emitter evidenced in (d).

at 986 nm so that the light-matter interaction is highly non-resonant both at this fundamental and the corresponding 493-nm second-harmonic wavelength, where KTP remains highly transparent.(Hansson et al., 2000)

The previously observed domain (see Figures 3b and 3c) is raster scanned and the SHG signal is collected in the backward direction with the same microscope objective and is detected by avalanche photodiodes operated in the photon-counting regime (Figure 3d). SHG is detected from a large fraction of the scatterers revealed by the white-light observation of the sample, as evidenced by the one-to-one correspondence between the spots that appear in Figure 3b and Figure 3d. Nearly all these emitters can also be unambiguously identified in the AFM image (white circles in Figure 3c). The AFM measurement determines the height of the nanocrystals, a value taken as an estimate of the nano-KTP size. A shape and phase analysis of the AFM scan indicates that other non-emitting nano-objects observed in the AFM image are most likely small polymer residues left after the etching process of the polymer during the structuration procedure.

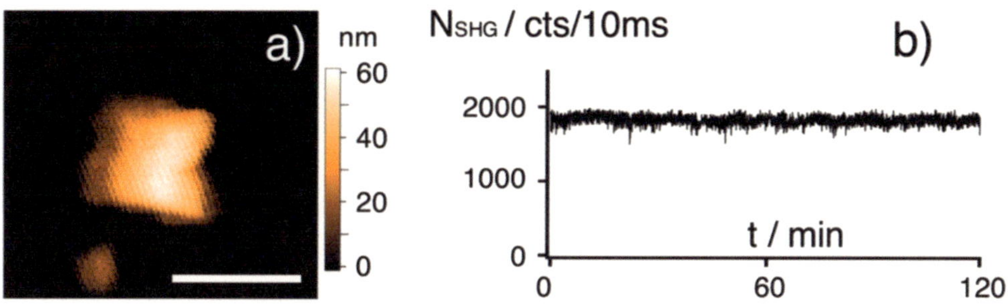

Fig. 4. Characterization of an isolated nano-KTP: (a) AFM image (scale bar: 200 nm), (b) photostability of the second-harmonic emission by the nano-KTP with average incident power 8 mW. Quadratic power dependence of the signal, as expected for SHG, has also been checked (not shown).

We now focus on the study of an isolated nano-KTP, which corresponds to the spot labeled #1 in Figure 3c and 3d. This single particle is representative of the properties of a large number of similar nanocrystallites. A zoom of the AFM image shown in Figure 4a reveals a single particle of almost rectangular shape, with a 60-nm height and a size of 120 nm in its transverse dimensions. We note that the value of the transverse size is over-estimated, due to convolution of the nanocrystal shape with the AFM tip and to residual polymer that might have stuck to the nano-KTP sides. The corresponding SHG image in Figure 3d is a diffraction-limited spot as expected for an emitter with sub-wavelength size. The emitted photon flux at the center of the spot leads to 2.5×10^5 detection counts per second for an average incident power of 8 mW. It leads to a measured signal-to-background ratio as large as 250. Most remarkably, Figure 4b shows the photostability of the emitter under ambient conditions (room temperature and direct exposure to oxygen) since a constant detection signal is recorded for more than 120 minutes. This feature is due to the highly non-resonant character of the interaction: indeed, there is no population in any excited level that could lead to the occurrence of photoinduced degradation processes.

3. Defocused imaging of second harmonic generation from a single nanocrystal

Under similar experimental conditions, we found that the second-harmonic reflection of a strongly focused infrared excitation beam at the surface of a macroscopic KTP crystal with known axis orientation can be accurately modeled by the radiation of a single nonlinearly induced dipole.(Le Xuan et al., 2006) To analyze the second-harmonic emission of a single nano-KTP with sub-wavelength size, we apply the same model which neglects depolarizing and propagation effects associated to the dielectric material.(Bohren & Huffman, 2004) Under an incident electromagnetic field at fundamental frequency ω, the nano-KTP is then equivalent to a single dipole oscillating at frequency 2ω, with components associated to the second-order susceptibility tensor coefficients

$$P_i^{2\omega} = \epsilon_0 \sum_{j,k} \chi_{ijk}^{(2)}(-2\omega;\omega,\omega)E_j^{\omega}E_k^{\omega} \tag{1}$$

where i, j, k span the Cartesian coordinates x, y, z in the laboratory frame. Describing the orientation of the crystal axes by the Euler angles , the nonlinear coefficients in the laboratory frame are related to the known nonlinear coefficients in its crystallographic X, Y, Z axes through simple rotations

$$\chi^{(2)}_{ijk} = \sum_{J,K} \chi^{(2)}_{IJK} \cos(\vec{i},\vec{I}) \cos(\vec{j},\vec{J}) \cos(\vec{k},\vec{K}) \quad (2)$$

The resulting nonlinear dipole radiates a second-harmonic field , which is predicted to have a specific polarization through equations 1 and 2. Following recently developed polarimetry techniques for nonlinear microscopy,(Brasselet et al., 2004; Le Floc'h et al., 2003) a polarizing beamsplitter analyzes the SHG orthogonal field components along x and y laboratory axes, located in the plane perpendicular to the microscope optical axis z. Each corresponding intensity is recorded as the excitation field is rotated with a half-wave plate in the path of the incident beam. Qualitatively, the polar graph of Figure 5a displays x and y second-harmonic emission similar to dipolar radiative patterns. They are related to the projection of the nonlinear emitting dipole in the (x,y) plane, i.e. Euler angle ϕ. In the case of a single nonlinear coefficient being non-null in Equations 1 and 2, we would have obtained two homothetic polar graphs. The different responses along x and y are due to the contribution of the five nonlinear coefficients in the KTP tensor.(Sutherland, 1996) Here, equations 1 and 2 allow us to retrieve $\phi = 70° \pm 5°$.
This polarimetry technique has however a low sensitivity to any dipole component along the longitudinal z-axis. To retrieve the orientation of the dipole in the three dimensions, we apply a defocused imaging technique.(Bohmer & Enderlein, 2003; Brokmann et al., 2005; Sandeau et al., 2007) For a given direction of propagation in the far field, the radiated field is proportional to

$$\vec{E}^{2\omega}_{FF} \propto \vec{k} \times \vec{k} \times \vec{P}^{2\omega} \quad (3)$$

while the intensity radiation diagram is given by, $I^{2\omega}_i(\vec{k}) \propto |\vec{E}^{2\omega}_{i,FF}(\vec{k})|^2$, where i stands for x, y, z. This radiation diagram is directly measured as a function of excitation polarization orientation by positioning a highly-sensitive CCD camera slightly out of the conjugated focal observation plane. The experimental defocused image is displayed in Figure 5b. Qualitatively, the outer "moon-shaped" structure is characteristic of an emitting dipole out of the (x,y) plane of the sample.(Bohmer & Enderlein, 2003; Brokmann et al., 2005; Sandeau et al., 2007) Compared to the case of single-chromophore fluorescence,(Bohmer & Enderlein, 2003; Brokmann et al., 2005) the image exhibits specific features associated to the nonlinear coupling described by equation 1 between the nonlinearly induced dipole and the polarization state of the excitation field . This allows us to determine the full set of Euler angles of the observed nanocrystal with a good accuracy while relying only on the SHG signal.(Sandeau et al., 2007) For the chosen nano-KTP #1 of Figure 3c, the model developed in (Sandeau et al., 2007) gives a single set of angles: Figures 5c and 5d display the numerical simulations corresponding respectively to Figures 5a and 5b with the determined Euler angles of the crystal axes.
For more insight on the method, we focus in the next part on a more detailed 3D orientation determination of one typical single KTP nanocrystal. The experimental setup is described in Figure 2. For defocused imaging, the SHG signal is directed towards a sensitive CCD camera (CCD_2 on the setup figure, having 512×512 pixels, 13×13 μm^2 pixel size, thermoelectrically cooled) which is translated by 15 mm towards the objective, so as to obtain a contrasted image of about 600×600 μm^2 size.

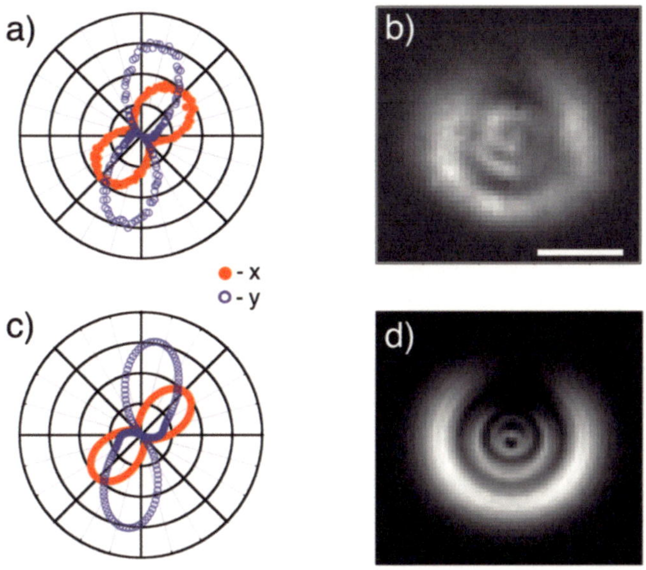

Fig. 5. 3D orientation of an isolated nano-KTP: (a) polar graph showing the measured SHG emission of the nanocrystal along x (red points) and y (blue points) axes, as a function of the direction of the linearly polarized excitation field, (b) corresponding experimental defocused image (scale bar: 250 μm), (c) and (d) numerical calculation of the polar graph and the defocused image respectively for a nano-KTP with Euler angles of the crystalline axis equal to ($70° \pm 5°$, $15° \pm 5°$, $60° \pm 5°$).

Figure 6 exhibits measured defocused images from a single KTP nanocrystal for different polarization orientations of the incoming fundamental field in the sample plane. Different features appear on these images. First, they exhibit a symmetry axis which orientation in the sample plane is related to the nanocrystal in-plane orientation. Second, the structure of the image is characteristic of a nanocrystal exhibiting out-of-plane orientation with $\theta \neq 0°$. Third, the symmetry axis of the structure rotates when changing the incident polarization direction. This is characteristic from a nonlinear coherently induced dipole emission as opposed to a single fluorescent dipole, since the KTP multipolar symmetry structure allows coherent nonlinear coupling with various incident field polarization directions.
The measured images are interpreted using a vectorial model calculation, in which the KTP nanocrystal is modeled by a cubic shape sampled by placing one individual dipole every nm^3 within this cube. The experimental imaging configuration is taken into account. Starting from the ring pattern observation for a first estimate of the 3D orientation parameter, the emission pattern calculation (Fig. 6(f-j)) is then implemented to manually adjust at best the measured images (Fig. 6(a-e)). This leads to a 3D orientation determination with error margins of $\pm$ 5° at maximum for θ and ϕ. This error is much larger for the ψ Euler angle (from $\pm$ 10° to $\pm$ 50°), which is specific to the case of KTP for which there is only a weak difference between nonlinear coefficients involving the "1" and "2" directions perpendicular to the main "3" axis.(Vanherzeele & Bierlein, 1992) In order to confirm the values for the obtained orientation angles, they were used as input parameters to calculate the in-plane polarimetric SHG response of the measured nanocrystal. The corresponding angular dependencies agree well

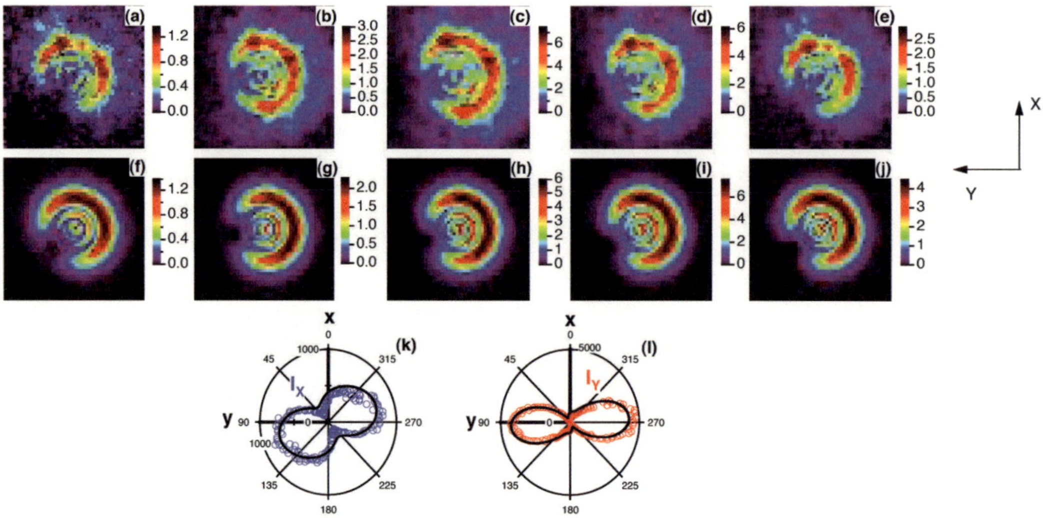

Fig. 6. (a-e) Experimental defocused images of a KTP nanocrystal for several incident polarization directions relative to the x-axis: (a) 0°, (b) 60°, (c) 90°, (d) 120° and (e) 150°. The images sizes are 600×600 μm^2. The integration time for each image is 5 s. (f-j) Corresponding calculated images, leading to the orientation parameters ($\theta = 30° \pm 5°, \phi = 115° \pm 5°, \psi = 90° \pm 20°$). (k,l) Experimental polarimetric analysis Ix and Iy of this nanocrystal (circle markers) and corresponding calculated polarization responses (lines) for the Euler set of angles deduced from the defocused imaging analysis.

with the experimental data (Fig. 6(k,l)). Note that a direct determination of the Euler angles from such polarimetric responses would have been more delicate since the KTP symmetry exhibits several ambiguous situations when projected in the sample plane. This points out the advantages of the defocused imaging technique, which encompasses a complete 3D coupling information in the observed pattern shapes.

Performing numerical simulation on nanocrystal of different sizes, we note that that small differences occur but in general the patterns of the defocused images have similar features, with a symmetry axis along the direction of the nonlinear induced dipole. The ring-structured shape appears to be in all cases representative of the tilt angle of the "3" axis of the crystal with respect to microscope objective axis z. The rough similarity among the observed pattern with respect to crystal size shows that radiation imaging can be used as a robust technique to determine orientation independently on the crystal size, provided that it lies below 150-200 nm. Above this size range, the fine structure of the observed rings is seen to clearly deviate from the single dipole model. For KTP nanocrystal of size smaller than 100 nm in diameter, the SHG emission can be modeled correctly in a dipolar approximation. In addition to 3D orientation determination, such a technique presents the unique property of providing a diagnostics for the crystalline quality of an isolated nano-object. Indeed the presence of nanocrystals of different orientations in a same focus point would strongly affect the defocused image properties described above, and allow a direct conclusion on the mono-crystallinity of the observed object. This is illustrated in Fig. 7, representative of a nanometric size KTP for which images are deprived from a symmetry axis and deviate from the previously observed ring structures. Such images are representative of the presence of at least two KTP nanocrystals in the focal volume of the microscope objective, which could not be revealed by the sole observation of the diffraction-limited spot size.

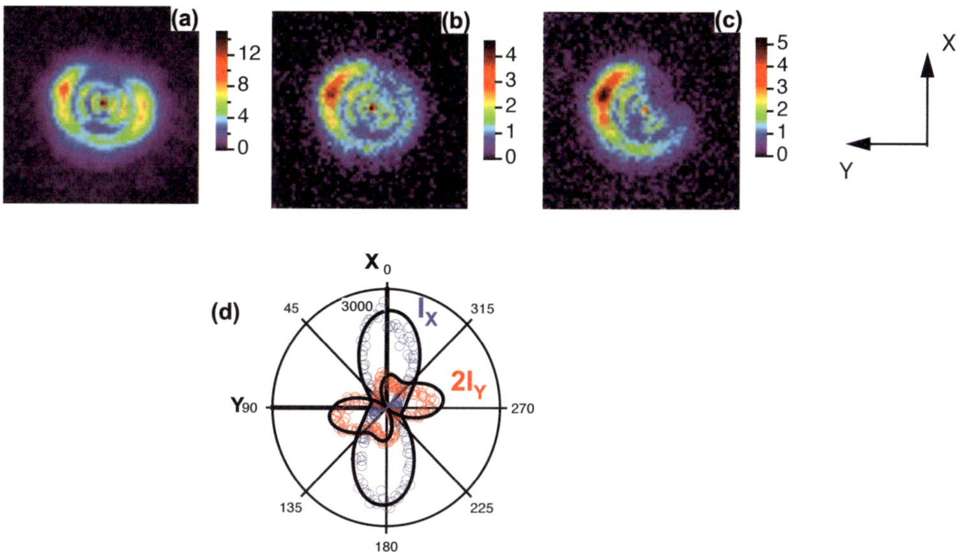

Fig. 7. (a-c) Defocused images of a nanocrystal for several incident polarization directions relative to the x (horizontal) axis: (a) 0°, (b) 306° and (c) 90°. The images sizes are 800×800 μm^2. This nano-object is visibly constituted of several distinct sub-domains, (d) Polarimetric analysis of this nano-cluster of nanocrystals. The adjustment of polarimetric responses (black line) is done using two nanocrystals of respective orientations $(\theta_1 = 20°, \phi_1 = 260°, \psi_1 = 0°)$ and $(\theta_2 = 45°, \phi_2 = 350°, \psi_2 = 0°)$.

A finer observation allows us to conclude that in this two-KTP nanocrystals picture, one of them is likely to lie close to the vertical x axis of the images, since the SHG pattern approaches a ring structure for a x-polarized excitation. The other KTP nanoparticle is preferentially excited for a polarization angle close to 45° (relative to x), and therefore manifests with larger signals in Fig. 7(b,c) as the excitation angle increases. This deviation from a mono-crystalline structure is confirmed in the polarimetric response of such a nano-object, which curve cannot be fitted using a single KTP unit-cell model, but rather by two or more nanocrystals orientations.

We have demonstrated in this section that defocused imaging of second harmonic generation radiation from a single KTP nanocrystal can provide a direct retrieval of its 3D orientation. This orientation information, based on the knowledge of the nano-object crystalline symmetry, can thus be used as a complete characterization of the excitation-dependent polarization state from a nonlinear nanosource. In addition, we show that this technique provides a unique possibility to inform on the mono-crystallinity nature of a nano-object. Finally, this analysis method is a promising tool of analysis of the vectorial electric-field character from an unknown electromagnetic incident radiation, taking advantage of the rich features appearing from the nonlinear coupling between matter at the subwavelength scale and complex optical polarization components.

4. Balanced homodyne detection of second harmonic radiation from single KTP nanocrystal

SHG is a coherent process thus interferometric detection schemes are well suited to the detection of the emitted field. Among possible techniques, coherent balanced homodyne is known to exhibits a sensitivity to extremely low photon flux rates (Yuen et al., 1983). Although high sensitivity with such a technique has been demonstrated to detect surface SHG emission from an air-semiconductor interface (Chen et al., 1998), to the best of our knowledge it had never been applied to the study of nano-objects. Since the optical phase of the SHG emitted field is directly related to the sign of the associated nonlinear susceptibility, measuring directly its value provides important information about the nonlinear material (Stolle et al., 1996). At nanometric scale, it allows to determine the absolute orientation of dipoles with nonlinear hyperpolarisabilities such as organic nanocrystals (Brasselet et al., 2004), to detect the polarity in cell membranes (Moreaux et al., 2000), or to study the boundary between microscopic crystalline domains (Laurell et al., 1992). Such application prospects strongly motivate the development of phase-sensitive SHG microscopy with highly sensitive detection compared to SHG interferometric schemes previously developed (Stolle et al., 1996). In this work coherent balanced homodyne detection is associated to SHG microscopy, allowing us to detect nano-objects with high signal-to-noise ratio and phase sensitivity.

The principle of the experiment is shown in Fig.8. Femtosecond infrared light pulses are injected in an inverted optical microscope and tightly focused. Backward second-harmonic emission by a nonlinear crystal placed on a glass cover slide at the microscope focus is collected by the microscope objective and transmitted through a dichroic mirror toward the detection set-up. A small fraction of the fundamental beam is also reflected from the cover-slide and follows the same optical path, providing the phase reference. Second-harmonic emission of the crystal, further referred as "signal", corresponds to an electrical field $E^{(2\omega)} = |E^{(2\omega)}| \exp[i\, \Phi_{\text{obj}}^{(2\omega)}]$, where $\Phi_{\text{obj}}^{(2\omega)}$ is the phase shift of the SHG field emitted by the object as compared to the one of the incident fundamental field $E_{\text{in}}^{(\omega)}$ on the object. For a macroscopic nonlinear crystal, the SHG field is $E^{(2\omega)} \propto \chi^{(2)} {E_{\text{in}}^{(\omega)}}^2$. Thus, $\Phi_{\text{obj}}^{(2\omega)}$ reflects the phase of the nonlinear susceptibility $\chi^{(2)}$. For a nonlinear crystal in its spectral transparence window, $\chi^{(2)}$ is real with a positive or negative value depending on the absolute orientation of the nonlinear response in respect to the crystal axis. Therefore determination of $\Phi_{\text{obj}}^{(2\omega)}$ allows to infer the absolute orientation of the nonlinear crystal.

For the coherent optical homodyne detection of the SHG signal, the local oscillator (LO) is generated by sending part of the incident infrared beam into a bulk BBO nonlinear crystal (Fig.8). A Glan prism ensures linear polarization of the local oscillator along the x-axis and x-polarized signal and local oscillator are recombined by a non-polarizing 50/50 beamsplitter cube. The 180° out-of-phase interferences at the two output ports of the beamsplitter are detected by photodetectors D_1 and D_2. In balanced detection mode, the two resulting photoelectric signals S_1 and S_2 are subtracted [1], thus canceling out their DC component. The interferometric signal is then

$$\Delta S^{(2\omega)} = K \times 2V \sqrt{P_{\text{LO}}^{(2\omega)} P_{\text{sig}}^{(2\omega)}} \cos\left(\Phi_{\text{sig}}^{(2\omega)} - \Phi_{\text{LO}}^{(2\omega)}\right) \tag{4}$$

[1] We use an autobalanced photoreceiver (Nirvana, Model 2007, New Focus, San Jose CA) which achieves 50 dB common noise rejection

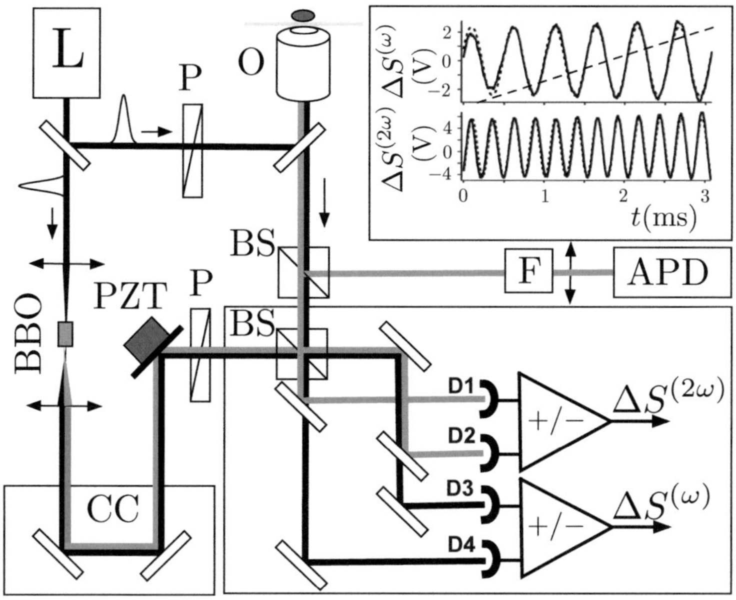

Fig. 8. Schematic of the experimental set-up. L: femtosecond laser (86 MHz repetition rate, $\approx$ 100 fs pulse duration, $\lambda = 986$ nm); BBO: nonlinear $\chi^{(2)}$ crystal; P: Glan prism; O: microscope objective (N.A. = 1.40, $\times$ 100) leading to a $\approx$ 300 nm FWHM diameter focal spot; BS: non-polarizing beamsplitter; CC: corner-cube; PZT: mirror mounted on a piezoelectric transducer; KTP: macroscopic $KTiOPO_4$ crystal or single sub-wavelength size crystal; F: SHG filter; APD: avalanche photodiode in photon counting regime; D_1 and D_2: p-i-n Si photodetectors of the balanced receiver recording SHG at 2ω (a SHG filter, not shown, is put in front of the detectors); D_3 and D_4: the same for the balanced receiver at fundamental optical frequency ω (an IR filter, not shown, is put in front of the detectors). *Insert*: full lines : detected signals as a function of time for a forward translation of the mirror. Up: signal $\Delta S^{(\omega)}$ at fundamental frequency with a sinusoidal fit at frequency $f = 1.9$ kHz, and bottom: $\Delta S^{(2\omega)}$ at SHG frequency with a sinusoidal fit at frequency $2f$ = 3.8 kHz, dashed line: voltage applied to the PZT on half a period (vertical scale: a.u.).

where $P_{LO}^{(2\omega)}$ and $P_{sig}^{(2\omega)}$ denote the mean optical power of local oscillator and signal, V is the fringe visibility, and K includes the quantum efficiency of the photodiodes. The phase shift of the signal at frequency 2ω , $\Phi_{sig}^{(2\omega)} = \Phi_1^{(2\omega)} + \Phi_2^{(2\omega)}$ contains the phase shift $\Phi_1^{(2\omega)}$ of the x-polarized radiation emitted by the nonlinear object and the phase shift $\Phi_2^{(2\omega)}$ due to propagation in the interferometer.

In order to measure the phase shift $\Phi_{sig}^{(2\omega)}$, temporal interference fringes are created by displacing a mirror mounted on a piezoelectric transducer (PZT) driven with a triangular shape voltage. The resulting fringes are shown in the inset of Fig.8. To extract the phase shift information, fringes at 2ω are compared to reference fringes resulting from fundamental beam. More precisely, we perform with detectors D_3 and D_4 a separate balanced homodyne detection of the residual fundamental beam reflected by the sample by mixing it with the residual light at ω

which follows the reference arm. A reference interference signal (see inset of Fig. 8) is then obtained, equal to $\Delta S^{(\omega)} = K' \times 2V' \sqrt{P_{LO}^{(\omega)} P_{sig}^{(\omega)}} \cos\left(\Phi_{sig}^{(\omega)} - \Phi_{LO}^{(\omega)}\right)$ with same notations meaning as in Eq. 4. The quantity to be extracted is then $\Delta\Phi = \left(\Phi_{sig}^{(2\omega)} - \Phi_{LO}^{(2\omega)}\right) - \left(\Phi_{sig}^{(\omega)} - \Phi_{LO}^{(\omega)}\right)$ which can be written as $\Delta\Phi = \Phi_1^{(2\omega)} + \Phi_0$, where Φ_0 is an "instrumental" phase shift. Although Φ_0 can be measured with a reference crystal its value is not required if we are only concerned in variation of $\Phi_1^{(2\omega)}$ while keeping Φ_0 constant. Finally, the value of $\Delta\Phi$ is extracted using a numerical procedure.

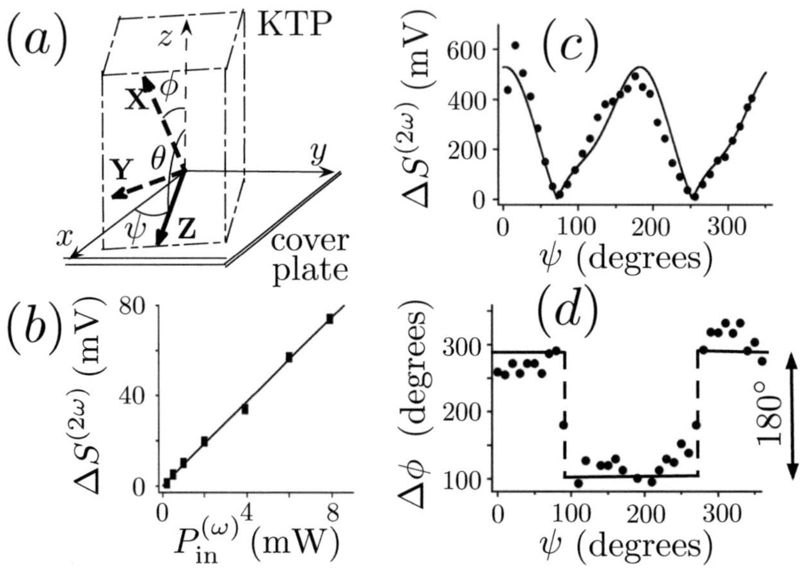

Fig. 9. Results from the balanced homodyne detection. (a) Laboratory axes (x, y, z) with x-polarized incident fundamental beam propagating along z-axis. $(\mathbf{X}, \mathbf{Y}, \mathbf{Z})$ KTP crystal principal axes ($\theta = 90°$ and $\phi = 23.5°$). Full lines denote axes in the (x, y) plane. ψ : rotation angle of the $\chi^{(2)}$ nonlinear crystal around the z-axis. (b) Linearity of $\Delta S^{(2\omega)}$ as a function of fundamental incident power. Points : experimental data, full line : best linear dependence fit. (c) : $\Delta S^{(2\omega)}$ as a function of ψ (in degrees). (d) : Relative SHG phase shift $\Delta\Phi$ as a function of ψ.

A bulk KTP crystal with axes as shown in Fig.9 is used to test the setup. The Z-axis is parallel to the crystal input face. With a strongly focused pump beam, a SHG backward emitted beam from the interface is observed (Boyd, 2003). We first check in Fig.9b that $\Delta S^{(2\omega)}$ associated to this surface SHG grows linearly with the fundamental incident power on the crystal $P_{in}^{(\omega)}$, as expected from Eq.4. If the crystal is rotated by 180° around the microscope optical axis (z-axis), the nonlinear coefficient $\chi^{(2)}$ is transformed in $-\chi^{(2)}$ and a 180° SHG-phase-shift is expected. Results are shown in Fig.9c. The amplitude shows a maximum of x-polarized SHG when the linear polarization is aligned along the Z-axis of the crystal (ψ = 0° or 180°) which exhibits the highest nonlinear coefficient. The full line is the theoretical result assuming an x-polarized incident plane wave normal to the interface and a single nonlinear emitting dipole which takes into account all nonlinear coefficients of the KTP $\chi^{(2)}$ tensor. It yields good agreement with experimental points. The phase-shift $\Delta\Phi$ (Fig. 9d) remains constant while the nonlinear dipole has a positive projection along the x-axis ($0° < \psi < 90°$). It then endures a 180ą-phase-shift when it becomes negative ($90° < \psi < 180°$). The small deviation to the constant value is attributed to drift of the ϕ_0 parameter during the step-by-step rotation of the KTP crystal.

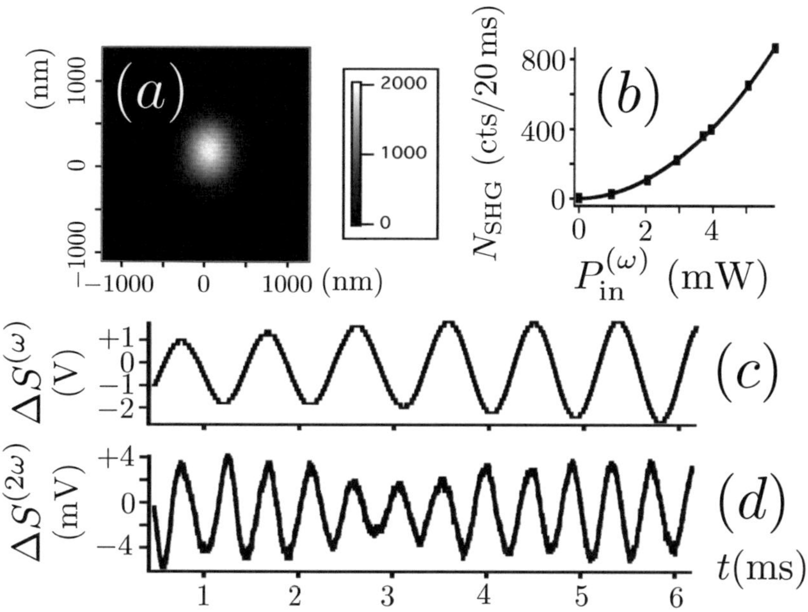

Fig. 10. Application of the balanced homodyne technique to sub-wavelength-size nonlinear crystals. (a) Raster-scan image of SHG signal from a KTP nano-crystal with an avalanche photodiode. The FWHM diameter of SHG spot intensity is about 300 nm, very close to the theoretical two-photon microscope resolution (320 nm FWHM). (b) Number of detected SHG photons as a function of incident power with a best square-law dependence fit. (c) Fundamental interference signal $\Delta S^{(\omega)}$ as a function of time for a single nano-crystal and (d) corresponding SHG interference signal $\Delta S^{(2\omega)}$.

We then apply the method to the detection of SHG from KTP nanocrystals. To first locate the crystals, the sample is raster-scanned in x and y directions with piezoelectric translators and SHG is detected with the avalanche photodiode as shown in Fig.10. A SHG signal image of such a nano-crystal is shown in Fig.10a. A quadratic dependence of the detected SHG intensity is observed upon varying $P_{in}^{(\omega)}$ as expected (see Fig.10b). After positioning the focused infrared excitation beam at the center of the detected emission spot of a single nano-crystal, we switch to the coherent balanced homodyne detection set-up. SHG interference fringes (Fig.10d) associated to fundamental ones (Fig.10c) are clearly visible, giving evidence for the coherence of the nano-crystal SHG emission.

Since the balanced homodyne detection method consists in the projection of the signal electric field on the spatio-temporal mode of the local oscillator, mode-matching between signal and LO beams determines maximal amplitude of the fringes. This can be quantified by the fringe visibility V of Eq.4 being equal to unity for perfect mode-matching. With a 2 mW LO average power, the smallest SHG fringe amplitude measured with 1 s duration averaging is equivalent to a signal power $P_{\rm sig}^{(2\omega)} \approx (7.2 \pm 5.0) \times 10^{-19}$ W assuming $V = 1$. The uncertainty evaluated for 95 % confidence interval is equivalent to 3.2 detected photons/s, close to the shot noise limit. However a unity fringe visibility is practically difficult to achieve. In an auxiliary experiment using equal powers for signal and LO beams, we measure $V = 0.21$, leading to an actual sensitivity of 80 photons/s. Such a reduced value for the visibility factor is attributed to imperfect mode-matching between signal and LO modes, including polarization mismatch, wave-front distortion, and frequency chirp on the SHG-emitted femtosecond pulses.

To conclude this section, we have demonstrated a SHG phase-sensitive microscope with coherent balanced homodyne detection, showing a sensitivity at the photon/s level. The high-spatial resolution and the sensitivity of the technique is well-adapted to the study of SHG from nano-crystals.

5. Coherent nonlinear emission from a single KTP nanoparticle with broadband femtosecond pulses

As a coherent process, the number of SHG photons emitted by a noncentrosymmetric nanoparticle scales as the square of the number of "oscillators" in the nanocrystal, i.e., as the sixth power of the nanoparticle average size. As a two-photon process the SHG intensity is also expected to scale, for a constant average laser power, as the inverse of the pulse time duration. It is therefore tempting to reduce the pulse duration from the standard 100 fs to 10 fs available from broadband ultrafast lasers in order to enhance the second harmonic photon emission rate by an order of magnitude, thus reducing the limit size of detectable nanoparticles. Nevertheless, this downscaling has to be done with care since high-order phase dispersion of the microscope objective and other optical components induce temporal aberrations in the excitation pulse interacting with the nanoparticle at the focus of the microscope (Guild et al., 1997). Using recently developed techniques for temporal characterization at the microscope focus and careful precompensation (Lozovoy et al., 2004), SHG from large objects has indeed been shown to scale as the inverse of the pulse time duration (Xi et al., 2008). Additionally, the broadband excitation corresponding to a 10-fs pulse offers the possibility of spectral manipulation for coherent control (Warren et al., 1993; Weiner, 2000) of two-photon absorption and non-resonant second harmonic generation (Broers et al., 1992; Meshulach & Silberberg, 1998; Wnuk & Radzewicz, 2007).

While it has been recently shown that a SHG-active nanoparticle of a size about a few hundreds of nanometers can be used for pulse detection in a nanoscopic version of the FROG technique (Extermann et al., 2008), the effects of manipulation of the excitation beam on SHG emission, i.e.,decreasing the time duration of a precompensated pulse as well as structuring its broadband spectrum, have not yet been investigated at the single nanoparticle level with a size well below the wavelength of light. Moreover in this size range phase matching conditions are automatically fulfilled, improving the coherent emission. Here we have shown that the use of broadband ultrashort laser pulses, precompensated using an automatic genetic algorithm,improves the second harmonic emission from a single nanoparticle of size about 100 nm. It results in a contrast enhancement of the SHG image obtained by raster scanning the sample. In this process, smaller nanoparticles are revealed for a given background. We also manipulate the broadband incident spectrum in a very simple manner to obtain a non-degenerate sum-frequency generation from a single nanoparticle.

The experiment is realized using a titanium doped sapphire (Ti:Sa) femtosecond laser with a 100 nm bandwidth. The laser spectrum is sufficiently broad to support pulses of approximately 10 fs duration. A genetic algorithm (Baumert et al., 1997) is used to search for the best phase correction on the incident pulse. A bulk KTP crystal with one face polished served as reference sample. Under femtosecond beam illumination, the second harmonic signal emitted by the crystal and detected with an avalanche photodiode operated in photoncounting regime served as the feedback information in the genetic algorithm procedure. The bulk crystal, instead of a nanocrystal, was selected because it provides a high signal to noise ratio, which speeds up the search for the optimal phase of the excitation pulse.
The phase correction was achieved with two consecutive systems: a two-prism compressor used for the rough compensation of (mostly quadratic) phase, and a compact pulse shaper using a diffraction grating and a spatial light modulator (SLM) (Weiner, 2000). The SLM-based pulseshaper was used for the genetic algorithm search for the optimal phase. To evaluate the result of this search we estimated the duration of the excitation pulse at the focus of the microscope by recording the interferometric autocorrelation of the pulse using the second harmonic field generated in the KTP nonlinear crystal. The result of the genetic algorithm search leads to a pulse of, approximately, 13 fs duration. Once the optimal pulse shape was found its duration can then be increased to any value by adding, with the SLM, a controlled amount of the second-order phase. The corresponding SHG photon number was then measured as a function of the corresponding pulse duration.

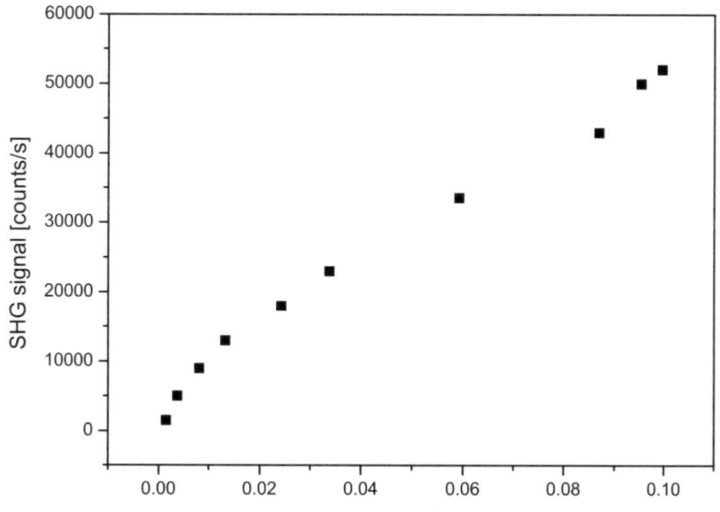

Fig. 11. Detected rate of the SHG from a 100 nm KTP nanocrystal as a function of pulse time duration showing the improvement of the SHG signal with the inverse of the pulse duration.

For a constant average incident power, the SHG signal is expected to scale as the inverse of the duration of the pulse at the microscope focus. Indeed, we observed a linear increase of the SHG signal with the inverse of the time duration, as shown in Fig. 11. However, the measured average slope is approximately 0.7 instead of unity. This is due in part to the broadband spectrum which cannot be perfectly compressed by the SLM due to pixelization of the correction signal and imperfections in spectral wings compensation. Analysis of the data shown in Fig. 11 reveals that the slope is indeed close to unity for long pulses, while it decreases with the pulse duration. Such a behavior is consistent with imperfect phase compensation: even small phase errors become important when the pulse duration approaches the Fourier transform limit. Despite this imperfect phase compensation, the SH count rate is nevertheless improved by almost an order of magnitude, as compared to excitation by standard 100 fs pulses with the same average intensity. With precompensated pulses, we recorded raster-scan maps of the SH signal originating from well dispersed KTP nanoparticles deposited by spin-coating a colloidal solution on a glass cover-slip. Figure 12 shows maps of the same area of the sample recorded with different pulse durations ranging between 200 fs and 13 fs. Firstly, for nanoparticles already visible at 200 fs, a clear increase of the signal-to-noise ratio is observed for shorter pulses. Secondly, Fig.12 reveals that a higher number of nanoparticles appear in the maps acquired with shorter pulses, since smaller objects can be detected due to enhanced second harmonic emission. This contrast enhancement is crucial for many practical applications of nonlinear microscopy.

The pulse autocorrelation measured on an individual KTP nanocrystallite in the focus of the microscope is close to the Fourier limit, which confirms the efficiency of the phase compensation and gives evidence that all spectral components of the precompensated pulse are converted by the nanoparticle. We note that the bandwidth of the SHG is, for bulk nonlinear crystal, limited by phase matching conditions (Boyd, 2003). However, because of the sub-wavelength dimension of the nanoparticle spectral filtering associated to phase-matching becomes negligible even for ultrashort femtosecond laser pulse.

The broadband nonlinear response opens the way to the coherent control of the different spectral components of the single incident beam in order to manipulate the second-order nonlinear emission (Dudovich et al., 2002).

As a proof-of-principle of such spectral manipulation, we used a two-band incident spectrum, containing well-separated but coherent "red" and "blue" parts (see inset of Fig. 13),in order to excite and subsequently filter out a non-degenerate sum-frequency (SF) signal. If we simply assume two average excitation frequencies "R" and "B" in the single incident beam,the SF signal is due to a nonlinear polarization of the form (Boyd, 2003):

$$P^{\omega_R+\omega_B} = \epsilon_0 \sum_{j,k} \chi^{(2)}_{ijk}(-\omega_R - \omega_B; \omega_R, \omega_B) E_j^{\omega_R} E_k^{\omega_B} \tag{5}$$

where i, j, k stand for x, y, z, the laboratory axes. With two well-separated spectral bands, Eq.5 shows that we have access to the different polarizations j, k of the driving fields, then select a specific nonlinear coefficient $\chi^{(2)}_{ijk}$, and thus control the direction i of the nonlinear dipole at the generated sum frequency. This can be applied e.g. to enhance the recognition of the nanoparticle-marker in a complex environment, and to investigate the shape dependence of the nonlinear optical response of an individual nanoparticle. Yet another possible future application concerns the measurement of the polarization properties of a femtosecond pulse in the focus of a microscope objective. In principle, the spatial resolution of such a measurement is limited by the minimum size of the nanocrystallite that produces a measurable signal. With sub 100-nm detection limit demonstrated in our experiment we hope to obtain in the future a detailed 3-D map of the polarization in the vicinity of the focus.

A simple two-band excitation is performed, as an experimental proof of principle, by blocking a part of the incident pulse spectrum. A bandpass filter is then placed in the detection path to transmit $\omega_R + \omega_B$, and reject both $2\omega_R$ and $2\omega_B$. We checked that excitation with any single spectral band, i.e., ω_R or ω_B, does not produce SF signal. Varying the power of the blue band

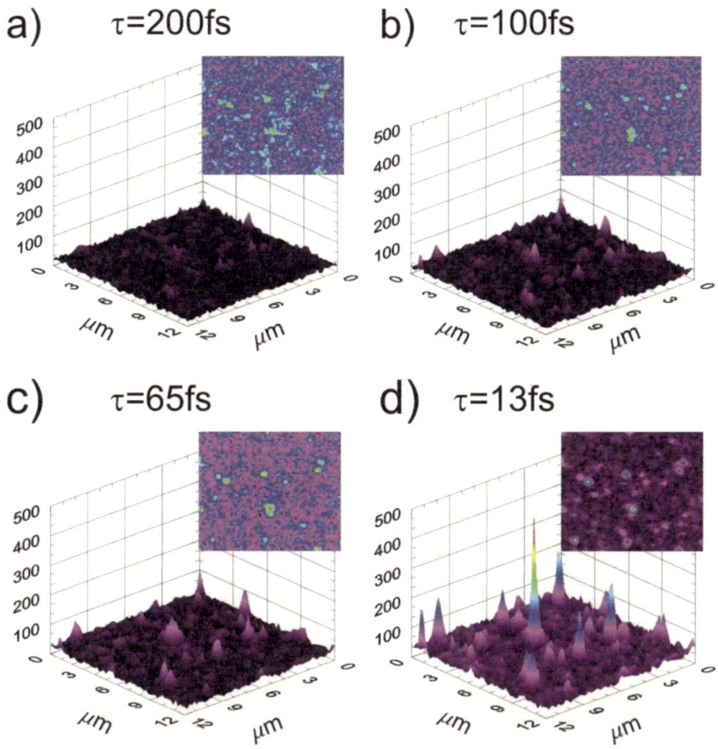

Fig. 12. SHG maps of the same area of the nanoKTP sample. The maps correspond to excitation with pulse durations of (a) 200 fs (SNR=2.5), (b) 100 fs (SNR=4), (c) 65 fs (SNR=4.5) and(d) 13 fs (SNR=16). For each pulse duration both a three-dimensional graph of the SH intensity as well as a normalized two-dimensional map are shown.

of the spectrum while keeping fixed the intensity of the red band leads to a linear increase of SFG vs. blue band power (Fig. 13), as expected from Eq. 5. Due to the intrinsic absence of ensemble averaging, this result can be considered as an improved version of Kurtz powder method (Kurtz & Perry, 1968), now applied on a single 100-nm size nonlinear nanoparticle.
To conclude this section, we have shown that phase-compensated 13-fs infrared pulses generated through the application of a genetic algorithm allow one to gain about one order of magnitude (as compared to standard 100-fs pulses with the same average intensity) in the SHG emission rate of a non-centrosymmetric single KTP nanocrystal. This leads to efficient optical detection of smaller nanocrystals which appear only when shorter pulses are used. Since there is no phase-matching limitation on the bandwidth of the excitation pulses, even shorter (sub-10 fs) pulse scan can be used to even further enhance the second harmonic signal. Optical autocorrelation from a nanocrystal with an AFM measured size of about 100 nm has been recorded, showing that pulse characterization can be obtained even on SHG-active nanoparticles of sub-wavelength size.

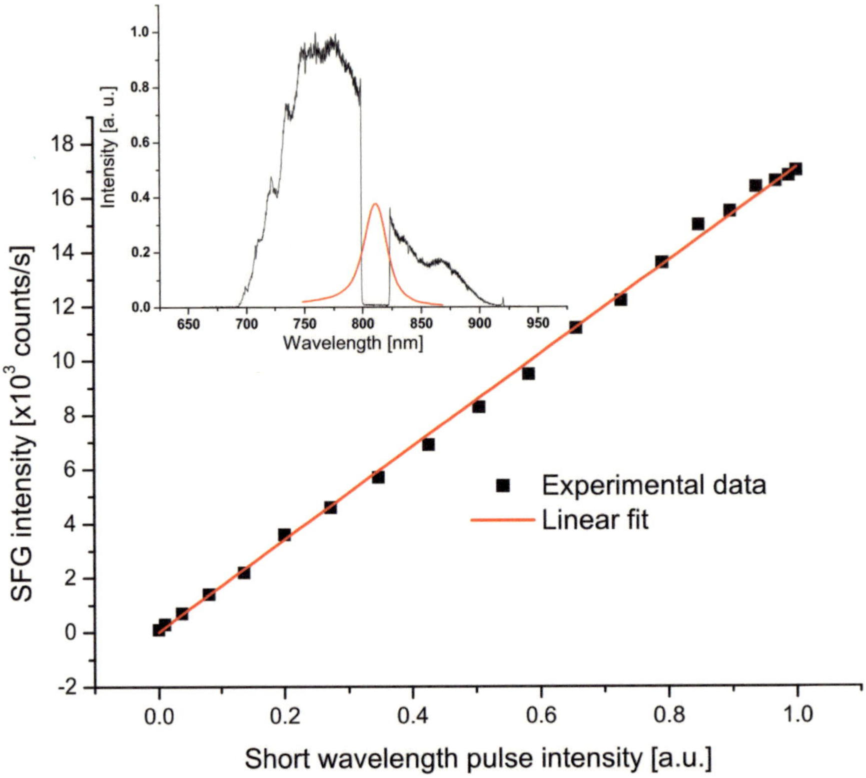

Fig. 13. Coherent sum frequency generation from a single nano-KTP vs the intensity of the blue band. Inset: corresponding pulse spectrum with two distinctive parts "red" and "blue". The nonlinear response is detected in the sum-frequency band with an interference filter transmission (spectral transmission shown in red). Note that in the latter case the wavelength scale has been multiplied by two.

6. Conclusion

In summary, we have reported the observation and the characterization of nanometric-sized crystals extracted by centrifugation from KTP powder. For a well-isolated single nanocrystal, in situ AFM analysis of its size and analysis of its second-harmonic emission properties have been performed. The highly efficient nonlinear response leads to the emission of a large number of SHG photons in a photostable and blinking-free manner due to non-resonant coherent interaction. By recovering the radiation pattern from the recorded defocused images, we retrieve the in situ three-dimensional crystal orientation. Solution-based chemical synthesis of KTP nanocrystals with a monodisperse size controlled by capping agents,(Biswas et al., 2007) now under way, should lead to optimized KTP nanocrystallites and to a more accurate estimate of the size-detection threshold. It also opens the way to surface functionalization of these nanocrystals. Fully characterized nano-KTPs are attractive for the development of novel schemes of nonlinear microscopy. Moreover, due to the non-resonant interaction, they can be envisioned for probing the localized electromagnetic field enhancement that appears at the apex of a metallic tip(Bouhelier et al., 2003) or is randomly generated by granular metallic structures(Stockman et al., 2004) while avoiding any detrimental quenching effect(Buchler et al., 2005; Carminati et al., 2006) induced by the metallic interface. The coherent character of the second harmonic emission from a single KTP nanocrystal have also been demonstrated in a balanced homodyne detection scheme, which permit nanoparticle detection with high signal-to-noise ratio and phase sensitivity. Furthermore, broadband femtosecond pulses have been successfully employed for studying of a single KTP nanoparticle, which open a bright perpective for spectral manipulation for coherent control of the nonlinear process at the nanoscale.

Acknowledgements

The authors are grateful to Dominique Lupinski and Philippe Villeval from Cristal Laser for initial nanoKTP material, Thierry Gacoin, Sandrine Perruchas, Cédric Tard and Géraldine Dantelle for the fabrication of the nanocrystal solution, Joseph Lautru for sample preparation. We acknowledge Jean-Pierre Madrange, André Clouqueur for technical assistance. This work is a fruitful collaboration with Nicolas Sandeau, Sophie Brasselet, Véronique Le Floc'h, Joseph Zyss, Chunyuan Zhou, Yannick de Wilde, Yannick Dumeige, François Treussart, François Marquier, Jean-Jacques Greffet, Rémi Carminati. This work is supported by AC Nanosience research program, CNano IdF research.

7. References

T. Baumert, T. Brixner, V. Seyfried, M. Strehle, and G. Gerber, "Femtosecond pulse shaping by an evolutionary algorithm with feedback," Appl. Phys. B 65, 779-782 (1997).

S. K. Biswas, A. Pathak, P. Pramanik, "Synthesis of Nanocrystalline $KTiOPO_4$ Powder by Chemical Method" J. Am. Ceram. Soc. 90, 1071 (2007)

M. Bohmer, J. Enderlein, "Orientation imaging of single molecules by wide-field epifluorescence microscopy", J. Opt. Soc. Am. B 20, 554 (2003).

C. F. Bohren, D. R. Huffman, *Absorption and Scattering of Light by Small Particles* ; Wiley-VCH: Weinheim, Germany, 2004.

L. Bonacina, Y. Mugnier, F. Courvoisier, R. Le Dantec, J. Extermann, Y. Lambert, V. Boutou, C. Galez, J.-P. Wolf, "Polar Fe(IO3)3 nanocrystals as local probes for nonlinear microscopy", Appl. Phys. B 87, 399 (2007).

A Bouhelier, M. Berverhuis, and L. Novotny, "Near-Field Second-Harmonic Generation Induced by LocalField Enhancement," Phys. Rev. Lett. 90, 013903.1-013903.4 (2003).

R. W. Boyd, in *Nonlinear Optics*, Academic Press, 2nd ed., (2003).

S. Brasselet, V. Le Floc'h, F. Treussart, J.-F. Roch, J. Zyss, E. Botzung-Appert, and A. Ibanez, "In situ diagnosticsof the crystalline nature of single organic nanocrystals by nonlinear microscopy" Phys. Rev. Lett. 92, 207401-204405 (2004).

B. Broers, L. D. Noordam, and H. B. van Linden van den Heuvall, "Diffraction and focusing of spectralenergy in multiphoton processes," Phys. Rev. A 46, 2749-2756 (1992).

X. Brokmann, M.-V. Ehrensperger, J.-P. Hermier, A. Triller M. Dahan, "Orientational imaging and tracking of single CdSe nanocrystals by defocused", Chem. Phys. Lett. 406, 210 (2005).

B. C. Buchler, T. Kalkbrenner, C. Hettich, V. Sandoghdar, "Measuring the Quantum Efficiency of the Optical Emission of Single Radiating Dipoles Using a Scanning Mirror",Phys. Rev. Lett. 95, 063003 (2005).

R. Carminati, J.-J. Greffet, C. Henkel, J.-M. Vigoureux, "Radiative and non-radiative decay of a single molecule close to a metallic nanoparticle", Opt. Commun. 261, 368 (2006).

J. Chen, S. Machida, and Y. Yamanoto, "Simultaneous measurement of amplitude and phase in surface second-harmonic generation", Opt. Lett. 23, 676 (1998).

T. A. Driscoll, H. J. Hoffman, R. E. Stone, P. E. Perkins, "Efficient second-harmonic generation in KTP crystals", J. Opt. Soc. Am. B 3, 683 (1986).

N. Dudovich, D. Oron, and Y. Silberberg, "Single-pulse coherently controlled nonlinear Raman spectroscopy and microscopy," Nature 418, 512-514 (2002).

J. Extermann, L. Bonacina, F. Courvoisier, D. Kiselev, Y. Mugnier, R. Le Dantec, C. Glez, and J. -P. Wolf,"Nano-FROG: Frequency resolved optical gating by a nanometric object," Opt. Express 16, 10405-10411(2008)

J. B. Guild, C. Xu, and W. W. Webb, "Measurement of group delay dispersion of high numerical apertureobjective lenses using two-photon excited fluorescence," Appl. Opt. 36, 397-401 (1997).

G. Hansson, H. Karlsson, S.Wang, F. Laurell, "Transmission Measurements in KTP and Isomorphic Compounds", Appl. Opt. 39, 5058 (2000).

C. Hsieh, R. Grange, Y. Pu, D. Psaltis, "Three-dimensional harmonic holographic microcopy using nanoparticles as probes for cell imaging," Opt. Express 17, 2880 (2009)

J. C. Johnson, H. Yan, R. D. Schaller, P. B. Petersen, P. Yang, R. J. Saykally, "Near-Field Imaging of Nonlinear Optical Mixing in Single Zinc Oxide Nanowires", Nano Lett 2, 279 (2002).

A. V. Kachynski, A. N. Kuzmin, M. Nyk, I. Roy and P. N. Prasad, ?Zinc Oxide Nanocrystals for Nonresonant Nonlinear Optical Microscopy in Biology and Medicine,? J. Phys. Chem. C, 112, 10721 (2008)

K. Kemnitz, K. Bhattacharyya, J. M. Hicks, G. R. Pinto, K. B. Eisenthal, T. F. Heinz, "The phase of second-harmonic light generated at an interface and its relation to absolute molecular orientation", Chem. Phys. Lett. 131, 285 (1986).

S. Kurtz and T. Perry, "A powder technique for the evaluation of nonlinear optical materials," IEEE J.Quantum Electron. 4, 333-333 (1968).

F. Laurell, M.G. Roelofs, W.Bindloss, H.Hsiung, A. Suna, and J.D. Bierlein, "Detection of ferroelectric domain reversal in KTiOPO4 waveguides," J. Appl. Phys. 71, 4664 (1992)

V. Le Floc?h, S. Brasselet, J.-F. Roch, J. Zyss, "In Situ Diagnostics of the Crystalline Nature of Single Organic Nanocrystals by Nonlinear Microscopy" J. Phys. Chem. B 107, 12403 (2003).

L. Le Xuan, S. Brasselet, F. Treussart, J.-F. Roch, F. Marquier, D. Chauvat, S. Perruchas, C. Tard, T. Gacoin, "Balanced homodyne detection of second-harmonic generation from isolated subwavelength emitters", Appl. Phys. Lett. 89, 121118 (2006).

L. Le Xuan, C. Zhou, A. Slablab, D. Chauvat, C. Tard, S. Perruchas, T. Gacoin, P. Villeval, and J. -F.Roch, "Photostable Second-Harmonic Generation from a Single KTiOPO4 Nanocrystal for NonlinearMicroscopy," Small 4, 1332-1336 (2008).

J. P. Long, B. S. Simpkins, D. J. Rowenhorst, P. E. Pehrsson, "Far-field Imaging of Optical Second-Harmonic Generation in Single GaN Nanowires", Nano Lett. 7, 831 (2007).

V. V. Lozovoy, I. Pastirk, and M. Dantus, "Multiphoton intrapulse interference. IV. Ultrashort laser pulsespectral phase characterization and compensation," Opt. Lett. 29, 775-777 (2004).

D. Meshulach and Y. Silberberg, "Coherent quantum control of two-photon transitions by a femtosecondlaser pulse," Nature 396, 239-242 (1998).

L. Moreaux, O. Sandre, J. Mertz, "Membrane imaging by second-harmonic generation microscopy", J. Opt. Soc. Am. B 17, 1685 (2000).

Y. Nakayama, P. J. Pauzauskie, A. Radenovic, R. M. Onorato, R. J. Saykally, J. Liphardt, P. Yang, "Tunable nanowire nonlinear optical probe", Nature 447, 1098, (2007).

G. H. Patterson, D.W. Piston, "Photobleaching in Two-Photon Excitation Microscopy", Biophys. J. 78, 2159 (2000).

P. Rejmankova, J. Baruchel, P. Villeval, C. Saunal, "Characterization of large KTiOPO4 flux grown crystals by synchrotron radiation topography", J. of Crystal Growth 180, 85 (1997).

N. Sandeau, L. Le Xuan, C. Zhou, D. Chauvat, J.-F. Roch, S. Brasselet, "Defocused imaging of second harmonic generation from a single nanocrystal", Opt. Express 15, 16051 (2007).

Y. Shen, P. Markowicz, J.Winiarz, J. Swiatkiewicz, P. N. Prasad, "Nanoscopic study of second-harmonic generation in organic crystals with collection-mode near-field scanning optical microscopy", Opt. Lett. 26, 725 (2001).

Y. R. Shen, "Surface properties probed by second-harmonic and sum-frequency generation", Nature 337, 519 (1989).

M. I. Stockman, D. J. Bergman, C. Anceau, S. Brasselet, J. Zyss, "Enhanced Second-Harmonic Generation by Metal Surfaces with Nanoscale Roughness: Nanoscale Dephasing, Depolarization, and Correlations", Phys. Rev. Lett. 92, 057402 (2004).

R. Stolle, G. Marowsky, E. Schwarzberg, G. Berkovic, "Phase measurements in nonlinear optics", Appl. Phys. B, 63, 491 (1996) and Refs. therein.

R. L. Sutherland, *Handbook of Nonlinear Optics* , Dekker, New York, 1996 pp. 267.

F. Treussart, E. Botzung-Appert, N. T. Ha-Duong, A. Ibanez. J.-F. Roch, R. Pansu, "Second Harmonic Generation and Fluorescence of CMONS Dye Nanocrystals Grown in a Sol-Gel Thin Film", Chem. Phys. Chem. 4, 757 (2003).

F. Treussart, V. Jacques, E Wu, T. Gacoin, P. Grangier, J.-F. Roch, "Photoluminescence of single colour defects in 50ănm diamond nanocrystals", Physica B 376, 926 (2006).

H. Vanherzeele, and J. D. Bierlein, "Magnitude of the nonlinear-optical coefficients of KTiOPO4 ", Opt. Lett. 17,982-985 (1992).

W. S. Warren, R. Rabitz, and M. Daleh, "Coherent Control of Quantum Dynamics: The Dream Is Alive,"Science 259, 1581-1589 (1993).

A. M. Weiner, "Femtosecond pulse shaping using spatial light modulators," Rev. Sci. Instrum. 71, 1929-1960 (2000).

P. Wnuk and C. Radzewicz, "Coherent control and dark pulses in second harmonic generation," Opt.Commun. 272, 496-502 (2007).

P. Wnuk, L. Le Xuan, A. Slablab, C. Tard, S. Perruchas, T. Gacoin, J-F. Roch, D. Chauvat, and C. Radzewicz, "Coherent nonlinear emission from a single KTP nanoparticle with broadband femtosecond pulses", Optics Express, 17, 4652 (2009)

P. Xi, Y. Andegeko, L. R. Weisel, V. V. Lozovoy, and M. Dantus, "Greater Signal and LessPhotobleaching in Two-Photon Microscopy with Ultrabroad Bandwidth Femtosecond Pulses," Opt.Commun. 281, 1841-1849 (2008).

H. P. Yuen, V. W. S. Chan, "Noise in homodyne and heterodyne detection," Opt. Lett. 8, 177 (1983).

F. C. Zumsteg, J. D. Bierlein, T. E. Gier, "$K_xRb_{1-x}TiOPO_4$: A New Nonlinear Optical Material," J. Appl. Phys. 47, 4980 (1976).

Zeolite nanocrystals - synthesis and applications

Teruoki Tago and Takao Masuda
Division of Chemical Process Engineering, Faculty of Engineering, Hokkaido University N13W8, Kita-Ku Sapporo, Hokkaido, 060-8628, Japan

1. Introduction

Zeolites, crystalline aluminosilicate materials, possess 3-dimensionaly connected framework structures constructed from corner-sharing TO_4 tetrahedra, where T is any tetrahedrally-coordinated cation such as Si and Al. These framework structures are composed of n-rings, where n is the number of T-atoms in the ring (e.g. 4-, 5-, and 6-rings), and large pore openings of 8-, 10-, and 12-rings are framed by these small rings. Figure 1 shows the pore sizes and framework structures of typical zeolites. The sizes of the intracrystalline pores and nanospaces, depending on the type of zeolite providing the framework, are close to the molecular diameters of lighter hydrocarbons. Moreover, strong acid sites exist on the nanopore surfaces, enabling the zeolites to be used as shape-selective catalysts in industrial applications such as fluid catalytic cracking of heavy oil, isomerization of xylene and synthesis of ethyl-benzene. However, compared to the sizes of micropores exhibiting a molecular-sieving effect, the crystal sizes of zeolites are very large, approximately 1–3 μm. When the zeolite is used as a shape-selective catalyst, the diffusion rates of reactant molecules within the zeolite crystals are lower than the intrinsic reaction rates. This resistance to mass transfer limits the reaction rates and low selectivity of intermediates. Moreover, since effective active sites (acid sites) for catalytic reactions are distributed on the internal surfaces of the main channels and the external surfaces of the crystal, the pore mouths are easily plugged due to coke deposition, leading to short lifetimes for the catalysts.

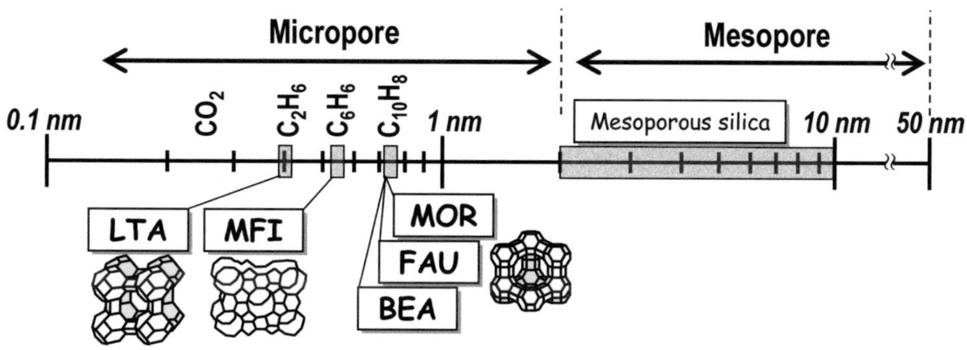

Fig. 1. Zeolite structure, pore size and molecular diameter of hydrocarbons.

Faster mass transfer is required to avoid these serious problems, and two primary strategies have been proposed; one is the formation of meso-pores within zeolite crystals (Groen et al., 20004a, 2004b; Ogura et al., 2001), and the other is the preparation of nano-crystalline zeolites (Tsapatsis et al., 1996; Mintova et al., 1999, 2002, 2006; Ravishankar et al., 1998; Grieken et al., 2000; Hincapie et al., 2004; Song et al., 2005; Larlus et al., 2006; Morales-Pacheco et al., 2007; Kumar et al., 2007).

In nano-crystalline zeolites, the diffusion length for reactant hydrocarbons, assignable to the crystal size, decreases, and the external surface area of the crystal increases as the crystal size decreases. The increase in external surface area and the decrease in diffusion resistance are effective for improving catalytic activity in gas-solid and liquid-solid heterogeneous catalytic reactions. Because of these favorable properties of nano-crystalline zeolites for catalytic reactions, the preparation methods for several types of zeolite nanocrystals have been reported and reviewed (Tosheva & Valtchev, 2005; Larsen, 2007).

In this chapter, a method for preparing nano-crystalline MFI and MOR zeolites in a solution consisting of a surfactant, organic solvent, and water is introduced (denoted as the emulsion method hereafter). Nano-crystalline zeolite is expected to be a promising material for increasing the outer surface area as well as decreasing the diffusion resistance of the organic reactant within the micropores, thereby improving the catalytic activity and lifetime.

2. Synthesis of nano-crystalline zeolites in water/surfactant/organic solvent

Usually, zeolites are prepared in an alkaline water solution containing Si and Al sources, and alkaline metal ions (sodium or potassium). In the synthesis of some types of zeolite, an organic structure directing agent (OSDA), such as an ammonium alkyl cation, is also necessary to form the zeolite framework. The water solutions containing these inorganics and OSDA are poured into a Teflon-sealed stainless steel bottle and heated to a desirable temperature.

In some types of zeolites (e.g. MFI and MOR), crystal nucleation occurs first, followed by the growth of the zeolite, with the nucleation simultaneously continuing to occur during the growth stage. One method to prepare nano-crystalline zeolite is the stoppage of the hydrothermal treatment when the nucleation occurs. However, because unreacted Si and Al species still remain in the solution, an amorphous phase tends to form on the surface of the zeolite crystals. Another method is increasing the nucleation rate by increasing the concentrations of the Si and Al sources. With increasing nucleation rate, the crystal size of the zeolite decreases. However, the zeolite thus obtained has a broad size distribution due to simultaneous nucleation and crystal growth during the hydrothermal treatment. Accordingly, in order to obtain nano-crystalline zeolites, it is important to separate the nucleation and growth stages. Recently, there has been growing interest in the synthesis of nano-crystalline zeolites with the addition of surfactants, including water/surfactant/organic mixture (Kuechl et al., 2010; Naik et al., 2002; Lee et al., 2005; Tago et al., 2004, 2009a, 2009b) (emulsion method), with research focused on size and morphology control, because these parameters affect the performance in applications, such as separation and catalytic reactions. Because surfactant molecules that have hydrophobic and hydrophilic organic groups in the molecules are adsorbed on a solid surface in a solvent, the interface energy difference between the solid surface and solvent can be reduced, leading to

enhancement of nucleation of metal and/or metal oxide nano-particles. In the emulsion method, the surfactant adsorption effect is applied to prepare nano-crystalline zeolites.
In general, there are four types of surfactants, anionic, cationic, nonionic, and bipolar surfactants. A nonionic surfactants; Polyoxyethylene(15)oleylether (O-15), polyoxyethylene(15)nonylphenylether (N-15), and poly(oxyethylene)(15)cethylether (C-15), and ionic surfactants, sodium bis(2-ethylhexyl) sulfosuccinate (AOT) and cetyltrimethyl ammonium bromide (CTAB), are employed. (Figure 2) Cyclohexane or 1-hexanol is used as an organic solvent.

nonionic surfactant

polyoxyethylene(15)nonylphenylether :(NP-15)

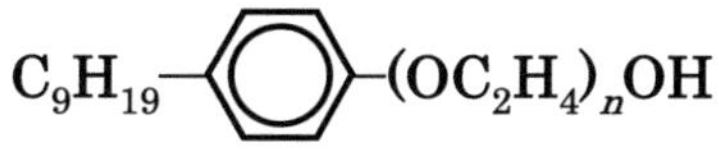

Polyoxyethylene(15)oleylether : (O-15)

$$C_9H_{18}= C_9H_{17} -(OC_2H_4)_nOH$$

poly(oxyethylene)(15)cethylether : (C-15)

$$C_{16}H_{33}-(OC_2H_4)_nOH$$

ionic surfactant

cetyltrimethyl ammonium bromide : (CTAB)

$$C_{16}H_{33}N(CH_3)_3Br$$

sodium bis(2-ethylhexyl) sulfosuccinate : (AOT)

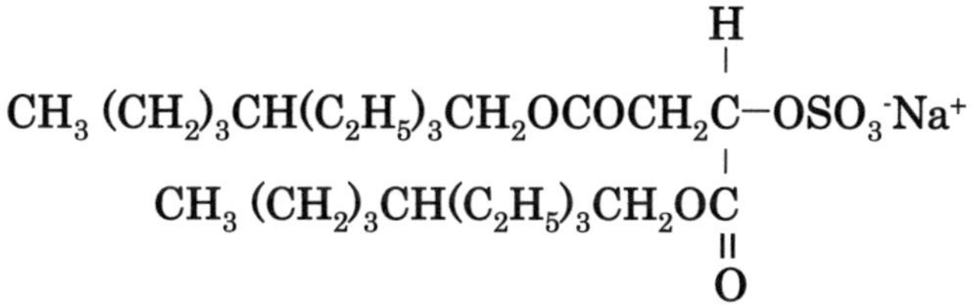

Fig. 2. Typical molecular structures of surfactants.

In the emulsion method, two solutions are prepared, one is a water solution containing Si and Al sources obtained by hydrolyzing metal alkoxide with a dilute OSDA/water solution, and the other is a surfactant/organic solvent. The concentrations of the Si and Al sources in the water solution and the molar ratio of Si to the OSDA as well as types of the OSDA are very important factors in preparing zeolite crystals. In the case of MFI, tetrapropylammoniumhydroxide (TPAOH) is used as an OSDA, and tetraethylammoniumhydroxide (TEAOH) is used for the preparation of MOR zeolite.

Moreover, two additional parameters, the concentration of surfactant in the organic solvent and the molar ratio of water to surfactant, are important in the emulsion method. These parameters affect the morphology and crystal sizes.

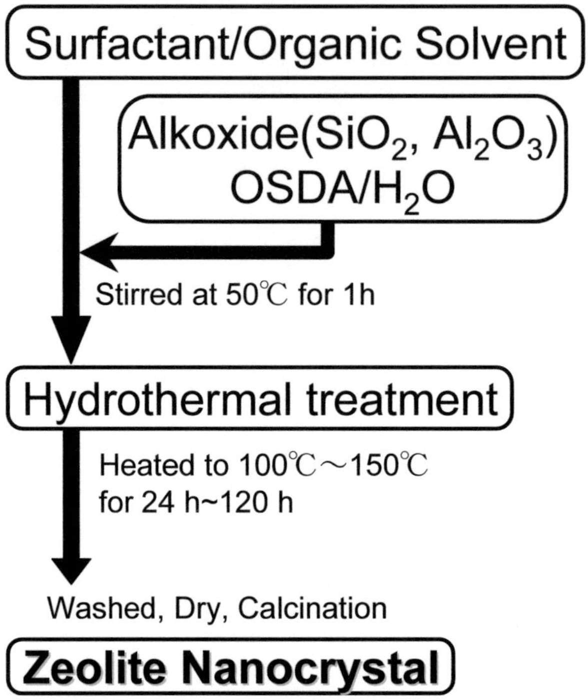

Fig. 3. Preparation procedure for nano-crystalline zeolite by the emulsion method.

The water solution thus obtained is added to the surfactant-organic solvent, and the mixture is magnetically stirred at 323 K for 1 h. The water/surfactant/organic solvent mixture is then placed in a Teflon-sealed stainless steel bottle, heated to 373 ~ 423 K, and held at the desired temperature for 12 ~ 120 h with stirring to yield zeolite nanocrystals. The precipitate thus obtained is centrifuged, thoroughly washed with propanol, dried at 100 °C overnight, and calcined under an air flow at 500 °C to remove the surfactant and the OSDA molecules. After the air calcination, the weight of the sample is measured to calculate the zeolite yield (Figure 3).

2.1 Effect of ionicity of surfactant on preparation of MFI zeolite nanocrystals

The morphologies of the silicalite-1 (MFI zeolite) nanocrystals prepared using various surfactants were investigated under conditions of TOES concentration of 0.63 or 2.73 mol/L, hydrothermal temperature of 140 or 100 °C and concentration of surfactant in organic solvent (solution B) of 0.50 mol/L. Figures 4 and 5 show X-ray diffraction (XRD) patterns and FE-SEM images of the obtained samples, respectively.

In the XRD pattern for the sample prepared in the AOT/cyclohexane solution, peaks corresponding to sodium sulfate rather than silicalite-1 are seen. Moreover, the sample showed an irregular morphology. In this case, the TPA-OH molecules cannot act as an OSDA in the synthetic solution. This result is ascribed to the fact that the surfactant of AOT and the OSDA of TPA-OH have opposite ionic charges.

When using CTAB/1-hexanol, the silicalite-1 crystals, which are approximately 1.0 μm in size, are embedded in the amorphous SiO_2 as seen from SEM observations. The coexistence of silicalite-1 crystals and amorphous SiO_2 is revealed by the X-ray diffraction analysis. Since the pH of the synthetic solution is alkaline, the surface of the SiO_2 produced by hydrolysis of TEOS has a negative charge. Accordingly, it is considered that CTAB and TPA-OH are independently adsorbed on the surface of SiO_2 because of their cationic ionicity. Therefore, the SiO_2 species that adsorb TPA-OH and CTAB change into silicalite-1 crystals and amorphous SiO_2, respectively.

In the case of the nonionic surfactants, C-15, NP-15 and O-15 (the nonionic surfactant/cyclohexane system), the XRD patterns of the samples showed peaks corresponding to pentasile-type zeolite (the reference silicalite-1), and mono-dispersed silicalite-1 nano-crystals were obtained, as seen from SEM observations. In contrast, the silicalite-1 crystal prepared in water (without a surfactant) shows a heterogeneous structure with smaller crystals (diameter of approximately 30 nm) on a larger one (approximately 120 nm), indicating that nucleation, crystallization and crystal growth occurred simultaneously. These results indicate that the ionicity of the hydrophilic groups in the surfactant molecules plays an important role in the formation and crystallization processes of the silicalite-1 nanocrystals. Since the aggregation of the silicalite-1 nuclei are inhibited by the adsorbed surfactants on the surface during hydrothermal treatment, mono-dispersed nanocrystals can be prepared.

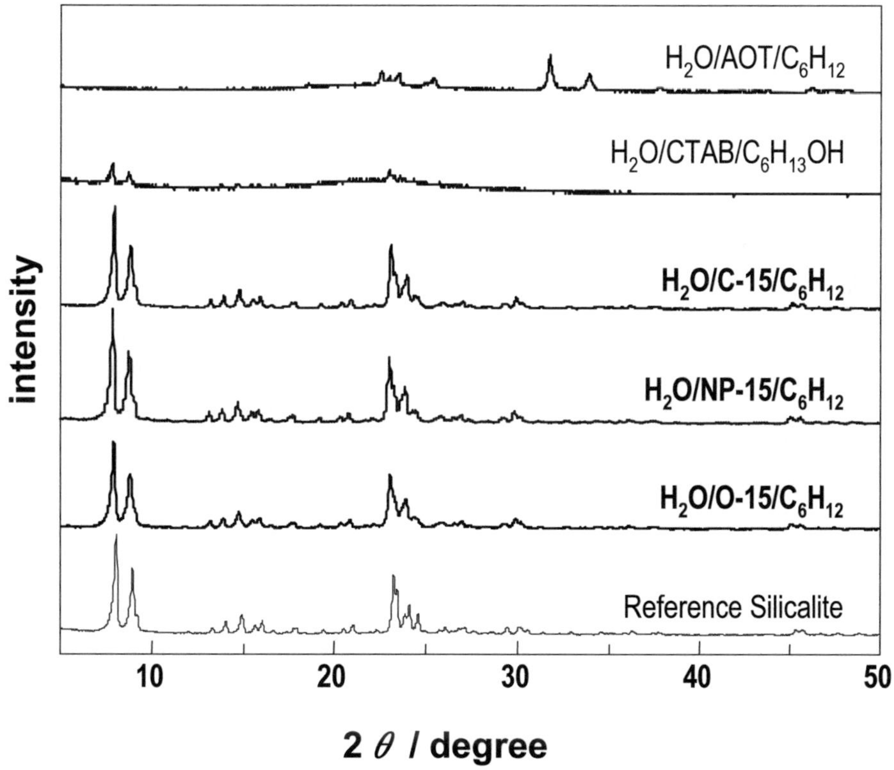

Fig. 4. X-ray diffraction patterns of samples prepared in water/surfactant/organic solvent. Effects of ionicity of the surfactant on crystallinity of MFI zeolite (Silicalite-1).

Nano-crystalline zeolite has a high external surface area, which is very important for use as heterogeneous catalysts and as seed crystals. To evaluate the crystallinity of the nanocrystals, X-ray diffraction patterns and Raman spectroscopy can be used. The former is based on elastic scattering of X-rays from structures that have long range order (zeolite crystal structure, types of zeolite), and the latter method can elucidate middle-range order (types of Si-O rings). Accordingly, the Raman spectra measurements are useful to evaluate the zeolite crystallinity in detail (Dutta et al., 1991; Li et al., 2000, 2001). Figure 6 shows the Raman spectrum of the silicalite-1 nanocrystals. The peaks in the range of 300 - 650 cm^{-1} are indicative of the type of silicon-oxygen rings present in the structure of zeolite. The spectrum also shows peaks around 380, 430 and 470 cm^{-1}, which correspond to the five-, six- and four-member rings, respectively. These peaks are in good agreement with the peaks of the reference silicalite-1 prepared in water. In the structure of the MFI-type zeolite, continuous chains of five-member rings are connected by the four- and six-member rings. These results indicated that the surfaces of the nanocrystals with a diameter of approximately 120 and 80 nm are also well-crystallized without amorphous SiO_2.

Fig. 5. FE-SEM micrographs of samples prepared in water/surfactant/organic solvent. Effects of ionicity of the surfactant on crystallinity of MFI zeolite (Silicalite-1).

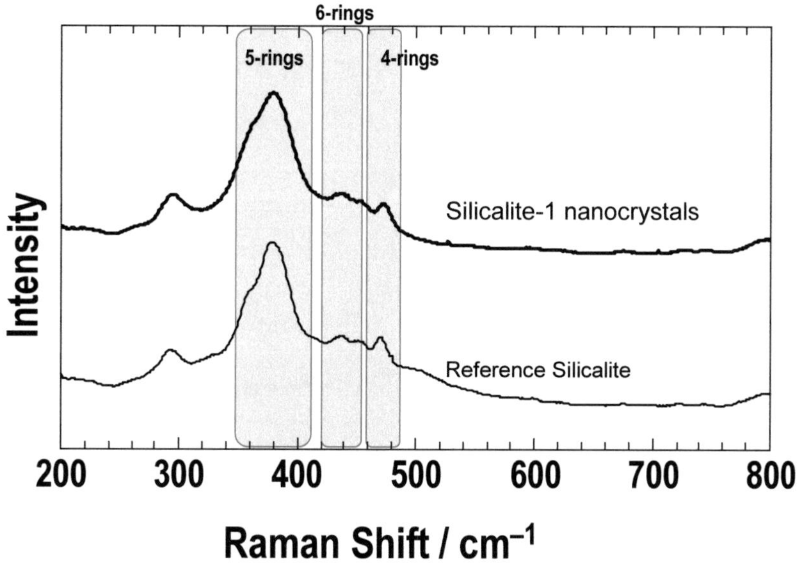

Fig. 6. Raman spectrum of the nanocrystals.

2.2 Preparation of MOR zeolite by the emulsion method

In the preparation of MFI zeolite nanocrystals by the emulsion method, since the non-ionic surfactant is revealed to be the appropriate surfactant to prepare nano-crystalline zeolite, this method is applied to prepare MOR zeolite nanocrystals as well. First, the effects of the hydrothermal time on the crystallinity of MOR zeolite prepared in water/O-15/cyclohexane are examined. The concentrations of Si in a water solution and a surfactant in cyclohexane are 0.75 and 0.5 mol/L, respectively, and the hydrothermal temperature is 423 K. Figure 7 shows the X-ray diffraction patterns of MOR zeolite prepared using the emulsion method with different hydrothermal times. The X-ray diffraction patterns of samples prepared in the water solution without a surfactant (conventional method) are also shown in the figure for comparison.

The samples prepared in the water solution show a broad halo pattern of SiO_2 at hydrothermal times of 72 and 96 h, and show peaks corresponding to MOR zeolite as well as BEA zeolite (beta-type zeolite) at 120 h. In contrast, as compared to the sample prepared in the water solution, though an amorphous pattern is observed from the sample prepared at a hydrothermal time of 72 h, the samples at hydrothermal times of 96 and 120 h show peaks corresponding to MOR zeolite. Moreover, peaks ascribable to other types of zeolite, e.g. MFI and BEA, cannot be detected. Because hydrolysis of the Si (tetraethylorthosilicate) and Al (aluminum-tri-isopropoxide) sources should be completed during the preparation of water solution containing Si, Al, and the OSDA molecules (TEA-OH), the appearance of the diffraction peaks corresponding to MOR zeolite after 96 h indicates that the MOR zeolite precursors are prepared in the solution until approximately 72 h, followed by nucleation and crystal growth of MOR zeolite. Accordingly, MOR and MFI zeolites can be prepared in water/surfactant/organic solvent. Sizes control by the emulsion method is next examined.

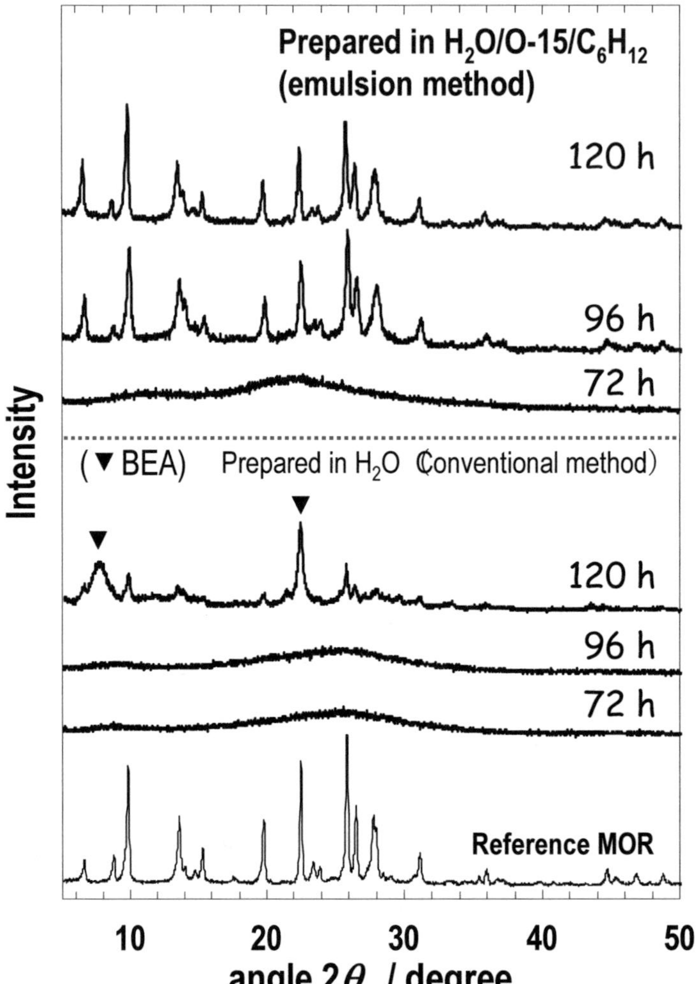

Fig. 7. X-ray diffraction patterns of MOR zeolite prepared in the emulsion method with different hydrothermal times.

2.3 Crystal size control

In a catalytic reaction using a zeolite, the reaction proceeds on acid sites located on the outer surface of the crystal as well as the inside pore surface. The crystalline zeolite affects the outer surface area and the diffusion length, assignable to the crystal size, for reactant hydrocarbons within the crystal. Accordingly, it is very important to develop preparation methods for zeolite nanocrystals with different crystal sizes.

In the emulsion method, it is considered that non-ionic surfactants adsorbed on the surfaces of zeolite precursors induce the formation of zeolite nuclei, enhancing the nucleation of zeolite. Moreover, the surfactant concentration in the solution can be easily changed. Accordingly, the effects of varying surfactant concentration [O-15] on the morphology of MOR zeolite were examined. Figure 8 shows FE-SEM images of the obtained samples. The Si/Al ratio is 12.5. The surfactant concentrations were changed in the range from 0.5 to 0.75

mol/L. Interestingly, the crystal size and morphology depended on the surfactant concentration, regardless of the same concentrations of the Si and Al sources, and template in the water solution. MOR zeolite nanocrystals with an average size of approximately 80 nm were obtained at a surfactant concentration of 0.5 mol/L. As the concentration increased to 0.65 mol/L, growth of MOR zeolite with column-like morphology is observed, and the crystal size reached ~1.0 µm at 0.75 mol/L. Figure 9 shows NH_3-TPD profiles and N_2 adsorption-desorption isotherms of the obtained samples, respectively, at surfactant concentrations of 0.5 and 0.75 mol/L. The NH_3-TPD profiles of these sample shows NH_3 desorption peaks above 600 K, ascribable to desorption of NH_3 from strong acid sites of MOR zeolites. Moreover, these MOR zeolites show almost the same peak area. The MOR zeolites also show almost the same N_2 adsorption-desorption isotherms, indicating that these zeolites exhibit almost the same surface area. Accordingly, these MOR zeolites possess almost the same number of acid sites, regardless of the crystal morphology.
As discussed above, it is revealed that the surfactant in the synthetic solution induces the nucleation and crystallization of zeolite as compared to the conventional preparation method (without a surfactant). Moreover, the surfactant concentrations in the solution were found to influence crystal growth. The crystal morphology changed from the nano-crystalline to large columnar crystals with increasing surfactant concentration.

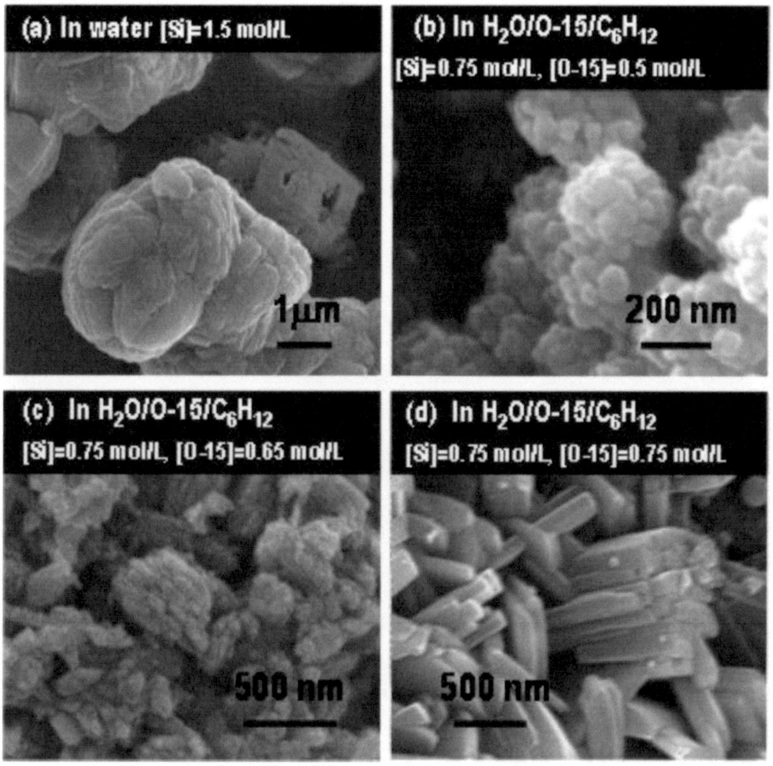

Fig. 8. FE-SEM micrographs of MOR zeolite with different crystal sizes. (Tago *et al.*, (2009b))

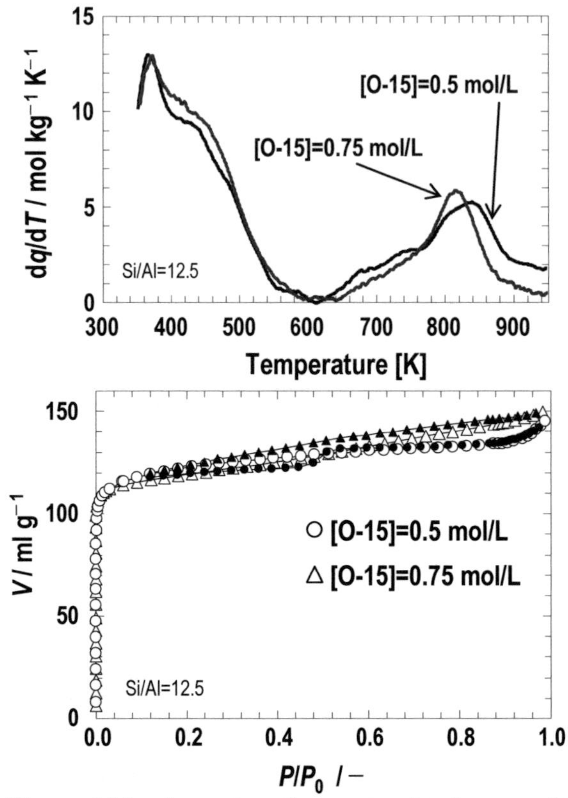

Fig. 9. NH_3–TPD profiles and N_2 adsorption–desorption isotherms of the MOR zeolite with different crystal sizes. (Tago *et al.*, (2009b))

2.4 Mechanism

In order to consider the mechanism of zeolite nanocrystals formation in water/surfactant/organic solvent, it is necessary to investigate the relationship between the ionicity of the surfactant and the template molecule, and the ionic charges of the surface of SiO_2 and/or zeolite precursor in detail.

In the method, the cyclohexane used as the organic solvent contributes to stabilization of the hydrophobic group of the surfactant. The effects of the ionicity of the hydrophilic group of the surfactant on the morphology and crystallinity of the obtained silicalite-1 zeolite nanocrystals are described above.

In an anionic surfactant (e.g. sodium bis(2-ethylhexyl) sulfosuccinate, AOT), because the surfactant of AOT has opposite ionic charges to the OSDA molecules, the OSDA cannot act as a structure-determining agent in the synthetic solution. Accordingly, in the relationship between the ionicity of the surfactant and OSDA molecules, a surfactant without electrostatic affinity to the OSDA molecules is needed for the preparation of zeolite crystals. In a cationic surfactant (e.g. cetyltrimethyl ammonium bromide, CTA-Br), the molecular structure is similar to that of the OSDA. Moreover, since the surface charge of the SiO_2 and/or zeolite precursor is negative, the surfactant molecules of CTA^+ with cationic hydrophilic group can be adsorbed on the surface, followed by the formation of amorphous SiO_2. In contrast, in the case of nonionic surfactants, MOR as well as MFI zeolite nanocrystals can be obtained.

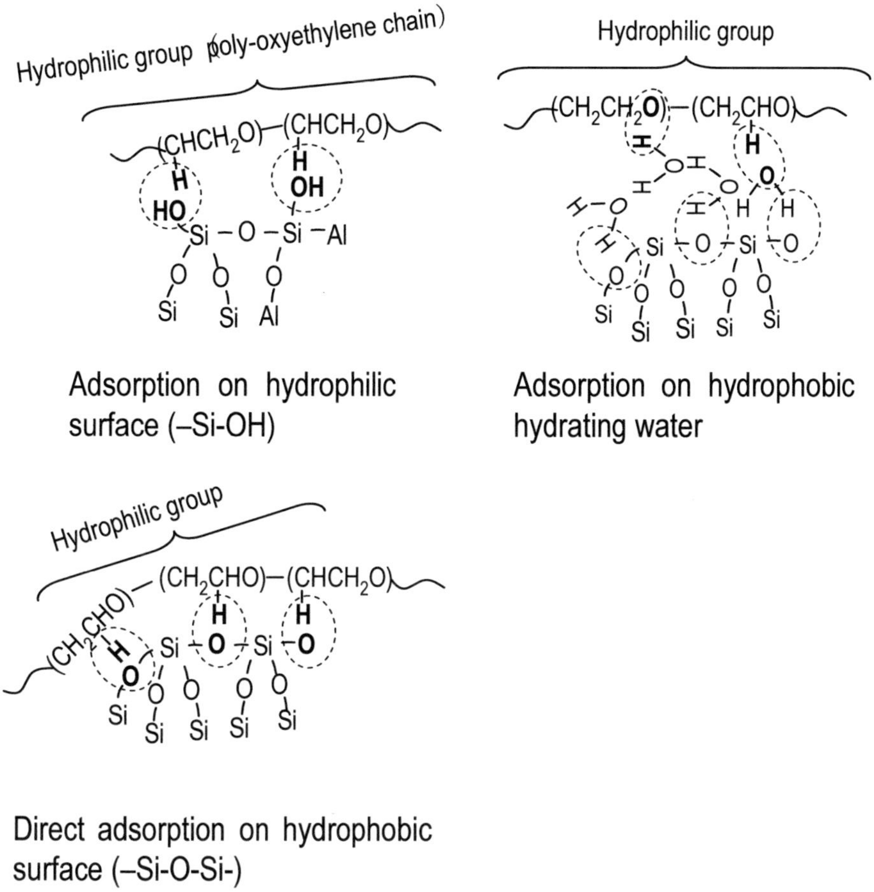

Fig. 10. A schematic showing the possible relationship between the surfactant and the zeolite surface. (Tago *et al.*, (2009b))

A schematic figure showing the possible relationship between the surfactant and the zeolite surface is shown in Fig. 10. The hydrophilic group of the surfactant O-15 is composed of poly-oxyethylene chains, making it a non-ionic surfactant. The surface of the zeolite precursor and crystal is composed of a hydrophobic surface (-Si-O-Si-) and hydrophilic silanol groups (-O-Si-OH). The hydrophilic groups of the surfactant can be adsorbed on the hydrophilic silanol groups, leading to stabilization of the silanol groups. On the hydrophobic surface, hydrophobic hydration occurs in the conventional preparation method without a surfactant (Burkett & Davis, 1994, 1995; De Moor, 1999a, 1999b). It is considered that the hydrophilic groups of the surfactant are adsorbed onto the water molecules hydrophobically-hydrated on the zeolite surface. Moreover, because the hydrophilic property of the poly-oxyethylene chains in the surfactant is much smaller than that in water, it is considered that the hydrophilic poly-oxyethylene chains can be directly adsorbed onto the zeolite surface due to their higher affinity to the zeolite surface than water molecules. These adsorbed surfactants can contribute to the stabilization of the surface of zeolite precursor and crystals, which will influence the nucleation rate of zeolite, so that the MOR zeolite nanocrystals can be obtained.

Three-dimensional cylindrical networks of the micellar structure composed of water and surfactant molecules likely exist in water/O-15/cyclohexane solvent, wherein silica-based cylindrical (Matsune et al., 2006) and spherical (Tago et al., 2002, 2003; Takenaka et al., 2007) materials including metal and/or metal-oxide nano-particles are obtained at 323 K. These networks lead to an increase in the viscosity of the solvent. Though the reaction temperature of 323 K is different from the hydrothermal temperature, the viscosity of the zeolite synthetic solution is thought to be high. It is considered that the viscosity of the solvent affected the nucleation rate and diffusion of silicate species. However, since the yields of zeolite are above 80%, the diffusion of silicate species in the solution is likely not a rate-limiting step to form zeolite nanocrystals, and thus the viscosity affects the nucleation rate. Moreover, as the surfactant concentration is increased, the excessive stabilization of the precursors and the high viscosity of the solution lead to a decrease in the nucleation rate of zeolite. Since the growth rate of the crystal is much higher than the nucleation rate, crystalline growth of the MOR zeolite is dominant at high surfactant concentrations, so that column-like crystals with a large size are obtained. Accordingly, the surfactant can play an important role in the nucleation and crystal growth of the MFI and MOR zeolite prepared in the water/surfactant/organic solvent (emulsion method), and the crystal size and morphology can be controlled by the surfactant concentration.

3. Applications of nanocrystals

3.1 Heterogeneous catalytic reaction over zeolite nanocrystals

The acid sites in zeolites are located on the external and internal surfaces. In nano-crystalline zeolite, the external surface area of the crystal increases and the diffusion length for reactant hydrocarbons, assignable to the crystal size, decreases as the crystal size decreases. Due to these favorable properties, nano-crystalline zeolites have been used as heterogeneous catalysts.

Zeolites exhibit a molecular sieving effect for lighter hydrocarbons, hence, the hydrocarbon molecules with sizes larger than the pore mouth size are mainly adsorbed on the external surface. Accordingly, an increase in the external surface area is effective for increasing catalytic activity. Song *et al.* (2004a, 2004b) reported the preparation of nano-crystalline MFI zeolites and they evaluated the effects of crystal size on external surface area and adsorption properties of toluene on the zeolite. Since the molecular size of toluene is almost the same as the pore mouth size in MFI zeolite, the external surface area, which is a function of the crystal size, affects the adsorption capacity for toluene. Zeolites with crystal sizes less than 100 nm have a higher adsorption capacity for toluene. Botella *et al.* (2007) have reported the Beckmann rearrangement reaction over nanosized BEA zeolite where the reactions mainly proceed on the external surface of zeolite. Serrano *et al.* (2005) have reported catalytic cracking of polyolefins over nano-crystalline MFI zeolite. Because the molecular size of polyolefins is larger than the pore size, cracking of polyolefins mainly occurs over the acid sites on the external surface, while the formation of lighter olefins proceeds on the internal pore surface. Jia *et al.* (2010) have reported dehydration of glycerol to produce acrolein over nano-sized MFI zeolite. Since the hydrophilic glycerol molecules are strongly adsorbed on the external surface of the zeolite, nano-crystalline MFI zeolite is effective for this reaction due to the high surface area. In these reports listed above, the increase in the external surface area of the zeolite nanocrystal is important to improve the catalytic activity and lifetime.

In contrast, Sakthivel *et al.* (2009) have reported the isomerization/cracking of hexane over nano-sized BEA zeolite. Since the molecular size of hexane is smaller than the pore size of MFI zeolite, the cracking reaction proceeds mainly on the acid sites located inside pores. Moreover, the small crystal size of BEA zeolite leads to low diffusion resistance of the reactant hexane and the products. Serrano *et al.* (2010) have reported epoxide rearrangement reactions over MFI zeolite, where the high external surface area and low diffusion resistance within the micropores due to the small zeolite crystal size enhance the rearrangement reaction without diffusion restriction of the epoxide molecules.
MFI zeolites are effective catalysts to synthesize lighter olefins by the methanol-to-olefin (MTO) reaction and acetone conversion to olefin. In these olefin syntheses, selective formation of lighter olefins occurs due to the molecular sieving effect of the zeolite. However, when the reaction of the hydrocarbon occurs over acid sites located on the external surface of zeolite crystals, this results in a non-shape-selective reaction as well as coke deposition, leading to short lifetimes for the catalyst. These undesirable phenomena will be accelerated in nano-crystalline zeolites because of their large external surface areas. Accordingly, a method for selective deactivation of acid sites located on the external surface of zeolite crystals is desired in order to prevent these undesirable reactions. Masuda *et al.* (2001) have reported a method called the "catalytic cracking of silane method", in which SiO_2 units are formed selectively on the acid sites of the zeolite using organic silane compounds. Tago *et al.* (2009c) have reported that the yields of lighter olefins such as ethylene and propylene are increased over the selectively de-activated MFI zeolite.

3.2 Structured materials composed of zeolite nanocrystals

Since the nano-crystalline zeolites have large external surface areas, they are potential candidates for seed crystals from which a structured material can be prepared. Moreover, the micro- and meso-pores formed in nano-crystalline zeolites are characteristic of the structured materials. Accordingly, design of the micro- and meso-pores is important.
Serrano *et al.* have reported the preparation of micro-, meso-, and macroscopic hierarchical materials composed of nano-crystalline zeolites. They prepared hybrid zeolitic-mesoporous materials with the properties of both MCM-41 and MFI zeolite using zeolite seeds by assembling around cetyltrimetyl ammonium bromide micelles. Moreover, they have also developed other methods to fabricate hierarchical materials using organosilane bonded zeolite nanocrystals, where the degree of aggregation of the zeolite can be controlled.
Nano-crystalline zeolites have been used as seed crystals for zeolite membranes. Since the pore spaces and pore sizes in the crystals represent a very important factor, affecting membrane performances (Hasegawa *et al.*, 2007), nano-crystalline zeolite crystals are useful as seed crystals. In the preparation of zeolite membranes, after the deposition and/or seeding of the nano-crystalline zeolites on the base materials, secondary growth of the zeolite was carried out by a hydrothermal treatment (Kita *et al.*, 1995; Kondo *et al.*, 1997). Wang *et al.* (2002) have reported nano-structured zeolite 4A membranes, wherein nano-crystalline zeolites 4A were seeded on an alumina filter by a dip-coating method. Hasegawa *et al.* (2006) and Tago *et al.* (2008) have reported Silicalite-1 nanocrystal-layered membranes, where the Silicalite-1 nanocrystals are piled up on an alumina filter, followed by hydrothermal synthesis to form a Silicalite-1 layer on the nanocrystal layer. In these membranes, the thickness of the nanocrystal layer (amount of nano-crystalline zeolite used as seed crystals) can be easily changed, so that the separation properties of these membranes

can be controlled (Hedlund, *et al.*, 1999; Pera-Titus *et al.*, 2005). Moreover, because the zeolite crystals are seed crystals for the membranes, the separation properties of the membranes are improved with decreasing crystal sizes of the zeolite.

4. Conclusions

Zeolites are being used extensively in industrial processes such as heterogeneous catalysts and adsorbents. Nano-crystalline zeolites have large external surface areas as well as low diffusion resistance. Accordingly, control of the crystal size and the tailoring the mesopores and macropores among the crystals are important factors for applications of these nanocrystals. In order to obtain nano-crystalline zeolites, it is important to separate the nucleation and growth stages. The addition of a surfactant into synthetic solutions of zeolites is a promising method to control the crystal size and improve the crystallinity due to adsorption of the surfactant on the zeolite surface.

5. References

Aguadoa, J.; Serrano. D. P. & Rodriguez, J. M. (2008). *Micropor. Mesopor. Mat.*, 115, 3, 504-513, 1387-1811

Botella, P.; Corma, A., Iborra, S., Montón, R., Rodríguez, I. & Costa, V. (2007). *J. Catal.*, 250, 1, 161-170, 0021-9517

Burkett S.L. & Davis M.E. (1994). *J. Phys. Chem.*, 98, 17, 4647-4653, 0022-3654

Burkett S.L. & Davis M.E. (1995). *Chem. Matter.*, 7, 5, 920-928, 0897-4756

De Moor P. P. E. A.; Beelen T. P. M., Komanschek B. U., Beck L.W., Wagner P., Davis M. E. & Santen R. A. (1999a). *Chem. Eur. J.*, 5, 7, 2083-2088, 0947-6539

De Moor P. P. E. A.; Beelen T. P. M. & Van Santen, R. A. (1999b). *J. Phys. Chem. C.*, 103, 10, 1639-1650, 1089-5647

Dutta, P. K.; Rao, K. M. & Park, J. Y. (1991). *J. Phys. Chem.*, 95, 17, 6554-6656, 0022-3654

Grieken, R. V.; Sotelo, J. L., Menendez, J. M. & Melero, J. A. (2000). *Micropor. Mesopor. Mat.*, 39, 1-2, 135-147, 1387-1811

Groen, J. C.; Peffer, L. A. A., Moulijn, J. A. & Perez-Ramirez, J. (2004a). *Micropore. Mesopor. Mat.* 69, 1-2, 29-34, 1387-1811

Groen, J. C. ; Jansen, J. C., Moulijn, J. A. & Perez-Ramirez, J. (2004b). *J. Phys. Chem. B*, 108, 35, 13062-13065, 1520-6106

Hasegawa, Y.; Ikeda, T., Nagase, T., Kiyozumi, Y., Hanaoka, T. & Mizukami, F. (2006). *J. Membr. Sci.*, 280, 1-2, 397-405, 0376-7388

Hasegawa, Y.; Nagase, T., Kiyozumi, Y. & Mizukami, F. (2007). *J. Membr. Sci.*, 294, 1-2, 186-195, 0376-7388

Hedlund, J.; Noack, M., Kolsch, P., Creaser, D., Caro, J.& Sterte, J. (1999). *J. Membr. Sci.*, 159, 1-2, 263-273, 0376-7388

Hincapie, B. O. ; Garces, L. J., Zhang, Q., Sacco, A. & Suib, S. L. (2004). *Micropor. Mesopor. Mat.*, 67, 1, 19-26. 1387-1811

Jia, C. J.; Liu, Y., Schmidt, W., Lu, A. H. & Schuth, F. (2010). *J. Catal.*, 269, 1, 71-79, 0021-9517

Kita, H; Horii, K., Ohtoshi, Y., Tanaka, K. & Okamoto, K. (1995). *J. Mat. Sci. Lett.*, 14, 3, 206-208, 0261-8028

Kondo, M.; Komori, M., Kita, H. & Okamoto, K. (1997), *J. Membr. Sci.*, 133, 1, 133-141, 0376-7388
Kuechl, D. E.; Benin, A. I., Knight, L. M., Abrevaya, H., Wilson, S. T., Sinkler, W., Mezza, T. M. & Willis, R. R. (2010). *Micropor. Mesopor. Mat.*, 127, 1-2, 104-118, 1387-1811
Kumar, S.; Davis, T. M., Ramanan, H., Penn, R. L. & Tsapatsis, M. (2007). *J. Phys. Chem. B*, 111, 13, 3393-3403, 1520-6106
Larlus, O.; Mintova, S. & Bein, T. (2006). *Micropor. Mesopor. Mat.*, 96, 1-3, 405-412, 1387-1811
Larsen, S. C. (2007). *J. Phys. Chem. C*, 111, 50, 18464-18474, 1932-7447
Lee, S.; Carr, C. S. & Shantz, D. F. (2005). *Langmuir*, 21, 25, 12031-12036, 0743-7463
Li, Q. H.; Mihailova, B., Creaser, D. & Sterte, J. (2001). *Micropor. Mesopor. Mat.*, 43, 1, 51-59, 1387-1811
Li Q. H.; Mihailova, B., Creaser, D. & Sterte, J. (2000). *Micropor. Mesopor. Mat.*, 40, 1-3, 53-62, 1387-1811
Masuda, T.; Fukumoto, N., Kitamura, M., Mukai, S. R., Hashimoto, K., Tanaka, T. & Funabiki, T. (2001). *Micropor. Mesopor. Mat.*, 48, 1-3, 239-245, 1387-1811
Matsune, H.; Tago, T., Shibata, K., Wakabayashi, K. & Kishida, M. (2006). *J. Nanoparticle Research*, 8, 6, 1083-1087, 1388-0764
Mintova, S.; Olson, N. H., Valtchev, V. & Bein, T. (1999). *Science*, 283, 5405, 958-960, 0036-8075
Mintva, S. & Valtchev, V. (2002). *Micropor. Mesopor. Mat.* 55, 2, 171-179, 1387-1811
Mintova, S.; Valtchev, V., Onfroy, T., Marichal, C., Knozinger, H. & Bein, T. (2006). *Micropor. Mesopor. Mat.*, 90, 1-3, 237-245, 1387-1811
Morales-Pacheco, P.; Alvarez, F., Del Angel, P., Bucio, L. & Dominguez, J. M. (2007). *J. Phys. Chem. C*, 111, 6, 2368-2378, 1932-7447
Naik, S.P., Chen, J.C. & Chiang, A. S. T.(2002). *Micropor. Mesopor. Mat.*, 54, 3, 293-303, 1387-1811
Ogura, M.; Shinomiya, S., Tateno, J., Nara, Y., Nomura, M., Kikuchi, E. & Matsukata, M. (2001). *Appl. Catal. A. Gen.*, 219, 1-2, 33-43, 0926-860X
Pera-Titus, M.; Llorens, J., Cunill, F., Mallada, R. & Santamaria, J. (1995). *Catal. Today*, 104, 2-4, 281-287, 0920-5861
Ravishankar, R.; Kirschhock, C., Schoeman, B. J., Vanoppen, P., Grobet, P. J., Storck, S., Maier, W. F., Martens, J. A., DeSchryver, F. C. & Jacobs, P. A. (1998). *J. Phys. Chem. B*, 102, 15, 2633-2639, 1089-5647
Sakthivel, A.; Iida, A., Komura, K., Sugi, Y. & Chary K. V. R. (2009). *Micropor. Macropor. Mat.*, 119, 1-3, 322-330, 1387-1811
Serrano, D. P.; Aguado, J., Escola, J. M. & Rodriguez, J. M. (2005). *J. Anal. Appl. Pyrol.*, 74, 1-2, 353-360, 0165-2370
Serrano, D. P.; Aguado, J., Escola, J. M., Rodriguez, J. M. & Peral, A. (2006). *Chem. Mat.*, 18, 10, 2462-2464, 0897-4756
Serrano, D. P.; Garcia, R. A. & Otero, D. (2009). *Appl. Catal. A-Gen.*, 359, 1-2, 69-78, 0926-860X
Serrano, D. P.; Van Grieken, R., Melero, J. A., Garcia, A. & Vargas, C. (2010). *J. Mol. Catal. A: Chem.*, 318, 1-2, 68-74, 1381-1169
Song, W.; Justice, R. E., Jones, C. A., Grassian, V. H. & Larsen, S. C. (2004). *Langmuir*, 20 (2004) 4696
Song, W., Justice, R. E., Jones, C. A., Grassian, V. H. & Larsen, S. C., *Langmuir*, 20 (2004) 8301
Song, W.; Grassian, V. H. & Larsen, S. C. (2005). *Chem. Comm.*, 23, 2951-2953, 1359-7345

Tago, T.; Hatsuta, T., Miyajima, K., Kishida, M., Tashiro, S. & Wakabayashi, K. (2002). *J. Am. Ceram. Soc.*, 85, 9, 2188-2194, 0002-7820

Tago, T.; Tashiro, S., Hashimoto, Y., Wakabayashi, K. & Kishida, M. (2003). *J. Nanoparticle Research*, 5, 1-2, 55-60, 1388-0764

Tago, T.; Nishi, M., Kono, Y. & Masuda, T. (2004). *Chem. Lett.*, 33, 8, 1040-1041, 0366-7022

Tago, T.; Nakasaka, Y., Kayoda, A. & Masuda, T. (2008). *Micropor. Mesopor. Mat.*, 115, 1-2, 176-183, 1387-1811

Tago,T.; Iwakai,K., Nishi, N. & Masuda,T. (2009a). *J. Nanosci. Nanotechnol.*, 9, 1, 612-617, 1533-4880

Tago, T.; Aoki, D., Iwakai, K. & Masuda, T. (2009b). *Top. Catal.*, 52, 6-7, 865-871, 1022-5528

Tago, T.; Sakamoto, M., Iwakai, K., Nishihara, H., Mukai, S. R., Tanaka, T. & Masuda, T. (2009c). *J. Chem. Eng. Jpn.*, 42, 1, 162-167, 0021-9592

Takenaka, S.; Orita, Y., Matsune, H., Tanabe, E. & Kishida, M. (2007). *J. Phys. Chem. C*, 111, 21, 7748-7756, 1932-7447

Tosheva, L. & Valtchev, V. P. (2005). *Chem. Mater.*, 7, 10, 2494-2513, 0897-4756

Tsapatsis, M.; Lovallo, M. & Davis, M. E. (1996). *Micropor. Mat.*, 5, 6, 381-388, 0927-6513

Wang, H.; Huang, L., Holmberg B. A. & Yan, Y. (2002). *Chem. Comm.*, 16, 1708-1709, 1359-7345

9

Complex nature of charge trapping and retention in NC NVM structures

A. Nazarov and V. Turchanikov
Lashkaryov Institute of Semiconductor Physics, National Academy of Sciences of Ukraine
Ukraine

1. Introduction

In conventional silicon floating gate (FG) non-volatile memory (NVM) charge responsible for the information stored is accumulated and retained in a bulk of polysilicon bar consisting of tens of thousands Si atoms where surface to bulk atoms ratio is negligibly small and interface nearly perfect due to the high temperature technological processes. Electrons, usually exploited for charge accumulation, are mainly the same quasiparticles as in the bulk Si isolated by macroscopic potential barrier on the bar boundaries. Typical structure of NVM memory transistor used in embedded and standalone systems is shown on the Fig.1a. Let alone the details of addressing, the main memory structure can be presented by stacked system of substrate, tunnel SiO_2, floating gate, control SiO_2 and program/access gate in split gate devices (Van Houdt et al., 1993) (Fig.1b). In this case programming is realized by hot-electron injection from drain region of MOS transistor and erasing by Fowler-Nordhaim (FN) tunnelling of electrons from floating Si-poly gate through the lowered barrier into silicon. There are a lot of problems that must be solved, but, despite of the efforts spent on development more and more reliable and capacious NVM elements and circuits, some fundamental problems persist. Here we mention only two of them.
Firstly, tunnelling oxide through which the floating gate is charged must not be very thin to prevent excessive random occurrence of leakage channels in dielectric, this, in turn, leads to SiO_2 overstressing by high electric fields needed to reach conditions of FN tunnelling, as contrasted to the direct tunnelling. To disclose the complexity further let us take into the account that the more perfect dielectric we create the more abrupt breakdown we obtain. That is: the transition between safe charging current and irreversible breakdown converges. As a consequence, more and more intricate cell structures are introduced for reprogramming substituting, for example, FN electrons tunnelling by hot holes injection (US Patent, 2004) and embedding additional charge balance controlling circuitry on the chip (US Patent, 2007).

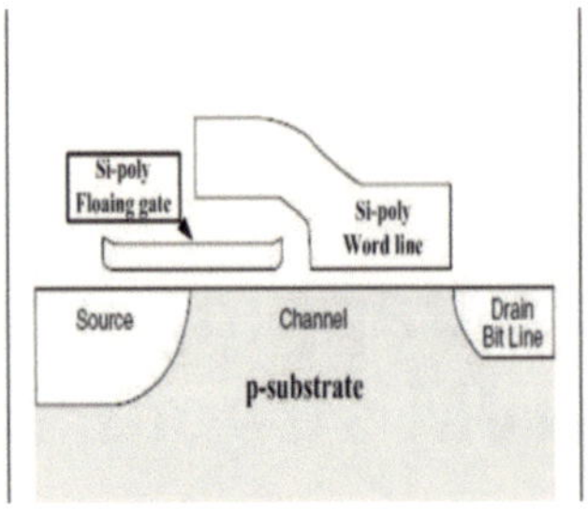

Fig. 1. (a) Typical structure of conventional floating gate NVM cell

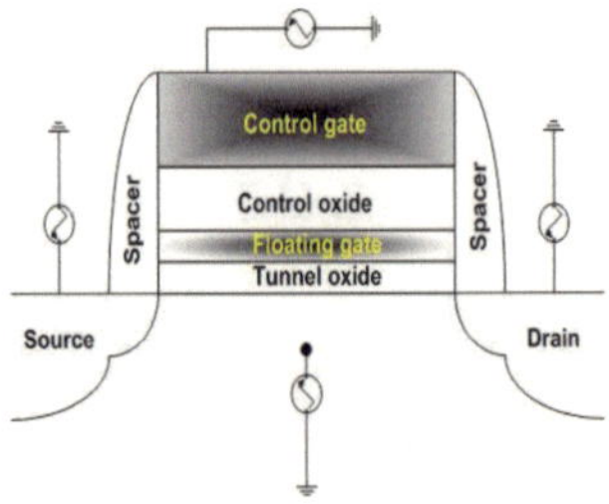

Fig. 1. (b) Split gate NVM cell with voltage sources needed to Write, Erase and Read.

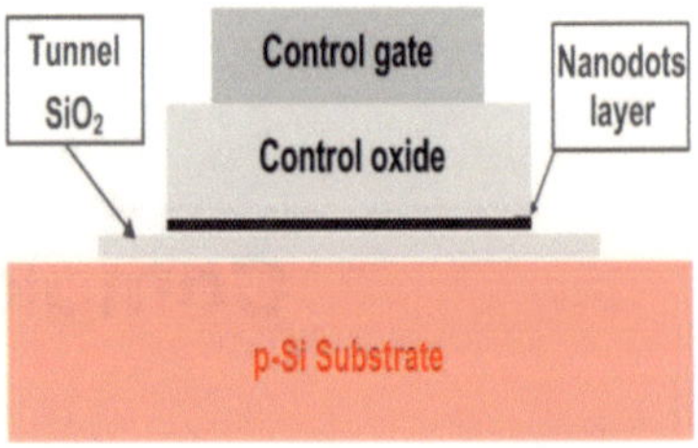

Fig. 1. (c) NVM capacitance cell with nanocluster inclusions

Secondly, scaling down of storage elements accompanied by high W/E potentials reaching 20V and more along with multylevel cell (MLC) technology spreading give rise for inappropriate programming risk through cross-programming forcing special measures to be taken for protection of adjacent cells (Kuchubatla, 2010).
Recent years activities in the field of research and development of alternative to conventional floating poly-Si NVM systems take into view the solution of these problems as a part of a new embedded and standalone NVM systems generation industry deployment. Here we concentrate on charge trapping and retention in silicon nanoclusters NVM structures as directly derived from bulk Si floating gate predecessors. Typical NC NVM structure presented on Fig.1c oddly enough resembles conventional NVM (Fig.1b), except the gate which is formed by array of nanoinclusions.
Qualitative differences in the electrical properties of conventional and NC systems have its roots in quantitative difference of Si atoms number in both systems. To illustrate this we present in Table1 quantity of Si atoms concentrated in a bar 20x50x50 nm^3 and nanoclusters with diameters 3 nm and 2 nm. Mean number of surface atoms is also presented in this table.

Type/Size	Bar 20x50x50 nm^3	Sphere of 3 nm diameter	Sphere of 2 nm diameter
Volume	5E-17 cm^3	1,41368E-20 cm^3	4,18867E-21 cm^3
No of Si atoms	2500000	706	209
No of Si surface atoms	18420	317	141
Surface to bulk	0.007368	0.449	0.675

Table 1. Atomic comparison of Floating Gate and NanoClusters NVM elements

From this we can see, that under the transition from bulk Si to NC the role of boundary atoms will grow. Also must grow the role of the whole NC medium including interclusters and boundary SiO_2.

Effect produced by bulk → NC transition	Electrical affected
(1) Bandgap, conduction and valence bands modifications 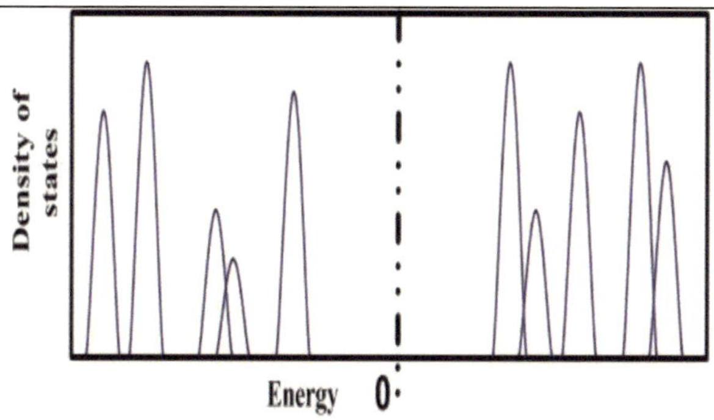	(2) • Bandgap widening due to quantum confinement effect • Electrons emission from NC instead of holes trapping • Nonexponential, staircase-like, charge relaxation (Coulomb blockade effect)
(3) Simultaneous accumulation of opposite charges 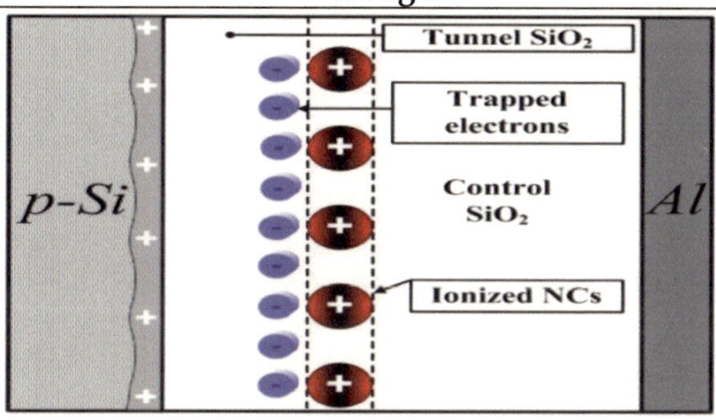	(4) • The possibility of unipolar reprogramming • Charge heterogeneities generation • Nonexponential retention characteristics • Strong dependence of retention on the gate material work function
(5) Stable current flow in pre-breakdown region. Soft breakdown 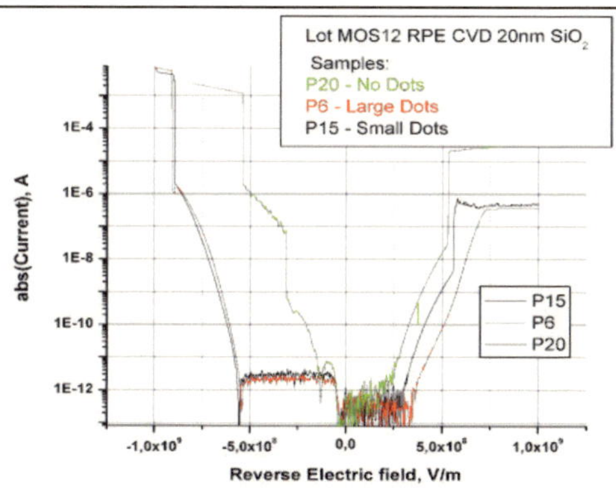	(6) • Wide range of reprogramming fields • Reduced necessity of charge balance control • The possibility of unipolar reprogramming • The possibility of unipolar and antipolar reprogramming of the same cell

Table 2. Qualitative effects introduced under conventional NVM to NC NVM transition.

The main feature of the transition is adding or removing electrons to the block only slightly affects their position in energy bands. Situation radically changes when nanoclusters are included in a gate dielectric as a storage media. In this case only dozens to hundreds silicon atoms participate in charge storage processes of which considerable part are boundary, forming microscopic traps with properties strongly dependent on local fields in turn, depending upon their population. Program/Erase procedures are not reduced to simple electrons injection/extraction but rather to trapping/recombination/emission. Complex charge interchange processes are involved during every stage of NC NVM cell operation: write, erase and storage, affecting the whole set of their characteristics, including window formation, charge relaxation and retention. As an orientation to the classification guidelines of this work we selected three main streamlines in which most evident divergence between conventional NC NVM cells occurred. These are tabulated in Table2.

2. Experimental

European Union sponsored consortium "Nanoforum" that provides a comprehensive source of information on all areas of nanotechnology to the business, scientific and social communities in its eighth report "Nanometrology"(2006) characterised capacitance spectroscopy as one of the main nanometrology techniques (Nanometrology, Eighth Nanoforum Report, 2006). We widely use such kind measurements for full electrical diagnostics of the NVM capacitance cell, which are first stage fabrication of the NVM transistor cell, but totally determines memory characteristics of the last device.

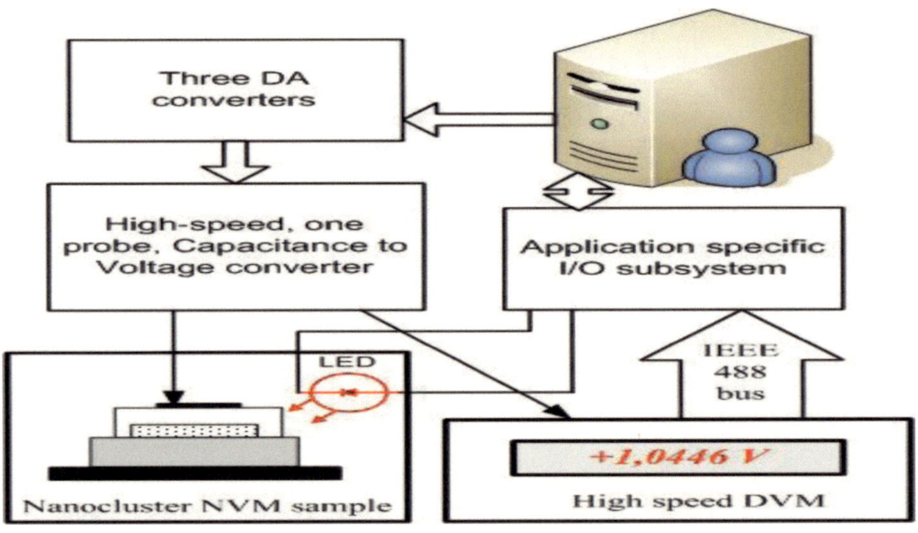

Fig. 2. A simplified block diagram of the computer aided nanocluster NVM cell diagnostic system.

The main characteristic of the W/E ability of the NVM capacitance cell is the window width under the positive and negative reprogramming biases on the gate. This window we calculate as a difference:

$$WINDOW = abs\left(V_{fb}(+) - V_{fb}(-)\right) \quad (1)$$

Where $V_{fb}(+)$ and $V_{fb}(-)$ are the flatband voltages of the capacitance-voltage (CV) characteristics calculated for positive and negative gate biases correspondingly. Respective simplified block diagram of the data measurement and acquisition system is presented on Fig. 2:

Main parameters of the system components are:

- DACs ranges are: ±5.12V(±2.56V), ±10.24V and ±20.48V with max bias of ±35.84V
- DVM max speed - 5000 meas./s
- CV converter time constant 0.2 ms

In our development of the experimental DAQ system on the first stage we foresaw three main measurement procedures (Turchanikov et al., 2005b): window cycling measurements; and W/E relaxation measurements; charge retention measurements.

2.1 Window cycling measurements

Application specific data acquisition system (DAQ), software was specially designed for window testing of write/erase cycles in nanomemory structures with the algorithm illustrated by Fig.3.

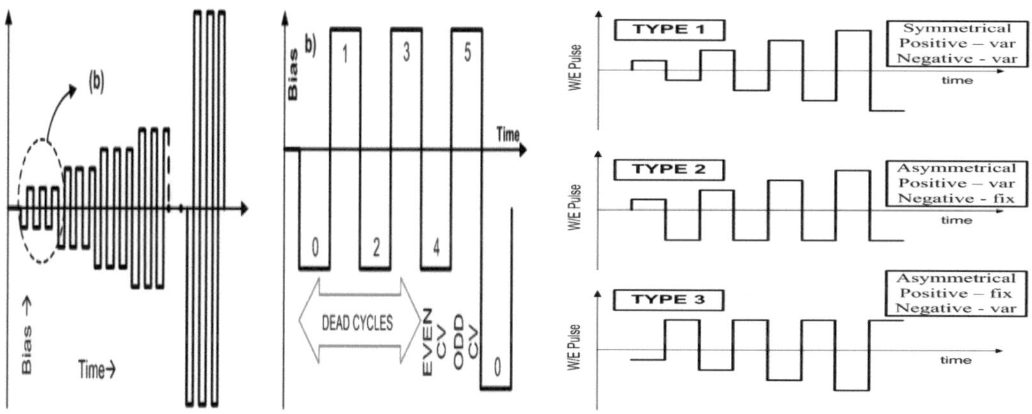

(b) (a) (c)

Fig. 3. a Algorithm of structure biasing under write/erase window testing of NVM structures with shifting bias, b. - Enlarged view of first BS series, c. - Window testing three biasing types initial potential levels are arbitrary.

First, CV characteristic of the virgin structure is measured. Next, antipolar symmetrical bias is applied to the structure, negative in the even cycles and positive in the odd for fixed time (in our case 20s). There were 6 such bias-stress (BS) cycles. After the first 4 cycles (dead cycles) no CV measurements were made with the goal to provide transient-free window formation. After the last two cycles CV characteristics were measured and stored and V_{fb} determined. Then bias was incremented (in our case by 1V) and above described series repeated, and so on up to 10 times. Thus we got 21 files with CV data and 2 files with V_{fb} data for each bias polarity with subsequent determination of window width (1). Each CV curve was measured in both directions: from accumulation to inversion and vice versa and contains 2048 points (1024 for each ramp direction). Measurement time was near 5 s, the V_{fb}

was obtained after simple smoothing "on the fly" for the second ramp branch to prevent distortions caused by discharging of the shallow trapping sites (Nikollian & Brews, 1982). Next, CV data were smoothed by 80-point Fast Fourier Transform (FFT), the dC/dV values were extracted and maximum of absolute (dC/dV) determined. The height of this maximum was used as the criteria of surface potential heterogeneities on the dielectric/silicon interface, which directly associated with heterogeneous charge trapping in the nanoclusters in the dielectric. Summarizing graphical presentation of the results obtained is given on Fig.4 (Turchanikov et al., 2005a).

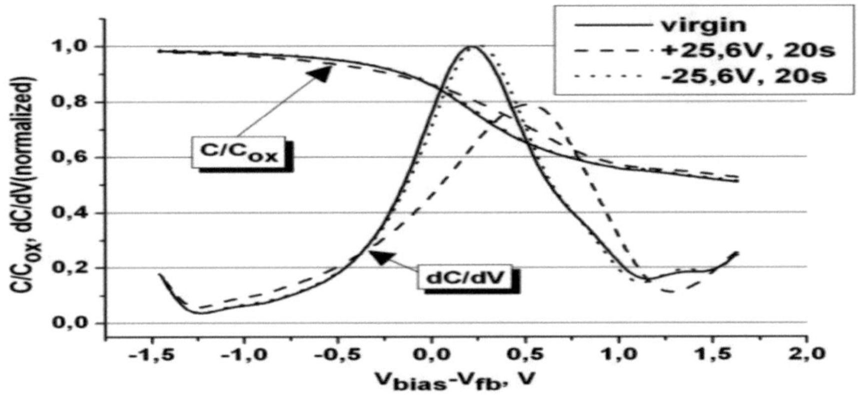

Fig. 4. Normalized C-V and dC/dV-V characteristics for virgin and subjected BS structures.

In each bias cycle the sample can be illuminated by LED to supply minority carriers in space charge region (SCR) with the goal to prevent nonequilibrium depletion of Si substrate that does not take place in the NVM MOSFET cell. Additionally the experimental system can measure window formation with one fixed bias and second - variable, ascending or descending (Fig.3.c). This is especially useful when breakdown is asymmetrical or for experiments with unipolar recharging.

2.2 W/E relaxation measurements

Window cycling gives us information on W/E window formation between bias cycles but not inside them. The latter may be extremely useful in estimating physical processes of nanoclusters recharging in the presence of external electric field (Tsoi et al., 2005). Algorithm of the cell charging for relaxation measurement is presented by Fig.5. In these experiments we used two identical series of bias time cycles with different biases in the first and second series. Measuring V_{fb} after each biasing period we obtain a relaxation dependence of V_{fb} on the cumulative time of biasing, t_{cum}, where t_{cum} is defined by:

$$t_{cum1} = t_1, t_{cum2} = t_{cum1} + t_2 \ldots t_{cumn} = t_{cumn-1} + t_n \quad (2)$$

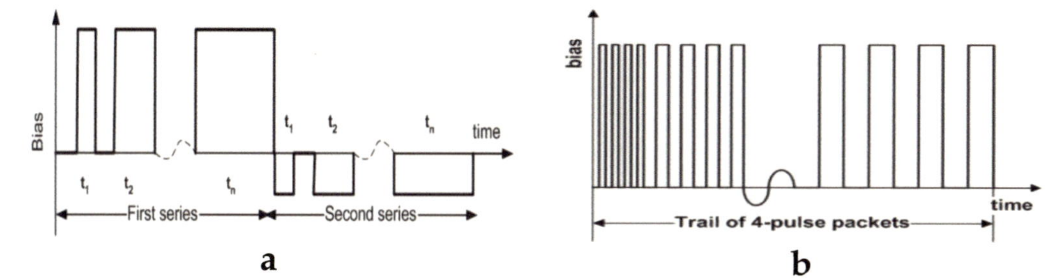

Fig. 5. Relaxation testing algorithm: (a) single and (b) multiple pulses packets.

Several relaxation cycles can be glued together, to obtain data on non-stationary processes developing in nanocluster NVM cells during W/E cycling. It should be noted, that sometimes pulses were grouped in pulse packets consisting of pulses with identical width which differs from packet to packet. This was introduced with the goal to access maximum resolution for low injection times and, simultaneously, overlap wide time range avoiding huge excessive data accumulation.

2.3 Charge retention measurements

We used successive approximations register (SAR) algorithm to measure charge retention after write or erase pulses. Time for SAR to reach desired precision was ~140 ms and the sequence of measurement intervals was regular in contrary to commonly used logarithmic scale. This was made starting from next reasons: 1) time needed to measure one V_{fb} point is small sufficiently as not to interrupt the whole retention process; 2) resulting data array is compact; 3) long interval between measurements may lead to misinterpretation of the data obtained. The latter is illustrated by Fig.6.

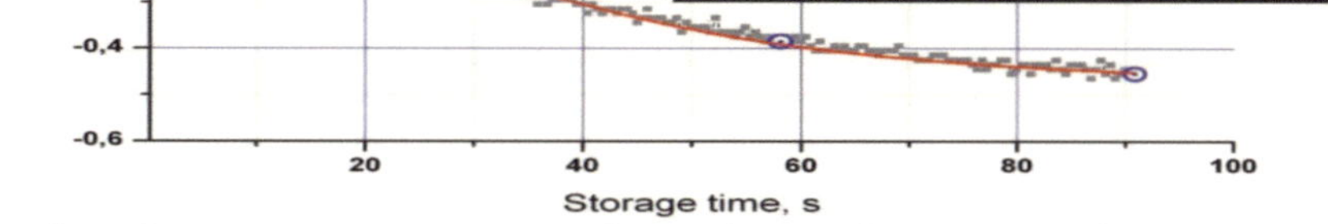

Fig. 6. Complex charge retention process (points) mistaken by simple exponential decay (line) deduced from logarithmic in time sampling (circles).

On this figure experimental points present complex charge dissipation in NC NVM cell consisting of two interdepending processes transforming to one another in the encircled region. Provided we have conventional sampling in the logarithmic time scale (circled points) relaxation process could be misinterpreted as a single exponential decay with a very good fitting degree following parameters placed in the inserted table.

3. Samples

In our work we used samples prepared on p-Si substrates using next technological processes for NC formation:

a) Low energy Si implantation in thermal SiO_2 (EME, Departament d'Electronica, Universitat de Barcelona, Barcelona, SPAIN) (Turchanikov et al., 2005c);

b) Ultralow energy Si implantation in thermal SiO_2 (Institute of Microelectronics, National Centre of Scientific Research IMEL/NCSR Demokritos, Athens, GREECE; Nanomaterials Group, CEMES-CNRS, Toulouse, France) (Coffin et al., 2005);

c) LPCVD amorphous Si deposition with subsequent solid-state recrystallization and oxidation (Institute of Microelectronics, National Centre of Scientific Research IMEL/NCSR Demokritos , Athens, GREECE) (Turchanikov et al., 2007);

d) LPCVD poly-Si deposition with RPECVD SiO_2 as a control gate dielectric (Institute of Semiconductor Electronics, RWTH Aachen University, GERMANY) (Turchanikov et al., 2005d).

The data on these samples are summarized in Table3.

Such variety of NC NVM samples allowed us to perform comparison of properties of the Si nanoclusters of different size and natures and to extract common features in the charge/discharge behaviour of such type systems.

Source	RWTH, Germany (d)	IMEL, Greece. (c)	IMEL, Greece. (b)	Barcelona Univ. (a)
Equivalent total oxide thickness (nm)	25-35 (calculated)	17	5+2; 10+2; 35+2	43 (nc layer 20nm)
Control Oxide thickness (nm)	20	10	3-5 (const)	12
Tunnel oxide thickness (nm)	2÷3 (good quality thermal oxide)	3.5 (good quality thermal oxide)	2; 7.5;	11 (with small nanocrystals)
Si nanocrystals size. Diam. (nm)	5÷10 (TEM)	3.3	2.5; 3	2.7±0.2
Si nanocrystals density (cm-2)	$(2\div4)x10^{11}$	10^{12}	$1,7x10^{12}$	$>10^{13}$
N atoms per NC	3270-26100	940	409; 706	515
Ns atoms on surface of NC	881-3524	384	220; 317	257

Table 3. Summarized characteristics of the experimental samples used.

XTEM and scanning microscopy pictures of the formed nanoclusters are presented in Fig.7.

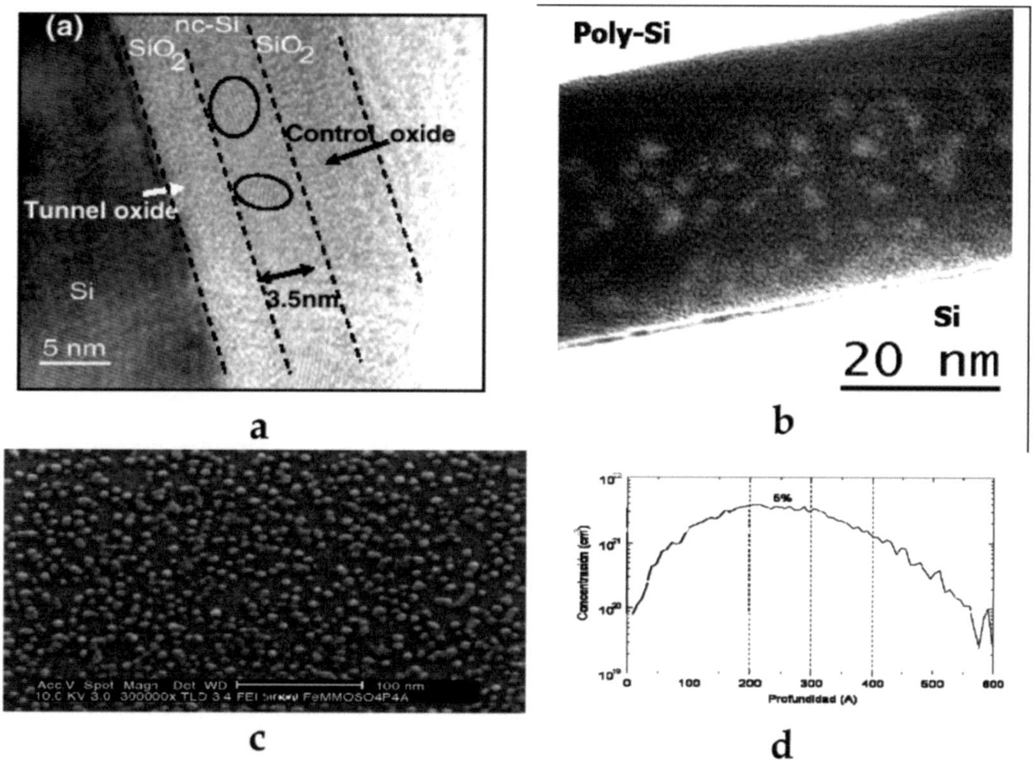

Fig. 7. a - XTEM picture of crystalline Si nanoclusters fabricated by technology (c); b – XTEM picture of Si nanoclusters fabricated by low-energy ion implantation with technology (a); c - SM picture of amorphous Si nanoclusters fabricated by technology (d); d - profile of Si ions implanted in silicon dioxide by technology (a).

4. W/E window formation

4.1 Nanoclusters' ionization

At the beginning let us devote some attention to conventional NVM. Formation of W/E window in these cells is realized through adding or subtracting negative electrons to the floating gate. Processes involved usually are direct tunneling, FN tunneling or hot carriers injection, sometimes hot holes injection is used to neutralize negative charge. But the latter process is rarely used due to the heavy damages produced by hot holes in SiO_2 and complexity in implementation (US Patent , 2006). Thus in both states (program and erase) the MOS transistors are in positive threshold voltages range.

The situation changed with nanoclusters embedded in the gate dielectric. In (Turchanikov, 2005a) we analyzed reprogramming cycles for NC NVM cells fabricated by technology of ion beam synthesis (IBS) but similar situation was observed on the samples with nanoclusters fabricated by different technologies. The write-erase cycles duration of 20 s for negative and positive polarity of start voltage pulses are depicted by Fig.8a and Fig.8b, correspondingly. It should be noted that during the first voltage pulse only positive charge build-up in the oxide is observed for both voltage polarities.

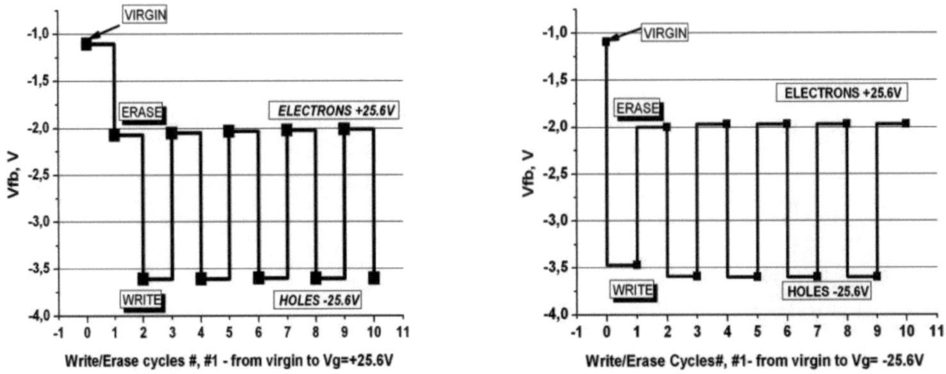

Fig. 8. Window write-erase cycling from virgin samples with first positive (a) and negative (b) poly-Si bias

On the next step of our study the transient processes of initial charge trapping in the virgin samples were examined. We applied voltage pulses of different duration from 10 ms to 1 s. Fig.9 presents the process of charge trapping under the voltage cycling on the gate electrode. It can be seen that the initial storage process has two stages: first, the positive charge accumulation and second, the positive charge dissipation. Under the subsequent bias stress (BS) with positive voltage, two transients had been also observed. It should be noted that full recovery of the C-V characteristics in the second BS cycle was not achieved. Detailed analysis of the transient process related to the hole trapping in first BS cycle, with both positive and negative voltage applied to the gate, reveals two stages of charge build-up in the oxide (see inset in Fig.9 for negative voltage applied). The transient process can be described by the following:

$$\Delta V_{FB}(t) = \Delta V_{FB}^{fast}(0) \times \exp(-t/\tau_{fast}) + \Delta V_{FB}^{slow}(0) \times \exp(-t/\tau_{slow}), \tag{3}$$

where $\Delta V_{FB}^{fast}(0)$, $\Delta V_{FB}^{slow}(0)$ and τ_{fast}, τ_{slow} are the initial (t=0) amplitudes and time constants for fast/slow trapping processes respectively. Density of the trapped charge (N_t) can be calculated using the following expression:

$$N_t = \frac{C_{ox} \Delta V_{FB}}{qS} \tag{4}$$

where C_{ox} is the oxide capacitance; q is the charge of electron; S is the area of the gate electrode. Parameters of charge storage kinetics for the first BS cycle for both positive and negative voltage on the gate are presented in Table4.

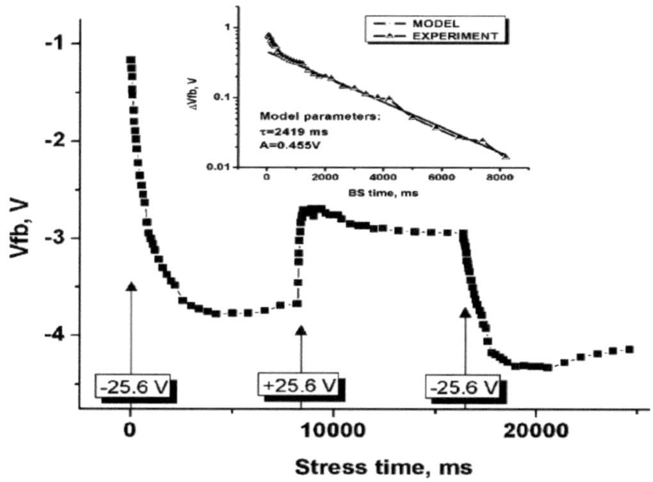

Fig. 9. Write-erase cycling starting from hole injection in the dioxide. Inset: experimental charge build-up modeling for first hole accumulation process with a single exponential decay.

The important point is the close coincidence of the time constants and the trapped charge density for the "fast" process for both positive and negative voltage applied during the first step of the charging. This implies the similar charging process in both cases.

APPLIED VOLTAGE	"Fast" process		"Slow" process	
	τ(ms)	N_t (10^{11} cm^{-2})	τ(ms)	N_t (10^{11} cm^{-2})
- 25.6 V	275 ± 57	4.43	1264 ± 163	10.25
+ 25.6 V	219 ± 22	3.50	2419 ± 235	2.34

Table 4. Relaxation process parameters for Fig. 9.

Together with flat-band voltage measurements after charge injection the slope of the C-V characteristics was determined. It's worthy to note that *main decrease of the slope is observed during "fast" stage of hole trapping both for positive and negative applied voltages* (see Fig.10). Special control experiments at liquid nitrogen temperature (similar to (Nikollian & Brews, 1982)) show that the slope decrease of the C-V characteristic is associated with the increase of fluctuations of the surface potential at the SiO2/Si interface, but not with the surface state generation.

Taking into account the above results one can conclude that *the "fast" process of the positive charge build-up in the oxide results in increase of the potential fluctuations at the SiO2/Si interface, that can be associated with electron emission from the silicon nanoclusters located near the SiO_2/Si interface*. Fig.11 presents the schematic picture of spatial distribution of the surface potential in the SiO_2/Si interface after the "fast" positive charge trapping process.

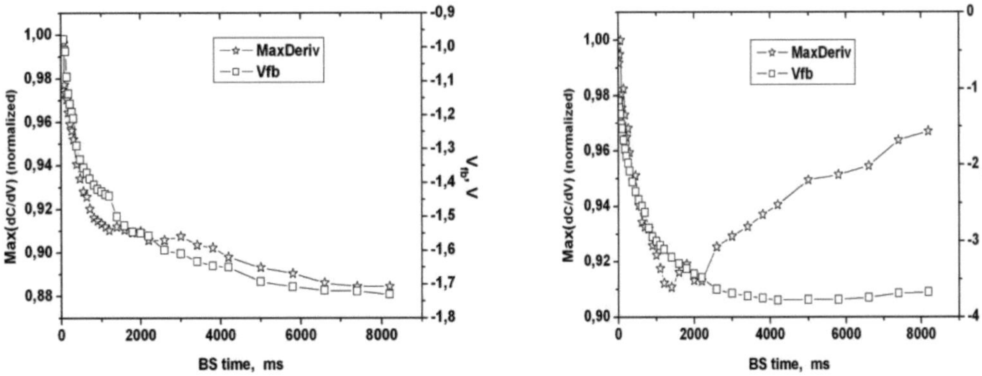

Fig. 10. Vfb and maximum of dC/dV vs. bias stress time for positive (+25.6V) (a) and negative (-25.6V) (b) bias on poly-Si electrode

The following build-up of the positive charge during the "slow" process smoothes the potential fluctuations in the SiO_2/Si interface at the positive voltage applied (Fig.10a) results in the restoration of the C-V characteristics slope for negative voltage (Fig.10b). The latter case can be also related to the positive charge dissipation process, which is observed at the last stage of the relaxation, i.e. the electrons trapping in the oxide. Indeed, if we apply positive voltage to the gate just after the first cycle with negative voltage, the total restoration of the C-V characteristic slope was observed. This is the evidence of nanocrystals charge neutralization.

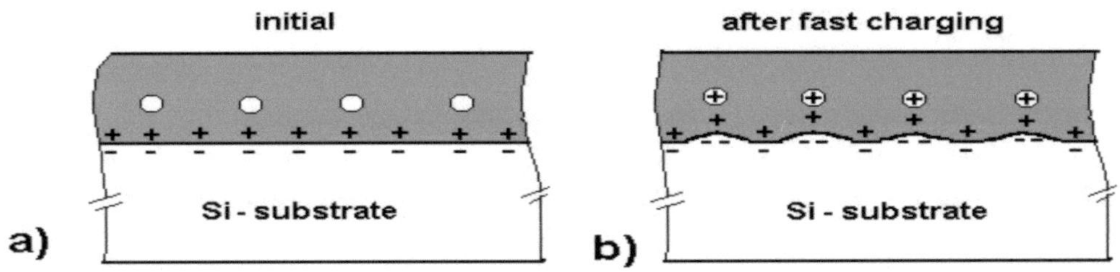

Fig. 11. Schematic picture of NCM device charging: initial conditions of the device (a) and after fast stage of charging (b) starting from virgin sample.

"Slow" process of positive charge trapping has a different set of parameters for negative and positive applied voltages (see Table4), so it is the argument in favour of different processes of trap charging at positive and negative voltages. At the same time, along with background positive charge incorporation in dielectric, there is the question of write/erase window formation. This process, as illustrated by Fig.9 at the second switching, is characterised by the following parameters: τ=63.7±3,3ms and trap concentration of N_t= (7.8±0.1)x10^{11}cm^{-2}. Window formation may be due to two processes: holes or electrons trapping/detrapping in dielectric. The latter is preferable because of the comparatively short time constant, as compared to positive charge build-up presented in Table4. Thus, we can conclude that independently on the polarity of the voltage applied on the first stage of the charging there are two main processes of the charge storage in Si nanocrystals rich SiO_2: first, *the ionisation of the Si nanocrystals in SiO_2 and their positive charging*; second, *uniform positive charge trapping and neutralization of the defects in dioxide matrix*. The evidence of fast recovery of the V_{fb} during negative-positive bias switching proves that the process of the forming write - erase window is, presumably, related to the electron trapping/compensating in modified SiO_2.

Detailed examination of complex relaxation processes (Turchanikov et al., 2005c) led us to the conclusion that silicon behavior in nanoclusters is similar to intrinsic silicon - no free carriers, no charge injection. But except for field ionization, there is an alternative way to get free carriers in the nanoclusters, namely, to raise the temperature of the sample. In paper (Turchanikov et al., 2005c) series of the above described experiments under the elevated temperatures in the range 20 C - 300 C were performed. Discharding relaxations involving high-field ionization along with intricate processes that took place under simultaneous nanoclusters ionization and neutralization by electrons injected from electrodes a family of relaxation curves presented on Fig.12 were obtained. If we guess some analogy between nanoclusters and bulk Si, assuming the model presented in Fig.13, when in common this process can be described as following:

$$\Delta N_p(t,T) \sim N_0(T)\left[1-\exp\left(-\frac{t}{\tau(T)}\right)\right], \tag{5}$$

where ΔN_p is the charge stored in nanocluster, N_0 is the free electrons concentration in nanocluster, τ is the relaxation time constant. Here we must not forget, that in real situation, according Table4, there are two processes and, consequently, two exponents. Assuming that Si nanocluster can be considered as intrinsic semiconductor, $N_0(T)$ can be presented as following:

$$N_0(T) \equiv n_i(T) = \sqrt{N_c N_v}\,\exp\,(-\frac{E_G}{2kT}), \tag{6}$$

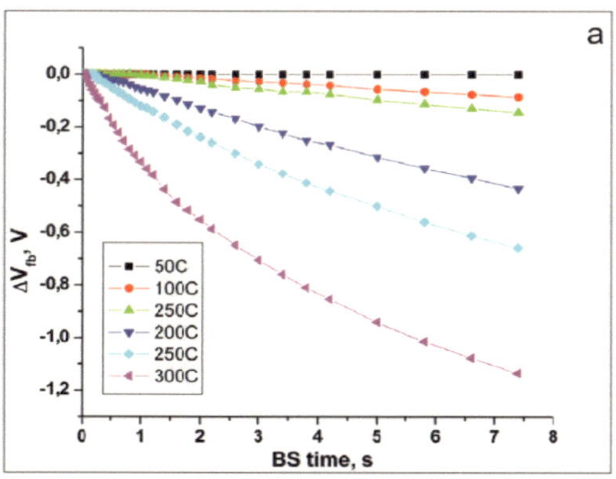

Fig. 12. Vfb relaxation at elevated temperatures.

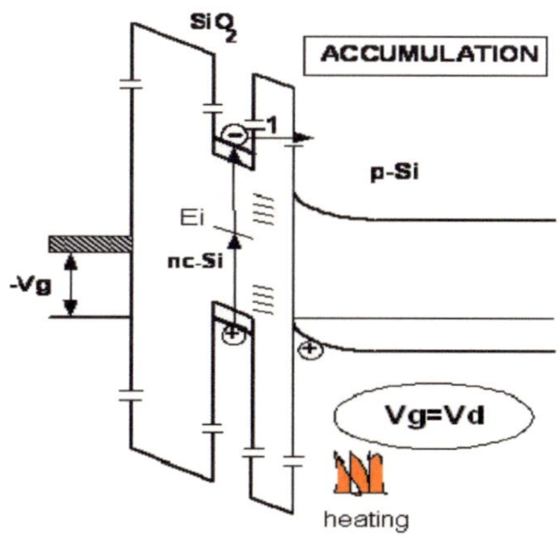

Fig. 13. Energy diagram for NC NVM structure for Si ionization.

where N_C, N_V and E_G – analogs of the bulk density of states in conductance and valence bands and bandgap correspondingly. Calculated activation energy for τ was in the range of 0.05÷0.22 eV.

This is a very low value for experimental temperatures over 150 ^{0}C but allows us to neglect second exponent in equation (5) and estimate E_g of the nanoclustes from expression (6). The obtained value of E_g=1.50±0.12 eV is in the good agreement with one found for the band gap for Si nanoclusters with size of near 3.5 nm (Garrido-Fernandez et. al., 2002).

4.2 Unipolar window formation in NC NVM structures

Usually window in NC NVM structures is formed by application of two opposite polarity pulses (Tsoi et al., 2005; Carreras et al., 2005) but occurrence of electron injection from nanocluster under medium electric field can be mirrored in terms of the applied programming bias polarity. Namely, there is no need to change bias polarity. The same result can be attained under the positive to more positive gate bias switching and vice versa. This phenomenon was called as "unipolar bias recharging" (Turchanikov et al., 2005d). Corresponding processes are illustrated in Fig.14.

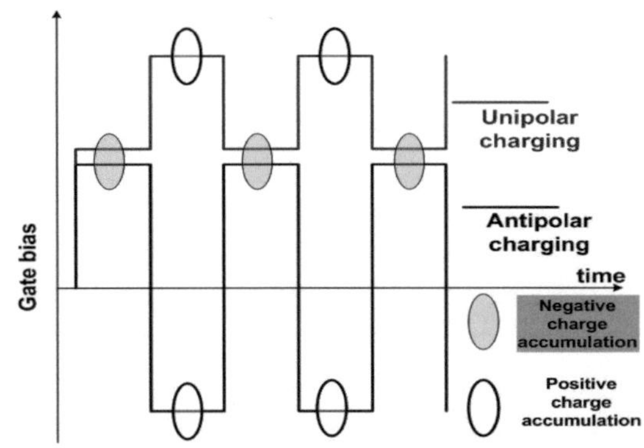

Fig. 14. A comparison of the W/E procedure in NVM cells under the usual (antipolar) and unipolar conditions.

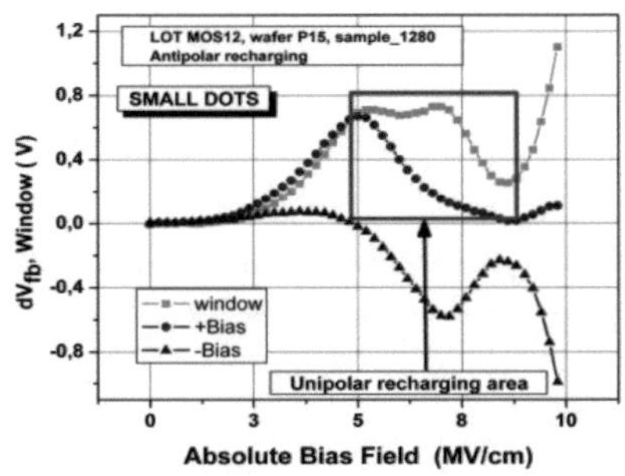

Fig. 15. Window formation under the antipolar charging and symmetrical bias for NVM structures with small nanodots. The zone selected for unipolar recharging experiments, due to the opposite charge polarity stored in a structure via biasing, is marked. Charging pulse duration is 1280 ms

The phenomenon is observed in all studied samples with nanoclusters floating gate. On example of the structures possessing of Si nanodots with size near 5 nm, which were obtained by LP CVD and low-temperature RPE CVD silicon dioxide (Winkler et al., 2004), some features of this process can be demonstrated. With the goal to understand the unipolar recharging phenomenon, at first, the NVM structures with silicon nanodots for charge accumulation under the symmetrical antipolar biasing were tested. As it can be seen from Fig.15, positive charge accumulation appears under the more positive bias in a case of unipolar charging, in contrary to the antipolar charging, when positive charge is accumulated under the most negative structure biasing.

The experimental data for positive bias may be explained by the model presented in Fig.16. First, under the low fields, negative charge, as a part of a through electron flow (not presented on this figure for the simplicity) is trapped on the limited number of sites associated with nanodots (i), second, with the field rising, negative trapped charge is compensated by Si ionization in nanodots (i.e. electron injection from the nanodots) (ii) and, third, after the emptying again, electron trapping prevails (iii). The latter inevitably invokes positive feedback with respect to the electric field in the main bulk of dielectric and subsequent breakdown of the available Si sites. The more Si in nanodots the more is the breakdown field, so *for the large nanoclusters the breakdown field is greater than for small.* This fact we verified experimentally by the measurement of current-voltage characteristics (see Table1).

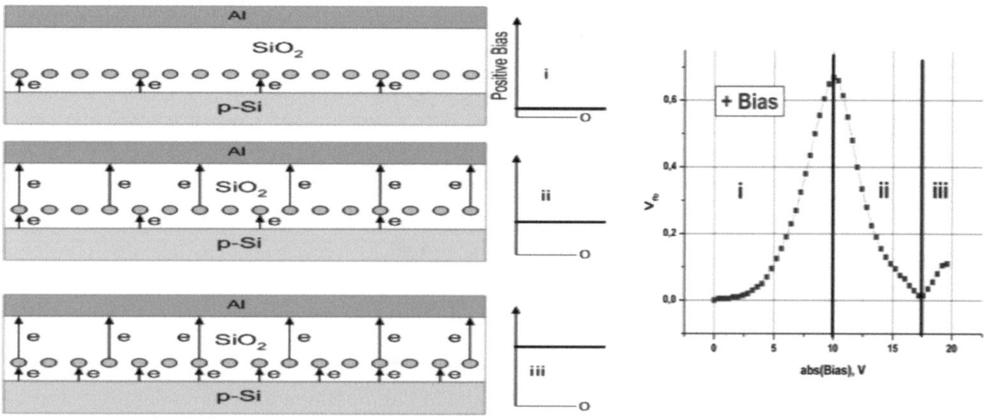

Fig. 16. Model of charge accumulation in NVM NC memory cells under the positive bias

Under the negative biasing of the NVM structure charge accumulation processes are rather different as presented by Fig.17. The main difference consists in the fact, that the charge accumulation in the region of low (i), intermediate (ii) and medium field (iii) is due to the carriers' trapping by the sites associated with nanodots in contrast to their ionization. In this

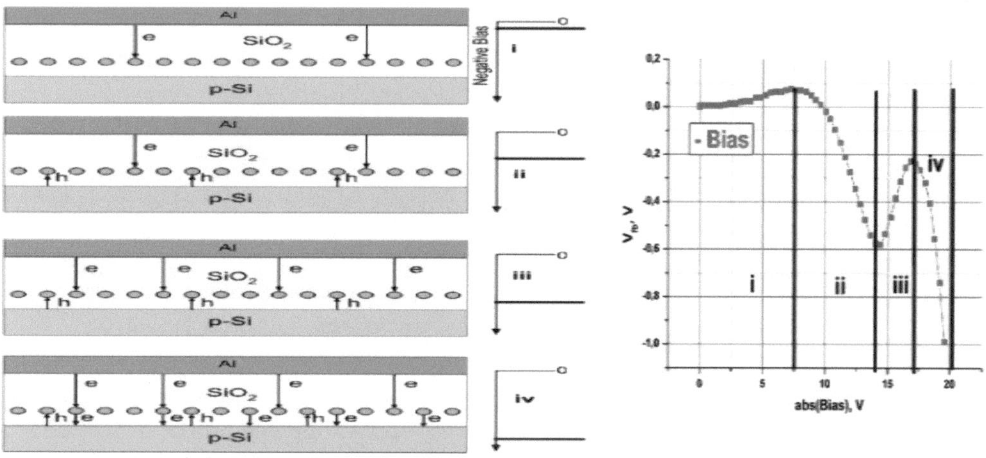

Fig. 17. Model of charge accumulation in NVM nanocluster memory cells under the negative bias.

situation (i ÷ iii) charge accumulation depends upon competitive processes of electron flow through imperfect RPECVD control oxide and holes tunneling through tunnel oxide. Under the high field (iv) nanodots ionization, with subsequent positive charge incorporation, prevails.

The possibility of the unipolar recharging process is dictated by two competitive processes in the singled out zone of Fig.15 (Turchanikov et al., 2005c):

1. Negative charge trapping under medium fields
2. Nanodots ionization (electron injection from nanodot) - positive charge incorporation.

5. Relaxation

5.1 Relaxations under unipolar and antipolar charging

Because of existing of electron injection process from nanoclusters during unipolar charging it's possible to expect differences in the relaxation processes under unipolar and antipolar charging. Above mentioned research was performed in paper (Turchanikov et al., 2005b), where algorithm of the cell charging for relaxation measurement was also described. It must be noted, that in this study cumulative charging time, i.e. the sum of biasing time for each positive and negative bias was the same and equals ~24 s.

Fig.18 represents the V_{fb} relaxation process sequencing, i.e. V_{fb} establishment with an increase of cumulative charging time in a series of bias pulses, under anti- and unipolar biasing for 7 W/E cycles for nanodots with size near 5 nm.

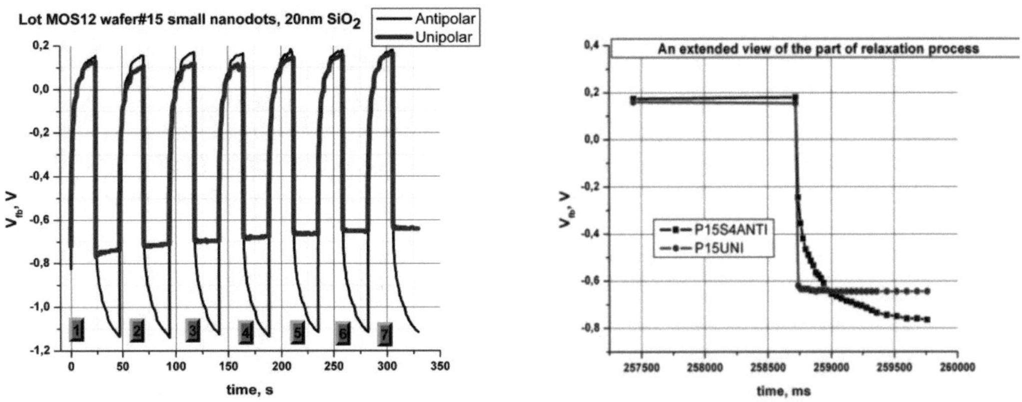

Fig. 18. a Anti- and unipolar relaxation charging process cycling in the structures with small nanodots. Reprogramming bias field for antipolar W/E cycling is +4.2x10^6 V/cm (first half-cycle in each cycle) and -4.2x10^6 V/cm (second half-cycle) , for unipolar - +4.2x10^6 V/cm (first) and +6.4x10^6 V/cm (second); b - a comparison of the anti- and unipolar relaxation process (an expanded view of the 7-th W/E cycle from Fig.18a).

The relaxation process of Vfb establishment obtained in paper (Turchanikov et al., 2007) was simulated on the basis of two relaxation time constants, discarding very short relaxation (<5 ms), which cannot be reliably processed by used DAQ system:

$$WINmodel(t) = A1*\exp(-\tau 1/t) + A2*\exp(-\tau 2/t), \tag{7}$$

where A1, A2, $\tau 1$, $\tau 2$ are simulated components of slow relaxation processes.
Thus it was experimentally determined the sum W/E window width Wexperimental, by simulated components of slow relaxation with time constants of $\tau 1$, $\tau 2$, and it was simulated maximum window WINmodel including the fast process with $\tau 3 < 5$ ms. The difference of this magnitudes ΔWIN has to characterize the fast component of the relaxation process with time constant $\tau 3 < 5$ ms. It was shown that under the **antipolar** recharging conditions a comparison of the experimental data with model parameters demonstrates a difference (±30%) in the predicted and observed results for positive charge trapping (Table5.). This difference may be understood on the basis that some fast high field electron emission (with time constant <5 ms) processes from the nanodots take place. In any case, *under the antipolar recharging, the formation of the W/E window is dictated, in the main part, by slow trapping of the free carriers*, electrons or holes, on the sites associated with nanodots.

In a case of **unipolar** biasing the situation is rather different. The positive charge incorporation, more than 95%, via switching to more positive bias potential is much faster (time constant < 5 ms) than negative charge trapping (time constant > 0.5 s) as illustrated by Fig.18b and Table5. Thus, under unipolar recharging *the formation of the W/E window is dictated mainly (the switching to more positive polarity) by fast trapping of the free carriers associated with electron injection from nanodots.*

	Single pulse duration 80 ms, cumulative charging time 24 s (300 pulses)			
	ANTIPOLAR		UNIPOLAR	
	Neg. charge accum.	**Pos. charge accum.**	**Neg. charge accum.**	**Pos. charge accum.**
$WIN_{experimental}$	**1.3 V**		**0.82 V**	
$\tau 1$, s /A1, % of experiment	**4.0/ 46%**	**7.4/ 31%**	**4.9/ 54%**	
$\tau 2$, s /A2, % of experiment	**0.33/ 41%**	**0.15/ 41%**	**0.56/ 41%**	**Σ <3%**
$WIN_{moel}(\infty)$ % of experiment	**87%**	**72%**	**95%**	
$\tau 3$, s / ΔWIN, % of experiment	**<0.005/ 13%**	**<0.005/ 28%**	**<0.005/ 5%**	**>97%**
Table denotations are in accordance with (1) and (2)				

Table 5. Relaxation parameters for the NVM samples

5.2. Staircase charging

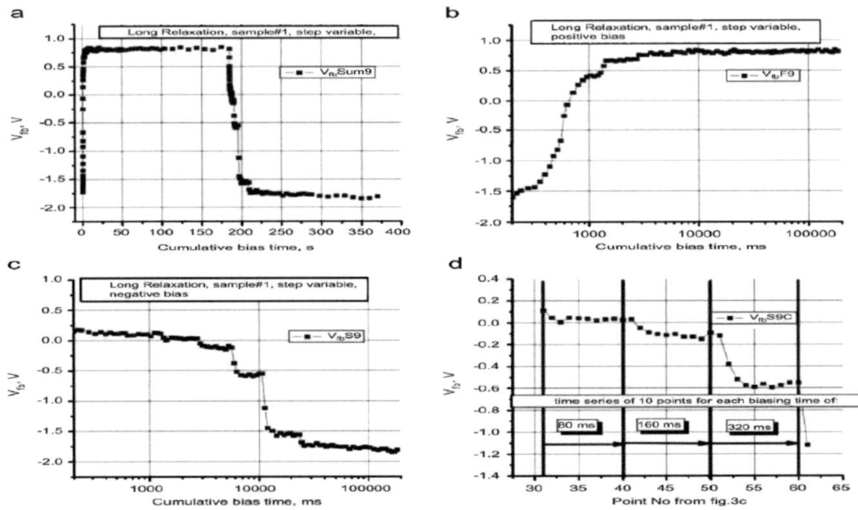

Fig. 19. Vfb relaxation under the irregular bias pulse sequence of 10 pulse trails with duration incremented by doubling in the range 20÷10240 ms, (a) summary relaxation curve; (b) positive bias of +7.52 V; (c) negative bias of -7.52 V; and (d) V_{fb} as a function of the number of BS attempts (reshaping of results of Fig.19c).

Performing experiments on charge relaxation with trail sequences of bias impulses (Fig.5) it was found that this relaxation was of staircase type (V. Turchanikov et al., 2007b). From Fig.19a, it can be concluded that during the positive BS cycle net negative charge is incorporated in the NVM structure, while during the negative BS cycle the situation is reversed, as expected. Detailed examination of the relaxation process, as presented by Fig.19b and c, reveals a staircase V_{fb} behavior in equilibrium during BS charging, that depends mostly on the injection time in each sequence rather than on the cumulative BS time. To illustrate this, the dependence of V_{fb} was rebuilt in coordinates of the number of BS attempts under the same bias time (see Fig.19d). It is evident that the stair appears upon the transition from one time sequence (10 points) to another. To verify the assumption that the W/E window width depends on the single bias width rather than on the cumulative charging time, the experiment under a constant bias with same cumulative BS charging time but different durations of a single bias shot was performed. These results are presented in Fig.20 and a summary of the corresponding W/E window is shown in Fig.21.
Using computer simulations, it was demonstrated that for a positive bias the window width follows the single decay time constant relation:

$$\Delta V_{fb} = \Delta V_{fb1}^{0(+)} \exp\left(\frac{-t}{\tau_1^{(+)}}\right), \tag{8}$$

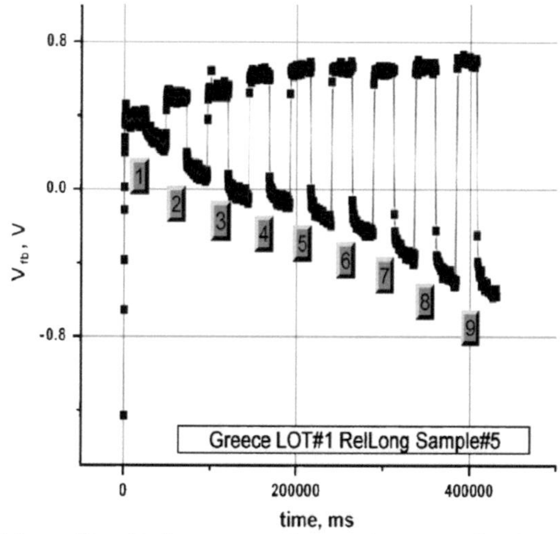

Fig. 20. Relaxation dependence of the write/erase window with fixed bias of +7.52 and -7.52 V, fixed cumulative charging time of 24 s and different durations of a single charging pulse. Numbers correspond to different single charging pulse durations in a cycle: 1.80, 2.160, 3.240, 4.320, 5.400, 6.480, 7.560, 8.640, and 9.720 ms. Cumulative charging time for each cycle was constant and equal to 24 s for both positive and negative cycles.

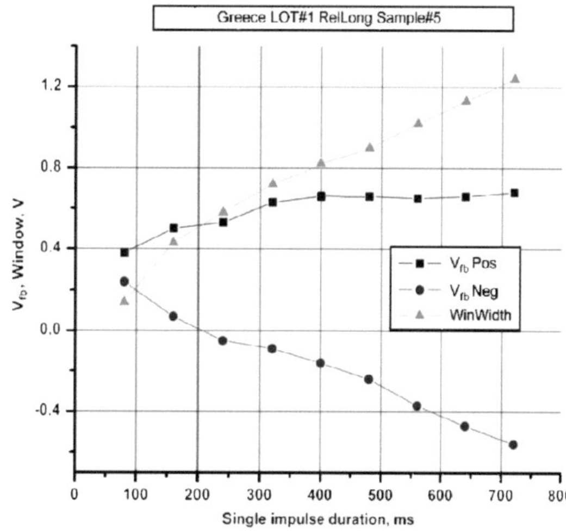

Fig. 21. Write/erase window formation under a fixed bias of +7.52V (V_{fb}Pos), -7.52V (V_{fb}Neg) and fixed cumulative charging time of 24 s, as a function of the duration of single charging pulse. Full triangles correspond to positive pulse application, full circles to negative, while full triangles show the width of the memory window.

where $\Delta V_{fb1}^{0(+)}$ represents the total flat band voltage shift in V for the given process for an applied positive voltage pulse and $\tau_1^{(+)}$ is the time constant of the relaxation process, with mean values of $\tau_1^{(+)}$= 235±35 ms and $\Delta V_{fb1}^{0(+)}$= 0.716±0.18 V. This means *that in every charging sequence the same single negative charge trapping mechanism prevails* (Fig.21). The density of charge trapping sites was found to be equal to N_t= 1×10^{12} cm^{-2}.
The process of V_{fb} relaxation under negative bias is more complex than that occurring under positive bias. Instead of a single relaxation process, in this case *at least two processes take place and a large dispersion of time constants was observed.* A fast time constant ($\tau_1^{(-)}$) varying from 200 to 350 ms was measured with $\Delta V_{fb1}^{0(-)}$ changing from 0.2 to 0.3 V, while a slow time constant ($\tau_2^{(-)}$) was also measured in the range of 11–25 s, with $\Delta V_{fb2}^{0(-)}$ in the range of 0.05–0.12 V. If to compare the modelling results for positive ($\Delta V_{fb}^{(+)} = \Delta V_{fb1}^{0(+)} \approx 0.72\ V$) and negative ($\Delta V_{fb}^{(-)} = \Delta V_{fb1}^{0(-)} + \Delta V_{fb2}^{0(-)} \approx (0.25\ /\ 0.42)\ V$) biases with the corresponding results from measurements $\Delta V_{fbmes}^{(+)} = 0.72\ V\ and\ \Delta V_{fbmes}^{(-)} = 0.72\ V$ it is possible to understand that an additional very fast process of positive charge accumulation (τ<5ms and N_{t3}≈(3.5×10^{11}÷15 5.8×10^{11}) cm^{-2}) that cannot be detected by our relaxation measurements setup takes place,

that contributes to NVM sample recharging. Moreover, we have detected a very slow process of positive charge accumulation under the negative gate bias.
Thus, it was concluded that *several single or distributed processes take place under the negative biasing* of the structure (positive charge accumulation), that can be associated either with hole trapping in the nanocrystals or alternatively with hole trapping in deep traps (slow process) located in the vicinity of the Si nanocrystals or electron emission from similar amphoteric defects. Possibly *the fastest process of positive charge trapping is associated with shallow states at the interface of silicon nanocrystals with SiO_2 or electron injection from nanocrystals.* Returning to the nature of staircase window formation (Fig.20), a computer simulation via C–V measurements of charge accumulation under charge leakage from the nanoclusters was made, from which it was proposed that charge draining from defect states at the interface of silicon nanoclusters with SiO_2 was possibly at the origin of the staircase characteristic of V_{fb} as a function of time. However, this process would have ended in the collapse of the W/E window in a short time (s), which does not happen in our structures, as it was deduced from charge retention studies (Turchanikov et al., 2005a). On the other hand, Coulomb blockade effects associated with carrier trapping/detrapping from nanoclusters of sizes below few nanometers (Carreras et al., 2005), as those involved in our samples, cannot be neglected. *So, finally, a superposition of two effects was proposed for the explanation of the staircase window formation: charge draining from defects in the vicinity of nanoclusters and Coulomb blockade effects associated with carrier trapping/detrapping in nanocrystals.*

6. Charge retention

The Fig.22 presents results of the retention experiments in the moderate time range (up to ≈3500 s), which was performed with the goal of estimation of the charge dissipation processes in NVM cells and window width formation possibilities. At the first glance there is nothing unusual in V_{fb} relaxation processes except for the facts that both retentions (for +8 V and +14 V biases) are cymbate and charge dissipation after +8 V pulse is very swift. The latter leads to the narrowing of W/E window from ≈0.75 V at t=6s (first measured point) to ≈0.11 V (t=3500 s) that can be correctly depicted by two exponential decays with time constants of $\tau 1$=75 s (fast) and $\tau 2$=1000 s (slow). Taking into account huge reduction of W/E window width in the course of relaxation the short time retention experiments were performed with increased time resolution (0.45 s). These results are presented by Fig.23. Let us describe the retention curve after +8 V pulse more thorough.
First, the W/E windows appears to be even greater then estimated from Fig.22 and exceeds 1.2 V. Second, accumulation of negative charge in NVM cell is obvious, because at the first moments of relaxation Vfb>0, hence there is no question about simple accumulation and dissipation of accumulated positive charge. And, third, retention cannot be presented by simple superposition of exponential decay processes.

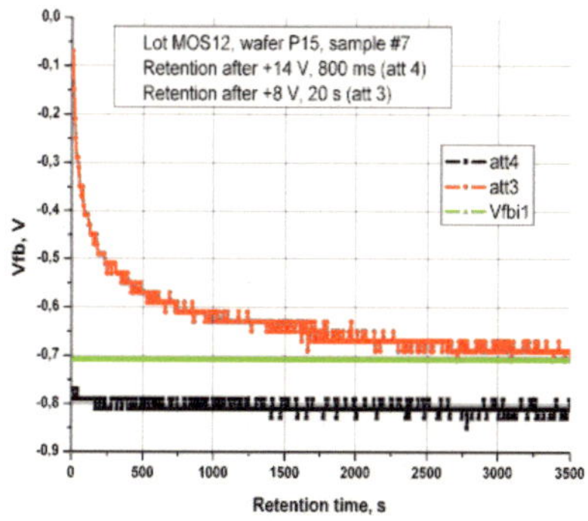

Fig. 22. Moderate time charge retention characteristics of NVM sample with nanodots under unipolar bias conditions. V_{fbi} – flatband voltage of the virgin sample.

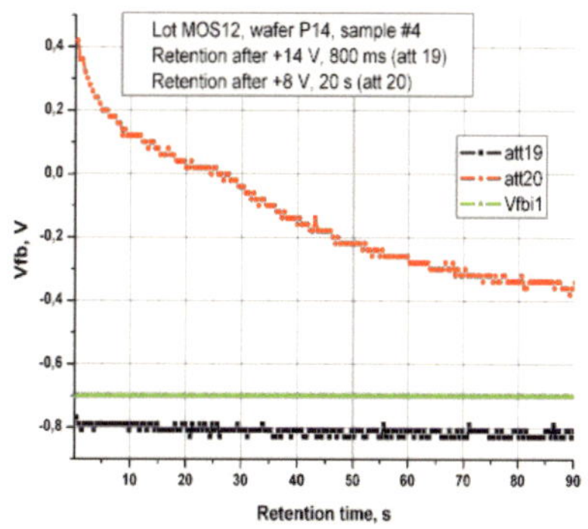

Fig. 23. Short time charge retention characteristics of NVM sample with nanodots under unipolar bias conditions. V_{fbi} – flatband voltage of the virgin sample.

The last approval signifies that the process has to be described the following expression

$$\Delta V_{fb}(t) = \Delta V_{fb1}(0) \times \exp\left(-\frac{t}{\tau_1}\right) + \delta(t-\tau_d) \times \Delta V_{fb2}(\tau_d) \times \exp\left(-\frac{\tau_{d-t}}{\tau_2}\right), \quad (9)$$

where τ_d is a "dead" time for the second relaxation process and $\delta(t-\tau_d)$ is a delta function. The process of this type physically means that retention constituents are not independent, i.e. the first charge dissipation process serves as a trigger turning on the second. In reality there must exist a third process that prevents the triggering of the second simultaneously retarding the first.

Basing on the above stated experimental data next model of charge redistribution during retention in nanodots NVM system was proposed (Fig.24). After positive applied biasing (small field, long time) two types of charge are accumulated in gate dielectric of NVM structure, one negative, localized in a plane near Si-SiO_2 interface, second - positive localized deeper in SiO_2 closer to metal gate. Naturally, there are distributions of such charges, we present them as concentrated in planes for the simplicity.

In accordance with our model the whole retention process splits into three stages:

- Stage 1 - under the field E1 electrons are tunneling from the trapping sites through tunnel SiO_2 into the silicon substrate. Direct recombination of positive and negative charges in dielectric is very slow which follows from the retention curves after +14 V bias (Fig.22, Fig.23) and does not substantially affect the process. Other possible processes are retarded either by E1, E2 or E3 fields directions. This process leads to decreasing of the whole set of electric fields in dielectric. For charge transport on this stage only the reduction of E1 is substantial. When this field drops to zero flow of electrons in the direction of Si stops - no field, no current, Stage 2 begins;
- Stage 2 - unstable equilibrium takes place, when field E3 reverse its direction and electrons emission from metal becomes possible. This emission leads to the recombination of some positive charge, in turn field E1 drops rising electron emission from the negative plane and restoring E1 to zero and so on, V_{fb} is stabilized on zero level - intermediate part on

Fig.23. This situation will continue up to the whole recombination of positive charge. Then charge dissipation proceeds to Stage 3. Naturally, the Stage 2 duration depends upon initial

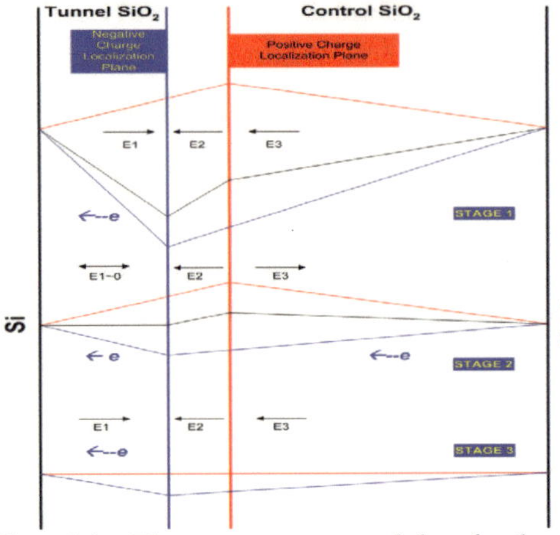

Fig. 24. Three stages model of charge dissipation in NVM structure with nanocluster inclusions after +8 V (20 s) biasing. Potential distributions induced by positively (red), negatively (blue) charged planes and sum distribution (black) as well as the field and charge flow directions are shown.

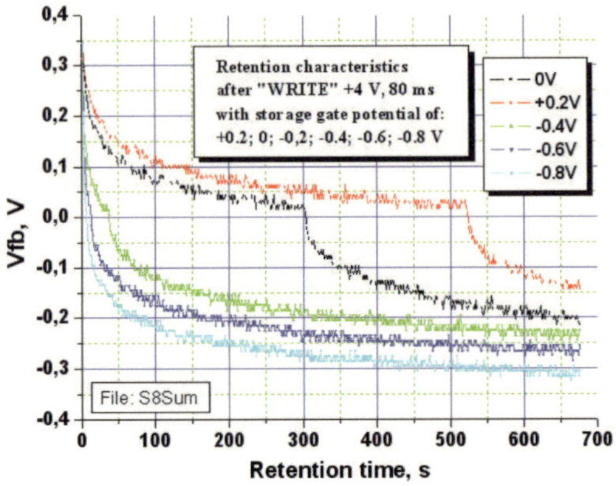

Fig. 25. Retention characteristics of IBS samples obtained by ultra low energy Si implantation with gate storage biased.

(after biasing) charges distribution in NVM structure. This, in turn depends upon the charging field, work functions of metal and Si, prehistory of the sample and technological details (cell structure) of the fabrication process. Stage 2 may be very long when field E1 drops to zero, but field E3 is not reversed and may be altogether absent if negative trapped charge is small;

- Stage 3 - only captured electrons remain in dielectric determining middle time Vfb relaxation process.

But the most evident confirmation of the crucial role that can play fields redistribution during storage period in NC NVM structures we obtained studying charge retention in samples with nanoclusters obtained by ultralow energy Si ion implantation (Coffin H. et al. 2005) with varying control gate storage potential (Ievtukh V. et al., 2008).

Slightly varying gate potential in the range +0.2 V ÷ -0.8 V we simulated possible gate material work function changes from Al to Pt (negative potentials are equivalent to work function increasing and positive to decreasing, gate material was Al) (*CRC handbook of Chemistry and Physics* (2008). Obtained results are presented on Fig.25. Thus, application of the positive voltage to the gate during retention process can considerably increase information window.

7. Conclusions

- Reduced quantity of Si atoms in storage media qualitatively affects all the electrical characteristics set of NVM cells with nanoclusters: window formation, W/E relaxation and charge retention.
- Specialized DAQ system, hardware and software, for studying the trapping/retrapping processes in NC NVM cells were designed.
- It was found that positive charge accumulation due to the nanoinclusions ionization in many cases is preferable for information storage in contrary to conventional structures.
- This process creates charge heterogeneities in dielectric mirrored in substrate space charge region that corresponds to local charge trapping in the dielectric.
- Effect of opposed charge accumulation under identical bias sign was found. It was called "unipolar recharging/reprogramming".
- In some cases W/E relaxation was of distinctly pronounced discrete ("staircase") nature that may be due to the interdependent redistribution of the different set of traps in dielectric separated by energy and/or space.
- This opens the way for alternative development of Multy Level NVM Cell (MLC) on the basis of injection duration in contrary to charge control.
- Temperature dependences of relaxation time constant enabled us to determine band gap in nanoclusters which was 1.50±0.12 eV for samples used.
- Charge retention processes in NC NVM cells are of complex nature, including parasite short-time specific for structures with nanoinclusions. Peculiar feature - to trap charge of opposite sign simultaneously - can lead to nonexponential decay, when completion of one process triggers the beginning of the other.
- Gate material work function variations directly affect retention characteristics of NC NVM cell and may be used for improvement of storage characteristics.

8. References

Carreras J., Garrido B., Arbiol J., Morante J. R. (2005) Reliable, Fast and Long Retention Si Nanocrystal Non-Volatile Memories in *Materials and Processes for Nonvolatile Memories*, edited by A. Claverie, D. Tsoukalas, T-J. King, and J.M. Slaughter, pp.229-234, (Mater. Res. Soc. Symp. Proc. 830, Warrendale, PA, 2005)

Coffin H., Bonafos C., Schamm S., Carrada M., Cherkashin M., Ben-Assayag G., Dimitrakis P., Normand P., Raspaud M., Claverie A. (2005) Si nanocrystals by ultra-low energy ion implantation for non-volatile memory applications *Materials Science and Engineering B*, Vol. 124–125, (2005), pp.499–503, ISSN 0921-5107.

CRC handbook of Chemistry and Physics (2008), ed. D.R. Lide, 89-th edition, 2008-2009, pp.2496, ISBN-10: 0849304792.

Garrido-Fernandez B, Lopez M, Garsia C, Perez-Rodriguez A, Morante J R, Bonafos C, Carrada M and Claverie A (2002) Influence of average size and interface passivation on the spectral emission of Si nanocrystals embedded in SiO2 *J.Appl.Phys.* Vol.91, No2, 2002, p.p.798-807, ISSN 0021-8979.

Ievtukh V., Turchanikov V., Nazarov A., Lysenko V., Normand P., Dimitrakis P. (2008) Processes of window formation and dissipation in nanocluster nonvolatilememory structures formed by ultra low energy ion implantation, *Europhysics Conference Abstract Volume 32F*, ISBN: 2-914771-54-1, 22nd Gen. Conference of the Condensed Matter Division of the European Physical Society, Rome, Italy 25÷29 Aug. 2008.

Kuchubatla, R. (2010) IMFT 25-nm MLC NAND: technology scaling barriers broken, *Electronic Engineering Times Europe*, April, 2010, pp. 16-18.

Nanometrology, Eighth Nanoforum Report (2006), *nanoforum.org European Nanotechnology Gateway*, July 2006, p. 60.

Nicollian E.H., Brews J.R.(1982) *MOS Physics and Technology* Wiley, ISBN 0-471-08500-6, NewYork,1982

Tsoi E., Normand P., Nassiopoulou A., Ioannou-Soulgeridis V., Salonidou A. and Giannakopoulos K. (2005) Silicon nanocrystal memories by LPCVD of amorphous silicon, followed by solid phase crystallization and thermal oxidation, *Journal of Physics: Conference Series*, Vol.10, 2005, pp.31–34, ISSN 1742-6588.

Turchanikov V.I., Nazarov A.N., Lysenko V.S., Carreras J., Garrido B. (2005a) Charge storage peculiarities in poly-Si-SiO2-Si memory devices with Si nanocrystals rich SiO_2 *Microelectronics Reliability*, Vol.45, No. 5-6, May-June 2005, pp. 903-906, ISSN 0026-2714.

Turchanikov, V., Nazarov, A., Lysenko, V., Winkler, O., Spangenberg, B., Kurz, H. (2005b) *Computer as a tool for nanocluster NVM cells diagnostics, EUROCON 2005.The International Conference on "Computer as a Tool" Serbia & Montenegro, Belgrade, November 22-24, 2005*

Turchanikov V.I., Nazarov A.N., Lysenko V.S., Carreras J., Garrido B. (2005c) Charge trapping in Si-implanted SiO2-Si memory devices at high electric fields and elevated temperatures *Journal of Physics: Conference Series*, Vol.10, 2005, pp.409-412, ISSN 1742-6588.

Turchanikov V.I., Nazarov A.N., Lysenko V.S., Winkler O., Spangenberg B., Kurz H. (2005d) Study of the unipolar bias recharging phenomenon in the nonvolatile memory cells containing silicon nanodots *Materials Science and Engineering: B*, Vol.124-125, 5 December 2005, pp.517-520, ISSN 0921-5107.

Turchanikov V., Nazarov A., Lysenko V., Ostahov V., Winkler O., Spangenberg B., Kurz H. (2007a) Charge accumulation in the dielectric of nanocluster NVM MOS structures under anti- and unipolar W/E window formation *Microelectronics reliability*, Vol.47, No4-5, 2007, pp.626-630, ISSN 0026-2714.

Turchanikov V., Nazarov A., Lysenko V., Tsoi E., Salonidou A., Nassiopoulou A. G. (2007b) Charging/discharging kinetics in LPCVD silicon nanocrystal MOS memory structures *Physica E: Low-dimensional Systems and Nanostructures*, Vol.38, No.1-2, April 2007, pp.89-93, ISSN 1386-9477.

US Patent (2004), No US 6,760,270 B2 Jul. 6, 2004.

US Patent (2006), No US 7,075,828 B2, Jul. 11, 2006.

US Patent (2007), No US 7,164,603 B2 Jan. 16, 2007.

Van Houdt, J., Haspelslag, L., Wellekens, D., Deferm, L., Groeseneken, G. and Maes, H.E. (1993). HIMOS - A High Efficiency Flash E PROM Cell for Embedded Memory Applications, *IEEE Trans.on Electron Devices*, Vol.40, No.12, pp. 2255

Winkler O., Baus M., Spangenberg B., Kurz H. (2004) Floating nano-dot MOS capacitor memory arrays without cell transistors, *Microelectronic Engineering*, vol. 73-74, 2004, pp. 719-724, ISSN: 0167-9317.

10

Carrier storage in Ge nanocrystals grown on silicon oxide by a two step dewetting / nucleation process

A. El Hdiy, M. Troyon, K. Gacem* and Q.T. Doan
LMEN, EA 3799, Université de Reims Champagne Ardenne, UFR sciences, 51687 Reims cedex 2, France
**At the present time, at CEA Saclay, France*

1. Introduction

In microelectronics, dimensionality miniaturization theoretically leads to reliability increase. However, an important concern in silicon technology is the effective reliability of MOS (metal-oxide semiconductor) devices such as MOSFETs (MOS field-effect transistors) and memory cells as they are scaled to smaller dimensions. Indeed, as the silicon dioxide (SiO_2) in CMOS technology is thinned below 2 nm for higher density and performance, limitations associated with the poly-silicon gate become increasingly important. These limitations include increasing poly-depletion effects, high gate resistance, doping impurity penetration,...As a consequence, some problems affecting the CMOS devices reliability appear. Among the main problems, the exponentially increased gate leakage current, the reduced threshold for dielectric breakdown and oxide charging which results in voltage shifts. To avoid these difficulties, group IV nanocrystals embedded in a SiO_2 matrix have been studied extensively because of their potential for integrated optoelectronic devices on silicon substrates. The use of silicon (Si) nanocrystals (NCs) instead of standard polycrystalline silicon floating gate was proposed and many studies have been made describing the NC capabilities in different devices (Tiwari et al., 1995; Park et al., 2003; Conibeer et al., 2006). However, due to their large bandgap variations and their potential for bringing a strong quantum confinement, germanium NCs have attracted more interest than Si-NCs as reported in many studies (Choi et al., 1999; Skorupa et al., 2003; Chatterjee et al., 2008).

For non-volatile memory device application, a long charge retention time at room temperature is the most important. This is the reason why the study of the NC specific properties is of principal importance. Indeed, device reliabilities are strongly affected by the variation of the NC structural parameters; namely, their density, their size, and consequently the NC spacing. The NC spacing controls the energetic potential recovery of the NCs and promotes the carrier hopping between them. Some of these effects cannot be evaluated by the standard techniques, because the evolution of the curves contains a global response; all the NCs are involved in the transport and storage, so one obtains just average information which is affected by the NC size fluctuation and density.

It is interesting to obtain information on an isolated NC (or a limited number of NCs) in a given structure to optimize the device functionalities. Indeed, electric transport through NCs and carrier storage in NCs depend on the NC structural properties. It appears that the use of the atomic force microscopy (AFM) is very appropriate if the properties of individual NCs or quantum dots (QDs) are to be exploited with nm-scale resolution (Okada et al., 2002; Stomp et al., 2005; Smaali et al., 2006, Smaali et al., 2010, Gacem et al., 2010). So, in this work, we present two kinds of results; results obtained with standard methods in which one brings average information, and results obtained by use of the C-AFM technique. The standard techniques consist of the use of high frequency (1 MHz) capacitance - voltage (*HFC-V*) and current - voltage (*I-V*) methods usually applied in microelectronic industry. Note that the kind of carriers involved in the electric transport and the charge storage depends on the nature of the electrode/sample contacts. With the standard methods, for which a metallic contact is used, electrons are concerned. However, conduction by holes dominates when working with a conductive AFM using a *p*-doped diamond probe, which leads to formation of an artificial nano-heterojunction at the contact with a large conduction band discontinuity. Obviously, if the C-AFM probe was a metallic probe (*Pt-Ir, Cr-Co,…*), electric conduction could be governed by electrons. Indeed, in a recent study, we studied the carrier transport through quantum dots by the C-AFM technique using two kinds of conductive AFM probes (*p*-doped diamond and metallic probes), and showed that the type of the carriers flowing through the nanostructure depended on the used AFM probe (Smaali et al. 2010).

HFC-V and *I-V* were used to study the carrier storage capabilities of the Ge-NCs in a large temperature range [77 - 300K]. For these techniques, the studied sample is similar to a MOS (metal - oxide - silicon) capacitor. This work allowed us to study not only the carrier transport and storage but also to evidence at room temperature a Coulomb blockade effect for nanocrystals of 3.5 nm average diameter. The latter appears for increasingly low voltages when the temperature decreases.

The C-AFM method was used to study carrier storage in a single NC in vacuum and at room temperature. With this technique, electric images of NCs of any arbitrary positions on the sample surface can be studied using the same conductive AFM probe. However, this technique requires that the NCs are not capped with an additional thick layer. In our case, the studied samples were elaborated in the same conditions than those used for standard techniques but without layer covering the NCs. The C-AFM brought interesting information on the NCs, especially on conduction characteristics and carrier storage. This study was performed with a home made AFM working inside a scanning electron microscope (SEM) (Troyon et al., 1997).

2. Sample elaboration

The crystalline silicon substrate used to elaborate samples is (100) oriented, boron doped and thermally oxidized to ~ 5 nm thick. The amorphous germanium layer was deposited by Molecular Beam Epitaxy (MBE) in ultra-high vacuum (10^{-11} Torr) at room temperature. The Ge-NCs were formed after 30 min in situ annealing at 700°C by the combination of unwetting and crystallization processes (Karmous et al., 2006; Szkutnik et al., 2008). The mean diameter (D) of Ge-NCs is uniquely controlled by the nominal thickness (t) of the amorphous layer following a relation $D \propto t$. The thickness was varied between 0.5 and 5 nm leading to NC mean diameters between 3.5 and ~34 nm. For all samples, Ge-NCs present a unique aspect ratio (height over diameter) of about 0.8 which strongly differs from the aspect ratio (about 0.15) of Ge quantum dots in epitaxy on Si substrate (Berbezier et al., 2002).

In the present work, our samples are labelled *A* and *B*. For the *A* samples, the silicon substrate was boron doped to ~10^{15} cm^{-3}, and a 18 nm thick amorphous silicon (a-Si) was deposited as a capping layer. The use of a-Si layer leads to an electrical conduction between one electrode and Ge-NCs for very weak biases. This situation could not be obtained if we had an insulating layer, as in memory devices, where the carrier exchange through the oxide makes damages and finally leads to an oxide breakdown. From an energetically point of view, the a-Si bandgap is higher than that of the Ge-NC, which reinforces the spatial confinement of the carriers in Ge-NCs, and at the same time remains small to have an electronic transport under biases as weak as those applied in this study. Weak biases avoid oxide damages in order to have reproducible measurements. In other words, it is not necessary to apply a high bias to the sample for characterization measurements. The samples of this series are labelled $A_{3.5}$, A_{17}, A_{21} and A_{35} and characterized by their average size and density (i.e., 3.5 nm and 2.4×10^{12} cm^{-2} for $A_{3.5}$, 17 nm and 1.5×10^{11} cm^{-2} for A_{17}, 21 nm and 6.7×10^{10} cm^{-2} for A_{21}, 35 nm and 2.4×10^{10} cm^{-2} for A_{35}). Transmission electron microscopy (TEM) cross-section image shown in Fig. 1(a), gives a typical example of the stack layers for Ge-NCs with an average diameter of 3.5 nm. It can be clearly seen that Ge-NCs are monocrystalline ({111} plans are identified) and free of extended defects as shown in Fig. 1(b).

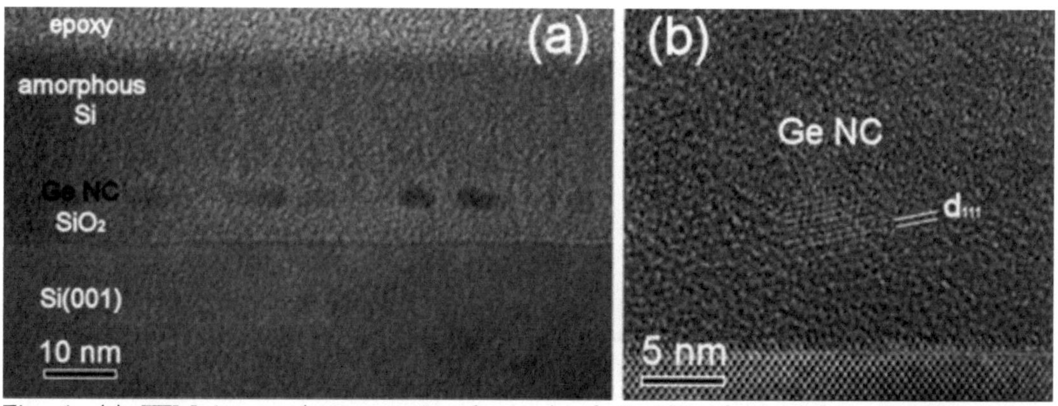

Fig. 1. (a) TEM image (cross sectional view) of Ge-NCs formed by annealing of an amorphous Ge layer (0.5 nm thickness) deposited on an ultrathin SiO_2 (5 nm thickness) for 30 min at 700 °C and capped by a 18 nm of amorphous silicon. (b) High resolution TEM image (cross sectional view) of a Ge-NC where the distance between {111} plans are evidenced (Szkutnik et al., 2008).

For the *B* sample, the silicon substrate was *n* doped to ~5×10^{18} cm^{-3} and the Ge-NCs were not capped in order to allow electrical AFM measurements. This sample contains 6×10^{9} Ge-NCs per cm^2 the average diameter of which is equal to 70 nm.

3. Experiments

3.1 *C-V* and *I-V* techniques

For the study of the *A* samples, electrical contacts were made using silver bond on the two faces of each sample. The structure is similar to a MOS capacitor (see Fig. 2). The area of the silver bond electrode is approximately circular and its diameter is about 2 mm. *HFC-V* and *I-V* measurements are obtained by limiting the gate bias to relatively low values (≤ 3 V) in order to prevent the silicon dioxide damaging and are carried out at various temperatures from 300 K down to 77 K. *I-V* measurements were made with a HP4140B picoammeter and *HFC-V* characteristics were performed with a HP4284A RLC-meter. We only focus here, on the *HFC-V* in order to have a direct response of the carrier exchange (trapping/detrapping) effect. The *HFC-V* measurements were made from inversion (positive gate bias) to accumulation (negative gate bias) regimes and immediately back to the inversion regime with the same voltage sweep rate. The carrier trapping/detrapping process was shown by the occurrence of the hysteresis cycle in the *C–V* curves.

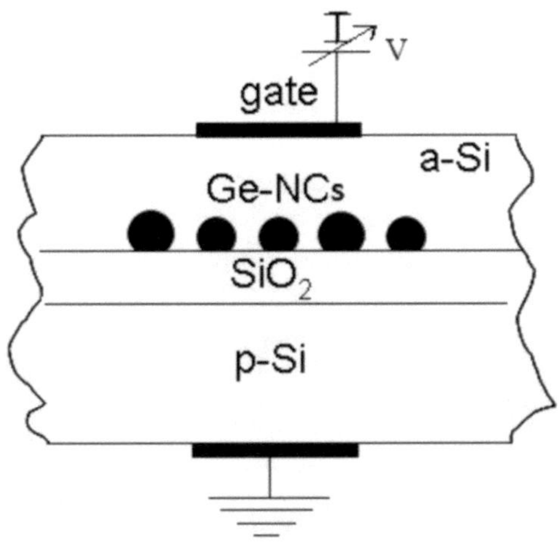

Fig. 2. A schematic cross section of the studied samples of the *A* series. *V* is the bias polarization: the structure is at inversion regime if V > 0 and at accumulation regime if V < 0.

3.2 Conductive-AFM/SEM technique

The sample *B* was studied by using a home-made AFM combined to a scanning electron microscope (SEM) (Troyon et al., 1997) equipped with a field emission gun (LEO Gemini). Our measurements were made with this AFM/SEM combined instrument because working inside the SEM allows one to get rid of the influence of the water layer covering the surface in ambient atmosphere which hinders the electric measurements. Indeed, working in wet atmosphere may create local oxidation of the surface at the tip/surface contact (Okada et al., 2001). Furthermore, this hybrid instrument allows obtaining three simultaneous images; electric, topographic and secondary electron images. For example, the secondary image shown in Fig. 3 allows the verification of the probe geometry and check the state of the probe and of the sample surface. Ge-NCs are well seen and are randomly distributed at the surface of the sample. Note that precise analysis of the Ge-NC structural (average size and density) properties, the size distribution and the density have been previously studied and it was shown that the distribution of the NC size is very narrow (Szkutnik et al., 2008).

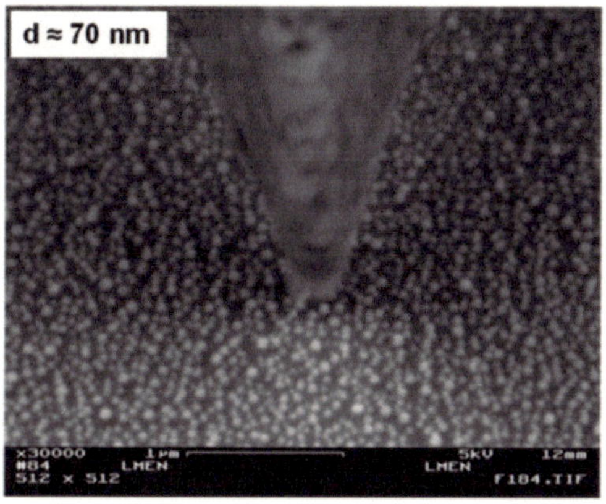

Fig. 3. Secondary electron image showing a *p*-doped diamond AFM tip in contact with Ge-NCs of 70 nm in diameter.

Another important reason to perform measurements inside the SEM is that this particular sample requires a preliminary electron beam irradiation in the condition for obtaining an electron beam induced current (EBIC) image to obtain afterwards (electron beam off) a measurable conduction current. Indeed, due to the thickness of the SiO_2 layer it was impossible to get a C-AFM image for polarization voltages smaller than ± 8 V. The reason why a preliminary electron beam irradiation is needed is explained in section 4.3.

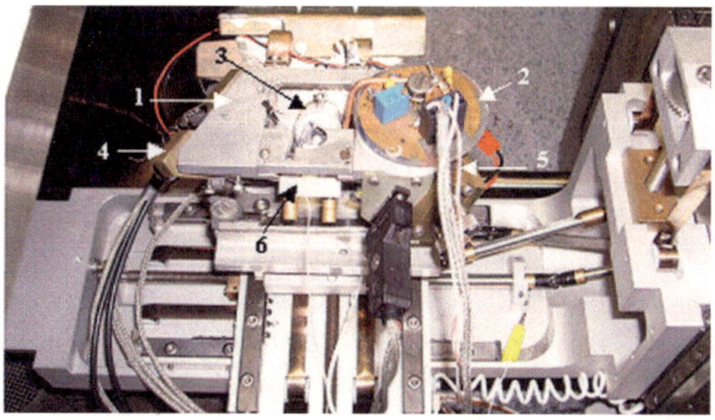

(1) AFM Head.
(2) Preamplifier.
(3) Piezo-tube supporting the sample.
(4) Position of the laser diode.
(5) Photodiode adjustment system.
(6) AFM cantilever support.

Fig. 4. View of the AFM mounted inside the specimen chamber of the SEM.

Fig. 4 gives an image showing the various components of the AFM and Fig 5 gives a schematic representation of the hybrid system used in the present work. The detected current remains very low. So, we installed a low noise (~5pA RMS, 1 MHz bandwidth) current/voltage amplifier (10^7 gain) close to the AFM tip, inside the SEM, to limit the electric noise. An ohmic contact was made on the back side of the samples by depositing an aluminium layer. The sample surface (the top side) was probed in the contact mode while a sample bias V was applied to the substrate with respect to the tip, which is itself referenced to the ground through the preamplifier. In our system, the sample is scanned with respect to the AFM probe. The electron-beam irradiation of the samples was performed in the conditions used for obtaining a nano-EBIC image with the e-beam focused in a fixed position just in front of the AFM probe and as close as possible to it (Smaali et al., 2008); the electron primary energy was 5 keV, the primary current was 1 nA, the sample bias was V = 0 volt and the images (256×256 pixels) were acquired at a line scanning frequency of about 0.5 Hz.

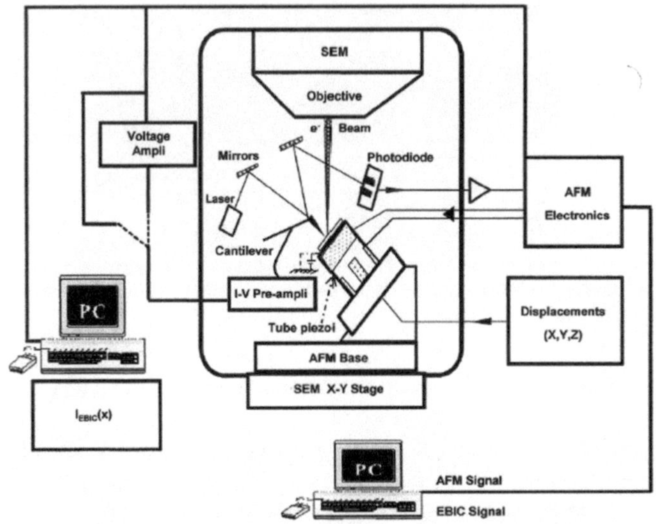

Fig. 5. Schematic representation of the experimental combined AFM/SEM set-up.

4. Results and discussion

4.1 Capacitance – voltage characteristics

HFC-V and *I-V* measurements cycling on *A* samples show the hysteresis loops representative of the carrier storage. The question is: where does the carrier storage occur? Does it appear in NCs and/or in the oxide bulk? To give an appropriate response, Fig. 6(a) shows typical *HFC-V* curves with arrows indicating the sense of the measurement. The same cycle is obtained whatever the beginning of the measurement direction from positive or negative gate bias. Note that the silicon substrate is *p*-doped. The *HFC-V* curves measured from positive to negative gate bias are shifted towards negative biases compared to the *HFC-V* curves measured in the opposite direction. This highlights the electron trapping/detrapping in Ge-NCs and clearly indicates that this process is either non-existent or negligible in oxide.

Let us describe the electron trapping process. In inversion regime (when the silicon substrate is negatively polarized), electrons in the substrate accumulate close to the Si/SiO_2 interface and holes (the majority carriers) move away from this interface towards silicon bulk. On the other side of the sample, the trapped electrons can be emitted from Ge-NCs to the gate via a-Si. Hence, the *HFC-V* curve is shifted to negative bias (this curve is taken as reference to calculate trapped charge due to *HFC-V* shift). If the bias is reversed, electrons can be injected in Ge-NCs and the *HFC-V* curve is moved to positive bias as shown by the arrows in Fig. 6(a). These electrons do not come from the substrate because of the electric field orientation, but from the other side via the a-Si. The contribution of the current coming from electron trapping in a-Si layer, which is expected due to the presence of defects (such as dangling bond and vacancies), is not considered here since carrier emission and capture in these defects are very fast in comparison to the time measurement. Indeed, since the energy levels of the defects are close to the conduction and valence bands they produce very fast conduction paths for electrons and holes. Note that the electron trapping is enhanced at low temperature.

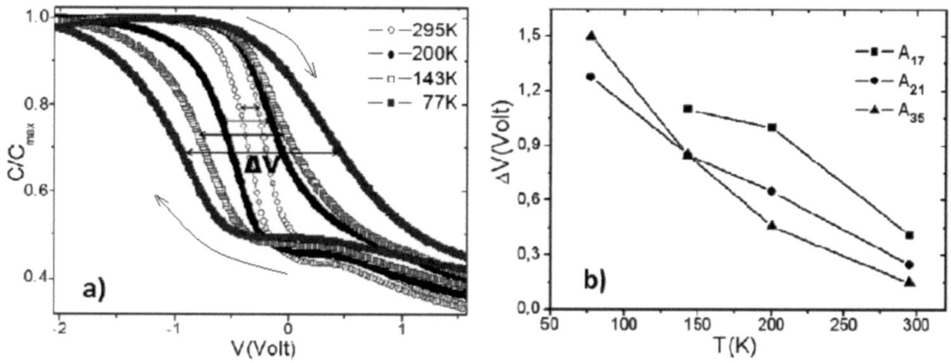

Fig. 6. (a) Electron storage in nanocrystals and its effects on the capacitance - voltage cycles, the temperature effect is also shown, (b) electron storage amplitude for three samples (A_{17}, A_{21} and A_{35}).

In Fig. 6(b), we present the hysteresis width (ΔV) deduced from the *HFC-V* cycles for three samples of the *A* series as a function of the temperature. ΔV is reduced when the temperature increases and is also reduced for samples containing NCs with high average diameter as shown in Fig. 6(b). Note that ΔV represents the global stored charge. The analytic expression of ΔV is given by Tiwari et *al.* (Tiwari et al., 1995), which shows that ΔV is directly related to the Ge-NC density, to the number of electrons stored by NC and to the NC mean diameter. Two main parameters control the ΔV amount: the Ge-NC size and density, but their effects can be different. If the NC size controls the trapped charge density (represented by ΔV) the trapping process will be stronger in NCs with high average diameter. This seems true at low temperature (< 150 K) as shown in Fig. 6(b); the total charge trapping is much more efficient in large Ge-NCs than in smaller ones. But at high temperature (> 150 K), the Ge-NC density controls the stored charge amount. The temperature plays thus a particular role in Ge-NC size and density effects on the global stored charge.
As it has been quoted above, eventual carrier storage can occur in the oxide after flowing through the oxide interfaces barrier height. Nevertheless, this is screened by the effective charge stored in NCs. Finally, the carrier exchange is made between the Ge-NCs and the gate as quoted in Fig. 7 giving a schematic and qualitative situation for electron storage in NCs when the gate is negatively polarized by reference to the Si substrate (on the left of Fig 7) and NCs discharging when the bias polarization is inverted (on the right of Fig. 7).

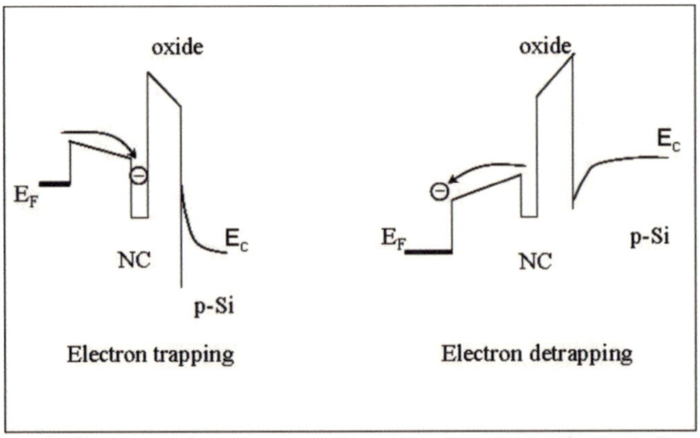

Fig. 7. Energetic band diagram showing carrier trapping in NCs and detrapping according to the bias polarization.

4.2 Current – voltage characteristics

The carrier trapping phenomena are also observed in *I-V* curves as shown in Fig. 8 where the measurements are performed from negative (-3 V) to positive (+3 V) and back to negative gate voltage. Note that the charge storage in standard MOS capacitors and its effects on the *I-V* curves is well known (Balland & Barbottin, 1989; El Hdiy et al., 1993). The shift of the *I-V* curve along the voltage axis is controlled by the type of the effective oxide charge; a positive oxide charge causes a negative voltage shift of the *I-V* curve, while a negative one leads to a positive voltage shift. Moreover, if the stored charge is located near the interfaces (gate/oxide or *p*-Si/oxide), the *I-V* curves show a distortion without shift (or with a very weak shift) along the voltage axis (El Hdiy et al., 1993).

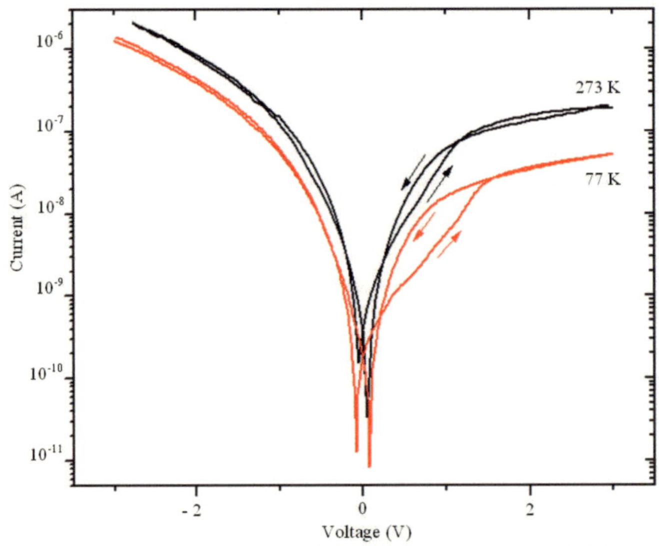

Fig. 8. Typical result showing occurrence of the hysteresis in the *I-V* curves obtained in the sample containing NCs of 21 nm diameter (A_{21}). The enlargement of the cycle is due to the thermo-electronic effect reduction.

Let us remark that only the *I-V* characteristics in inversion regime present a hysteresis cycle. And this hysteresis increases when the temperature is reduced indicating the increase of the carrier storage as previously reported (Kuo & Nominanda, 2006). We also notice that the *I-V* curves are not shifted indicating that the trapping/detrapping phenomena do not occur in the oxide bulk but more in the Ge-NCs as explained in the following.
The measurement first begins at negative bias (-3 V). At negative bias, holes move from the *p*-Si valence band to the gate by tunnelling effect, namely Fowler Nordheim tunnelling (FNT) effect, while electrons move from the gate to Ge-NCs, but their energy and their density are affected in the a-Si. Hole (majority carriers) density is higher than that of electrons (minority carriers). It is expected that trapping process (electrons and/or holes trapping) appears in both oxide bulk and in Ge-NCs; holes in the oxide and electrons in NCs. Indeed, when the bias voltage increases from -3 V to zero, becomes positive and reaches + 3V, hole density is reduced but the density of electrons coming from Si substrate becomes more and more important showing a current increase at positive bias voltage (from 0 to +3 V). When the applied voltage turns back from +3 V to 0 V, the current presents higher values than the first scan. In this case, one has two possibilities leading to increase of current; the first process is the hole detrapping from the oxide, and the second process is the electron detrapping from the Ge-NCs. Both processes can appear during the scan of the bias from zero to +3 V. However, if the hysteresis was related to the hole trapping/detrapping in the oxide, the sense of the hysteresis cycle, indicated by the arrows in the figure would be inverted. Because hole trapping reduces the electron barrier height and the tunnelling distance, leading to a higher current when the bias voltage varies from 0 to +3 V than when it varies from +3 to 0 V. *I-V* measurements were also performed at low temperatures. The results show that the hysteresis size depends on the temperature; the hysteresis width expends when the temperature decreases suggesting that the carrier trapping is enhanced as also shown by the HFC-V technique. Finally the same carrier trapping process is revealed by both *I-V* and *HFC-V* characteristics.
Let us note that the use of the Fowler-Nordheim tunnelling standard expression can qualitatively lead to determine the density of the carriers trapped in Ge-NCs, or at least it can inform on the trapping efficiency versus the temperature. This approach was previously made in the case of the standard MOS capacitor (DiMaria et al., 1993; El Hdiy & Ziane, 2001) to extract the trapped charge density from *I-V* measurements. We carefully use it according to the following assumptions: (a) there is no trapped carriers in the oxide, or the eventual stored carriers do not affect the initial oxide field E_0, (b) the charge contained in the NCs is supposed to be distributed in a thin layer at the NC/oxide interface and (c) the supplementary field caused by the carrier storage process remains lower than E_0.
The standard FNT expression (Fowler & Nordheim, 1928) is:

$$I = AE^2 \exp(-B/E) \tag{1}$$

Where, I is the tunnelling current, A and B are the FNT parameters considered to be constant. E is the electric field near the injecting electrode (the interface which is negatively polarized). Note that the expression given in Eq. (1) is valid only for high voltages; at low voltages, we must take into account the total expression of the tunnelling current (FNT and direct tunnelling) (Schuegraf & Hu, 1994). The oxide electric field can be expressed as:

$$E \approx E_0 - \frac{q\Delta N}{\varepsilon_{Ge}} \quad (2)$$

Considering that the contribution of the stored electrons in NCs to the electric field E variations remains relatively negligible, the density of the trapped electrons is given by:

$$\Delta N(T) = -\frac{E_0^2 \varepsilon_{Ge}}{qB} Ln\left(\frac{I}{I_0}\right) \quad (3)$$

Where, E_0 is the field when the NCs are totally empty, q is the elementary charge, ε_{Ge} is the Ge permittivity (ε_{Ge} is used instead of ε_{oxide} because the charge is stored in Ge-NCs) and ΔN is the density of the stored electrons. I_0 and I are the current before charge storage ($\Delta N = 0$) and after carrier storage ($\Delta N \neq 0$), respectively.
During the measurement under negative gate bias, electrons are trapped by NCs because of their weak energy at the injecting electrode. When the gate bias is inverted, the negative charge stored in NCs affects the tunnelling current leading to a current I lower than I_0. The use of Eq. (3) to data of Fig. 8, allows us to show that the stored charge density at 77 K is twice higher than that trapped at 300 K; $\Delta N(77\text{ K}) / \Delta N(300\text{ K}) \approx 2$. On the other hand, ΔN can be measured by taking the current (I and I_0) values at bias voltage corresponding to the large width of each cycle hysteresis.

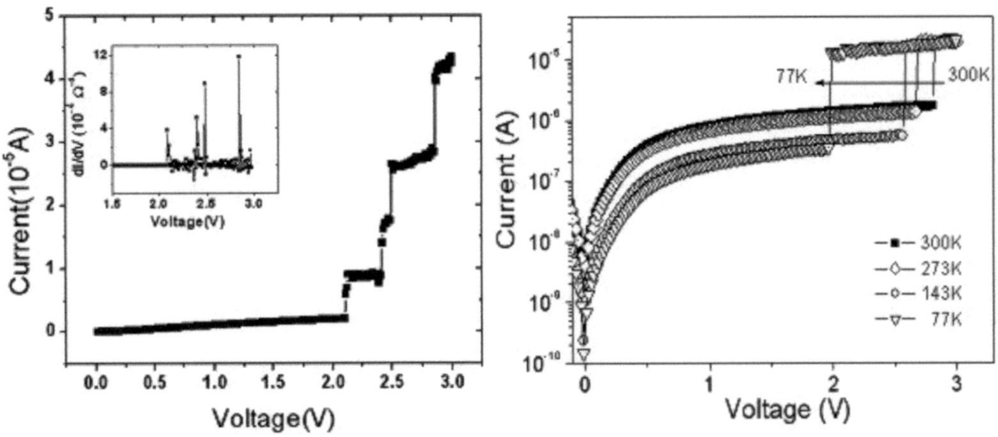

Fig. 9. (a) Current-voltage characteristic at a positive gate bias (inversion regime). The steps observed in the reverse I-V curve are related to electron resonant tunnelling via Ge NC discrete levels. The inset contains the corresponding conductance as a function of gate voltage. (b) Evolution of current-voltage characteristics with temperature. Current jumps appear for increasingly weak voltages when the temperature decreases.

In addition to the carrier trapping evidence, we notice the remarkable behaviour exhibited by NCs which have a mean diameter of 3.5 nm and an average density of 2.5×10^{12} cm^{-2}. Their I-V curves present steps in the reverse part as shown in Fig. 9(a). The observed jumps are not linked to the oxide breakdown for the following reasons. The maximum value of the oxide field is lower than 6 MV/cm (value from which damage in the oxide can occur), and to prevent defect creation in the oxide, the bias was scanned rapidly for higher values (from 2 to 3 V). Moreover, to be sure that the observed steps do not correspond to the oxide breakdown, the *I-V* characteristics were repeated many times indicating approximate reproducible results. This little difference in terms of the plateau voltage position and height can be linked to both the temperature and size dispersion effects. Moreover, other measurements made at low temperatures have shown that the plateaus became larger. These jumps are representative of a resonant tunnelling of electrons into the discrete energy levels of Ge-NCs. Such tunnelling resonance behaviour has been attributed to the Coulomb blockade effect in Si quantum dots embedded in Si-rich SiO_2 deposited in plasma phase (Kim, 1998). Other recent studies have also shown the Coulomb blockade at room temperature in roughly spherical nanocrystalline silicon dots the main diameter of which was 5 nm (Wu et al., 2004). .

Let us now explain this resonant tunnelling current. It is well known that in tunnelling resonance, electrons accumulate between double barriers (Goldman et al., 1987; Kim, 1998, Kareva et al., 2002; Vexler et al., 2006). In the case of our sample, electrons can temporarily accumulate at the Si/SiO_2 interface quantum well and between the double barrier formed by Ge-NCs, *a*-Si, and oxide. Under a positive bias, electrons are injected from the Si/SiO2 quantum well through the oxide to the subband levels of the Ge-NCs where they fill the same energy levels as in the Si/SiO_2 quantum well. A simple description of the tunnelling resonance process can then be given: if we consider that electrons have the energy level of the silicon conduction band edge at the Si/SiO_2 interface (E_{c0}), when the gate is positively biased, current flows only if E_{c0} is swept through the first quantum energy level (E_1) of Ge-NCs resulting in a current peak in the *I-V* curve. Since current continues to flow after E_{c0} is swept past E_1, one observes current steps rather than peaks in the *I-V* curves. Two main conditions for observing the Coulomb blockade effect are (a) the tunnelling resistance between Ge NC and the electric contacts must be higher than the quantum resistance ($Rt >> h/e^2 \approx 26$ kΩ) and (b) single electron charging energy ($E=e^2/2C$)>> k_BT, where $C \approx 4\pi\varepsilon_0\varepsilon_{a-Si}r = 2.277$ aF is the self-capacitance of an approximately spherical Ge NC and $\varepsilon_{a-Si} = 11.7$ (Ref. Orwa et al.'2005) (this expression of C gives an overestimation of the capacitance value, since the spherical form of the Ge-NC is truncated at the contact with the oxide). This capacitance gives a single electron charging energy E of about 35 meV, supporting the existence of the Coulomb blockade effect. The first criterion can be met by weakly coupling the NC to two electrodes which is the case in our sample. The NCs are confined between SiO2 and non doped amorphous silicon. The second criterion depends on temperature. At room temperature, the Coulomb blockade condition occurs when the capacitance is smaller than the thermal one, $C_{th} = 3.1$ aF (3.1×10^{-18} F), for a single quantum dot (or a diameter ≤ 5 nm). In the situation investigated here, the capacitance values are smaller than C_{th}. Consequently, the charge energy (e^2/C) is higher than the thermal energy (k_BT). On the other hand, Fig. 9(b) shows that current jumps appear for increasingly weak voltages when the temperature decreases (Gacem et al., 2007). Note that this Coulomb blockade effect appears only for the NCs with the smallest main diameter (3.5 nm), and not in the other studied samples which probably would require temperatures lower than 77 K. This is related to the strong quantum confinement in 3.5 nm Ge-NC than in the others. Indeed, figure 10(a) gives a schematic energetic band diagram of the studied samples showing the bandgap of each element and corresponding to a thermal equilibrium situation characterized by the alignment of the Fermi level. The conduction and valence band bending are not shown for convenience. The minimum difference between the a-Si and Ge-NC conduction bands is about 0.15 eV (here a-Si gap is equal to 1.7 eV), because it is well known that the bandgap of thin amorphous silicon may be substantially increased as compared to bulk a-Si owing to the confinement effect. This means that the bandgap value of the a-Si could by higher than 1.7 eV. Fig. 10(b) shows the NC gap as a function of the mean diameter obtained from calculation developed by Niquet et al. (Niquet et al., 2000) for spherical Ge nanocrystals. The Ge-NC band gap weakly increases from the sample A_{35} to A_{21} and then A_{17}, but takes a high value (~1.55 eV) for the sample $A_{3.5}$. This behaviour is remarkable and its consequence on electric characteristics is to lead a stronger quantum confinement than for the other samples, which is even revealed at room temperature.

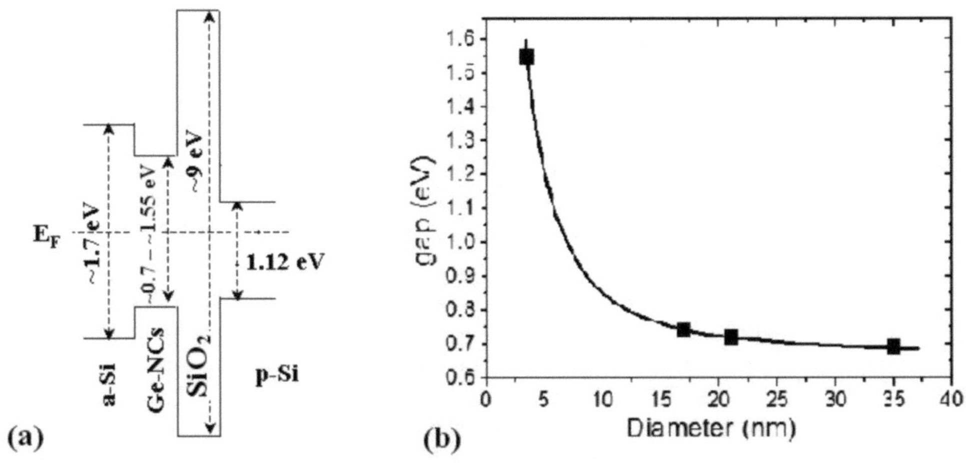

Fig. 10. (a) Schematic band diagram of the sample structure. The conduction and valence band bending are not shown here. (b) Energy bandgap of spherical Ge nanocrystals versus NC size; the line corresponds to theoretical calculation extracted from (Niquet et al., 2000).

4.3 Conductive-AFM measurements

I-V and *HFC-V* measurements show that the electron trapping process occurs in Ge-NCs. But these techniques cannot allow the study of what happened in a single NC. They give average information which must be completed by supplementary data on single NC. This is the reason why we used the AFM technique to study the transport and the trapping/detrapping process in a NC.

The study was done by conductive atomic force microscopy (C-AFM). The measurements were performed on sample *B* at room temperature. The tip that we used was *n*-doped silicon tip covered with *p*-doped diamond-like carbon (resistivity ≈ 0.003-0.005 Ω cm) and the nominal cantilever stiffness was 2 N/m. In preliminary experiments, we first collected C-AFM images in which the measured current was a few pA which is not enough to obtain good resolution images. For detecting measurable current at a reasonable bias voltage (lower than ± 8 V) we were obliged to preliminary irradiate the sample with the electron

beam in the condition for obtaining a nano-EBIC image. Note that this phenomenon has also been observed for InAs/GaAs quantum dots (Troyon & Smaali, 2008).
First of all, let us explain what is the nano-EBIC technique. Its principle is schematically represented in Fig. 11.

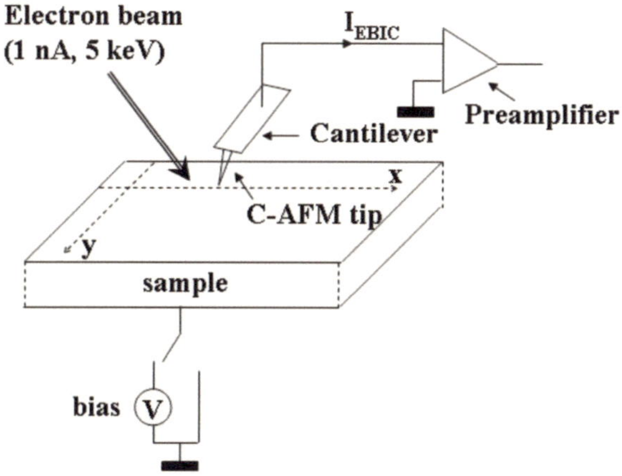

Fig. 11. Schematic representation of the set up used for the nano-EBIC technique. The AFM tip and the electron beam focused near this one are both immobile, whereas the sample is scanned in the (x, y) plane.

The AFM and EBIC images are simultaneously obtained by scanning the sample with respect to the fixed AFM probe by using the piezoelectric tube of the AFM. During image acquisition the electron beam is focused in a fixed position just in front of the AFM probe and as close as possible to this one. The back side of the sample is connected to the ground and the current is punctually collected by the conductive AFM tip near the electron probe during scanning the sample. The tip-sample contact constitutes a nano-heterojunction with formation of a very small depletion zone which is able to collect the electron beam induced current. The size of the depletion zone determines the resolution. We have demonstrated that a resolution of the order of 20 nm can be obtained (Smaali et al., 2008; Troyon & Smaali, 2008).

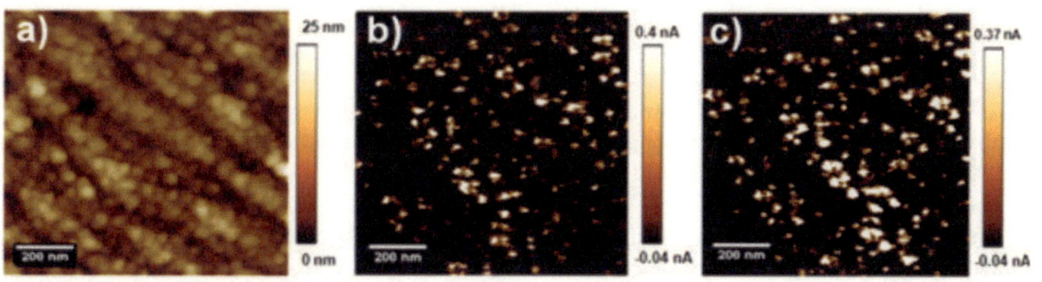

Fig. 12. Topographic image a), b) electric image obtained at 0 V just after acquisition of an e-beam induced current image (e-beam off) and c) second electric image obtained two hours after in the same irradiated zone.

After electron beam irradiation in the condition for obtaining electron beam induced current, C-AFM measurements without electric biasing (0 V) were performed at different times with the electron beam off. Typical results are shown in Fig. 12. This figure gives a comparison between three kinds of images acquired on the same area of a sample containing NCs of 70 nm diameter: (a) is an AFM topographic image, (b) is a C-AFM image obtained without sample bias (e-beam off) just after acquisition of a nano-EBIC image and (c) is obtained in the same condition as image (b) but 120 minutes later. Although the tip radius is large (~150 nm), the NCs are well resolved; this can be probably explained by the presence of small diamond nanocrystals or grains formed at the extremity of the tip during the fabrication process. Note that even at 0 V, 0.4 nA of current is obtained. The fact that a current could be collected without bias is explained by the charge transfer resulting from the alignment of the Fermi levels of the sample and probe during scanning.

The C-AFM images afterwards obtained on the previously irradiated zone show that the current flowing through the NCs remains of the same order of magnitude as the image taken two hours before, although the sample is not polarized. This process shows that charges have been trapped in the NCs during the electron beam irradiation and produce a current, which is a hole current. Indeed, the contact between the *p*-doped diamond AFM tip and the sample surface forms a nano-heterojunction. This kind of nano-contact leads to the transport and trapping of holes, but prevents electrons to flow due to the high conduction bands discontinuity. This mechanism is explained by the energetic band diagrams of Fig. 13. Fig. 13(a) shows the band diagram of the contact between the AFM diamond tip and the oxide and Fig. 13(b) shows the band diagram when the AFM tip is in contact with the NC. In both cases, electrons cannot flow because of the large conduction band discontinuity.

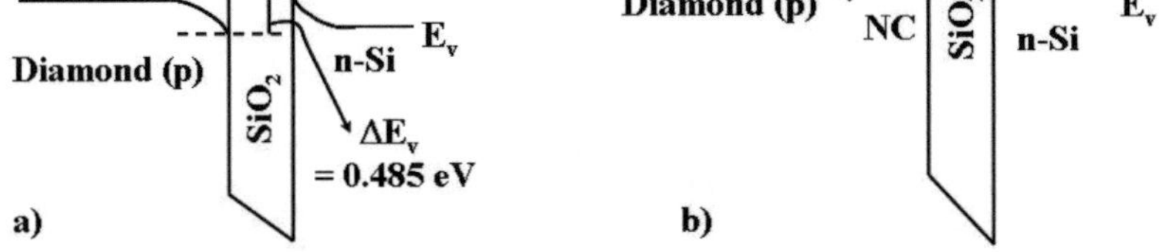

Fig. 13. Energy band diagrams of the probe-sample nano-contact when (a) the probe is positioned in contact with the oxide and (b) when it is positioned above the NC the bandgap of which (0.7~1.5 eV) is size-dependent.

During scanning, holes can move from the tip to the NC because of the weakness of the band bending at the tip/NC interface. Coming from the substrate by tunnelling through the oxide, some of holes are trapped in the quantum well (in the NC) and could accentuate the NC energy band bending. With no contact between the AFM tip and the sample, holes remain stored in the NCs because of the presence of a native oxide covering the NCs and also of the vacuum which creates a barrier confining holes in the NCs. The hole detrapping begins when the AFM tip is in contact with the NC in two situations; unbiased and negatively polarized tip. In the unbiased tip case, a part of stored holes can tunnel through the little barrier from NCs to the AFM tip. While a negative polarization to the tip (which does not affect the barrier height at the NC/tip interface) accentuates the band bending at each interface enhancing the hole current through the structure. Whatever the sign of the electric polarization, electrons cannot move through the system (tip/NC/oxide/silicon) because of the strong conduction band pseudo-discontinuity added to the barrier height due to the presence of the oxide.

As quoted above, electric conduction results from hole detrapping from NCs. This discharging effect of Ge-NCs has been clarified by the analysis of a series of C-AFM images recorded on one single NC just after the acquisition of a NF-EBIC image, without polarizing the sample. Five images have been recorded every 30 min. Between two images the C-AFM probe is retracted to avoid contact with the sample. The successive electric images show a progressive reduction of the intensity with a total extinction of the NC at the fifth image when the NC is completely discharged. Fig. 14 shows the evolution of the maximum current measured in the NC versus time and on the right part the corresponding current images and the simultaneously acquired topographic images which were extracted from larger images. These images reveal a detrapping phenomenon. The C-AFM images obtained without sample bias, demonstrate the capacity of the NCs to trap charges and also to release charges after each passage of the C-AFM tip on the NCs in the condition of no sample bias. Losing charge during acquisition of an electric image reduces the current in the next image. Indeed, during contact between the AFM tip and an isolated NC under 0 volt bias, the Fermi levels alignment leads to hole emission from previously charged NC because of the low effective barrier height at this contact which facilitates the current flow. Obviously, silicon hole tunnelling effect through the oxide is unlikely because of its thickness (~5 nm). So, after the acquisition of the following electric images the hole current is more and more reduced.

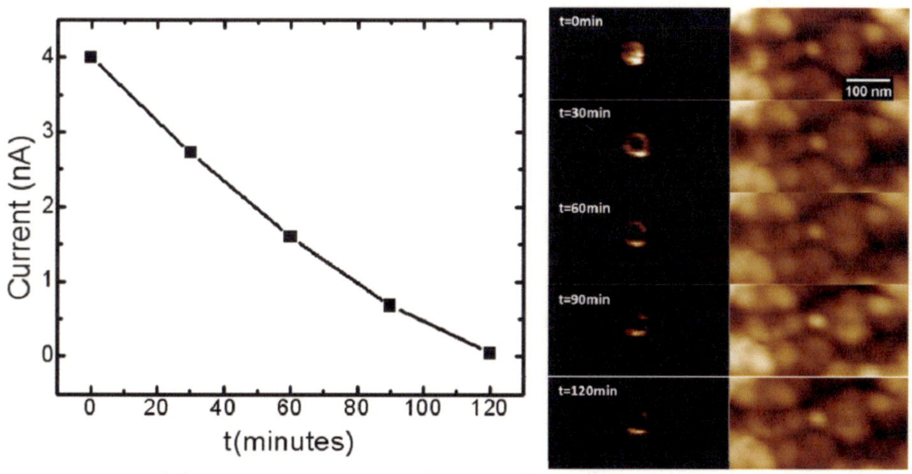

Fig. 14. Variation of the maximum current flowing through an NC versus time in a series of five images (V = 0) illustrating a hole detrapping process. The electric images recorded every 30 min after irradiation, are presented on the right as well as the simultaneously acquired topographic images.

The charge retention is an important and crucial parameter which requires attention in both elaboration procedure and reliability characterization. The use of NCs to replace the standard floating gate in memory devices is necessary since many problems can be avoided. It is interesting to make comparison between both characterization techniques used in this work even if two kinds of nanostructures (containing capped or no capped NCs) have been differently studied. Both methods give information on carrier transport and storage. They also underline dominant role of the Ge-NCs in the carrier capture and storage. However, it is clearly seen that the C-AFM technique is required if information on an isolated NC (or a limited number of NCs) in a given structure is needed. It would be interesting to study the stored charge amount as a function of the NC size. This could be performed by Electrostatic Force Microscopy technique which is our future ambition.

5. Conclusion

Ge nanocrystals grown by a two step dewetting / nucleation process on an oxide/*p*-(or *n*-) doped silicon substrate have been electrically studied by standard methods (capacitance - voltage and current - voltage) and by conductive atomic force microscopy technique. The kind of the carriers involved in the electric transport and the charge storage depends on the nature of the electrode/sample contacts. With the standard methods, for which a metallic contact was used, electrons were concerned. The study was performed over a wide range of temperature varying from 300 down to 77 K and results evidenced an electron storage phenomenon in nanocrystals. The temperature effect in the enhancement of the electron retention was revealed by the enlargement of the hysteresis cycles obtained on *HFC-V* and *I–V* measurements. Indeed, the reduction of the measurement temperature reduces the thermo-electronic emission from NCs and facilitates carrier storage. Resonant tunnelling effect through germanium nanocrystals with large voltage gaps was observed at room temperature in ultradense Ge nanocrystals of ~ 3.5 nm mean size. It appeared for increasingly low voltages when the temperature decreases. All these results are consistent with a Coulomb blockade effect in ultrasmall Ge nanocrystals. Conditions for the occurrence of the Coulomb blockade have been reminded.

On the other hand, the C-AFM technique using a *p*-doped diamond probe was used to study uncapped Ge-NCs. The current flowing through the conductive probe was measured for imaging local conductance, while the deflection of cantilever was optically detected for revealing geometrical structure. The heterojunction resulting from the AFM tip/sample contact allowed hole transport through the structure and hole storage in Ge-NCs. Preliminary electron beam irradiation in the conditions for obtaining a nano-EBIC image was needed for performing the C-AFM study. Without this e-beam irradiation, no noticeable current could be detected.

6. Acknowledgments

Isabelle BERBEZIER and her group are acknowledged for providing the samples. This work was supported in part by the European Community FP7 program in the scope of the FIBLYS project under grant CP-TP 214042-2.

7. References

Balland, B. & Barbottin G. (1989). The trapping and detrapping kinetics impact on C(V) and I(V) curves. In: *Instabilities in Silicon Devices Vol.2*, Barbottin G. & Vapaille A. (Ed.), (19-69), North-Holland, Amsterdam.

Berbezier, I.; Ronda, A. & Portavoce. (2002). SiGe nanostructures: new insights into growth processs. *A. J. Phys. Condens. Matter.*, Vol. **14,** (8283-8331).

Conibeer G. *et al.* (2006). Silicon nanostructures for third generation photovoltaic solar cells. *Thin Sol. Films,* Vol.**511–512,** (654-662).

Chatterjee, S. (2008). Titania-germanium nanocomposite as a photovoltaic material. *Solar Energy* **Vol. 82,** (95-99)

Choi, W. K.; Kanakaraju, S.; Shen, Z. X.; & Li, W. S. (1999). Characterization of Ge nanocrystals in a-SiO_2 synthesized by rapid thermal annealing. *Appl. Surf. Sci.* Vol. **144-145**, (697-701).

DiMaria, D. J.; Cartier, E. & Arnold, D. (1993). Impact ionization, trap creation, degradation, and breakdown in silicon dioxide films on silicon. *J. Appl. Phys.* Vol. **73**, No.7 (3367-3384)

El hdiy, A.; Salace, G.; Petit, C.; Jourdain, M. & Vuillaume, D. (1993). Study of defects induced by high-electric-field stress into a thin gate oxide (11 nm) of metal-oxide-semiconductor capacitors. *J. Appl. Phys.* Vol. **74**, No.2 (1124-1130).

El Hdiy, A. & Ziane, Dj. (2001). Electric bidirectional stress effects on metal-oxide-silicon capacitors. *J. Appl. Phys.* Vol.89, No.2 (1405-1410).

Fowler, R.H. & Nordheim, L. (1928). Electron emission in intense electric fields. *Proc. Roy. Soc. London.* **Vol. 119**, No. (173-181).

Gacem, K.; El Hdiy, A.; Troyon, M.; Berbezier, I.; Szutnik, P.D.; Karmous A. & Ronda A. (2007). Memory and Coulomb blockade effects in germanium nanocrystals embedded in amorphous silicon on silicon dioxide. *J. Appl. Phys.* Vol.102, (093704/1-4).

Gacem, K.; El Hdiy, A.; Troyon, M.; Berbezier, I. & Ronda, A. (2010). *Nanotech.* Vol.21, (065706/1-6).

Goldman, V. J.; Tsui, D. C. & Cunningham, J. E. (1987). Observation of Intrinsic Bistability in Resonant-Tunneling Structures. *Phys. Rev. Lett.* Vol.58, No.12 (1256-1259)

Karmous, A.; Berbezier, I. & Ronda A. (2006). Formation and ordering of Ge nanocrystals on SiO_2. *Phys. Rev. B* Vol. 73, No.7 (075323/1-5).

Kareva, G. G.; Vexler, M. I.; Grekhov, I. V. & Shulekin, A. F. (2002). Electron Tunneling through a Double Barrier in a Reverse-Biased Metal-Oxide-Silicon Structure. *Semiconductors*, Vol.36, No.8 (889-894).

Kim, K. (1998). Visible light emissions and single-electron tunnelling from silicon quantum dots embedded in Si-rich SiO_2 deposited in plasma phase. *Phys. Rev. B,* Vol.57, No.20 (13072-13076)

Kuo, Y. & Nominanda, H. (2006). Nonvolatile hydrogenated-amorphous-silicon thin-film-transistor memory devices. *Appl. Phys. Lett.* Vol. 89 (173503/1-3)

Niquet, Y. M.; Allan, G.; Delerue, C. & Lannoo, M. (2000). Quantum confinement in germanium nanocrystals. *Appl. Phys. Lett.* Vol. **77**, No.8, august 2000, (1182-1184).

Okada, Y.; Miyagi, M.; Akahane, K.; Iuchi, Y. & Kawabe, M. (2001). Self-organized InGaAs quantum dots on GaAs (311)B studied by conductive atomic microscope tip. *J. Appl. Phys.* Vol. 90, No.1, (192-196).

Okada, Y.; Miyagi, M.; Akahane, K.; Kawabe, M. & Shigekawa, H. (2002). Self-organized InGaAs quantum dots grown on GaAs (311)B substrate studied by conductive atomic microscope technique. *J. Cryst. Grow.*, Vol.245, (212-218).

Orwa, J. O.; Shannon, J. M.; Gateru, R. G. & Silva, S. R. P. (2005). Effect of ion bombardment and annealing on the electrical properties of hydrogenated amorphous silicon metal-semiconductor-metal structures. *J. Appl. Phys.*,Vol.97, No. (023519/1-5)

Park, N.M.; et al. (2003). Size-dependent charge storage in amorphous silicon quantum dots embedded in silicon nitride. *Appl. Phys. Lett.* Vol.83, No.5, august 2003, (1014-1016)

Schuegraf, K. F. & Hu, C. (1994). Hole injection SiO_2 breakdown model for very low voltage lifetime extrapolation. *IEEE Trans Electron Dev.* Vol. 41, No.5 (761-767)

Skorupa, W.; Rebohl, L. & Gebel, T. (2003). Group-IV nanocluster formation by ion-beam synthesis. *Appl. Phys. A* Vol.76, (1049-1059).

Smaali, K.; Troyon, M.; El Hdiy, A. And Molinari, M. (2006). Imaging the electric properties of InAs/InP(001) quantum dots capped with a thin InP layer by conductive atomic force microscopy: Evidence of memory effect. *Appl. Phys. Lett.* Vol.89, (112115/1-3).

Smaali, K; Fauré, J; El Hdiy, A & Troyon, M. (2008). High-resolution scanning near-field EBIC microscopy: application to the characterisation of a shallow ion implanted p^{+}–n silicon junction. Ultramicroscopy 108 (605-612).

Smaali, K.; El Hdiy, A.; Molinari, M. & Troyon, M. (2010) Band gap determination of the native oxide capping quantum dots by use of different kinds of conductive AFM probes: example of InAs/GaAs quantum dots. *IEEE. Electron. Dev.* To be published

Stomp, R.; Miyahara, Y.; Schaer, S.; Sun, Q.; Guo, H. & Grutter P. (2005). Detection of single-electron charging i an individual InAs quantum dot by noncontact atomic-force microscopy. *Phys. Rev. Lett.* Vol.94, (056802/1-4).

Szkutnik, P. D.; Karmous, A.; Bassani, F.; Ronda, A.; Berbezier, I.; Gacem, K.; El Hdiy, A. & Troyon, M. (2008) Ge nanocrystals formation on SiO_2 by dewetting: application to memory. *Eur. Phys. J. Appl. Phys.* Vol. 41, (103-106).

Tiwari, S.; Rana, F.; Chan, K.; Hanafi, H.; Chan, W.; & Buchanan, D. (1995). Volatile and Non-Volatile Memories in Silicon with Nano-Crystal Storage. *IEDM Tech. Dig.* Vol. 95, (521-524).

Troyon, M.; Lei, H.N.; Wang, Z. & Shang, G. (1997). A Scanning force microscope combined with a Scanning electron microscope for multidimensional data analysis. *Microanal. Microsc. Microstr.* Vol.8 (393-402).

Troyon, M.; Smaali, K.; Molinari, M.; El Hdiy, A.; Saint-Girons, G.; & Patriarche, G. (2007). Local electronic transport through InAs/InP(001) quantum dots capped with a thin InP layer studied by an AFM conductive probe. *Semicond. Sci. Technol.* Vol.22 (755–762).

Troyon, M. & Smaali, K. (2008). Influence of electron irradiation on electronic transport mechanisms during the conductive AFM imaging of InAs/GaAs quantum dots capped with a thin GaAs layer. *Nanotech.* Vol. 19, (255709/1-7)

Vexler, M. I.; El Hdiy, A.; Grgec, D.; Tyaginov, S. E.; Khlil, R.; Meinerzhagen, B.; Shulekin, A. F. & Grekhov, I. V. (2006). Tunnel charge transport within silicon in reversely-biased MOS tunnel structures. *Microelectron. J.*, Vol.37, (114-120).

Wu, L.; Dai, M.; Huang, X.; Zhang, Y.; Li, W. Xu, J. & Chen, K. (2004). Room temperature electron tunneling and storage in a nanocrystalline silicon floating gate structure. *J. Non-Cryst. Sol.* Vol. 338-340, (318-321).

Nano-crystals for quadratic nonlinear imaging: characterization and applications

Sophie Brasselet
Institut Fresnel, Domaine Universitaire St Jérôme, 13397 Marseille Cedex 20
France

Joseph Zyss
Laboratoire de Photonique Quantique et Moléculaire, Institut d'Alembert, ENS Cachan,
61 av. du président Wilson, 94235 Cachan
France

1. Introduction

Nonlinear optics is a well established field today, covering a large and rich range of applications which expands every year. Its extension to sub-wavelength scale processes is however highly non trivial, and nano-scale nonlinear optics is only in its infancy. Downscaling high order light-matter optical interaction is however nowadays accessible using new technological tools such as near field techniques, high resolution microscopy and pulsed femtosecond lasers. As a consequence, a large amount of effort has been recently invested into the invention and engineering of nano-structures that exhibit nonlinear optical properties and lead to new optical functions. Among the promising routes followed by nonlinear nano-optics, the development of new nano-sources originating from frequency mixing processes is particularly successful. In this chapter, we describe how nano-crystals have been advantageously developed and used in Second Harmonic Generation (SHG) (or by extension Sum Frequency Generation : SFG), which is the lowest order nonlinear process, dependent on the square of the incident field (Boyd, 1992). Recent developments have shown that these nano-structures are potentially key elements in various fields, such as new nano-probes for bio-imaging, or nano-scale optical fields probing in the ultra-short pulses regime. They also combine the interesting properties of frequency mixing processes with the nano-scale regime: unlike the resonant fluorescence process, coherent harmonic generation is active in the non-resonant regime and therefore free from photo-bleaching which is otherwise generally considered a major drawback from fluorescent nano-probes today. The coherent nature of frequency mixing also allows generating different wavelengths and therefore creating nano-sources of a large range of possible frequencies. An additional advantage of these nano-sources is to avoid phase-matching constraints since their size is well below the coherent length of the underlying nonlinear process, at which nonlinear propagation suffers from destructive interferences between the propagating and induced nonlinear waves. The availability of a large emission frequency range finally allows imaging

bright nano-emitters over a dark background in complex environments sometimes contaminated by auto-fluorescence such as in living cells.
The coherent build up of an induced dipole at twice the frequency of the incident field demands a material exhibiting a non-centrosymmetric structure, which explains why crystals featuring such a structure, although not a majority, have been such determining materials in nonlinear optics to these days, both as a source of inspiration as well as towards applications (Boyd, 1992; Chemla & Zyss, 1987).

As shall be reviewed in the first section of this chapter, downscaling non-centrysymmetry to the nano-scale has been achieved following different possible routes, covering both bottom-up to top-down approaches, in organic as well as in inorganic structures. These nano-probes, which are the nano-scale counterparts of the traditional nonlinear active bulk materials, require specific attention with respect to material engineering. Indeed working at nanometric scales requires revisiting the basic rules of nonlinear optics such as phase matching and nonlinear coupling optimization, which lead to original light-matter interaction processes. Very much like in linear optics, working with nano-objects in nonlinear optics is therefore a research field in itself.
While the study of nonlinear bulk crystals has largely developed along the search for macroscopic nonlinear coupling directions and phase matching conditions, the investigation of nonlinear nano-crystals requires a complete different approach, furthermore constrained by the lack of control of the orientation of the object in the macroscopic framework. In this context, determination of the orientation of the nanocrystal with respect to the laboratory framework sets a problem by itself, thus in considerable contrast with the case of bulk crystals which can be in most cases cut and oriented at will. We will describe in the second section of this chapter the different techniques which have been advantageously developed, either based on a randomization of the nano-crystals orientation in solution, or by immobilization on a substrate by nonlinear microscopy.
Finally, the last section of this chapter will summarize the actual status of applications of nonlinear active nano-crystals, both for bio-imaging using nonlinear microscopy in living cells and tissues, and for nanophotonics, in particular for nano-scale optical fields probing.

2. Nano-crystals development for Second Harmonic Generation

The conception and fabrication of crystals with nanometric to sub-micrometric size exhibiting strong nonlinear quadratic optical responses is a topic of growing interest. In particular, Second Harmonic Generation (SHG) active nano-crystals offer a broad range of potential applications from nanoprobes for bio-imaging to nano-scale photonics with original optical properties. SHG, which provides the ability to transform an exciting optical frequency into its harmonic, when transposed to the nano-scale, could provide new types of light sources and labels with (i) an emission wavelength far away from its excitation (in particular in the infra-red range, which is better adapted to bio-imaging in tissues) (ii) a lower photobleaching probability due to the possibility to work far from resonances, (iii) possibility to tune the emission wavelength by tuning the incident one, or by performing frequency mixing of different excitations, (iv) a less stringent dependence, compared to bulk crystals, on phase matching conditions.

The first prerequisite condition when engineering a SHG-active nano-object is to obtain a non-centrosymmetric structure. As seen in the examples below, a large amount of researches have lead to new routes of fabrications of SHG-active nano-crystals, in order to meet this requirement: either by using crystal unit-cell structures deprived of center of symmetry in inorganics, or engineering a non-centrosymmetric packing of non-centrosymmetric molecules in molecular organic crystals (Chemla & Zyss 1987). As a positive consequence, the coherent emission from SHG-active nano-crystals is supposed to be weakly dependent on the shape of the nano-structure, contrary to metallic nanoparticles which rely on the non-centrosymmetry breaking at their interface. Nevertheless, such shape dependence remains a research topic of interest for nano-objects of the order of a significant fraction of the wavelength (typically from a few ten's to a few hundred's of nanometers). In addition, these structures are expected to exhibit a small size-dispersion, to be amenable to manipulation in solution, and ultimately to be bio-compatible, which imposes low chemical reactivity, low toxicity, stability in aqueous environment, and possibly surface functionalization. They are finally also expected to lend themselves via adequate biomineral or bio-organic linkage, to tethering to some biological material of interest, such as towards the SHG monitoring of intra or inter cellular traffic of a DNA fragment, a biotine-avidine linkage moiety, a protein, an antigen-antibody pair or a deactivated virus. As described in this section, various strategies have been developed to achieve stable and efficient SHG nano-emitters, either from a top-down or from a bottom-up approach.

2.1 Organic nano-crystals

Molecular based nano-crystals are the first structures developed as SHG-active nano-materials. Based on molecular nano-assemblies, they have been designed starting from SHG active asymmetric molecules as individual building blocks, and further grown from a bottom-up approach. The fabrication of these novel nanomaterials constitutes an alternative approach to the synthesis of multichromophoric architectures, which has been a trend over the last decade (Yokoyama et al. 2000, Rekaï et al. 2001, Le Bozec et al. 2001). Molecular-based assemblies present furthermore the advantage of a large flexibility with respect to the nanoparticle nature and the active molecule used. The fabrication procedure follows an essential requirement, which is to conserve the molecular non-centrosymmetry at the nanometric scale, in addition to a high molecular density packing, which ensures SHG efficiency. Assuming indeed a nano-object consisting of n molecules of induced nonlinear dipoles oriented in the same direction, the overall SHG signal is proportional to n^2, which can lead to measurable signals in nanostructure of tens of nanometer sizes (Brasselet & Zyss 2007) (see Section 3).

One of the first reports of second-harmonic emission for nanoscale molecular crystals (Shen et al., 2000) was referring to the emblematic NPP structure (NPP for N-(4-nitrophenyl)-(L)-prolinol)) whereby the usual trend towards centrosymetry had been fought by introduction into the molecular structure of a chiral amino-acid fragment (proline alcool derivative) (Zyss et al. 1984) and which had marked a milestone in nonlinear molecular crystals as the first infrared optical parametric oscillator (Josse et al. 1992). Another early example was based on the non-centrosymmetric CMONS molecule (α-[(4'-methoxyphenyl)methylene]-4-nitro-benzene-acetonitrile, a polar donor- acceptor nonlinear structure) (Figure 1a), from which nano-crystals have been fabricated by rapid nucleation of the molecules, confined in the nano-pores of a silica network sol-gel (1:1 tetramethoxysilane (TMOS) : methyltrimethoxysilane (MTMOS)) formed by spin-coating (Sanz et al. 2001). The size of the nano-crystals, ranging from 20 to 100nm with a remarkable monodispersity, was controlled by the growth kinetics and parameters such as temperature, matrix porosity, and the CMONS:alkoxides molar fraction (Sanz et al. 2001, Treussart et al. 2003). The study of individual nano-crystals by SHG microscopy has confirmed the non-centrosymmetric structure of the nano-objects, which was consistent with the bulk symmetry (Brasselet et al. 2004). Although these structures showed very efficient SHG activity in addition to two-photon fluorescence, they were embedded in a gel film and therefore not useable for future labelling and manipulation.

Developing isolated nano-objects which can be manipulated in solution has been achieved in $MnPS_3$/DAZOP particles of 10nm size, based on the hybrid bottom-up synthesis of an intercalation compound consisting of push-pull organic chromophores (DAZOP : (4-[2-(4-dimethylaminophenyl)azo]-1-methylpyridinium) embedded in a layered manganese hexathiohypodiphosphate ($MnPS_3$) inorganic host lattice (Yi et al. 2004) (Figure 1b). The DAZOP non-centrosymmetric ferroelectric-like arrangement is obtained within the layered $MnPS_3$ matrix taking advantage of ionic interactions operating during the particle formation within nanoreactors provided by inverse micelles (typically using Brij97 as a surfactant). This strategy has lead to particles with an average second order hyperpolarisability β of the order of 42.10^{-27} esu, measured by Hyper Rayleigh Scattering (HRS) (Yi et al. 2005) (see Section 3). This value is three to four orders of magnitude higher than the β coefficient typically measured in nonlinear molecules. Studies by SHG microscopy (see Section 3) on individual particles immobilized in PVA (polyvinylalcohol) spin coated thin films have shown that the particles tend to form aggregates of larger size (about 100nm), still preserving a quasi one-dimensional order (Delahaye et al. 2006, Delahaye et al. 2009). The stability of the SHG emission from these particles was seen to be strongly dependent on the surfactant used for the synthesis, suggesting an environment dependence of the particle photo-damage (Delahaye et al. 2009).

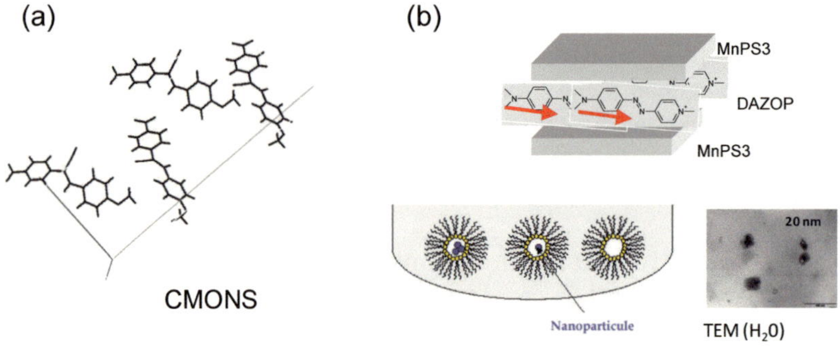

Fig. 1. (a) Unit-cell structure of the CMONS macroscopic crystal in its non-centrosymmetric form. (b) $MnPS_3$/DAZOP nano-crystals : scheme, fabrication procedure and Transmission Electron Microscopy Image.

Top-down approaches have also been attempted to reach molecular nano-crystals. Crystal grinding has been tested on the very efficient co-inclusion compound based on the

confinement of guest non-linear active molecules within the host polycyclic hydrocarbon perhydrotriphenylene (PHTP) channels (Komorowska et al. 2005). The photoactive molecules DANA (4-Dimethylamino-40-nitroazobenzene), used as guest systems, lead to typical bulk values of 340 pm/V. Particles of sizes ranging from 400 to 1000 nm (for which the goal of mono-dispersion had not been pursued at this stage) could be obtained after fine grinding in a PVA-water solution, followed by sonication. Adequate dilution before spin-coating allowed obtaining isolated particle active for SHG and Two Photon Fluorescence.
Another interesting structure, building-up on the concept of octupolar symmetry (Zyss & Ledoux 1994) has been developped based on TTB (1,3,5-tricyano-2,4,6-tris(p-diethylaminostyryl)benzene), a third-order symmetry molecule forming highly efficient nonlinear crystals (Le Floc'h et al. 2005). Micro- and nano-crystals have been investigated (Brasselet & Zyss 2007). 3D Octupolar symmetry evidences in principle an optimized structure whereby the SHG response does not depend neither on the incident polarization nor on the orientation of the object (Brasselet & Zyss 1998). However in the case of TTB, the planar geometry still led to a highly anisotropic nonlinearity (Le Floc'h et al. 2005). Resorting to a fully 3-D octupolar structure (Zyss & Ledoux 1994) would allow generating an orientation independant SHG response, which is of considerable interest for nano-crystal tracking for instance.

Among other possibilities, the fabrication of molecular aggregates from micelles structure based on the mixing of surfactant and active molecules has also been successful to generate non negligible SHG signals (Eisenthal 2006). These structures, which size can be easily engineered, are interesting model systems to study the contribution of surface effects in the nonlinear radiation from sub- micrometric size objects (Revillod et al. 2008).

2.2 Inorganic nano-crystals

While organic nano-crystals can exhibit a wide range of absorption wavelengths governed by chemical synthesis, inorganic nanostructures are often transparent from the visible to 12 μm. A large amount of studies is currently undertaken in order to fabricate robust, transparent, mono-disperse functionalizable inorganic nano-crystals, with typical sizes ranging from 10 to 300 nm. Since the nano-structures described below exhibit very similar nonlinear coefficients (typically in the range 10-30 pm/V in bulk), the goal of the report below is not to discuss their relative efficiencies but rather to give an overview of the fabrication and application routes exploited so far. Efficiency measurement issues are discussed in Section 3 of this chapter.

Zinc Oxide (ZnO) nano-crystals are among the first inorganic nano-crystals studied for nonlinear optics (Johnson et al. 2002). Made of a nontoxic and biocompatible semiconductor material, ZnO nano-crystals were later synthesized in organic media by using a nonhydrolytic sol−gel process and subsequently dispersed in aqueous media using phospholipid micelles, leading to successful encapsulation such as used for drug delivery (Kachynski et al. 2008, Kuo et al. 2009). Further incorporation with the biotargeting molecule folic acid makes this structure adapted to a wide range of targeted bio-imaging studies. ZnO nano-crystals of <100nm size have been synthesized, with additional X-ray diffraction and Transmission Electron Microscopy (TEM) ascertaining for their tetragonal (and pyramid shape) structures (Kachynski et al. 2008). ZnO exits also in the form of nano-wires of typical

1 μm lenths with hexagonal cross-sections, which have been studied also for bio-imaging applications (Nakayama et al. 2007).

The non-toxicity of Iron Iodate has motivated the fabrication of $Fe(IO_3)_3$ nano-crystals of typical size 20-300nm, by co-precipitation synthesis (Galez et al. 2006). The $Fe(IO_3)_3$ material belongs to the hexagonal space group $P6_3$ with nonlinear coefficients in bulk of about 10pm/V off resonance. Nano-crystals of 30 nm sizes have been seen to aggregate in larger clusters (80nm) when the solvent is evaporated on a surface (Bonacina et al. 2007).
Single-crystalline $KNbO_3$ nano-wires of rectangular cross-section of 40nm-400nm x 1-20 μm sizes (depending on the reaction time) could be obtained by a hydrothermal method, mixing at high temperature Potassium hydroxide and niobium pentoxide in deionized water (Magrez et al. 2006). Structural characterizations showed a conserved orthorhombic phase (*mm2*) with the growth axis of the wires parallel to the [011] direction. A rough estimate for the averaged nonlinear coefficient of $KNbO_3$ nano-wires lead to equivalent bulk values of about 10-30 pm/V, as expected in $KNbO_3$ crystals. These $KNbO_3$ nano-wires have been applied to trapping and bio-imaging in living cells (Nakayama et al. 2007).

Aside from these bottom-up techniques, nano-crystals have also been obtained from powders, with a good size dispersion ascertained by Dynamic Light Scattering (DLS) experiments.
$KTiOPO_4$ (KTP) nano-crystals, isolated from a flux-grown powder extracted from a bulk crystal, can be commercially provided (Cristal Laser). These nanocrystals are monodisperse in size (with sizes ranging between 10nm and 80nm). KTP belongs to the orthorhombic point group *mm*2 with typical nonlinear coefficients ranging from 2 to 17 pm/V off resonance. Thorough studies have been undertaken on isolated KTP nano-crystals, in order to measure small size crystals (Le et al. 2006), infer their three-dimensional orientation on substrates (Sandeau et al. 2007), and relate SHG and size information using combined atomic force microscopy and optical techniques (Le et al. 2008).
At last, $BaTiO_3$ nano-crystals of 30nm and 90nm sizes were obtained from commercial dry powders (Nanoamor (30nm) and Techpower (90nm)). Specific studies have allowed a good control of the dillution behavior of the particles in solution to further bio-imaging applications (Hsieh et al. 2009, Hsieh et al. 2010). By dispersion in aminomethylphosphonic acid and sonication of the colloidal solution, the phosphonic acid reacts on the particle surface, ending up with an amine group coating, which introduces an electrostatic repulsion between the particles. $BaTiO_3$ is of tetragonal structure with similar range of nonlinear coefficients as in the crystals reported above. Measured efficiency cross section on single nano-crystals has been evaluated and seen to surpass, by orders of magnitude, the typical fluorescence cross section from traditionally used molecules in two-photon fluorescence (Rodriguez et al. 2009b, Pu et al. 2008, Hsieh et al. 2009).
In all the examples above, considerable stability and efficiency of the measured signals could be observed, as compared to traditional nano-probes used in fluorescence. It is therefore not surprising that many of the previously described structures (ZnO, $Fe(IO_3)_3$, $BaTiO_3$ and $KNbO_3$) have been directly applied to imaging in biological media (see Section 4 of this chapter).

Semi-conductor quantum dots are now the smallest existing structures active for SHG with typical sizes ranging from 2 to 15 nm. While semiconductor quantum dots are widely known and used in fluorescence microscopy as bio-markers, their use for nonlinear

radiation has been much less studied. CdS and CdTe cadmium based materials are both based on a zinc blende crystal structure of cubic Tetrahedral Td symmetry. CdS nanoparticles of about 5 to 10 nm were first synthesized by co-precipitation reaction between an aqueous solutions of $Cd(NO_3)_2$ and Na_2S in the presence of hexametaphosphate (HMP) as a stabilizer (Fu et al. 2001, Wang et al. 2005). Measured hyperpolarizabilities in solution by HRS were found to be up to 7.24×10^{-26} esu, which is four orders of magnitudes higher than typical nonlinear molecules. This value was found to be higher than the one expected from a pure bulk response (typically 78pm/V off resonance), indicating possible surface effects (Fu et al. 2001).
CdTe/CdS nanocrystals with a diameter of 10 to 15 nm have been recently studied by SHG microscopy, immobilized in a PMMA (Polymetylmetacrylate) thin film (Zielinski et al. 2009). Their synthesis is based on a progressive crystalline growth in a noncoordinating solvent, by regular injection of cadmium oleate solution and trioctylphosphine/Te followed by trioctylphosphine/S. Polarization analysis of the second-harmonic emission confirms the expected zinc blende symmetry (see Section 3). The nonlinear efficiency of these structures is expected to come predominantly from the core (with a bulk CdTe nonlinear coefficient of about 150 pm/V), even though the shell is also non-centrosymmetric. In addition to the possibility to observe very small nano-crystals thanks to a high efficiency, intriguing wavelength dependence of the SHG efficiency indicated a probe of deep resonant levels mechanisms (Zielinski et al. 2009).

3. Optical characterization of SHG active nano-crystals

SHG is a multiple field tensorial coupling process, involving both the crystal symmetry and the incident fundamental as well as outgoing harmonic fields polarizations. This coupling has been accounted for by introduction of the nonlinear tensor field and its contraction with the molecular susceptibility tensor, a procedure that can be generalized to any order, leading to a convenient invariant scalar formulation (Boulanger et al. 1997, Brasselet & Zyss 1998). Furthermore, in objects smaller than the focal excitation volumes, the SHG efficiency grows with the object size, as expected from the higher number of active molecules/unit cells (Brasselet 2010). Therefore the analysis of a SHG signal from an isolated object, and in particular the measure of its efficiency, relies on the knowledge of its orientation, its symmetry, and its size relative to the excitation focal volume. While size information can possibly be known on average by techniques such as Dynamic Light Scattering (DLS), Transmission Electron Microscopy (TEM), or Atomic Force Microscopy (AFM), symmetry information can be inferred from bulk structures or more rarely from X-ray diffraction studies on an ensemble of nano-objects. Although these measurements provide averaged information, they can be still valuable as a first input for SHG investigations from single nano-crystals. In particular the nano-crystals symmetry will often be considered to be the same as for the bulk, thus preserved at the nano-scale.

In a planar wave approximation, the overall measured SHG intensity of a single nano-crystal originates from the coherent sum of n induced nonlinear dipoles, n being the number of molecules (or crystalline unit-cells) contained in the nano-crystal present in the focal volume of excitation (Delahaye et al. 2009, Brasselet 2010). The macroscopic components of these induced nonlinear dipoles, projected in the *(X,Y,Z)* macroscopic frame, are proportional to:

$$P_I^{2\omega}(\Omega) = n\sum_{J,K} \beta_{IJK}(\Omega) E_J^{\omega} E_K^{\omega} \tag{1}$$

with *(I,J,K)* = *(X,Y,Z)*, and $\beta_{IJK}(\Omega)$ being the SHG tensorial component of the crystal unit-cell (or unit molecule) oriented along the Euler set of angles $\Omega = (\theta, \phi, \psi)$. E_J^{ω} is the projection of the incoming field polarization at the ω fundamental frequency on the *J* macroscopic direction. From a microscopic point of view, $\beta_{IJK}(\Omega)$ originates from the projection of the microscopic unit-cell nonlinear hyperpolarizability β_{ijk} (*(i,i,k)*= *(1,2,3)* being the unit cell frame), projected in the macroscopic frame, and can be written (Zyss et al. 1982):

$$\beta_{IJK}(\Omega) = \sum_{i,j,k} (\vec{i}.\vec{I})(\Omega)(\vec{j}.\vec{J})(\Omega)(\vec{k}.\vec{K})(\Omega)\beta_{ijk} \tag{2}$$

with $(\vec{i}.\vec{I})(\Omega)$ the cosine projection factor between the*i* (microscopic) and *I* (macroscopic) axes. Note that $d_{ijk} = N\beta_{ijk}/2$ are the traditionally used nonlinear coefficients in a bulk crystal with *N* the molecular (or unit-cell) density. They can therefore be used as known numbers in this expression. Eq. (2) is inferred from the oriented gas model (Zyss et al. 1982), widely used in the understanding of nonlinear responses from organic materials. It assumes in particular that there is no significant interaction between the molecules which therefore behave such as uncorrelated nonlinear emitters, which has been a realistic approach in many examples, subject of course to adequate local field corrections, such as the classical Lorenz-Lorentz or Onsager terms.

As described in the next Sections, two measurements of the macroscopic information from the nonlinear induced dipoles radiation can be performed:

- An incoherent summation over a large collection of n_{nc} nano-crystals, all macroscopic dipoles $P_I^{2\omega}(\Omega)$ being oriented in random directions and positions, from either time or spatial (or both) fluctuations (see Section 3.1). In this case the overall intensity analyzed along the I polarization direction is proportional to:

$$I_I^{2\omega} \propto \left\langle \left| P_I^{2\omega}(\Omega) \right|^2 \right\rangle_{\Omega} = n_{nc} \sum_{J,K} n^2 \left\langle \beta_{IJK}(\Omega) \beta_{ILM}(\Omega) \right\rangle_{\Omega} E_J^{\omega} E_K^{\omega} E_L^{\omega *} E_M^{\omega *} \propto n_{nc}.n^2.\left\langle \beta^2 \right\rangle.\left(I^{\omega}\right)^2 \tag{3}$$

  With $\sqrt{\langle \beta^2 \rangle}$ being the averaged measured nonlinear coefficient. The incoherent nature of the process imposes in this case that the signal is proportional to n_{nc}, the number of nano-crystals in the focal volume.

- A direct coherent measurement of the macroscopic radiating dipole $P_I^{2\omega}(\Omega)$ from a single nano-crystal oriented in the Ω direction (see section 3.2):

$$I_I^{2\omega}(\Omega) \propto \left|P_I^{2\omega}(\Omega)\right|^2 = \left|n\sum_{J,K}\beta_{IJK}(\Omega)E_J^{\omega}E_K^{\omega}\right|^2 \propto n^2.\beta^2.\left(I^{\omega}\right)^2 \quad (4)$$

In both cases the signal is proportional to n^2, with n the number of unit-cells (or molecules) in the nano-crystal. Since $n = N.V$ with V the nano-crystal volume and N the crystal density, the measured signals are proportional to the sixth power of the nano-crystal diameter. Note that rigorously, V is the overlap between the excitation volume and the nano-crystal volume itself.

3.1 Averaging SHG information

In order to measure SHG efficiency from nano-crystals while circumventing the issue of their orientation knowledge, studies have been performed using orientation-averaging techniques. Averaging SHG information is typically performed in a powder measurement, which is an historical way to analyse SHG efficiency from molecular micro-crystals (Kurtz et al. 1968, Halbout et al. 1984). Powder analyzes have been performed for instance on ferroelectric Strontium barium niobate nanoparticles of 40 nm size, made from $Sr_xBa_{1-x}Nb_2O_6$ (SBN) (Rodriguez et al. 2009). The drawback of this technique is the lack of knowledge on the possible formation of macroscopic clusters.
Another traditional technique, Hyper Rayleigh scattering (HRS) (Terhune et al. 1965), relies on the measurement of the nonlinear scattering of the nano-objects in solution, which is more amenable to environments where the nano-crystals can be isolated from their neighbours by electrostatic repulsion. HRS, which has been a determining tool for molecular engineering in nonlinear optics (Zyss & Ledoux 1994) including molecular based nano-objects (Le Bozec et al. 2001), is a robust technique to infer averaged nonlinear coefficients from nano-crystals, using known solvents as references. Since the obtained signal is a result of time, spatial and orientational averaging over fast diffusing molecules or particles, this method does not require a preliminary orientation of molecules or particles before study. It has been applied to CdS particles (Fu et al. 2001, Wang et al. 2005), molecular aggregates (Eisenthal et al. 2006, Revillod et al. 2008), hybrid nanoparticles (Yi et al. 2005), $BaTiO_3$ and $PbTiO_3$ nano-crystals (Rodriguez et al. 2009b), and more recently $FeTiO_3$ nano-crystals (Mugnier et al. Submitted). $\sqrt{\langle\beta^2\rangle}$ averaged coefficients up to more than 10^{-26} esu have been measured, which is orders of magnitudes above single nonlinear molecules, due to their size extension into larger scale aggregates (as a comparison a DAZOP molecule efficiency is typically 220.10^{-30} esu (Yi et al. 2005)).

While these techniques are powerful to provide an averaged efficiency estimation of nano-crystals, they are however strongly limited when analysis is needed at the single nano-object level, which is the relevant one towards nano-crystal engineering. In particular, no particular knowledge can be inferred on the crystalline quality of isolated crystals, neither on their possible behaviour distributions.

3.2 SHG imaging of single nano-crystals

SHG microscopy, initially introduced as an imaging tool to visualize the microscopic crystal structure in polycrystalline ZnSe (Hellwarth, 1974), allows today investigating nano-crystals one by one, provided they are sufficiently distant from one another (typically by 1 µm) on a glass substrate or in a polymer thin film. SHG microscopy is based on the illumination of a

nano-object by focussing the light of a pulsed laser (for instance a Ti:Sa laser, tuneable between 690 nm and 1050 nm (150fs pulse width, 80MHz repetition rate), using a high numerical aperture objective (NA ranging between 0.9 and 1.4). The focused light forms a focal volume of typically 300 nm in the lateral dimension and 700 nm in the axial direction. In a typical SHG microscope, the SHG emission from nano-objects is collected backward through the same objective as used for the fundamental excitation, then filtered-out spectrally to possibly remove remaining laser light and fluorescence emission, and focused on avalanche photodiodes (Figure 2a). The final image of nano-crystals is generally obtained by scanning either the sample on a piezoelectric mount or the excitation beam by galvanometric scanners, over typically 20 μm to 100 μm dimensions (Figure 2a) (Brasselet et al. 2004, Brasselet & Zyss 2007). Other implementations have been performed based on the phase information from the SHG signal in single nano-crystals, taking advantage of the coherent nature of this optical process. Based on interferometric measurements with a reference signal coming from the SHG field of a macroscopic nonlinear crystal, these implementations have allowed either a direct phase measurement in a homodyne detection set-up (Figure 2b, Le et al. 2006), or 3D imaging using digital holographic image reconstruction (Figure 2c, Pu et al. 2008). The first scheme has allowed measuring KTP nano-crystals below 30nm size, benefiting from the homodyne detection sensitivity provided by the local oscillator (Le et al. 2006). The second provides a scheme for a scan-free 3D imaging. In this case sensitive CCD camera is placed away from the object plane, and the image is added coherently with a reference SHG plane wave. By digital propagation in the observed interference pattern, the field is reconstructed at any plane, with a diffraction limited resolution (Pu et al. 2008).
In the context of quadratic nonlinear microscopy, let us also mention a parent technique, also resorting to a frequency mixing process, the Pockels linear electro-optic effect that is leading to an $\omega \pm \Omega$ optical frequency modulation, where ω and Ω stand respectively for the higher frequency of the illuminating optical laser field and for the lower one attached to an externally applied electric voltage (herein ranging from the kHz to a few tens of MHz). In an interferometric read-out configuration using a sensitive homodyne detection to boost the modulated signal while minimizing the intensity actually shining the material, this configuration turns out to be compatible with low cw power (typically 100μW). This effect has been implemented and applied to poled polymers (Toury et al. 2006), doped artificial membranes (Hajj et al. 2009), and more recently to KTP nano-crystals down to 100 nm size (Hajj et al. Submitted).

Obviously a SHG image of many isolated nano-crystals exhibits a wide range of intensities (Figure 2a), for several reasons:

- This is first due to the size variations between nano-crystals, their efficiency scaling with the sixth power of their diameter, as detailed above. In addition an estimation of size variations in the 10-100 nm range is impossible: any size below the diffraction limit exhibits an image spot of roughly 300 nm size, and only crystals sizes above this limit can be discriminated.
- Even in a mono-disperse population, intensity variations occur from the wide range of possible orientations of nano-crystals in the sample, coupling with different efficiencies to the incident polarization as seen in Eq. (4).

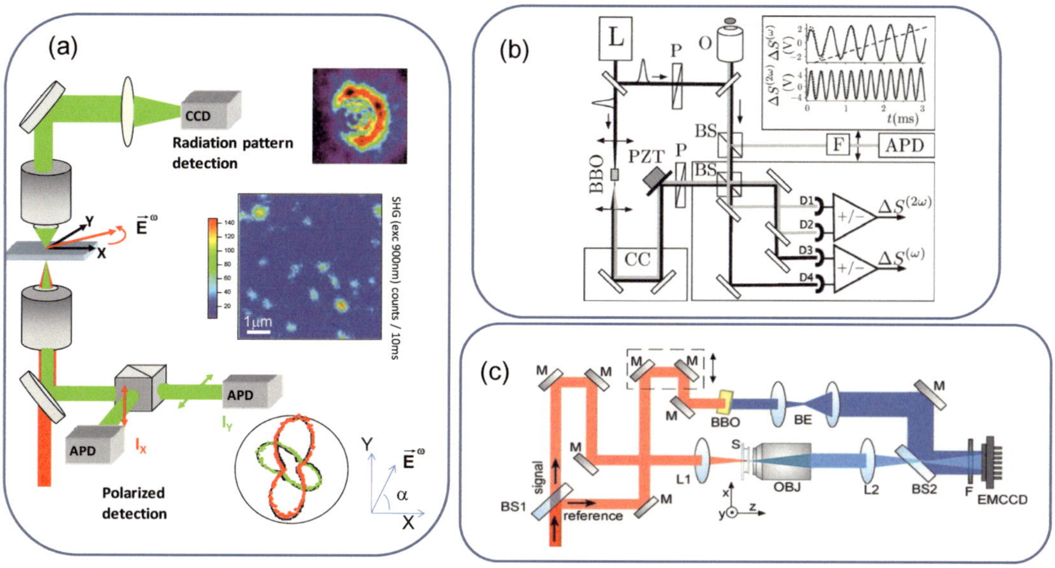

Fig. 2. (a) Imaging and polarization resolved nonlinear microscope. A polarizing beamsplitter separates the signal towards two avalanche photodiodes (APD). SHG polarimetry consists in recording the SHG emission while the incident linear polarization is rotated (angle α) on two orthogonal analysis directions (I_X and I_Y are represented in polar plots in red and green respectively for a given nano-crystal). A typical SHG image (from Brasselet 2010) is shown from a thin PVA polymer film containing MnPS3 particles ($MnPS_3$/DAZOP/Brij97, 10^{-4} mol/L DAZOP). Excitation wavelength : 920nm (a SHG spectral filter is placed in the detection path at 460 nm - excitation power: 1 mW-SHG signal scale : counts/50ms). A defocused imaging (above) is shown from a 80 nm size nano-KTP (Sandeau et al. 2007). (b) (from Le et al. 2006) SHG balanced homodyne detection set-up. (c) (from Hsieh et al. 2009) Harmonic Holography set-up. On both cases, a reference arm is created by SHG in a BBO (β-barium borate) crystal. While the homodyne set-up restores the information from interferences fringes observed by modulating a mirror position in the interferometer, the holography set-up images directly the phase information in a wide field interference pattern, which is used for further reconstruction imaging.

3.3 Determining orientation and crystallinity information from a single nano-crystal

Measuring a single nano-crystal nonlinear efficiency is therefore not direct and requires careful investigation of the SHG response. In order to approach this issue, knowledge of the SHG polarization coupling mechanism in the nano-crystals is required. In addition to SHG imaging, a full polarimetric analysis has been therefore developed based on the recording of the SHG emission signal under a continuous rotation of the incident linear polarization in the sample plane (Brasselet et al. 2004). The polarization rotation is performed by an achromatic half-wave plate mounted on a rotating step motor, while the signal is recorded on two polarization directions, using two avalanche photodiodes separated by a polarizing beamsplitter (Figure 2a). The orientation (Ω) and symmetry (β tensor) information in the nano-crystal are contained in the $\beta_{IJK}(\Omega)$ coefficients of Eqs. (2) and (4). The principle of polarization-resolved SHG is to tune the field contributions E_J^{ω} by rotating a linear

polarization in the *(X,Y)* sample plane (Figure 2a). Typical polarization responses projected on the *X* and *Y* directions are illustrated in Figure 2a and Figure 3 on a 1D symmetry nano-crystal. Very anisotropic polarization responses can be observed in the direction of the nano-object, as expected, whereas a symmetry containing different orientations, such as in a 1D nano-object cluster, would lead to more complex polarimetric response (Figure 3b).
It is therefore possible, providing that the nano-crystal symmetry or orientation is known, to discriminate mono-crystalline structures from non-crystalline ones (Brasselet et al. 2004, Komorowska et al. 2005, Brasselet & Zyss 2007, Brasselet 2010). A non-crystalline structure is seen in this case as a collection of nonlinear dipoles oriented in different Ω directions adding up incoherently. This information, which is not accessible from an ensemble measurement, is of paramount importance towards further nano-crystals synthesis. SHG polarimetry has been successful in investigating molecular ordered materials (Le Floc'h et al. 2003, Anceau et al. 2005). Recently, demonstrations have shown its potential to provide orientational and structural behaviour of single isolated CMONS nano-crystals (Brasselet et al. 2004), as well as hybrid (Delahaye et al. 2006, Delahaye et al. 2009) and inorganic (Zielinski et al. 2009) nano-crystals.

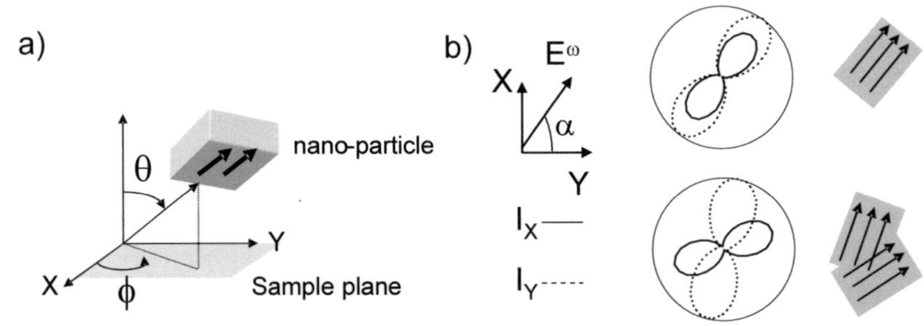

Fig. 3. (a) Representation of the Euler set of angles Ω (θ,ϕ,ψ) defining the orientation of a nano-crystal of high symmetry axis designated by the black arrows. In the case of a nano-object abiding to cylindrical (or higher) symmetry as here, one needs only two Euler angles, the third ψ rotation around the symmetry axis of the nano-object being then unnecessary. (b) Polarization-resolved SHG response from a single 1D symmetry nano-crystal (above) versus a collection of different orientations of 1D symmetry nano-crystals (below). The polarization response is represented as a polar graph in the projection sample plane (X,Y), projected along the X and Y polarization analysis directions.
Several observations must be added to the polarimetric analysis discussed above, including considerations which are crucial towards a proper analysis of polarization dependent SHG data from single nanocrystals.

- *Instrumentation distortions by reflection optics.* In most of the measurements a polarization analysis requires rotating a linear polarization before reflection on a dichroic mirror or other mirrors (Le Floc'h et al. 2003; Chou et al. 2008; Schön et al. 2008). It is know that these optics, although often corrected for a reliable *s/p* (or *X/Y*) intensity reflection ratio, can strongly distort any intermediate polarization angle by applying a phase shift between the *X* and *Y* components. The input polarization, becoming elliptic, needs to be correctly accounted for in the analysis of Eq. (4). This can be done by preliminary calibrations (Le Floc'h et al. 2003, Schön et al. 2008).

- *Effect of the high aperture excitation/collection*. Eq. (4) is written in the plane wave approximation. Observation of single nano-crystal however requires high numerical aperture (NA) objectives for both the excitation and collection of the SHG radiation which requires a more rigorous analysis. Indeed, under such conditions, an incident field polarization in the *(X,Y)* plane is known to contain also *Z* contributions in the focal plane, that can reach up to 45% of the incident amplitude at the edge of the focussing spot (Richards & Wolf 1959). Depending on the nano-crystal off-plane orientation, the SHG signal therefore also contains *Z*-coupling nonlinear coefficient contributions that can strongly alter its polarization response (Yew & Sheppard 2006, Schön et al. 2010). Such contributions are however located spatially at the border of the incoming Gaussian field amplitude, and may induce extra nonlinear coupling terms only for objects reaching 100 nm to 150 nm size. The high aperture collection also mixes-up polarization states of the emission radiation. Both effects can be fully taken into account by a complete model calculating the nonlinear induced dipole at every point of an object placed at the center of the focal spot of the objective (Sandeau et al. 2007, Schön et al. 2010). In general nano-crystals of sizes well below 100 nm will only experience slight modifications of their polarization responses under high NA conditions.

- *Effect of the spatial extension of the nano-crystal*. Another limitation to Eq. (4) is that it considers that the coherently added dipoles are all positioned spatially on the optical axis of the microscope objective, thus ignoring possible phase retardation effects in the volume of the object. Extensive studies accounting for the spatial extension, of a nano-crystal, have shown that for sizes above a few tens of nm, the radiation does not resemble that of a single macroscopic dipole volume (Sandeau et al. 2007), in particular the backward emission is considerably reduced comparing to the forward emission, and phase matching conditions have been re-written accounting for the Gouy phase shift occurring in the excitation volume (Sandeau et al. 2007). Whereas the "punctual object" assumption might be crucial for efficiency measurements in nano-objects of unknown size distribution, it does not however alter the polarization resolved microscopy analyses which are relative measurements.

- *Limitation to a 2D projection information*. The information provided by polarimetric SHG is limited to the projection of the nonlinear tensor in the sample plane *(X,Y)* and is therefore not a complete 3D information. Indeed the excitation field is defined in this plane, therefore only *X* or *Y* coupling directions are allowed in this geometry, except from high NA focusing effects as discussed above. Any information related to off-plane coupling requires complementary analysis. In some cases, as demonstrated for CMONS (Figure 4a), two-photon fluorescence can provide additional information which, combined with SHG, is sufficient to deduce such 3D orientation and crystallinity information from the nano-object (Brasselet et al. 2004). Pure 2D projection analysis in SHG may also lead to the missing 3D information in nano-crystals of complex symmetry involving non-diagonal coupling directions (Zielinski et al. 2009, Brasselet 2010). Another method has been proposed to deduce 3D information from nano-crystals, based on the defocused imaging of the nonlinear radiation from the nano-object (Figure 2a, 4b). This has lead to successful determination of 3D orientation information based on the image analysis of the projection of the nonlinear dipole radiation expansion in the *(X,Y)* sample plane (Sandeau et al. 2007).

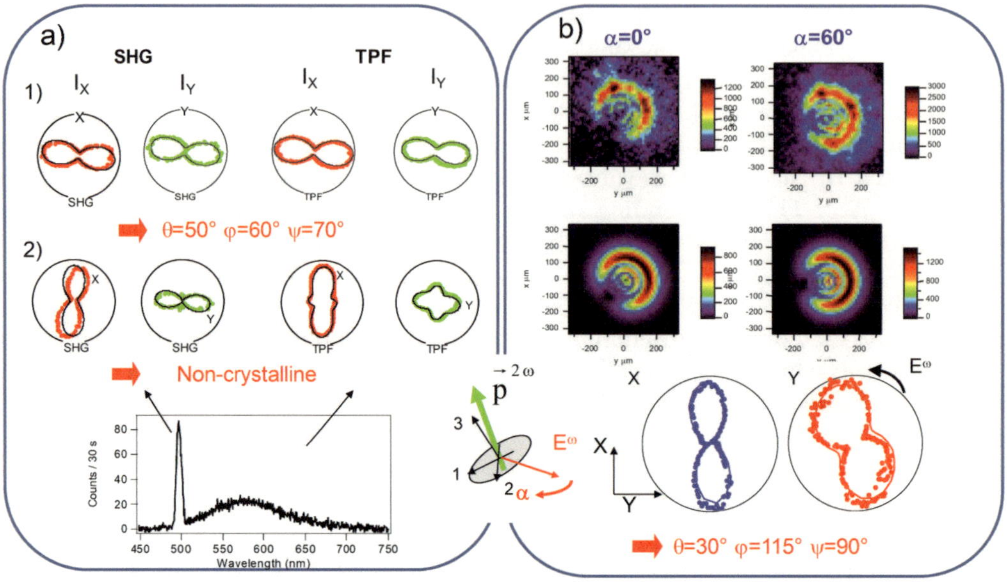

Fig. 4. (a) SHG and two-photon fluorescence (TPF) polarization dependence from two different CMONS nano-crystals embedded in a sol-gel film (Brasselet et al. 2004). The SHG and TPF emission wavelengths can be discriminated and therefore lead to a separated polarization analysis of the two signals. 1) : mono-crystalline nanoparticle which projection in the sample plane resembles that of a 1D nano-object. The 3D orientation of the nano-crystal could be deduced from the simultaneous fit of the X and Y projections of the SHG and TPF responses. 2) typical nano-crystal which polarization response cannot be fit by a pure mono-crystalline orientation. (b) Defocused imaging of 80 nm a KTP nano-crystal for two different input polarizations in the sample plane (α angles). The defocused images, compared to modelling accounting for the full propagation of nonlinear induced dipoles in the microscope, allowed retrieving a 3D orientation information, confirmed by polarimetry (Sandeau et al. 2007).

3.4 Determining the nonlinear efficiency from a nano-crystal

Based on Eq. (4) and assuming that the nano-crystal symmetry and 3D orientation are known, it is now possible to determine its nonlinear efficiency. This is generally done by measuring a reference nonlinear response, similarly as in HRS (Yi et al. 2005, Revillod et al. 2008), which is provided here by a macroscopic crystal. In known excitation/detection polarization coupling directions, the SHG signal from a crystal of known orientation is:

$$I_{ref}^{2\omega} \propto \left(\chi_{ref}^{(2)}\right)^2 . \left(I^{\omega}\right)^2 \tag{5}$$

which nonlinear coefficient $\chi_{ref}^{(2)}$ is taken as a reference. The proportionality coefficient contains the same collection and scaling factors as in Eq. (4), therefore SHG intensities can directly be compared, for a same input power, and a given (IJK) set of polarization conditions:

$$n^2 . \beta_{IJK}(\Omega)^2 = \chi_{IJK}^{(2)}(\Omega)^2 = \chi_{ref}^{(2)} . I_{ref}^{2\omega} / I_I^{2\omega}(\Omega) \tag{6}$$

The macroscopic nonlinear coefficients from a nano-crystal $\chi_{IJK}^{(2)}(\Omega)$ can therefore be measured for a given orientation. Note that this macroscopic coefficient is generated from an object of much smaller size than that of a bulk crystal. Since $\chi_{IJK}^{(2)}(\Omega)$ is proportional to the volume of the nano-object (Eq. (4)), and similarly $\chi_{ref}^{(2)}$ is proportional to the excitation volume V_{exc} in the microscope, the nano-crystal efficiency can be transformed into an "equivalent bulk" value which is the efficiency that the nano-crystal would exhibit if it were filling the whole focal volume: $\chi_{IJK,bulk}^{(2)}(\Omega)=\chi_{IJK}^{(2)}(\Omega).V_{exc}/V$. Typically V_{exc} ~ 60.10^6 nm^3, and V ranges between 10^3 nm^3 and 10^6 nm^3. Note that since the excitation takes place over about 300 nm in the sample plane, it is considered here as homogeneous over the nano-crystal size. More refinements are necessary when the nano-crystal size approaches the diffraction limit size (Delahaye et al. 2009). Values up to 1pm/V have been measured in 100 nm size aligned nano-clusters of $MnPS_3$/DAZOP hybrid nanoparticles, which reach up to 1000 pm/V in an equivalent bulk crystal (Delahaye et al. 2009). In addition of being stable, these nano-crystals are therefore very bright nano-sources.

3.5 Retrieving size information from a nano-crystal

Assuming that both the orientation Ω and efficiency $\chi_{IJK}^{(2)}(\Omega)$ of the nano-crystal can be measured, it is now possible to retrieve size information from this object. For this, the macroscopic to microscopic relation expressed in Eq. (2) is used and introduced in:

$$V=\chi_{IJK}^{(2)}(\Omega)/(N.\beta_{IJK}(\Omega)) \tag{7}$$

Where $\beta_{IJK}(\Omega)$ are deduced from the microscopic nonlinear coefficients β_{ijk} (Eq. (2)), known from the unit-cells crystals values: $d_{ijk}=N\beta_{ijk}/2$ and assuming that both bulk nonlinear efficiencies and density are the same as in the bulk crystal.

Such an analysis has been performed on a large collection of $MnPS_3$/DAZOP hybrid nano-crystals immobilized in PVA (Figure 5), which efficiency were used to evaluate a lower limit to the particles size distribution (Delahaye et al. 2009). The lower limit size given here is imposed by the lack of knowledge of the 3D off plane orientation angle of the nano-crystals (even though the 2D orientation of the dipoles in the plane could be determined by SHG polarimetry as described in the previous Section). The population of the deduced sizes was seen to be consistent with the sizes of the spots measured on the SHG scans, which most often were above the diffraction limit. This analysis allowed in particular evidencing an aggregation behaviour of the nano-particles in the polymer film, which lead to 100 nm size objects. Such statistical analysis, accounting for both SHG polarization dependence and image size measurements, can therefore provide a complete picture of the structural behaviour of nano-crystals.

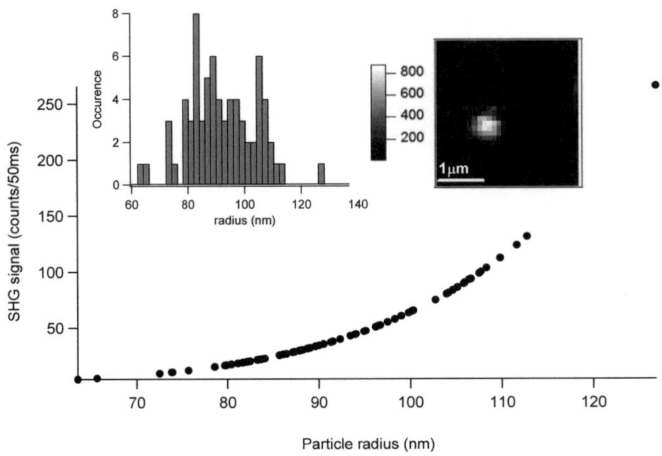

Fig. 5. Histograms of sizes of (MnPS3/DAZOP/Brij97, 10^{-4} mol/L DAZOP) nano-crystals in a PVA polymer film. These sizes were deduced from the SHG signal dependence on the particle radius R (as detailed in the text), depicted in the curve where each marker represents an isolated nano-crystal. In this analysis the DAZOP molecules were assumed to have the same hyperpolarizabtility as in solution, namely $\beta = 220\times10^{-30}\,esu$. The estimated size could be compared to the size measured for each nano-crystal on a SHG scan, which some times surpasses the diffraction limit (Delahaye et al. 2009).

4. Applications

4.1 Bio-imaging

While SHG microscopy, together with polarimetric analyses, is today particularly successful for bio-imaging benefiting from intrinsic responses from bio-molecular assemblies (Campagnola et al. 2002, Zipfel et al. 2003, Stoller et al. 2002, Strupler et al. 2007), biological applications still require efficient nano-emitters that can report biological relevant information at nano-scale in targeted regions of a sample of interest. Several issues such as toxicity, photo-bleaching, instability, lack of sharp contrast, are still considered as strong drawbacks for the fluorescence process which is more traditionally used in bio-imaging.

These issues can be advantageously circumvented by the use of SHG active bio-marker. SHG active nano-crystals are interesting alternatives due to the stability of their emission, the flexibility in the emission wavelengths and the subsequent possibility to observe them over a dark background, and at last their potential to no-toxicity. They can be additionally excited in the IR (700-1500nm), within wavelength ranges that are less scattered by biological media. They also do not exhibit phase-matching constraints contrary to more bulky media, being therefore able to efficiently emit SHG regardless of the phase properties of the excitation field.

ZnO nano-crystals have been successfully used for bio-imaging applications (Kachynski et al. 2008). Sum frequency, second harmonic, and non-resonant four wave mixing nonlinear signals have been obtained from stable dispersion of ZnO nanoparticles targeted to live tumor (KB) cells. Robust intracellular accumulation of the targeted (FA incorporated) ZnO nanocrystals could be observed without any indication of cytotoxicity. A water dispersion of well-separated single nanocrystals was obtained from an encapsulation technique, using

phospholipid micelles formed by the self-assembly of polymer-grafted lipids (Kachynski et al. 2008).
The holographic SHG imaging described in Section 3 allowed 3D imaging of $BaTiO_3$ particles of 30 nm size in HeLa cells, treated by incubation at 37°C with the particles uptaken by endocytosis (Hsieh et al. 2009). The particles were seen to be randomly distributed in the cells, possibly as clusters, into vesicles. The cells were seen to be alive after incubation, and fixed for imaging (Hsieh et al. 2009, Kuo et al. 2009, Hsieh et al. 2010).
At last, $KNbO_3$ was successfully incorporated and optically trapped for imaging in living cells (Nakayama et al. 2007).

SHG active nano-crystals are also of interest as punctual nano-sources for imaging through thick tissues, since the wide range of usable wavelengths can be extended to the IR, which is less scattered by natural media (Bohren & Huffmann, 1983). The wide frequency range available from SHG active nano-crystals has been exploited using $Fe(IO_3)_3$ nanoparticles illuminated at 800nm and 1550nm, in tissue phantoms made of submicrometric polystyrene beads, as well as in a section of a fixed mouse liver (Exterman et al. 2009). Using long working distance objectives (with lower NA objectives down to 0.6), better imaging quality was observed at large depths (180 μm in a phantom) for the larger IR wavelength.
At extensive depths, biological tissues are known to exhibit scattering properties which can also alter considerably the quality of optical focusing. SHG efficiencies can therefore be largely decreased by scattering when exploring large depths. A possible method to correct for the deformation of the optical wavefront brought by the medium is based on phase conjugation, which however requires a point source inside the medium. Indeed in this technique a source is used as a reporter of the local phase of the excitation field; its scattered field is then recorded in amplitude and phase in a hologram, and a phase-conjugated beam, reconstructed from this field, is sent back into the sample, thus providing automatic correction of phase distortions. SHG active nano-crystals (300 nm size $BaTiO_3$) have been used to play the role of the point source (Hsieh et al. 2010). This approach has shown successful digital phase conjugation in turbid media using off-axis digital holography to record the scattered SHG field from the nano-crsytals in a turbid medium made of a layer of parafilm fixed on the particles sample. The measured phase conjugated beam was generated in a spatial light modulator, and then sent back through the turbid medium. This observation of a nearly ideal focus on a nano-crystal, would not have been feasible in linear optics where all possible structures scatter the incident field without any spatial discrimination (Hsieh et al. 2010).

4.2 Nano-scale optical fields probing

The nonlinear mechanism taking place in SHG couples coherently the emission induced dipole to the excitation fields, therefore providing a possible way to probe excitation fields at the nano-scale in phase, polarization and amplitude. Thanks to their sizes, nano-crystals are unique reporters of optical fields properties at nano-scales. This approach has been advantageously used to apply Frequency Resolved Optical Gating (FROG) (Trebino et al. 1997) in single $Fe(IO_3)_3$ nano-crystals under ultra-short excitation (69 fs), bringing new prospective to monitor the phase and amplitude spectral distortions of such pulses (Exterman et al. 2008). This technique, usually performed on bulk samples, is a unique

demonstration that opens to the exploration of optical fields time properties at any place in a focal spot, since the size of the used particle is well below the optical resolution. The additional advantage of the use of nano-crystal is the absence of constraints imposed by nonlinear phase-matching in bulk crystals, which can be limiting in the case of extreme ultra-short pulses analysis.
SHG responses under ultra-short pulses are generally strongly dependent on the spectral and time phase profile of the pulses. In particular the optimum efficiency is reached for Transform Limited pulses, which exhibit no spectral/time dependence in phase over their whole bandwidth ("flat phase"). In order to maximize the SHG response from nano-crystals under short pulses excitations (13 fs to 200 fs), an optimization of the signal has been performed in KTP nano-crystals by directly adapting adequate optimization generic algorithms (Baumert et al. 1997) to their emission signals (Wnuk et al. 2009). As expected, an increase of the emission intensity has been observed when decreasing the duration of the excitation fs pulses.

5. Conclusion

This chapter has given an overview of the current state of research on SHG active nano-crystals. While this field is still in constant progress, many achievements have already allowed revisiting nonlinear optics in order to adapt to the scale of these new nano-objects. Their applications to both nanophotonics and biophotonics fields have shown their potential and are probably only at their infancy.

6. Acknowledgements

We thank Véronique Le Floc'h, Kasia Komorowska, Christelle Anceau, Nicolas Sandeau, Dominique Chauvat, Jean-François Roch, and Isabelle Ledoux for fruitful discussions and contribution to nonlinear microscopy developments in Laboratoire de Photonique Quantique et Moléculaire, ENS Cachan, France. Part of this work has been undertaken in collaboration with Alain Ibanez (Institut Néel, Grenoble, France), Emilie Delahaye and René Clément (Laboratoire de Chimie Inorganique, Institut de Chimie Moléculaire et des Matériaux d'Orsay, Université Paris XI, Orsay, France), S. Perruchas, C. Tard and T. Gacoin (Laboratoire de Physique de la Matière Condensée, Ecole Polytechnique, Palaiseau, France).

7. References

Anceau C., Brasselet S., Zyss J. (2005). Local orientational distribution of molecular monolayers probed by nonlinear microscopy. *Chem. Phys. Lett.* 411(1-3), 98-102

Baumert T., Brixner T., Seyfried V., Strehle M. et Gerber G. (1997). Femtosecond pulse shaping by an evolutionary algorithm with feedback. *Appl. Phys. B: Lasers and Optics*, 65, 6

Bohren C. F., and Huffmann D. R. (1983). *Absorption and Scattering by Small Particles.* Wiley Science paperback, Wiley & Sons, New York

Bonacina L., Mugnier Y., Courvoisier F., Le Dantec R. , Extermann J., Galez C., Boutou V., Wolf J.P. (2007). Polar Fe(IO3)3 nanocrystals as local probes for nonlinear microscopy. *Applied Physics B, Lasers and Optics*, 87(3), 399-403

Boulanger B. and Zyss J. (1997). *Physical Properties of Crystals, Vol. D of International Tables for Crystallography*, A. Authier, ed. Kluwer, Dordrecht, The Netherlands

Boyd R. W. (1992). *Nonlinear Optics*. Academic, New York

Brasselet S. and Zyss J. (1998). Multipolar molecules and multipolar fields: probing and controlling the the tensorial nature of nonlinear molecular media. *J.Opt.Soc.Am.B* 15(1), 257-288

Brasselet S. and Zyss J. (2007). Nonlinear polarimetry of molecular crystals down to the nanoscale. *C. R. Physique* 8, 165-179

Brasselet S. (2010). Second Harmonic Generation microscopy in molecular crystalline nano-objects. *Nonlinear Optics, Quantum Optics* 40 (1-4), 83-94

Brasselet S., Le Floc'h V., Treussart F., Roch J. F., Zyss J., Botzung-Appert E., and Ibanez A. (2004). In situ diagnostics of the crystalline nature of single organic nanocrystals by nonlinear microscopy. *Phys. Rev. Lett.* 92(20), 4

Chemla D.S. and Zyss J. (Eds.) (1987) *Nonlinear Optical Properties of Organic Molecules and Crystals* (Vol.1). Academic Press, Orlando

Chou C.-K., Chen W.-L., Fwu P. T., Lin S.-J., Lee H.-S., and Dong C.-Y. (2008). Polarization ellipticity compensation in polarization second-harmonic generation microscopy without specimen rotation. *J. Biomed. Opt.* 13(1), 014005

Delahaye E., Sandeau N., Tao Y., Brasselet S., Clément R. (2009). Synthesis and Second Harmonic Generation microscopy of nonlinear optical efficient hybrids nanoparticles embedded in polymer films. Evidence for Intra and Inter nanoparticles orientational synergy. *J. Phys. Chem. C*, 113 (21) 9092–9100

Delahaye E., Tancrez N., Yi T., Ledoux I., Zyss J., Brasselet S., and Clement R. (2006). Second harmonic generation from individual hybrid MnPS3-based nanoparticles investigated by nonlinear microscopy. *Chem, Phys. Lett.* 429, 533–537

Eisenthal Kenneth B. (2006). Second Harmonic Spectroscopy of Aqueous Nano- and Microparticle Interfaces. *Chem. Rev.* 106, 1462-1477

Extermann J., Bonacina L., Courvoisier F., Kiselev D., Mugnier Y., . Le Dantec R, Galez C., Wolf J.P. (2008). Nano-FROG: Frequency Resolved Optical Gating by a nanometric object. *Optics Express*, 16(14) 10405-10411

Extermann J., Bonacina L., Cuna E., Kasparian C., Mugnier Y., Feurer T., Wolf J.P. (2009). Nanodoublers as deep imaging markers for multi-photon microscopy. *Optics Express*, 17(17) 15342-15349

Fu D., Zhang Y., Liu J., Lu Z. (2001). Hyper-Rayleigh scattering of CdS nanoparticles stabilized by inorganic heteropolyanions. *Mater. Lett.* 51, 183-186

Galez C., Mugnier Y., Bouillot J., Lambert Y., and Le Dantec R. (2006). Synthesis and characterisation of Fe(IO3)3 nanosized powder. *J. Alloys Comp.* 416, 261-264

Hajj B., de Reguardati S., Hugonin L., Le Pioufle B., Osaki T., Suzuki H., Takeuchi S., Mojzisova H., Chauvat D., Zyss J. (2009). Electro-Optical Imaging Microscopy of Dye-Doped Artificial Lipidic Membranes *Biophys. J.* 97(11), L29

Hajj B., Gacoin T., Zyss J., Chauvat D. (submitted 2010). Electrooptical Pockels scattering from a single nanocrystal.

Halbout J.-M., Blit S., Chung T. (1984). Evaluation of the phase-matching properties of nonlinear optical materials in the powder form. *IEEE Journ. of Quant. El.* 17(4), 513-517

Hellwarth, R. Christensen, P. (1974). Nonlinear optical microscope examination of structures in polycrystalline ZnSe. *Opt. Commun*, 12, 318-322

Hsieh C. L., Grange R., Pu Y., and Psaltis D. (2009). Three-dimensional harmonic holographic microcopy using nanoparticles as probes for cell imaging. *Opt. Express* 17(4), 2880–2891

Hsieh C. L., Pu Y., Grange R., and Psaltis D. (2010). Digital phase conjugation of second harmonic radiation emitted by nanoparticles in turbid media. *Opt. Express* 18(12), 12283-12290

Hsieh C. L., Grange R., Pu Y., and Psaltis D. (2010). Bioconjugation of barium titanate nanocrystals with immunoglobulin G antibody for second harmonic radiation imaging probes. *Biomaterials* 31(8), 2272–2277

Johnson J. C., Yan H. Q., Schaller R. D., Petersen P. B., Yang P. D., and Saykally R. J. (2002). Near-field imaging of nonlinear optical mixing in single zinc oxide nanowires. Nano Lett. 2(4), 279–283

Josse D., Dou S.X., Zyss J., Andreaaza P., Périgaud A. (1992). Near-Infrared Optical parametric Oscillation in a N-(4-Nitrophenyl)-L-Prolinol Molecular Crystal. *Appl.Phys.Lett.* 61,121

Kachynski A. V., Kuzmin A. N., Nyk M., Roy I., and Prasad P. N. (2008). Zinc oxide nanocrystals for nonresonant nonlinear optical microscopy in biology and medicine. *J. Phys. Chem. C* 112(29), 10721–10724

Komorowska K., Brasselet S., Zyss J., Pourlsen L., Jazdzyk M., Egelhaaf H.J., Gierschner J., Hanack M. (2005). Nanometric scale investigation of the nonlinear efficiency of perhydrotriphynylene inclusion compounds. *Chem. Phys.* 318 (1-2), 12-20

Kuo T. R., Wu C. L., Hsu C. T., Lo W., Chiang S. J., Lin S. J., Dong C. Y., and Chen C. C., (2009). Chemical enhancer induced changes in the mechanisms of transdermal delivery of zinc oxide nanoparticles. *Biomaterials* 30(16), 3002–3008

Kurtz S. K., Perry T. T. (1968). A Powder Technique for the Evaluation of Nonlinear Optical Materials. *J. Appl. Phys.*, 39, 3798–3813

Le X. L., Zhou C., Slablab A., Chauvat D., Tard C., Perruchas S., Gacoin T., Villeval P., and Roch J. F. (2008). Photostable second-harmonic generation from a single KTiOPO4 nanocrystal for nonlinear microscopy. *Small* 4(9), 1332–1336

Le X. L., Brasselet S., Treussart F., Roch J. F., Marquier F., Chauvat D., Perruchas S., Tard C., and Gacoin T. (2006). Balanced homodyne detection of second-harmonic generation from isolated subwavelength emitters. *Appl.Phys. Lett.* 89(12), 121118

Le Bozec, H., Le Bouder, T., Maury, O., Bondon, A., Ledoux, I., Deveau, S., Zyss, J (2001). Supramolecular Octupolar Self-organization towards Nonlinear Optics. *Adv. Mater.*, 13, 1677-1681

Le Floc'h V., Brasselet S., Roch J.F., Zyss J. (2003). Monitoring of orientation in molecular ensembles by polarization sensitive nonlinear microscopy. *J. Phys. Chem. B*, 107, 12403-12410

Le Floc'h V., Brasselet S., Zyss J., Chao R., Lee S.H., Jeon S.J., Cho M., Min K.S., Suh M.P. (2005). High efficiency and quadratic optical properties of a fully optimized 2D octupolar crystal by nonlinear microscopy. *Adv. Mater.* 17(2),196-200

Magrez, A. , et al. (2006). Growth of single-crystalline KNbO3 nanostructures. *J. Phys. Chem. B* 110, 58–61

Mugnier Y., Houf L., El-Kass M., Le Dantec R. , Hadji R., Vincent B., Djanta G., Badie L., Eschbach J., Rouxel D. and Galez C. (submitted 2010). In situ crystallization and growth dynamics of acentric iron iodate nanocrystals in w/o microemulsions probed by Hyper-Rayleigh Scattering measurements.

Nakayama Y., Pauzauskie P. J., Radenovic A., Onorato R. M., Saykally R. J., Liphardt J., and Yang P. D. (2007). Tunable nanowire nonlinear optical probe. *Nature* 447(7148), 1098–1101

Oudar J.L., Chemla D.S. (1977). Hyperpolarizabilities of the nitroanilines and their relations to the excited state dipole moment. *J. Chem. Phys* 66, 2664-2670

Pu Y., Centurion M., Psaltis D. (2008). Harmonic Holography–a new holographic principle. *Appl. Opt.* 47 A103-A110

Rekaï, E. D., Baudin, J. B., Jullien, L., Ledoux, I., Zyss, J., Blanchard-Desce, M. *Chem.Eur. J.* 2001, 7, 4395

Revillod G., Duboisset J., Russier-Antoine I., Benichou E., Bachelier G., Jonin Ch., Brevet P.F. (2008). Multipolar Contributions to the Second Harmonic Response from Mixed DiA-SDS Molecular Aggregates. *J. Phys. Chem.* C 112, 2716

Richards B., and Wolf E. (1959). Electromagnetic diffraction in optical systems. 2. Structure of the image field in an aplanatic system. *Proceedings of the Royal Society of London Series a-Mathematical and Physical Sciences* 253, 358–379

Rodríguez E. M., Speghini A., Piccinelli F., Nodari L., Bettinelli M., Jaque D., and Sole J. G. (2009). Multicolour second harmonic generation by strontium barium niobate nanoparticles. *J. Phys. D Appl. Phys.* 42(10), 4

Rodriguez E. V., de Araujo C. B., Brito-Silva A. M., Ivanenko V. I., and Lipovskii A. A. (2009), Hyper-Rayleigh scattering from BaTiO3 and PbTiO3 nanocrystals. *Chem. Phys. Lett.* 467(4-6), 335–338

Sandeau N., Le X.L. , Chauvat D., Zhou C., Roch J. F., and Brasselet S. (2007). Defocused imaging of second harmonic generation from a single nanocrystal. *Opt. Express* 15(24), 16051–16060

Sanz N., Baldeck P. L., Nicoud J. F., LeFur Y., and Ibanez A. (2001), Solid State Sci. 3, 867.

Shen Y., Swiatkiewicz J., Winiarz J., Markowicz P., Prasad P. (2000), Second-harmonic and sum-frequency imaging of organic nanocrystals with photon scanning tunneling microscope. *Appl.Phys.Lett.* 77, 2946 (2000)

Schön P., Munhoz F., Gasecka A., Brustlein S., and Brasselet S. (2008). Polarization distortion effects in polarimetric two-photon microscopy. *Opt. Express* 16(25), 20891–20901

Schön P., Behrndt M., Ait-Belkacem D., Rigneault H., Brasselet S. (2010). Polarization and Phase Pulse Shaping applied to Structural Contrast in Nonlinear Microscopy Imaging. *Phys. Rev. A* 81, 013809

Terhune R.W., Maker P.D., SavageC.M. (1965). Measurements of Nonlinear Light Scattering. *Phys. Rev. Lett.* 14, 681-684.

Toury T., Brasselet S., Zyss J. (2006) Electro-optical microscopy: mapping nonlinear polymer films with micrometric resolution *Opt. Lett.* 31(10), 1468-1470

Trebino R., DeLong K.W., Fittinghoff D.N., Sweetser J.N., Krumbügel M.A., Richman B.A., and Kane D.J. (1997). Measuring ultrashort laser pulses in the time-frequency domain using frequency-resolved optical gating. *Rev. Sci. Instrum.* 68, 3277

Treussart F., Botzung-Appert E., Ha-Duong N.-T., Ibanez , Roch J.-F., Pansu R. (2003). Second Harmonic Generation and Fluorescence of CMONS Dye Nanocrystals Grown in a Sol-Gel Thin Film. *ChemPhysChem* 4, 757–760

Wang X., Zhang Y., Wang G., Zhang C., Fu D., Shen Y., Lu Z., Cui Y., Ochiai S., Uchida Y., Kojima K. and Ohashi A. Second-order nonlinear optical properties of centrosymmetric nanoparticles studied by hyper-Rayleigh scattering (HRS) technique. (2005). *Chinese Optics Letters*, Vol. 3, Issue S1, S125-S127

Wnuk P., Le L. X., Slablab A., Tard C., Perruchas S., Gacoin T., Roch J. F., Chauvat D., and Radzewicz C. (2009). Coherent nonlinear emission from a single KTP nanoparticle with broadband femtosecond pulses. *Opt. Express* 17(6), 4652–4658

Yew E. Y. S., and Sheppard C. J. R. (2006). Effects of axial field components on second harmonic generation microscopy. *Opt. Express* 14(3), 1167–1174

Yi T., Tancrez N., Clément R., Ledoux-Rak I., Zyss J. (2004). Organic-MPS3 nanocomposites with large second-order nonlinear optical response. *J. Lumin*, 110, 389-395

Yi T., Clément R., Haut C., Catala L., Gacoin T., Tancrez N., Ledoux I., Zyss J. (2005). J-Aggregated Dye-MnPS3 Hybrid Nanoparticles with Giant Quadratic Optical Nonlinearity. *Adv. Mater.* 17, 335-338

Yokoyama, S., Nakahama, T., Otomo, A., Mashiko, S (2000). Enhancement of the Molecular Hyperpolarizability in Multi-choromophoric Dipolar Dendrons. *J. Am. Chem. Soc.*, 122, 3174-3181

Zielinski M., Oron D., Chauvat D., and Zyss J. (2009). Second-harmonic generation from a single core/shell quantum dot. *Small* 5(24), 2835–2840

Zyss J. and Oudar J.L. (1982). Relations between microscopic and macroscopic lowest order optical nonlinearities of molecular crystals in one and two-dimensional units. *Phys Rev A* 26 (4), 2028-2048

Zyss J., Nicoud J.F., Coquillay M. (1984). Chirality and hydrogen-bonding in molecular crystals for phase-matched second-harmonic generation: N-(4-nitrophenyl)-(L)-prolinol (NPP). *J. Chem. Phys.* 81, 4160

Zyss J. and Ledoux I. (1994) Nonlinear Optics in Multipolar Media: Theory and Experiments. *Chem.Rev.*94,77-105

12

Hybrid colloidal nanocrystal-organics based light emitting diodes

Aurora Rizzo, Marco Mazzeo and Giuseppe Gigli
National Nanotechnology Laboratory (NNL), CNR-Nanoscienze, Università del Salento,
Italian Institute of Technology (IIT) via Morego, 30 I-16163 Genova, Italy
Via per Arnesano Km 5 I73100 Lecce, Italy

1. Introduction

In the last few years a new and promising field of research defined "nanoscience" has attracted an increasing interest. This new field involves the capability to manipulate, fabricate and characterize the material structure at a nanometer scale. Research dedicated to this field has an interdisciplinary character and covers physics, engineering, chemistry, material science and, molecular biology. The recent possibility to manipulate the matter with nanoscopic resolution revealed novel physical and chemical properties of low dimensional structures in between bulk materials and atomic sizes. Among all the nanostructured materials, semiconductor quantum dots (QDs) or nanocrystals have been used in a wide array of fields such as electronic and optoelectronic devices. The most recent techniques in the preparation of nanocrystals are the chemical methods and in particular the colloidal nanocrystals synthesis. These techniques allow for the growth of semiconductor nanocrystals in solution and consequently their integration in the organic electronic technology. The most effective technique is the synthesis by thermal decomposition of precursors in hot coordinating solvents. This allows the growth of semiconductor (II-VI, III-V), (Murray et al., 1993; Murray et al., 2001) metal (Murray et al., 2001; Sun et al., 1999) and metal oxide nanocrystals with high quality, (Rockenberger et al., 1999) almost free from defects and narrow size distribution (less than 5%).

In particular the mostly used material in optoelectronic application synthesized by this technique is the luminescent CdSe QDs. Usually the CdSe cores are surrounded by a wider band gap semiconductor (such as ZnS) that passivates the surface states of nanocrystals with a consequent increase in the photoluminescence. In order to ensure the solubility, the QDs are covered by an organic capping group.

These structures are smaller in size than the diameter of a Bohr exciton in a bulk crystal of the same material. By reducing the size of the nanocrystals core, the quantum confinement of the electron, hole and exciton is increased, with a consequent increase in the exciton energy. For instance in CdSe nanocrystals, the quantum confinement increases the exciton energy from a bulk bandgap of 1.7 eV (λ=730nm red edge of visible spectrum) to any value up to 2.7 (λ=450nm blue luminescence). The spectral tunability can access almost all the visible range with a single material set. Moreover the saturated color emission (linewidths of $<$35nm Full Width at Half Maximum), the high luminous efficiency even in the blue region,

the possibility to tailor the external chemistry without affecting the emitting core as well as the inorganic nature make these structures ideal candidates in the fabrication of hybrid organic/inorganic light emitting diodes (LEDs) potentially more stable and with a longer lifetime with respect to the fully organic counterpart.
In recent years, indeed the development of organic light emitting diode (OLED) showed several basic physical problems that are very difficult to overcome. One is the relatively short lifetime if compared with other electroluminescent devices. Lifetime have been extended using new chemistry to prevent crystallization and sophisticated packaging schemes to avoid the degradation of the organic emitting molecules by water and oxygen. The chemistry optimization must be repeated for each emitter and differences in the aging time of different molecules emitting in the red, green and blue region poses problems in the OLED color stability. Inorganic QDs are robust and high luminescence lumophores but they present poor charge conduction properties. Recently many efforts have been devoted in the successful integration of these nanocrystalline materials in OLED technology. However the studies on photocurrent and injection in thin disordered films of nanocrystals between metal electrodes revealed that transport and injection in these films cannot be described in the framework of the band model of semiconductor (Ginger et al., 2000; Leatherdale et al., 2000). Nanocrystals are particles separated one to each other by organic surfactant. The high degree of disorder in these films suggests that hopping and tunneling between localized states are the principal mechanism for the charge transport (Ginger et al., 2000). The hopping model is similar to the one described for the organic semiconductors, but in the case of nanocrystals the energetic disorder arises from the size distribution of the particles and the geometric disorder comes from their spatial separation, also induced by the surfactant. Moreover colloidal nanocrystals have poor charge transport properties especially in the case of thick multilayer films. To avoid the problem of poor charge conduction in LED fabrication, colloidal nanocrystals have to be merged with organic semiconductor hole and electron transporting/injection materials. In the last few years several hybrid device structures based on colloidal QDs and organic materials have been reported.
In this chapter, the recent development in the fabrication of Hybrid organic/QD-LEDs will be reviewed; In particular the diverse approaches to fabricate multilayer devices and white light emitting devices will be discussed in details.

2. Hybrid Light Emitting Diodes based on Polymer/QDs

The fist work on colloidal semiconductors nanocrystal LEDs demonstrated a device structure in which five monolayer of CdSe quantum dots were deposited on the top of spin coated and thermal converted semiconducting p-paraphenylene vinylene (PPV) layer. Nanocrystal layers were bound on the polymer with exane dithiol functionalization (Colvin et al. 1994). The light emission arises from the recombination of holes injected into the PPV polymer layer with electrons injected into the multilayer film of CdSe nanocrystals. The low external quantum efficiency of the devices (between 0.001 and 0.1%) can be attributed either to the low photoluminescence (PL) efficiency of the CdSe core QDs, or to the poor electron conduction through the five layer thick QD film (Morgan et al., 2002). Moreover, the use of the hexane dithiol for the device fabrication adds impurities that can trap the charges and quenching the QD excitons.

Subsequently electroluminescence has been obtained by incorporating CdSe QDs into a thin film (100 nm) of polyvinylcarbazole (PVK), a photostable hole conducting polymer, and an oxadiazole derivative (t-Bu-PBD), an electron transport species (Dabbousi, 1995). Three devices with different size of QDs (~32Å, ~40Å, ~60Å diameter) have been fabricated by spin coating nanocrystals with the mixture of PVK and t-Bu-PBD as an emitting layer (Figure 1). The maximum external quantum efficiency in air and at room temperature of these devices is quite low (~0.0005%) and it decreases with decreasing of the QD size.

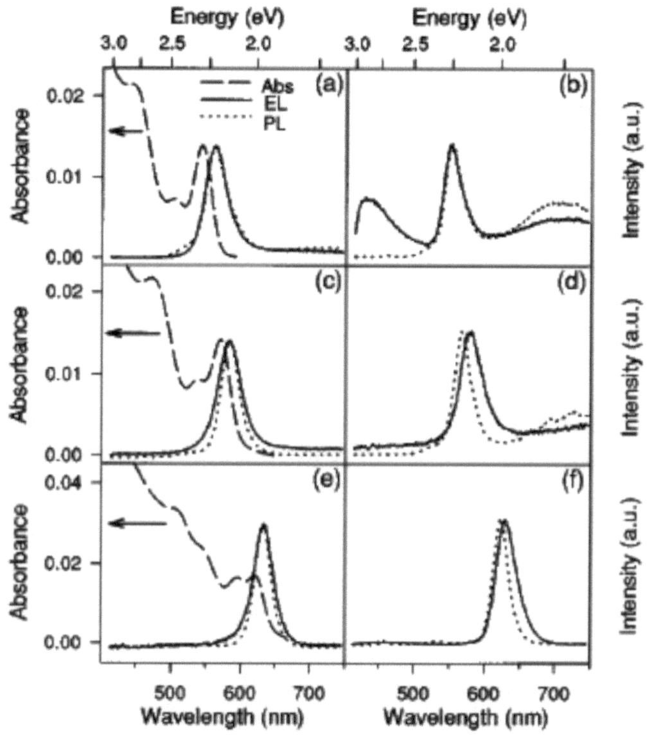

Fig. 1. Absorption (dashed lines), photoluminescence (exc. 457.9 nm) (dotted Lines), and electroluminescence (solid lines) spectra for 32 Å (a),(b), 40Å (c),(d), and 60 Å (e),(f) diameter CdSe nanocrystallite/PVK/PBD devices at room temperature (a), (c), (e) and at 77 K (b), (d), (f). (Reprinted with permission, Dabbousi et al. , 1995. Copyright 1995, American Institute of Physics).

A significant improvement over the previous devices has been obtained by using CdSe(CdS) core/shell QDs (Schlamp et al., 1997). These core/shell nanocrystals consist in a CdSe core surrounded by a shell of epitaxially grown CdS. The core and shell energy levels alignment confines the holes in the core while electrons are delocalized in throughout the structure. This results in a higher PL quantum yield (> 50% in solution at room temperature) and photo-oxidative stability of the CdSe(CdS) compared with the CdSe core samples. The active nanocrystal multilayer has been deposited on a thermal converted PPV polymer (Figure 2).

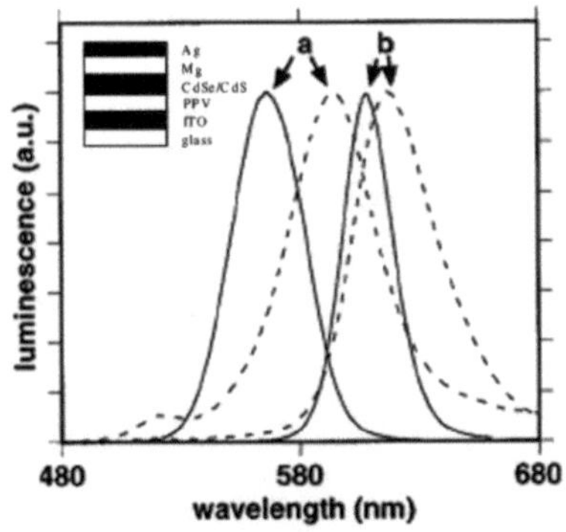

Fig. 2. PL (solid lines) and EL (dashed) of samples of (25, 5) CdSe(CdS) (a) and (34, 7) CdSe(CdS) (b) nanocrystals. The PL was collected in solution at room temperature. Inset: device configuration. (Reprinted with permission, Schlamp et al., 1997. Copyright 1997, American Institute of Physics).

The devices emit from red to green with external quantum efficiency of 0.22% at a brightness of 600 cd/m^2 and a current density of 1 A/cm^2. Despite the potential advantages of using QDs as emitters, all these QD-LED structures have efficiencies that are far below of the all organic LED technology ones.
The most limiting factor in the QD devices performances is the poor conductivity of the colloidal nanocrystals compared to semiconductor materials (Leatherdale et al., 2000) and in all the structures reported above the QDs act both as emitters and electron transport species. A multilayer device configuration can allow an independent optimization of materials for charge injection, transport and, emission.

3. High efficiency hybrid LED insulating QD layer function

3.1 QD-LED by phase segregated nanocrystal monolayer

In 2002 an innovative QD-LED structure incorporating a monolayer of CdSe/ZnS core/shell QDs sandwiched between hole and electron transporting organic layers was proposed (Coe et al., 2002). In this device structure colloidal QDs act only as a lumophores and they do not participate to the charge conduction process. The organic layers transport charge carriers to the vicinity of the QD monolayer from which the luminescence originates. In this way the problem of QD poor charge conductivity due to the insulating layer of surfactant, that coat their surface, was overcome. For the device fabrication a solution of QDs and organic small molecule material is spin coated on an Indium Tin Oxide anode. During the spin-coating process the QDs, covered by an aliphatic capping layer, phase separate from the aromatic small molecules, such as N,N'-diphenyl-N,N'-bis(3-methylphenyl)-(1,1'-biphenyl)-4,4'-diamine (TPD), and form a layer on the organic surface.
The thickness and the coverage of the nanocrystals layer depends on the QD concentration in the solution. For the device fabrication the QD concentration is optimized in order to obtain a single close packed monolayer. By exploiting the phase segregation technique they fabricated high luminance devices with emission in the red-green region achieving a maximum external quantum efficiency >2%.(Coe et al., 2002, Coe-Sullivan et al., 2003; Coe-Sullivan et al., 2005) (Fig. 3).

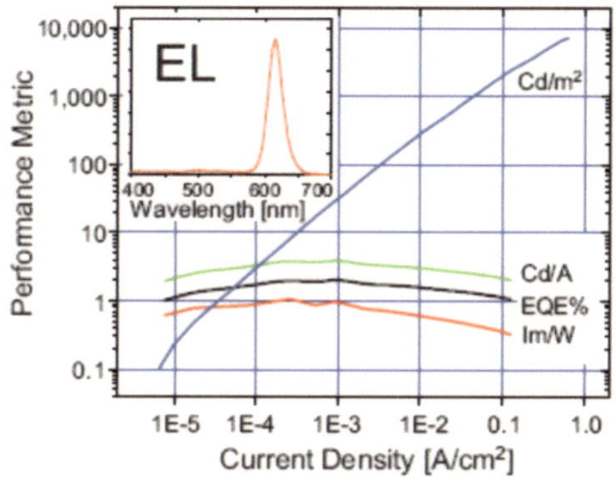

Fig. 3. Performance metrics of the QD-LED, with external quantum efficiency of in excess of 2% and a maximum luminescence over 7,000cd/m2. Inset: EL spectrum with a saturated color red emission peaking at 615nm , and full width at half maximum of 27nm (Coe-Sullivan et al., 2005. Copyright Wiley-VCH Verlag GmbH & Co. KGaA. Reproduced with permission.).

3.2 Multilayer devices by cross-linked hole transporting layer

Recently an alternative strategy for QD-LED fabrication by spin-coating CdSe/ZnS QDs onto a thermally cross-linked hole transporting layer, polystyrene (PS)- N,N′-diphenyl-N,N′-bis(4-n-butylphenyl)-(1,1′-biphenyl)-4,4′-diamine(TPD)-perfluorocyclobutane (PFCB) was reported (Zhao, 2006). Following the deposition of the QD layer, the electron transporting layer, 1,3,5-tris(N- phenylbenzimidazol-2yl)benzene (TPBI), was deposited by thermal evaporation. EL spectrum and device structure are reported in fig. 4.

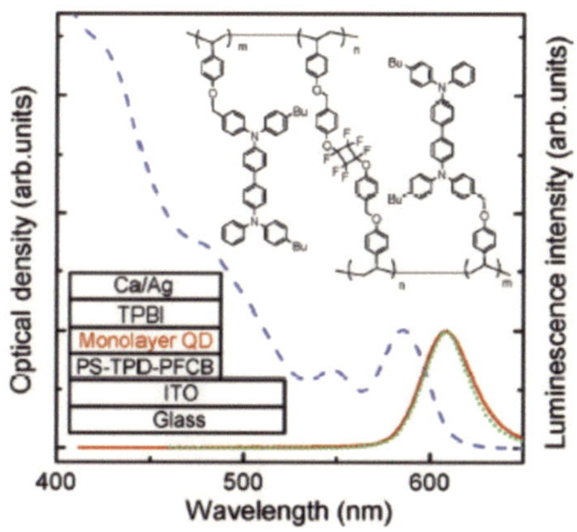

Fig. 4. Absorption (dashed line), photoluminescence (dotted line), and electroluminescence spectra (solid line) for CdSe/CdS QD-LEDs with a structure ITO/PS-TPD-PFCB (30 nm)/CdSe QDs/ TPBI (40 nm)/Ca (30 nm)/Ag (120 nm) at a voltage of 6.0 V. Structures of the multilayered CdSe/CdS QD-LEDs and cross-linked PS-TPD-PFCB are shown in the inset of the figure. (Reprinted with permission, Zhao et al., 2006.Copyright 2006, American Chemical Society).

The maximum external quantum efficiency of these devices was 0.8% at 100 cd/m^2 and a maximum brightness in excess of 1000 cd/m^2.
The same group improved the performance of these devices structure by using thermal annealing of the quantum dot layer reaching an external quantum efficiency of 1.6% at a brightness of 100 cd/m^2 (Niu et al., 2007).
Although QD-LEDs with QDs monolayer have the advantages of higher efficiency and lower turn-on voltages, their output power, maximum luminance and color purity are limited owing to the low chromophore quantity and the poor confinement of excitons in the active QD region.
Recently high-performance red, orange, yellow and green QD-LEDs based on QDs with a CdSe core and a ZnS or CdS/ZnS shell have been reported (Sun et al., 2007). Their maximum luminance reached 9,064, 3,200, 4,470 and 3,700cd/m^2 (Fig. 5).

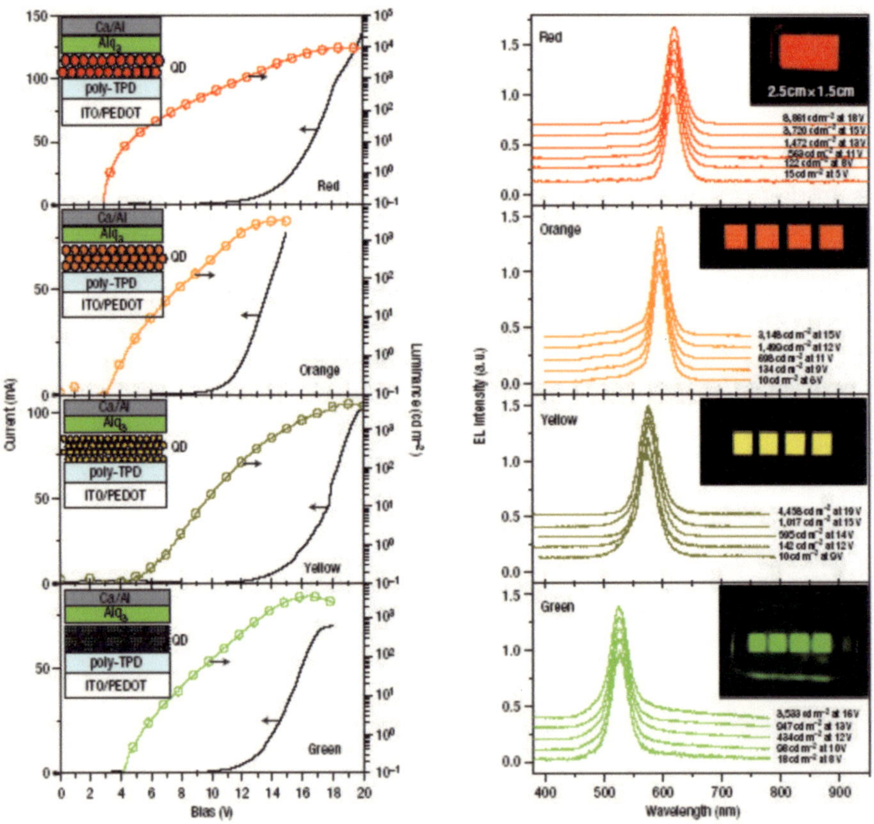

(a) (b)

Fig. 5. (a) Current/voltage and luminance/voltage characteristics of red-, orange-, yellow- and green-emitting QD-LEDs. The insets show schematic device configurations of the corresponding QD-LEDs. (b) EL spectra of the red-, orange-, yellow- and green- emitting QD-LEDs operating at different luminances voltages. The insets show images of the devices under operation. (Reprinted with permission, Sun et al., 2007. Copyright 2007, Nature Publishing Group).

The superior performances of the QD-LEDs arise from the careful preparation of highly purified, uniform and monodispersed colloidal core–shell QDs, and optimization of the thicknesses of the polymer HTL, the QD layer and the ETL. Indeed, for use in QD-LEDs, all the QDs were subjected to a multistep purification process to remove the organic ligands. After purification, the red, orange and yellow QDs were in solid powder form with a quantum yield (QY) higher than 30% in toluene, whereas the green QDs were still in liquid solution, with a QY of 10% in toluene, owing to their smaller size. During the purification process, the removal of organic ligands from the QDs produces surface defects that may trap charges.
A trilayer structure of indium tin-oxide(ITO)/poly(ethylenedioxythiophene):polystyrene sulphonate (PEDOT:PSS) (25 nm)/HTL (45 nm)/QDs (15–20 nm)/ETL (35 nm)/Ca (15 nm)/Al (150 nm) was used in the fabrication of the QD-LEDs, with ITO/PEDOT:PSS as the anode, poly(N,N0-bis(4-butylphenyl)-N,N0-bis(phenyl) benzidine) (poly-TPD) spin-coated from its chlorobenzene solution as HTL, QD layers spin-coated from their toluene solutions as emissive layers, tris(8-(hydroxyl-quinoline) aluminium (Alq_3) as ETL, and Ca/Al as the cathode. In the QD-LEDs, the HTL and ETL thicknesses were both optimized to confine the injected electrons and holes to recombine predominantly within the QD layer and provide optimal hole and electron transportation.

3.3 QD-light emitting diodes with metal oxides charge transport layers

For practical device application it is fundamental to improve the shelf-life robustness. Recently the use of sputtered amorphous inorganic semiconductors as robust charge transport layers was reported (Caruge, 2008). In particular 20nm thick film of NiO is deposited by radiofrequency (RF)-sputtering on the ITO electrode; the active layer consists of ZnCdSe alloyed QDs, deposited by spin-coating, with red emission peak at 638nm. The electron transporting layer is an optically transparent film of alloyed ZnO and SnO_2. NiO and ZnO:SnO_2 are p-type and n-type semiconductors respectively (Fig. 6).

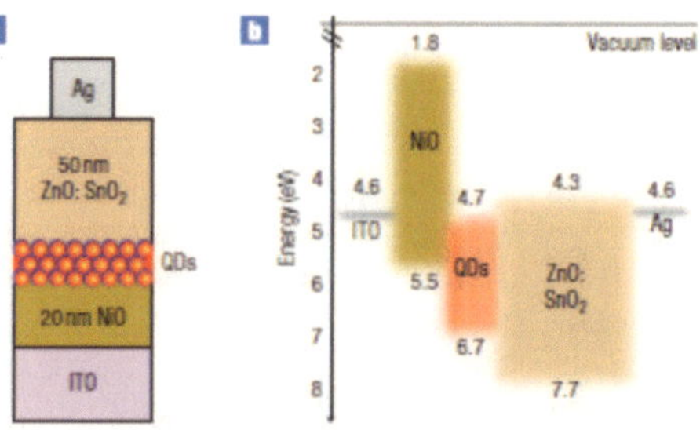

Fig. 6. (a) Device configuration: the Ni is used as hole transporting layer, the QD as active luminescent layer, the ZnO:SnO_2 as electron transporting layer. (b) Band diagram determined from UV photoemission spectroscopy and optical absorption measurements. (Reprinted with permission, Caruge et al., 2008. Copyright 2008, Nature Publishing Group).

A maximum EQE of nearly 0.1% was demonstrated, the peak brightness was measured to be 1950 cd m^{-2} at 19.5 V and 3.73 Acm^{-2}. Comparable brightness and J-V characteristics is claimed when the devices were tested after being stored for four days, in contrast to unpackaged organic QD-LEDs, which cannot withstand prolonged atmospheric exposure.

Recently high efficiency QD-LEDs by using a sol gel TiO_2 electron transporting layer was reported (Cho et al., 2009); the TiO_2 layer is deposited by spin-coating, enabling the QD-LED to be fabricated by means of an all-solution process (Fig. 7). The emissive layer is made of a crosslinked QD layer directly spin-coated on a poly[(9,9-dioctylfluorenyl-2,7-diyl)-co-(4,40-(N-(4-sec-butylphenyl)) diphenylamine)] (TFB) hole transporting layer. The linker molecule, 1,7-diaminoheptane, is attached to the QD through exchange with the pre-existing surfactants or by binding to empty sites on the QD surface. The crosslinking is obtained by a post-deposition thermal treatment. The crosslinking of the QD layer greatly improves the luminance and luminous efficiency.

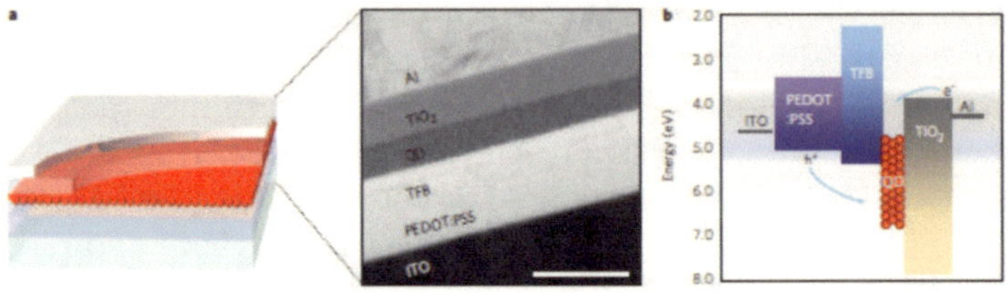

Fig. 7. (a) Device structure (left) and cross-sectional TEM image (right) of the QD-LED. TFB, poly[(9,9-dioctylfluorenyl-2,7-diyl)-co-(4,40-(N-(4-sec-butylphenyl))diphenylamine)]. Scale bar, 100 nm. (b) Energy band diagram. (Reprinted with permission, Cho et al., 2009. Copyright 2009, Nature Publishing Group).

It is demonstrated that the energy band offset between the QD and the HTL was reduced from 1.5 to 0.9 eV that leads to an increase by over a factor of 10 in the maximum luminous efficiency because of more efficient hole injection and enhanced charge balance. Moreover the metal oxide TiO_2 layer shows an improved electron injection superior if compared with the standard organic Alq_3; this is a consequence of the lower band offset (0.4 eV) for Al/TiO_2 compared to that for Al/Alq_3 (1.2 eV). And higher electron mobility of the sol-gel TiO_2 (1.7 × 10-4 cm2 V-1 s-1) compared to that for Alq_3 (~1.0×10^{-4} cm^2 V^{-1} s^{-1}). The turn-on voltage in the TiO_2-based QD-LED also significantly decreased to 1.9 V, which is smaller than the QD bandgap of 2.1 eV and much lower than that observed for the Alq_3-based device (~4.0 V). The overall device performances are improved, in particular the luminance reach the value of 12380 cd m^{-2} and a maximum power efficiency of 2.41 lm W^{-1}.

4. White Hybrid Light Emitting Devices

To obtain white light, all the three primary colors (red, green and blue) have to be produced simultaneously. Since it is difficult to obtain all primary emissions from a single molecule, excitation of more than one organic species is often necessary, thus introducing color stability problems. Due to the different degradation rates of the employed organic compounds, the emission color of the device can, in fact, changes with time. The CdSe semiconductor QDs exhibit a size-dependent color variation due to quantum confinement effect, which covers almost whole the visible range (Suzuki et al., 1996; Ginger et al., 2000; Coe-Sullivan et al., 2005). Additionally, the fluorescence efficiency and, in particular the stability of the nanocrystals, can be greatly improved by modifying the particle surface. These characteristics can be merged with peculiar properties of organic materials, such as flexibility and ease of processing, to give rise to a novel class of low cost hybrid white-LEDs with improved lifetime and color stability. We proposed two different white emitting device structures based on colloidal semiconductor QDs. In the first structure, red emitting CdSe/ZnS QDs are blended in a blue emitting polyfluorene (Li et al., 2005). The matrix polymer provides a simple device preparation process due to its high processability and a blue emitting component in the device emission. The function of green component in the white emission spectrum is given to a green emitting electron transporting material evaporated on top of the polymer/QDs film. In the second structure (Li et al., 2006), three different size of CdSe/ZnS QD samples emitting at wavelength of 490nm, 540nm, 620nm are dispersed in a high mobility hole organic matrix. White bright electroluminescence is obtained from the ternary nanocrystal composites.

4.1 Organic-Inorganic Hybrid White LEDs Based on Polymer and QDs

Host-guest systems are typically employed to obtain white light emission by exploiting two mechanisms, namely, Förster energy transfer (Förster, 1959; Lakowicz, 1983) and charge transfer (Utsugi et al., 1992; Suzuki et al., 1996). All these processes have to be accurately controlled in order to obtain white EL. We demonstrated that a balanced white emission is obtained in hybrid system poly[(9,9-dihexylfluoren- 2,7-diyl)-alt-co- (2-methoxy-5- {2-ethylhexyloxy} phenylen-1,4-diyl)](PFH-MEH):QDs/ Alq_3 when Förster energy transfer in the guest-host system is accomplished by charge transfer from PFH-MEH and Alq_3 to QDs during the electrical excitation (Li et al., 2005). The absorption and PL spectra of all components and possible energy/charge-transfer pathways were shown in fig. 8.

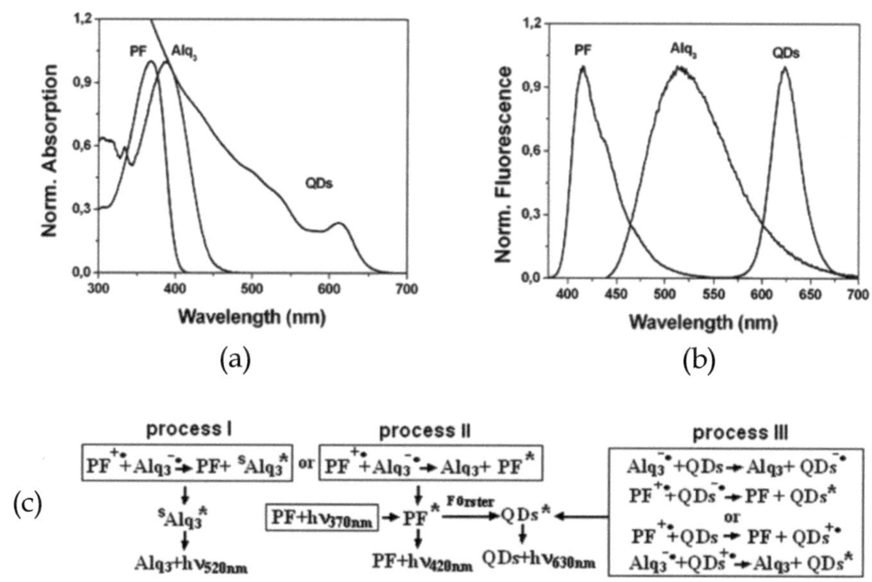

Fig. 8. (a) Absorption and (b) Photoluminescence (PL) spectra of device components. (c) Possible pathways leading to emissive states in device ITO//PEDOT-PSS// PFH-MEH (PF):CdSe/ZnS//Alq_3//Ca/Al. (Reprinted with permission, Li et al., 2005. Copyright 2005, American Institute of Physics).

Electroluminescence measurements have been carried out on ITO//PEDOT-PSS//PFH-MEH:CdSe/ZnS//Alq_3//Ca/Al structures. In fig. 9 the EL spectra for different PFH-MEH:CdSe/ZnS concentration ratios are reported.

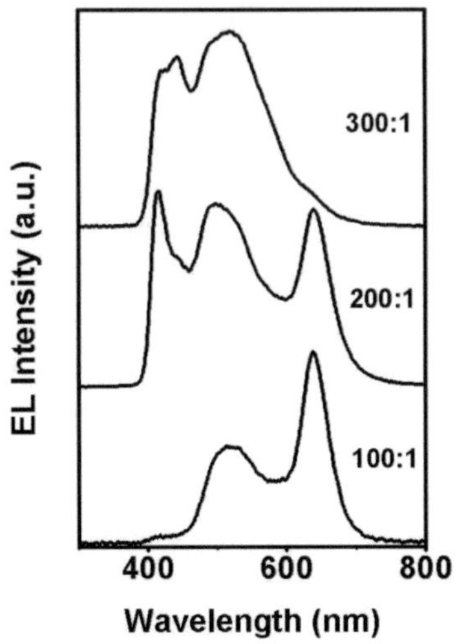

Fig. 9. EL spectra for the device ITO//PEDOT-PSS//PFH-MEH:CdSe/ZnS//Alq_3// Ca/Al with different ratio (PFH-MEH:CdSe/ZnS, c%) (Reprinted with permission, Li et al., 2005. Copyright 2005, American Institute of Physics).

In order to achieve white EL emission, the different color components have to be accurately balanced by controlling both the Förster energy transfer and charge transfer mechanisms. To this aim we fabricated devices with different PFH-MEH:QDs concentration ratios, namely 200:1 and 300:1. At low QDs concentration (300:1), the possible pathways are process I and II. Process I involves the transfer of a hole from a PFH-MEH cation radical ($PF^{+\bullet}$) to an Alq_3 anion radical ($Alq^{3-\bullet}$). Process II involves the transfer of an electron from an Alq_3 anion radical ($Alq^{3-\bullet}$) to a PFH-MEH cation radical ($PF^{+\bullet}$). Both the mechanisms result in excited Alq_3 and PFH-MEH molecules which can decay radiatively, originating blue and green EL emission. Negligible red emission is instead originated from the low amount of QDs. By increasing the concentration of CdSe/ZnS QDs (concentration ratio 200:1), the possible pathway are process I, II and III. A relevant additional role is assumed by the following processes: Förster energy transfer to QDs from excited PFH-MEH and Alq_3 molecules, sequential charge transfer of a hole from PFH-MEH followed by transfer of an electron from Alq_3 or charge transfer of an electron from Alq_3 followed by transfer of a hole charge from PFH-MEH (processes III). Efficient electroluminescence at the three primary colors is thus obtained from PFH-MEH, Alq_3 and QDs with a balanced white spectrum with CIE (0.30,0.33). Maximum External Quantum Efficiency of 0.24% is measured at $1mAcm^{-2}$ and 11V. In these QDs based white-LEDs, holes are considered to be injected from the ITO electrode through PEDOT:PSS layer into the polymer hole conductor and are eventually transported to the QDs. Similarly, the electrons are considered to be injected from Ca/Al cathode into the Alq_3 and are eventually transported to the QDs. Since the high electron affinity of QDs the electrons are better confined within the surface PFH-MEH:QDs/Alq_3, thus enhancing the balance between opposite carriers in the region where more efficient radiative exciton recombination can occur. In particular, charge transfer processes to CdSe / ZnS core-shell quantum dots are found to be the key element for well balanced white emission.

4.2 Bright White LEDs from Ternary Nanocrystals Composites

To obtain a high performance hybrid device in which QDs act as lumophores is the occurrence of efficient exciton recombination in the inorganic nanocrystals. This is usually inhibited by the poor electron conduction of the inorganic species which limits exciton formation (Leatherdale et al., 2000). Therefore low EL efficiency is observed in QDs/polymers blend based devices. (Colvin et al., 1994; Dabbousi et al., 1995; Schlamp et al., 1997). The phase-segregation technique proposed by Coe et al. allows for the fabrication of high efficiency hybrid monochromatic emission devices, but it involves a narrow QDs size distribution to form high-coverage monolayers, which is compatible only with monochromatic emitting QD-LEDs (Coe et al., 2002; Coe-Sullivan et al., 2003; Coe-Sullivan et al., 2005). To date, efficient QD-LEDs that emit white light only from nanocrystals are still a challenge due to the lack of proper fabrication techniques.

In order to obtain efficient simultaneous emission from different sizes of QDs composites, we proposed a novel device structures in which exciton formation in the inorganic QDs is not exclusively obtained by direct charge injection but by the accurate control of the energy/charge transfer mechanisms from the organic host (Li et al., 2006). We demonstrate the first efficient hybrid light-emitting device, with near white emission from chemically and optically stable ternary nanocrystal composites dispersed in an organic matrix. White bright emission is obtained from homogenous blends, without phase segregation between the active ternary QDs composites and the organic matrix 4,4,N,N-diphenylcarbazole (CBP), exploiting the energy transfer and charge-trapping properties of the different species. The proposed approach is a new general method for the fabrication of potential long operating lifetime, high efficiency white light-emitting devices. The emission mechanisms and the structure of the devices and are shown in the fig. 10a, 10b. In fig. 10c the Photoluminescence (PL) spectra corresponding to isolated lake blue, green and red quantum dots in solid state are shown. The EL spectrum at 13V is shown in fig. 11a. Spectral peaks at wavelengths of 490nm, 540nm and 618nm are attributed to the emission of lake blue, green and red quantum dots, respectively, in agreement with the PL spectra in the solid state (see fig. 10c).

A complete quenching of the CBP band is observed in the EL spectrum, resulting in efficient emission at the three primary colors from the QDs, giving rise to a balanced near white overall emission [CIE (0.32,0.45)] at 13V. Furthermore, the color coordinates do not remarkably change with increasing the operating voltage in the range 10V-28V. Possible pathways leading to emissive states are shown in fig. 10a. In our device, holes are injected from the ITO contact through the PEDOT:PSS layer into the CBP host matrix, and are transported towards the QDs. Similarly, electrons are injected from the Ca/Al cathode into the Alq_3 and are transported to the QDs. Then exciton generation on QDs occurs via two parallel processes: (I) charge-trapping and (II) Förster energy transfer from CBP and Alq_3. In the former process, electrons may be trapped in the QDs owing to the relative energy alignment of the LUMO levels of CBP, Alq_3 and QDs, thus, efficient exciton formation on the QDs can occur after recombination with high mobility holes. In the latter process, excitons form on organic molecules CBP (IIa) and Alq3 (IIb), then undergo Förster energy transfer to the lower-energy QDs sites, where they can recombine radiatively.

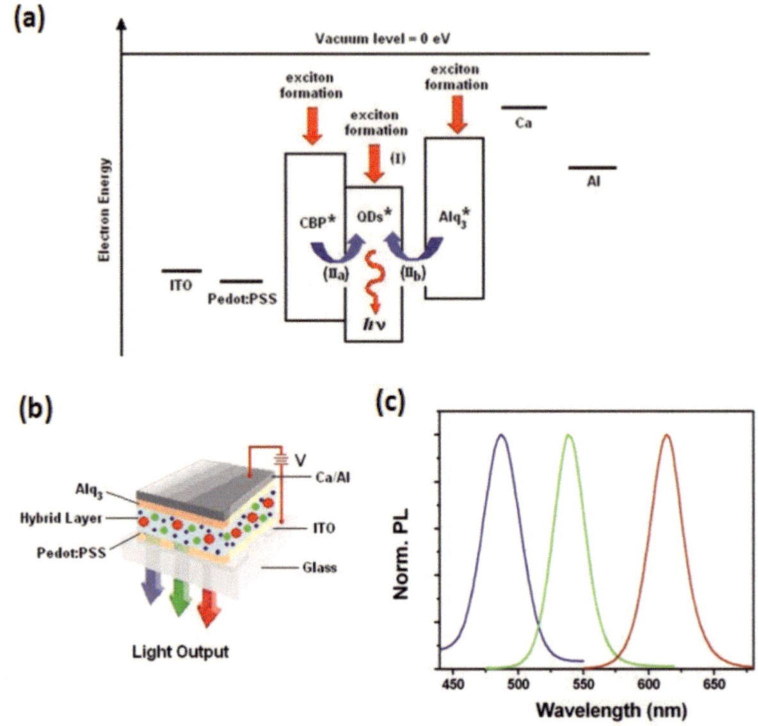

Fig. 10. (a) Proposed simplified energy level diagram of devices and possible exciton creation mechanisms on the QDs: (I) charge-trapping and (II) Förster energy transfer. (b) The structure of the device. (c) PL spectra corresponding to isolated lake blue, green and red quantum dots, measured in solid state. (Li et al. , 2006. Copyright Wiley-VCH Verlag GmbH & Co. KGaA. Reproduced with permission.).

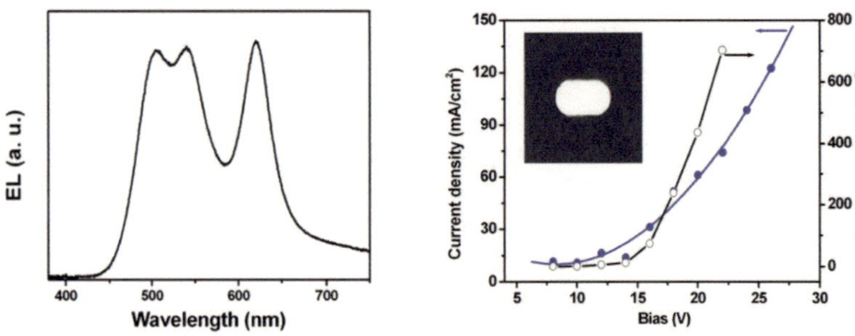

(a) (b)

Fig. 11. Electroluminescence (EL) spectra and characteristics of the ternary QDs device. (a) EL spectrum for the device: ITO//PEDOT:PSS//CBP:QDs (B,G,R:18,2,1) //Alq_3//Ca/Al. (b) V-I (dot), V-L (circle) characteristics of the device. Inset: A photo taken from the working device. (Li et al. , 2006. Copyright Wiley-VCH Verlag GmbH & Co. KGaA. Reproduced with permission.).

Possible pathways leading to emissive states are shown in fig. 10a. In our device, holes are injected from the ITO contact through the PEDOT:PSS layer into the CBP host matrix, and are transported towards the QDs. Similarly, electrons are injected from the Ca/Al cathode into the Alq_3 and are transported to the QDs. Then exciton generation on QDs occurs via two parallel processes: (I) charge-trapping and (II) Förster energy transfer from CBP and Alq_3. In the former process, electrons may be trapped in the QDs owing to the relative energy alignment of the LUMO levels of CBP, Alq_3 and QDs, thus, efficient exciton formation on the QDs can occur after recombination with high mobility holes. In the latter process, excitons form on organic molecules CBP (IIa) and Alq_3 (IIb), then undergo Förster energy transfer to the lower-energy QDs sites, where they can recombine radiatively. The maximum brightness of the device is 1050 cd m^{-2} at 58 mA cm^{-2}, which corresponds to a current efficiency of 1.8 cd A^{-1}, and a turn-on voltage of 6V are measured in air atmosphere fig. 11b. To our knowledge, this is the highest efficiency hybrid device with white emission only from ternary QDs composites, whose luminance satisfies lighting application requirements (i.e. 1000 cd m^{-2}).

5. Hybrid Light Emitting Diodes by Micro-Contact Printing

In previous sections different approaches for hybrid LEDs fabrication have been reported. All these device structures are limited by the chemical properties of the employed materials and by the deposition techniques. The main consequence is a restriction of the organic materials choice, especially for the holes transporting and injection layers, which is crucial for the optimization of the device emission. In fact, unlike organic small molecules, QDs cannot be deposited by thermal evaporation because of their high molecular weight and only wet deposition techniques, such as spin-coating and drop-casting are available. Consequently, the separation of the QD active layer from the organic transport layers and the fabrication of multilayer organic/inorganic structures are still challenging. The lack of an appropriate QD deposition technique strongly limits the implementation of Hybrid LEDs in the heterojunction devices technology. This technology could result in an improvement of the devices efficiency and lifetime.

In this frame we developed two totally dry, simple, and inexpensive deposition techniques to transfer colloidal semiconductor QDs onto organic substrates (Rizzo et al., 2008a; Rizzo et al., 2008b). These innovative techniques are modification of the standard microcontact printing (μCP). In the standard μCP, a poly(dimethylsiloxane) (PDMS) stamp is inked by the material which is then transferred onto the solid substrate by conformal contact of the elastomeric stamp and the substrate. However this technique cannot be used for the deposition of colloidal semiconductor nanocrystals. PDMS stamp, indeed, can be readily swelled by a number of non polar solvents such as toluene and chloroform, in which nanocrystals for optoelectronic applications are usually dispersed in.

The first technique we developed is named μCP-Double Transfer because it consists in a two step process. First the nanocrystals are deposited from a toluene solution on a float glass substrate, when the solvent is dried, the PDMS stamp pad is brought in conformal contact with the nanocrystal film, then the PDMS stamp is peeled away and the nanocrystals are transferred to the surface of the stamp. Finally, QDs are deposited by single step microcontact printing from the PDMS stamp to the surface of previously evaporated organic thin films. (Rizzo et al., 2008a).

The second technique consists in protecting the PDMS stamp pad with a SU-8 photoresist layer deposited by spin coating and cross linked in order to make it insoluble to toluene. (Rizzo et al., 2008b). The two techniques that we propose avoided the swelling of the PDMS stamp and combined the ease of wet deposition processes with the possibility to grow a heterojunction QDs-LED using thermal evaporation. White and red electroluminescence from colloidal QDs has been obtained demonstrating that these deposition techniques are fully compatible with OLED technology.

5.1 Hybrid Light-Emitting Diodes from μ-CP Double-Transfer of Colloidal Semiconductor CdSe/ZnS Quantum Dots onto Organic Layers

The sketch of the Double Transfer technique is reported in fig. 12. Drop-casting is the simplest technique for the deposition of a QDs solution on the glass substrate. The degree of order in the resulting assembly is mainly determined by the rate of solvent evaporation: slowing down the evaporation of solvent, in our case toluene, and preventing it from a rapid dewetting is crucial for a successful double transfer. A slow evaporation allows nanocrystals to diffuse and self assemble into large periodical (long range order) structures with fewer defects. In order to slow down the evaporation after the drop deposition, the substrates were placed in an over-saturated environment of toluene vapour.

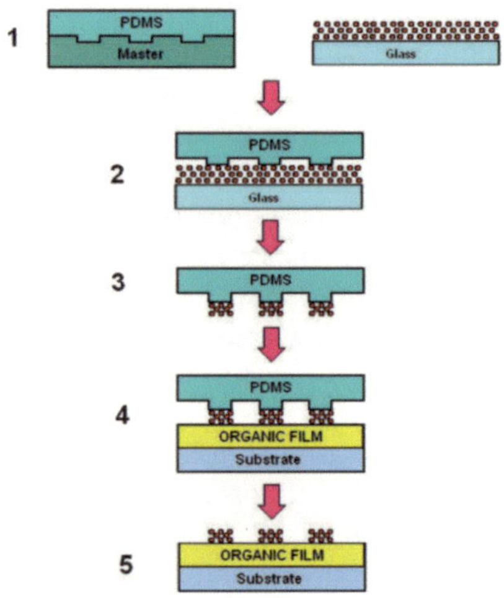

Fig. 12. A schematic illustration of the procedure used to transfer colloidal nanocrystals on organic thin films. 1. Colloidal QDs are deposited from toluene solution on a cleaned float glass substrate; 2. when the solvent is dried, a PDMS stamp pad is brought in conformal contact with the surface of the nanocrystals film; 3. The PDMS stamp is then peeled away and the nanocrystals are transferred to the surface of the stamp; 4. QDs are deposited by single step microcontact printing from the PDMS stamp to the surface of previously evaporated organic thin films; 5. Nanocrystals are transferred on the organic substrate (Rizzo et al., 2008a. Copyright Wiley-VCH Verlag GmbH & Co. KGaA. Reproduced with permission.).

The quality of the drop-casted films was verified by Transmission Electron Microscope (TEM) measurements. In fig. 13a and 13b we compare two samples realized depositing 20μL of the same colloidal solution on a TEM grid in over-saturated environment (fig. 13a) and under ambient condition (fig. 13b). After a slow dewetting of the solvent, (about 30 minutes) the nanocrystals form a compact layer on a scale of several microns. For a fast solvent evaporation (less than 5 min) the film presents the formation of multilayer domains and microscopic voids. Deposition of a uniform QDs layer on the top of the organic layers has a paramount importance for the correct functioning of the hybrid LED devices. The layer quality of QDs formed on the surface of PDMS reflects the morphology of the QD film on the glass substrate, i.e. it depends on the concentration of the nanocrystals solution and on their assembling on the glass surface.

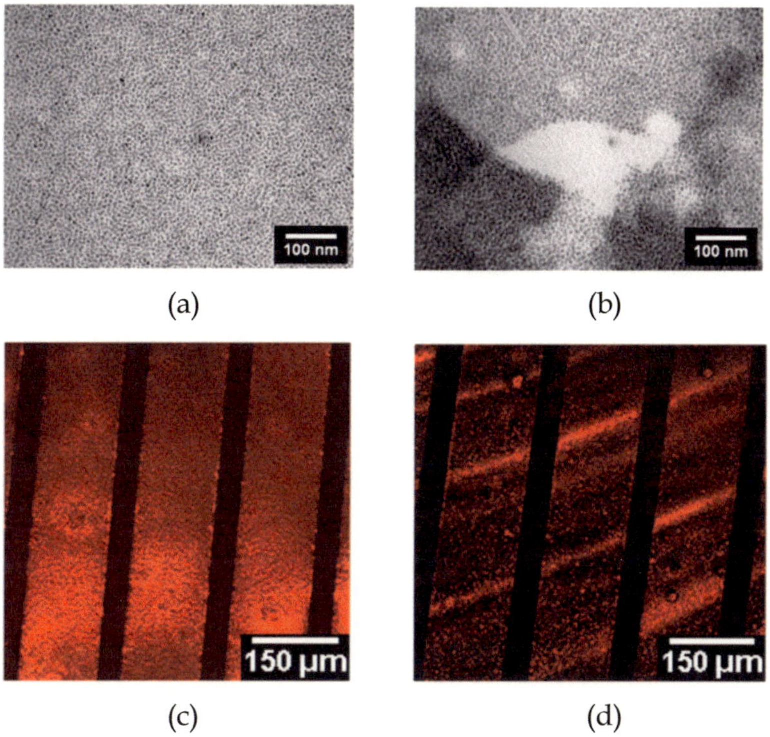

Fig. 13. a) TEM images of nanocrystals deposited by drop-casting on TEM grid after slow evaporation of the solvent and b) fast evaporation of the solvent. c) Confocal microscopy image of QDs transferred by double transfer µCP onto CBP organic substrates from a glass substrate after slow evaporation of the solvent; d) QDs on CBP transferred from a glass substrate after fast evaporation of the solvent. (Rizzo et al., 2008a. Copyright Wiley-VCH Verlag GmbH & Co. KGaA. Reproduced with permission.).

In fig. 13c and 13d we show the confocal microscope images of QDs transferred on CBP layer by µCP Double Transfer from glass substrates after slow (fig. 13c) and fast evaporation (fig. 13d) of the solvent. The morphology of 50×150µm QDs stripes results uniform for samples made after slow evaporation of the solvent and very dishomogeneous for the one made after a fast evaporation of the solvent. This is also a further proof of how effective slower solvent evaporation is in assuring highly homogeneous film of colloidal QDs at the micrometer scale. Increasing 15 times the concentration of QDs in toluene solution, the thickness of the transferred layer can vary from few layers to more than 100 nm.

The transferring mechanism of QDs from the glass substrate to the stamp can be explained by considering the strong interaction between the PDMS stamp surface and the colloidal nanocrystal hydrophobic surfactant. This can be attributed to the existence of highly mobile hydrophobic oligomers on the PDMS stamp surface. Moreover PDMS has a very low surface energy of 21.6dyn/cm thus allowing the transfer of the QDs on a higher surface energy substrate, such as the CBP. In this contest the organic material acts as glue for the nanocrystals.

In order to realize a hybrid LED that emits light from nanocrystals alone, a high coverage and compact QDs layer has to be deposited on the top of the organic material. For this reason in the device fabrication we decided to use a flat PDMS pad for the QDs transfer. The device structure (fig. 14a, inset) consists of a thermally evaporated N,N'-Bis(naphthalen-1-yl)-N,N'-bis(phenyl)benzidine (α-NPD) hole injection layer (HIL) doped with 2,3,5,6-tetrafluoro-7,7,8,8-tetracyano-quinodimethane (F4-TCNQ) and CBP hole transporting layer (HTL). A QDs active layer of 80 nm is transferred on the organic substrate by the Double Transfer technique reported above. The 2,9-dimethyl-4,7-diphenyl-1,10-phenanthroline (BCP) hole blocking layer (HBL), the Alq_3 electron transporting layer (ETL), and the electrodes are evaporated after the QDs deposition.

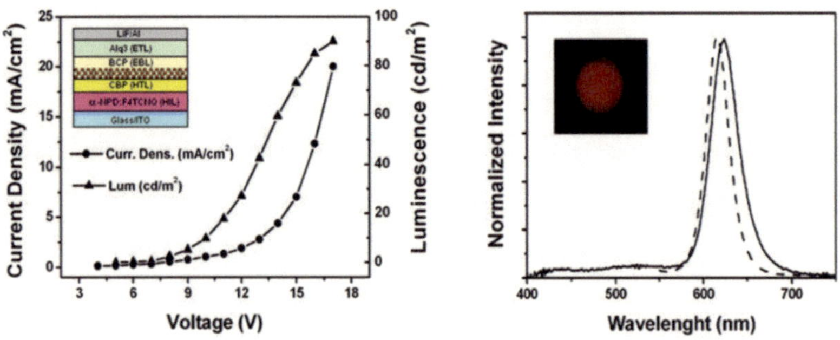

(a) (b)

Fig. 14. Structure and performance of a hybrid LED fabricated by µCP double transfer. a) Current-density–voltage and luminescence–voltage characteristics of the hybrid LED; inset: a schematic illustration of the device structure. b) Electroluminescence (solid line) and PL (dashed line) spectra; the inset shows a picture of a functional 1.5mmx1.5mm area device (Rizzo et al., 2008a. Copyright Wiley-VCH Verlag GmbH & Co. KGaA. Reproduced with permission.).

In the proposed device the holes and electrons are delivered to the nanocrystals layer through the injection and transport layers. The electroluminescence (EL) spectrum and the picture of the working device in fig. 14b demonstrate that the emission is dominated by QDs band edge emission. Photoluminescence (PL) spectrum of the nanocrystals is reported in dashed line in the same figure.

In fig. 14a a current density versus voltage and luminescence versus voltage characteristics are reported. The maximum brightness of the device is 90 cd m^{-2} at 20 mA cm^{-2}, which corresponds to an external quantum efficiency of 0.15%.

5.2 White Electroluminescence from a μ-CP Deposited CdSe/ZnS Colloidal Quantum Dot Monolayer

In Figure 15 a schematic sketch of the second approach for QD deposition is displayed (Rizzo et al., 2008b).

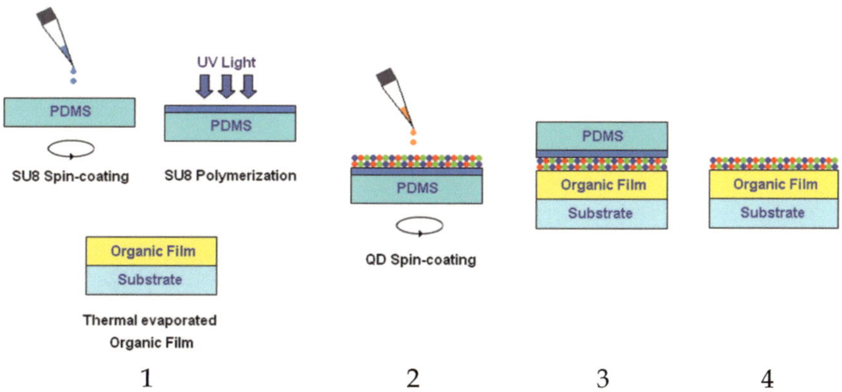

Fig. 15. A schematic sketch of the modified μCP technique used to deposit colloidal nanocystals on organic thin films. 1. The SU-8 is spin-coated on the top of a flat PDMS stamp pad, then exposed to UV radiation and subsequently thermally cross-linked in order to make the film insoluble to toluene; 2. a mixed blue, green and red CdSe/ZnS QD solution is spin-coated at on the SU-8 protected PDMS; 3. the inked stamp is then inverted on the surface of a previously evaporated organic thin film and peeled away after 30 second; 4. nanocrystals are transferred on the organic substrate. (Rizzo et al., 2008b. Copyright Wiley-VCH Verlag GmbH & Co. KGaA. Reproduced with permission.

In fig. 16 we report atomic force microscopy (AFM) images of the same film. Fig. 16a was taken at the interface between the CBP and QDs layer at the edge of a pinhole defect. By the profile (fig. 16b) we can estimate QD film thickness of about 10nm. In fig. 16c, 16d, 16e, and 16f AFM images of the QD film at different resolutions are reported. Pictures in fig. 16a, 16c, 16d, point out the uniform distribution of the QDs on the organic material with an average roughness of about 5nm. Moreover in the higher resolution images in fig. 16e and 16f is possible to discriminate the two phases: QDs from 4-10 nm in diameter on the top and organic CBP as underlayer. The film such prepared was used for the fabrication of white emitting hybrid LED. In Fig. 17 the electroluminescence spectra at different applied voltages are reported, showing an increase of the green and blue components relative to the red one when voltage is increased from 5V to 10V. This is consistent with the larger barrier for holes injection in the smaller QDs. Two dominant QD excitation mechanisms are present in the hybrid device: the Förster energy transfer and the charge trapping. In the charge trapping process the electrons may be trapped in the QD owing to the absence of any energy barrier between the organic Alq_3 and BCP layers and CdSe/ZnS one. For these charged QDs the barrier to holes injection and transfer from the α-NPD/CBP bilayer is reduced. Thus, exciton formation and consequent radiative recombination occur on QD sites upon an acceptance of high mobility holes from CBP. The presence of a α-NPD (420nm) small band emission is due to the presence of pinhole defects, grain boundaries, interstitial spaces in the QD thin layer that allow the formation of exciton on organic sites. The HOMO level alignment of α-NPD/CBP suggests an accumulation of holes in the α-NPD at the HIL/HTL interface.

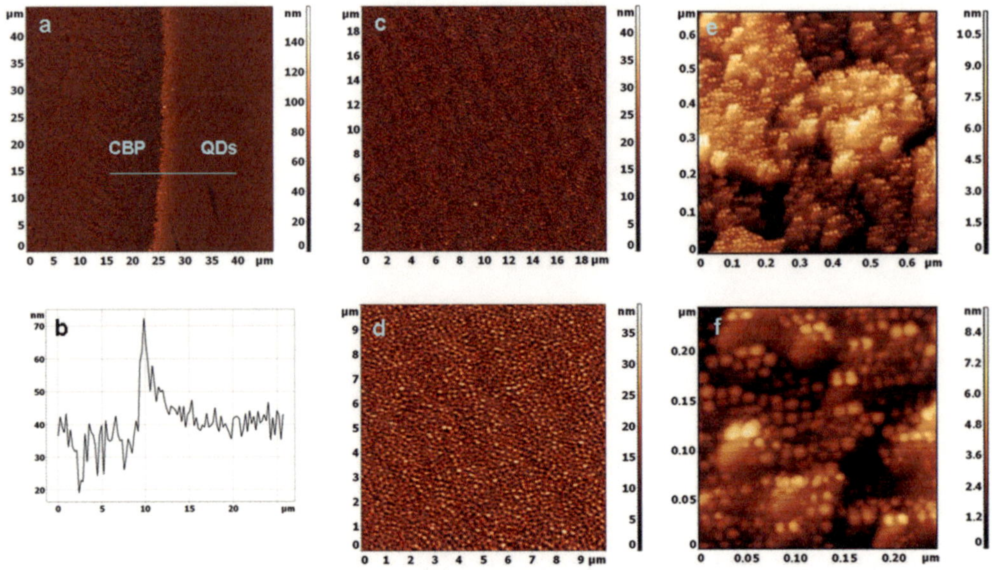

Fig. 16. Atomic force microscopy (AFM) characterization of the transferred nanocrystal layer. a) AFM image of the layer at the edge of a pinhole defect and b) profile corresponding to a); c), d), e), f), AFM images of the QD film at different scanning sizes. (Rizzo et al., 2008b. Copyright Wiley-VCH Verlag GmbH & Co. KGaA. Reproduced with permission.).

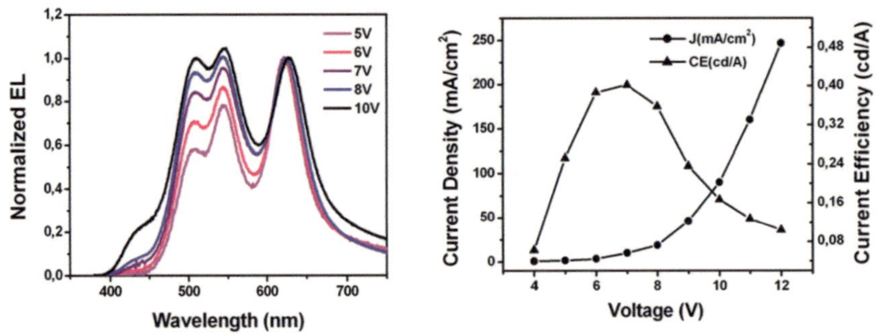

(a) (b)

Fig. 17. a) EL spectra of the device at different applied voltages normalized at the red peak. b) Current-Voltage I-V characteristic and current efficiency (CE) for the hybrid device. . (Rizzo et al., 2008b. Copyright Wiley-VCH Verlag GmbH & Co. KGaA. Reproduced with permission.).

By increasing the voltage and, consequently the current, the fraction of emission coming from the organic layer increases. The development of these two techniques enables us to separate the deposition of the organic materials from the QDs deposition and the fabrication of multilayer complex device structures without any restriction in the organic material choice. By taking advantage from the independent processing of QD we fabricated red and white emitting multilayer LEDs with a doped holes injection layer.

6. Conclusion

Several approaches have been followed for the incorporation of inorganic colloidal semiconductor nanocrystals in organic light emitting diodes technology. As a results we first reported two methods for the fabrication of white emitting hybrid organic/inorganic LEDs wherein colloidal CdSe/ZnS QDs are blended in a blue emitting polymer or in a high mobility hole transporting organic small molecule. The major breakthrough has been reached with the fabrication of white hybrid LEDs from ternary nanocrystal composites. CdSe/ZnS QDs with different sizes simultaneously emit light at different wavelength (blue, red and green region) to achieve white emission. By accurately controlling the Förster energy transfer and the charge trapping on QDs sites we were able to fabricate the highest efficiency hybrid device with emission only from QD composites. The maximum satisfies the lighting application requirements. Moreover the development of two innovative micro-contact printing techniques allow the fabrication of multilayer hybrid LEDs structure without restrictions in the organic under-layers. These deposition methods permit the integration of colloidal QDs in the promising p-doping organic technology. The recent results demonstrate that the successful integration of colloidal nanocrystals in OLED technology could enable the creation of a new set of promising devices with a possible improvement in the color stability and the device lifetime.

7. References

Murray, C. B.; Norris, D. J.; Bawendi, M. G. *Journal of the American Chemical Society* **1993,** 115, (19), 8706-8715.

Murray, C. B.; Sun, S. H.; Gaschler, W.; Doyle, H.; Betley, T. A.; Kagan, C. R. *Ibm Journal of Research and Development* **2001,** 45, (1), 47-56.

Sun, S. H.; Murray, C. B. *Journal of Applied Physics* **1999,** 85, (8), 4325-4330.

Rockenberger, J.; Scher, E. C.; Alivisatos, A. P. *Journal of the American Chemical Society* **1999,** 121, (49), 11595-11596.

Ginger, D. S.; Greenham, N. C. *Journal of Applied Physics* **2000,** 87, (3), 1361-1368.

C. A. Leatherdale, C. R. Kagan, N. Y. Morgan, S. A. Empedocles, M. A. Kastner, and M. G. Bawendi, "Photoconductivity in CdSe quantum dot solids", Phys. Rev. B, **62**, 2669-2680, 2000.

V. L Colvin, M. C. Schlamp, and A. P. Alivisatos, "Light-emitting diodes made from cadmium selenide nanocrystals and a semiconducting polymer", Nature, **370**, 354-357, 1994.

Morgan, N. Y.; Leatherdale, C. A.; Drndic´, M.; Jarosz, M. V.; Kastner, M. A. & Bawendi, M. (2002). Electronic transport in films of colloidal CdSe nanocrystals, *Phys. Rev. B*, 66, 075339

Dabbousi, B. O.; Bawendi, M. G.; Onitsuka; O. & Rubner, M. F. (1995). Electroluminescence from CdSe quantum-dot/polymer composites, App. Phys. Lett., 66, 1316-1318

Schlamp, M. C.; Peng, X.; & Alivisatos, A. P. (1997). Improved efficiencies in light emitting diodes made with CdSe(CdS) core/shell type nanocrystals and a semiconducting polymer", *J. App. Phys.*, 82, 5837-5842

Coe, S.; Woo, W. K.; Bawendi, M.; & Bulovic, V. (2002). Electroluminescence from single monolayers of nanocrystals in molecular organic devices *Nature,* 420, 800-803

Coe-Sullivan, S.; Woo, W. K.; Steckel,J. S.; Bawendi, M. G.; & Bulovic, V. (2003). Tuning the performance of hybrid organic/inorganic quantum dot light-emitting devices, *Org. Electron.* 4, 123-130

S. Coe-Sullivan, J. S. Steckel, W. K. Woo, M. G. Bawendi, & V. Bulovic, (2005) Large-Area Ordered Quantum-Dot Monolayer via Phase Separation During Spin-Casting, *Adv. Func. Mater.* 15, 1117-1124

Zhao, J.; Bardecker, J. A.; Munro, A. M.; Liu, M. S.; Niu, Y. H.; Ding, I.-K.; Luo, J.; Chen, B; A. K.-Y., Jen; & Ginger, D. S. (2006). Efficient CdSe/CdS Quantum Dot Light-Emitting Diodes Using a Thermally Polymerized Hole Transporting Layer, *Nano Lett.* 6, 463-467

Niu, Y. H.; Munro, A. M.; Cheng, Y.-J.; Tian, Y. Q.; Liu, M. S.; Zhao, J.; Bardecker, J. A.; Plante, I. J.-L.; Ginger, D. S.; & Jen, A. K.-Y. (2007). Improved Performance from Multilayer Quantum Dot Light-Emitting Diodes via Thermal Annealing of the Quantum Dot Layer, *Adv. Mater.* 19, 3371-3376, 2007

Sun, Q.; Wang, Y. A.; Li, L. S.; Wang, D. Y.; Zhu, T.; Xu, J.; Yang, C. H.; & Li, Y. F. (2007). Bright, multicoloured light-emitting diodes based on quantum dots *Nature Photonics*, 1, 717-722.

Caruge, J. M. Halpert, J. E. Wood, V. Bulovic´, V. & Bawendi, M. G. (2008). Colloidal quantum-dot light-emitting diodes with metal-oxide charge transport layers *Nature Photonics* 2, 247-250

Cho, K.-S.; Lee, E. K.; Joo, W.-J.; Jang, E.; Kim, T.-H.; Lee, S. J.; Kwon, S.-J.; Han, J. Y.; Kim, B.-K.; Choi, B. L.; & Kim, J. M. (2009). High-Performance Crosslinked Colloidal Quantum-Dot Light-Emitting Diodes Nat. Photonics, 3, 341– 345

Suzuki, H.; & Hoshino, S. (1996). Effects of doping dyes on the electroluminescent characteristics of multilayer organic light-emitting diodes, *Journal of Applied Physics* 79, (11), 8816-8822

Li, Y. Q.; Rizzo, A.; Mazzeo, M.; Carbone, L.; Manna, L.; Cingolani, R.; & Gigli, G. (2005) White organic light-emitting devices with CdSe/ZnS quantum dots as a red emitter *Journal of Applied Physics,* 97, 113501

Li, Y. Q.; Rizzo, A.; Cingolani, R.; & Gigli, G. (2006). Bright white-light-emitting device from ternary nanocrystal composites *Advanced Materials,* 18, 2545-2548.

Förster, T. (1959). Transfer mechanisms of electronic excitation, *Discuss. Faraday Soc*, 27, 7-17

Lakowicz, J. R. Principles of Fluorescence Spectroscopy. Plenum (New York): 1983.

Utsugi, K.; & Takano, S. (1992). Luminescent Properties of Doped Organic EL Diodes Using Naphatalimide Derivative, *Journal of the Electrochemical Society,* 139, 3610-3615.

Rizzo, A.; Mazzeo, M.; Palumbo, M; Lerario, G.; D'Amone, S.; Cingolani, R.; & Gigli, G. (2008). Hybrid Light-Emitting Diodes from Microcontact-Printing Double-Transfer of Colloidal Semiconductor CdSe/ZnS Quantum Dots onto Organic Layers, *Adv. Mater.* 20, 1886-1891

Rizzo, A.; Mazzeo, M.; Biasiucci, M.; Cingolani, R.; & Gigli, G. (2008) White EL from Microcontact Printing Deposited Colloidal Quantum Dot Monolayer, *Small* 4, 2143-2147

13

Seasoning ferroelectric liquid crystal with colloidal nanoparticles for enhancing application flavors

Jung Y. Huang
Department of Photonics, Chiao Tung University
The Republic of China at Taiwan

1. Introduction

The properties of nanostructured materials can be significantly different from those of bulk materials. This prospect has attracted immense academic and industrial interests to develop new synthetic schemes of nano building blocks and self-assembling methodologies for efficiently manipulating a variety of building blocks into functional materials. Liquid crystal (LC) materials possess both spatial order and mobility at the molecular level, thus may serve suitable candidates for self-assembly of nanoscale materials. The LC-based nanoscience (Hegmann et al. (2007)) has recently made significant progress to cover the topics of the synthesis of nanomaterials using LCs as templates, the design of LC nanomaterials, self-assembly of nanomaterials in LC phases, and defect formation in LC-nanoparticle suspensions.

For the synthesis of nanomaterials and self-assembly of nano objects in LC, researchers had demonstrated the shape-selective synthesis of gold nanoparticles in a liquid crystal medium and assembled these nanoparticles to spontaneously form ribbonlike patterns (Mallia et al (2007)). For the design of LC nanomaterials, suspensions of ferroelectric nanoparticles in a nematic liquid crystal (NLC) host had been shown to possess an enhanced dielectric anisotropy and are sensitive to the sign of an applied electric field (Cheon et al. (2005)). Large colloidal particles in a LC-nanoparticle suspension can cause strong director deformations and that usually lead to defect formation in LC matrices. By controlling the particles' shapes researchers can disturb the uniform alignment of the surrounding nematic host to generate highly directional pair interactions (Lapointe et al (2009)). Therefore, researchers in the LC-based nanoscience are now able to tailor colloidal interactions to realize high-precision self-assembly in anisotropic nematic fluids.

Small particles dispersed in LC at high concentration create almost a rigid LC suspension. From simulation (Hung & Bale (2009)), a nanocube in a NLC medium with perpendicular anchoring at the nanocube's faces tends to align in such a way that none of its faces is parallel or perpendicular to far-field LC director n(r). A triangular nanoprism with homeotropic anchoring of the nematic at its surfaces tends to align with both its long axis and one of its rectangular faces perpendicular to n(r). For systems of nanoprisms in NLC, inverted parallel arrays (the long axes of the nanoprisms are parallel, and one of the prisms is inverted with respect to the other one) are thermodynamically more stable than linear arrays (the long axes of the particles are collinear and the particles have the same orientation), which in turn are more stable than parallel arrays (the long axes of the particles are parallel and the particles have the same orientation). The minima observed in the potentials of mean force curves for the inverted parallel and linear arrays are significantly deeper than that observed for the parallel array. These NLC-mediated, anisotropic interparticle interactions can make the particles bind together at specific locations, and thus could be exploited to assemble the particles into ordered structures with different morphologies.

At low concentrations, LC nanocolloids appear similar to a pure LC with unique properties (Ouskova et al. (2003)). The diluted suspensions are stable, because the small concentration of nanoparticles does not significantly perturb the director field in the LC, and interaction between the particles is weak. Most importantly, the nanoparticles can share their intrinsic properties with the LC matrix due to the anchoring with the LC. For engineering applications (Kobayashi (2009)), nanoparticle-embedded LCDs using nanoparticles of metals, semiconductors, inorganic oxide dielectrics and polymers had been shown to reduce both the threshold voltage and response time, to improve frequency modulation electro-optical response, and even to yield a new display mode.

Unlike NLC, which possesses only orientational order, ferroelectric liquid crystals (FLC) exhibit smectic layer structures in which the director in the smectic layers tilts with respect to the layer normal (Lagerwall (1999)). Clark and Lagerwall (Clark & Lagerwall (1980)) invented the concept of surface-stabilized ferroelectric liquid crystal (SSFLC) to realize the attractive properties of FLC materials, which include microsecond response time, bistability and wide viewing angle. However, to meet the industrial application criteria, further improvements on the properties of FLC are required. Modifying an existing FLC material by doping with appropriate nanoparticles could produce new FLC by blending instead of synthesizing new mesogenic molecules and therefore had attracted significant interest (Kaur et al. (2007); Li et al (2006); Reznikov et al. (2003); Shiraishi et al. (2002)). However, the behaviors of a dilute suspension of colloidal nanometer-scale particles in smectic medium remain elusive. Further study to yield insight into the interaction between nanoparticle and FLC species (Huang et al. (2008)) is crucial not only for designing better blending schemes but also for the understanding of this new material system.

This chapter aims to provide such an insight. By using FLC doped by ZnO nanocrystals (nc-ZnO) as a model system, we illustrate several important concepts such as what material properties can be modified by the doping and what physical processes likely be involved. We first applied the capacitance-voltage (CV) measurement technique to reveal the roles of multiple dipolar species in a typical SSFLC film (Li & Huang (2009)). We found that nc-ZnO doping into the SSFLC not only enhances the dynamic polarization of the beneficial dipolar species but also minimizes the polarization response of the species with counter effects. We then exploited two-dimensional infrared (2D IR) absorption correlation technique to verify that ZnO nanocrystals are dispersed uniformly into the FLC host and the uniform doping leads to an improved aligning structure of FLC molecules (Huang et al. (2008)). We proposed that the ZnO nanodots tie up surrounding FLC dipolar species and yield a molecular binding effect. Finally, dynamic light scattering (DLS) was employed to show that nc-ZnO doping increases the elastic constant of FLC and renders the mechanical response of the FLC film into under-damping.

2. Preparation and Characterization of Materials

2.1 Material Preparation

Zinc oxide nanocrystals were chosen for this study in view of the low-cost synthetic process with non-toxic chemicals. We synthesized colloidal ZnO nanoparticles with the procedure reported in the literature (Goyal et al. (1993); Meulenkamp (1998)). To obtain a stabilized nc-ZnO colloid, we coated the nc-ZnO with 3-(trimethoxysilyl) propyl methacrylate (TPM) and purified the colloid as reported in the literature. The TEM image reveals the nc-ZnO colloid to have a size distribution of a mean diameter of 3.2 nm and a full width-at-half maximum (FWHM) of 1 nm (Huang et al. (2008)).

FELIX 017/100 FLC (from Clariant) with spontaneous polarization of P_s =47 nC/cm2 at 25°C exhibits a thermotropic transition sequence of Cr $\overset{-28°C}{\longleftrightarrow}$ SmC* $\overset{73°C}{\longleftrightarrow}$ SmA* $\overset{77°C}{\longleftrightarrow}$ N* $\overset{85°C}{\longleftrightarrow}$ Iso. The material properties of the FLC related to this study are summarized in Table 1. To prepare doped FLC material, an appropriate amount of ZnO nano powder was added into the FLC and then homogenized with ultrasonic at 85°C for 40 minutes. The substrates of each SSFLC cell are indium tin oxide (ITO)-coated glass plates of 12 mmx15 mm in dimension, which were coated with SE7492 polyimide (from Nissan Chemical) and rubbed unidirectionally to produce uniform FLC alignment. We assembled SSFLC cells with two anti-parallelly rubbed substrates and maintained the cell gap with 2-μm diameter silica balls dispersed in a UV-curable gel NOA65. The nc-ZnO doped FLC in the isotropic phase was filled into a test cell and then cooled slowly to 35°C. All SSFLC cells prepared for this study exhibit a stable single SmC* domain under an optical conoscope.

2.2 Material Characterization

The characterization of nanoparticles requires highly sophisticated analytical tools. Such tools must be suited for the studies where the nanoparticles are dilutely doped into a complex medium. Generally speaking a combination of analytical techniques results in the best characterization. In this study, a variety of characterization techniques were used to establish (i) whether the nanoparticles were distributed uniformly in the ferroelectric liquid crystal medium. (ii) Was there any influence of nc-ZnO doping on the structural ordering and dynamic response of the SSFLC? (iii) What is the physical picture related to the interaction between the nc-ZnO and ferroelectric liquid crystal molecules? Combined capacitance-voltage measurement, two-dimensional infrared absorption correlation technique, and dynamic light scattering seem to be useful for characterizing the nc-ZnO doped ferroelectric liquid crystal in the smectic C* phase.

The C-V measurements of the SSFLC cells (Li & Huang (2009)) were carried out at 25°C with a computer-controlled HP4284A LCR meter, which provides a bias voltage range of 35 V and a sinusoidal wave of 1.0 V with a frequency range of 20 Hz?1 MHz. We first used C-V measurement technique to ensure each empty cell to have a voltage-independent capacitance about 717±35 pF at 1 kHz.

Figure (1) presents the schematic of a SSFLC cell in an infrared absorption spectrometer (Huang et al. (2008)). Fourier-transform infrared (FTIR) spectra from 900 to 3500 cm^{-1} with 4-cm^{-1} resolution were recorded with a liquid nitrogen-cooled HgCdTe detector and a home-made data acquisition electronics. For time-resolved FTIR measurement, bipolar square-wave pulses were used to drive the test cell. The waveform is comprised of a +10V pulse from 0 to 140 μsec, followed by a field-free period extending from 140 to 500 μsec. Then a -10V pulse extends from 500 to 640 μsec, followed by a field-free duration from 640 to 1000 μsec. For each polarization direction of the incident infrared beam, a total of 32 time-resolved interferograms were acquired. The 2D correlation analysis with varying infrared polarization angles were calculated based upon an algorithm developed by Noda (Noda (1993)), and was implemented in a software 2Dshige by S. Morita (Morita (2004)).

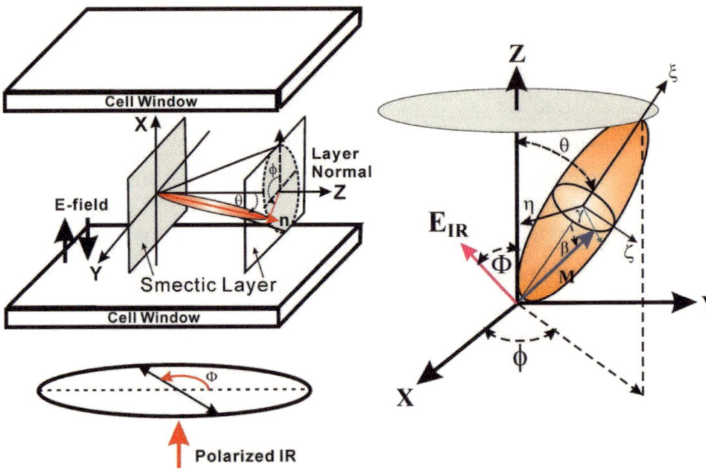

Fig. 1. Configuration of SSFLC in a FTIR apparatus. Z denotes the rubbing direction, which is also the layer normal of the smectic layers. **E** is the direction of the applied electric field. **M** on the right diagram denotes the IR dipole, which tilts from the molecular long axis ξ with an angle β and can rotate about the ξ-axis by γ.

Figure (2) presents the schematic diagram of the DLS apparatus used to probe thermal fluctuations of a SSFLC cell. A He-Ne laser with a wavelength of 632.8 nm was used as the excitation light source. The test cell was inserted in a setup with crossed polarizer and analyzer. We adjusted the incident light polarization to be along to the rubbing direction of the test cell. We adopted a scattering geometry with the scattering vector lying on the cell's substrates and being perpendicular to the rubbing direction. This experimental geometry ensures that the incident photons can only be scattered by the fluctuation of the azimuthal angle $\varphi(t)$ of FLC directors. Furthermore, a forward scattering geometry with a small scattering angle ($\sim 4^o$) was used to ensure only the fundamental mode of thermal excitation of the cell to be observed. The scattered photons were detected with a silicon avalanche photodiode (SPCM-AQR-15, PerkinElmer) and then fed into the Flex02-01D digital correlator (Correlator.com, NJ) to calculate the intensity autocorrelation function in real time. The SSFLC test cells in all measurements were kept at a specific temperature with $\pm 0.1^o$C precision.

Referring to the coordinate system described in Fig. 1, the intensity of the transmitted light through the setup can be expressed as (Huang et al. (2008))

$$I = I_0[\sin^2 2\theta \sin^2 \frac{\Gamma}{2}]\cos^2\varphi, \tag{1}$$

where $\Gamma = k\Delta n\ d/2$ denotes the phase retardation caused by the SSFLC film and d is the film thickness.

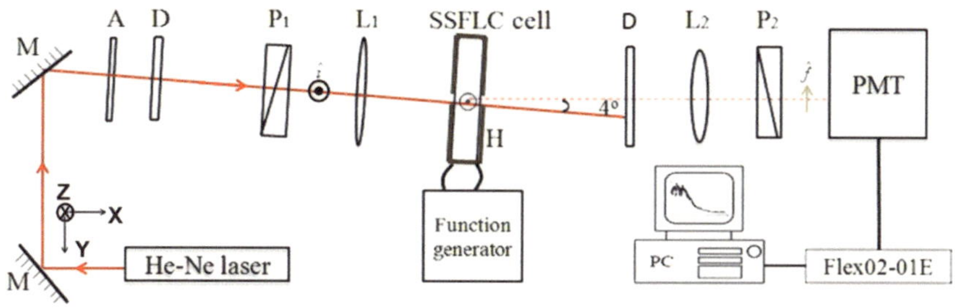

Fig. 2. Experimental configuration of DLS. M: mirror; P1 and P2: polarizers; L: Lens; D: diaphragm. Here the polarization of the incoming beam is set to be parallel to the rubbing direction of the SSFLC cell, but perpendicular to both the scattering plane and the polarization analyzing direction for the output scattering beam.

3. Experimental Results

Note that ZnO possesses a wurtzite structure that renders ZnO nanoparticle with a diameter of $2R_0$=3.2 nm to have a permanent dipole moment of μ_{ZnO}=50 Debye (Nann & Schneider (2004); Shim & Guyot-Sionnest (1999)). It is inspiring to first estimate the energy density of the dipolar interaction resulting from nc-ZnO doping. Being nano-size objects, the nc-ZnO particles shall behave like a molecular dopant in SSFLC. The dipole-dipole interaction energy between nc-ZnO and one of the surrounding dipolar species μ_{CO} reads $V = \mu_{ZnO}\mu_{CO}(1 - 3\cos^2\theta)/(4\pi\varepsilon_0 r^3)$. Because the dipole-dipole interaction depends on the relative orientations θ, the molecules can exert forces on one another and in fact can not rotate freely. The orientation with lower energy is favoured so there is a non-zero dipolar interaction energy. At a uniform doping concentration of one weight percent, we estimated one ZnO nanoparticle to be surrounded by about n_{FLC} =15000 molecules within a sphere of $R_{\max}$ =14 nm centered at the nc-ZnO. The thermal average of the dipole-dipole interaction energy density between nc-ZnO and surrounding FLC molecules becomes (Keesom (1921); Magnasco et al. (2006))

$$U_a^{ZnO-FLC} = -\frac{2\,\mu_{CO}^2\,\mu_{ZnO}^2\,n_{FLC}\,N_{ZnO}}{3\,(4\pi\varepsilon_0)^2 k_B T\,R_0^3\,R_{\max}^3}; \tag{2}$$

We estimated $U_a^{ZnO-FLC} \approx 1000J\cdot m^{-3}$ with $N_{ZnO} = 1.4\times10^{23}m^{-3}$, μ_{CO} =1.5 Debye, and $k_BT = 4\times10^{-21}J$. This interaction energy has an alignment effect similar to that FLC molecules will experience within a distance of 100 nm from an alignment surface of an anchoring strength $1\times10^{-4}J\cdot m^{-2}$ and shall produce an observable effect in our SSFLC cells.

Indeed, by inserting a doped SSFLC cell into a setup of crossed polarizer and analyzer and rotating the cell about the substrate normal, we found the resulting azimuthal pattern of the optical transmittance has a reduced light leakage in the dark state by 9 times while the optical throughput in the bright state is increased by 1.2 times, leading to an increased contrast by a factor of 10 (Li & Huang (2009)).

3.1 The Influence of nc-ZnO Doping on the CV Characteristics of SSFLC

The origin of the observed improved alignment of FLC molecules by nc-ZnO doping can be revealed in more detail with the C-V characterization technique. Figs. 3(a) and 3(b) present two typical CV characteristic curves of SSFLC obtained with a negative-to-positive voltage sweep. We can use Preisach model to derive the CV characteristic curve of a SSFLC film with multiple dipolar species (Li & Huang (2009))

$$C_{cell}(V_{ex}) = C_{const} + \sum_{i=1}^{n} \frac{P_{si} \cdot \delta_i}{\cosh^2[\delta_i \cdot (V_{ex} \pm V_{Ci}^{\mp})]} A. \quad (3)$$

where C_{const} denotes the voltage-independent part of the capacitance of FLC; A the cell area; P the polarization, and V_{ex} the driving voltage. P_{si} in Eq. (3) represents the spontaneous polarization as all dipoles of the ith species are fully aligned to an external field. V_{Ci} is the coercive voltage at which the electric polarization of the ith species vanishes during switching from one orientation to the other orientation (Blinov & Chigrinov (1993); Demus et al (1998)). $\delta_i = \log\{[1 + (P_r/P_s)_i]/[1 - (P_r/P_s)_i]\}/V_{Ci}$ is a constant related to species with P_r and P_s denoting the remnant and spontaneous polarizations, respectively. The $(+/-)$ sign refers to an increasing/decreasing V_{ex}. The CV curves of Fig. 3 can fit to Eq. (3) with four dipolar species being involved in the SSFLC films. From the data fitting, we found that by doping with nc-ZnO, the peak capacitance contributed by species 1 is increased while capacitances from species 2, 3 and 4 are reduced. ZnO nanocrystal doping also decreases V_c of species 1 by 1.2V, while increases V_c of species 2 and 4 by 1.5V and 0.5V, respectively.

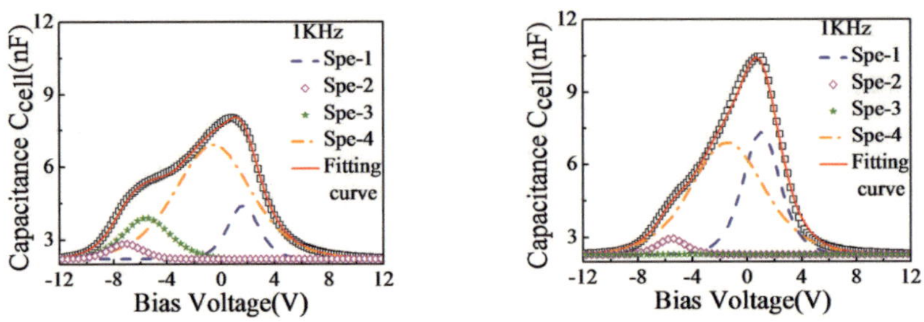

(a) (b)

Fig. 3. CV characterization of SSFLC. The CV characteristic curves (open squares) at 1 kHz and the corresponding fitting results (solid lines) from the four dipolar species in a SSFLC cell (a) without and (b) with nc-ZnO doping.

One of the major advantages of SSFLC is microsecond response time for true video rate applications. Therefore, it is also interesting to investigate how P_s of the four dipolar species behaves at high driving frequencies. Figure 4(a) shows that nc-ZnO doping results in an increase in P_s of species 1 while that from the remaining three species are diminished; the doping-induced spontaneous polarization change (ΔP_s) of the four species decays with frequency, reflecting more and more FLC molecules can not catch up the driving field at high frequency. These findings strongly support the notion that doping with nc-ZnO enhances the dynamic polarization of the beneficial dipolar species while it minimizes the polarization response of the species with counter effects.

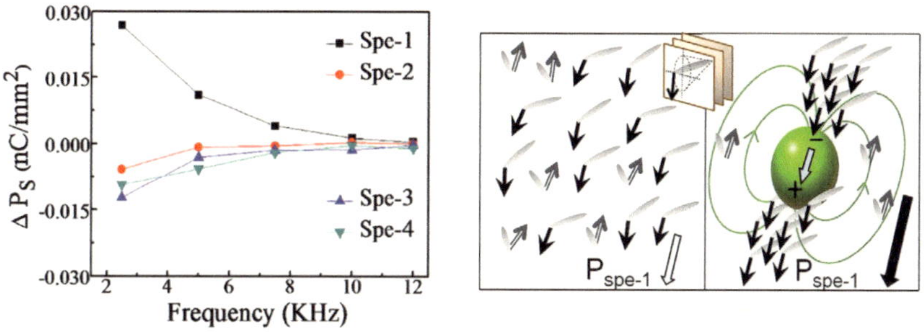

(a) (b)

Fig. 4. Polarization response from dipolar species.(a) The doping-induced spontaneous polarization changes (ΔP_s) of the four dipolar species in a SSFLC cell as a function of driving frequency. (b) Schematic drawing illustrating the dipolar interaction of species 1 in an undoped SSFLC cell (left) and the dipolar interaction of species 1 with nc-ZnO in a doped SSFLC cell (right). The inset figure shows the smectic layer structure of SSFLC and Pspe1 denotes the spontaneous polarization of species 1.

Figure 4(b) presents a schematic drawing illustrating our concept about the dipolar interaction of species 1 in an undoped and a doped SSFLC cell. Dipoles associated with species 1 could adjust their spatial distribution and likely aggregate on those regions near the north and south poles of the nc-ZnO, and accordingly align in the same direction with the nc-ZnO dipole. The number of the dipolar species 1 lying on the equatorial plane of the nc-ZnO becomes smaller. By this way, P_s from species 1 can therefore be increased by nc-ZnO doping.

3.2 2D IR Study on the Effect of nc-ZnO Doping in SSFLC

Two-dimensional IR correlation is a technique where the spectral intensity is plotted as a function of two independent spectral variables. By spreading spectral peaks along the second dimension, one can gain an advantage of sorting out complex or overlapped spectral features that normally cannot be resolved in a one-dimensional spectrum. Generalized 2D IR spectroscopy is an effective mathematical tool to elucidate spectral details of a dynamic system (Ozaki & Noda (2000)). The data yielded from the technique are usually presented with synchronous and asynchronous plots. In our case, the synchronous plot reveals the information about the in-plane order and the similarity in the IR azimuthal angular patterns, while the asynchronous plot offers the information about the dissimilarity in the azimuthal angular patterns. Therefore, the information about the alignment of submolecular species and their field-induced switching dynamics can be effectively probed with the 2D IR absorption correlation technique.

In Fig. 5, the field-free 2D synchronous plots in the 2820-3000 cm^{-1} region were shown on the left column and the asynchronous on the right-hand side for the undoped (first row) and the nc-ZnO doped (second row) SSFLC cells. The major auto-peak appearing at 2926cm^{-1} is assigned to the anti-symmetric stretching of CH2 group (*a*-CH2) and reflects an in-plane alignment order of *a*-CH2. The appearance of the cross-peaks A: 2926 (*a*-CH2) vs. 2840 cm^{-1} (*s*-CH2), B: 2926 vs. 2876 cm^{-1} (*s*-CH3), C: 2926 (*a*-CH2) vs. 2900 cm^{-1} and D: 2975 (*a*-CH3)

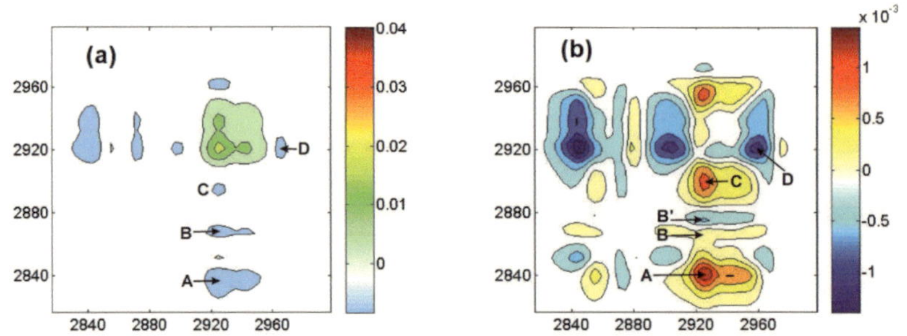

(a) (b) (c)

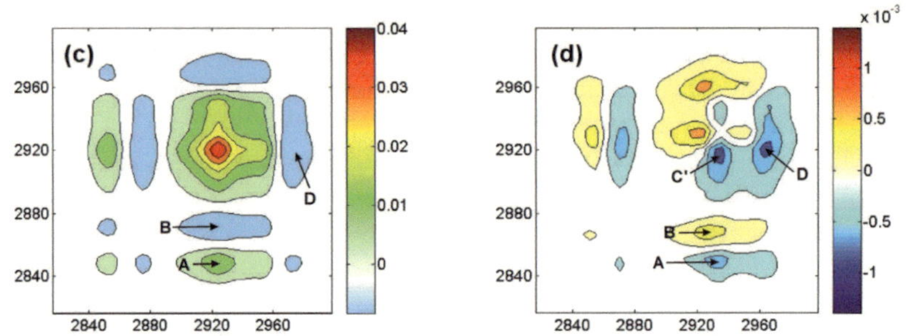

(d)

Fig. 5. 2D IR plots of CH stretching modes. Synchronous (a and c) and asynchronous (b and d) 2D IR correlation plots from the alkyl-chain modes of SSFLC cells without (a and b)and with nc-ZnO doping (c and d).

vs. 2926 cm^{-1} indicates the IR-active dipoles of CH2 and CH3 to be angularly correlated. Notice that the asynchronous cross-peak can reflect relative orientation difference of IR-active dipoles. Two cross peaks B and B′ in Figure 5(b) can therefore be attributed to be *s*-CH3 from different species. The cross peaks A, B and C were found to be positive while B′ and D to be negative. By defining Φ_0 as the apparent angle of the maximum IR absorbance, the asynchronous cross-peak can be expressed as $\Psi_a(\nu_1, \nu_2) = 0.11 \ \sin 2[(\Phi_0(\nu_1) - \Phi_0(\nu_2))]$ (Huang & Shih (2006)), indicating that IR-active modes associated with the cross peaks A, B, and C relative to that of the *a*-CH2 mode at 2926-cm^{-1} have smaller Φ_0 while peaks B′ and D have larger Φ_0.

In the synchronous plot, the major auto-peak of the SSFLC doped with nc-ZnO appearing at 2926 cm^{-1} is more intense than the undoped sample, indicating an improved alignment order with nc-ZnO doping. In the asynchronous plot, the cross peaks C and D of the doped sample become in phase, suggesting that CH2 and CH3 in the doped SSFLC are packed into a more ordered structure. This shall happen only when the ZnO nanodots are dispersed uniformly in the FLC medium.

In Fig. 6, the synchronous 2D IR plots from the core groups, the ring C=C stretchings of the FLC were presented. The cross-peaks of C(1514 vs. 1608 cm^{-1}), B(1584 vs. 1608 cm^{-1}) are more distinctive with nc-ZnO doping, indicating a more ordered structure of the core groups in the doped SSFLC cell.

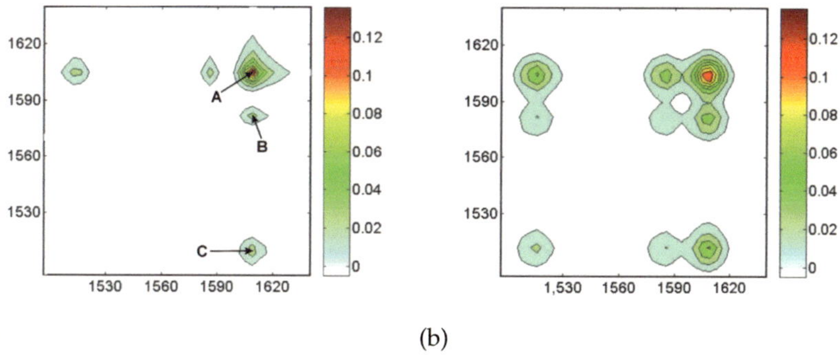

Fig. 6. 2D IR plots of core groups. Synchronous 2D IR correlation plots from the core groups of SSFLC cells (a) without and (b) with nc-ZnO doping.

By using time-resolved 2D IR correlation technique, the field-induced switching of SSFLC can be finely resolved into the reorientation of submolecular species. We had successfully exploited a global 2D phase defined as $\tan^{-1}[\Psi_a(\nu_1, t; \nu_1, 0)/\Psi_s(\nu_1, t; \nu_1, 0)]$ to reflect the orientational variations of IR-active modes (Huang et al. (2008)). The resulting 2D IR phase angles of the IR-active modes associated with the FLC cores and alkyl chains are presented in Fig. 7. The phase angles change sign with the polarity of the driving field, indicating that during the field-induced reorientation the IR-active molecular dipoles projected onto the substrate surface point to two opposite sides of their corresponding field-free directions. The CH2/CH3 stretching modes associated with the alkyl chains are loosely connected, resulting in larger angular spreads than that the core groups. By doping nc-ZnO into SSFLC, the orientational variations of the IR-active dipoles reveal smaller angular spreads, suggesting doping with nc-ZnO yields more concerted reorientation processes.

3.3 Dynamic Light Scattering from pure SSFLC and nc-ZnO Doped SSFLC

The experimental results described above clearly indicate that a dilute suspension of colloidal nanometer-scale particles in SSFLC film can significantly improve both the field-free alignment and the field-induced dynamic response of the film at the submolecular level. However, due to the unavoidable random impacts from thermal excitation, the SSFLC molecules behave stochastically. The random force constantly pushes FLC molecules away from the orientations of minimal elastic energy. As the FLC molecules return to their equilibrium orientations, they release the elastic energy into the rotational kinetic energy and through which, the kinetic energy is dissipated via the viscous dragging force. To advance our knowledge of the new ferroelectric liquid crystal material, we further conducted a series of dynamic light scattering measurements.

Dynamic light scattering (DLS) is an efficient technique to probe the stochastic behavior of a material. The methodology can yield an autocorrelation function $G(t) = \langle I_s(t) I_s(0) \rangle$ (Berne & Pecora (2000)) of the optical scattering intensity $I_s(t)$ resulting from the SSFLC film under study. Figure 8(a) presents the autocorrelation curves of undoped SSFLC acquired at varying temperatures. We can find several characteristic features from the measured DLS curves. First, autocorrelation at short delay time $G(t \to 0)$ increases with temperature of the

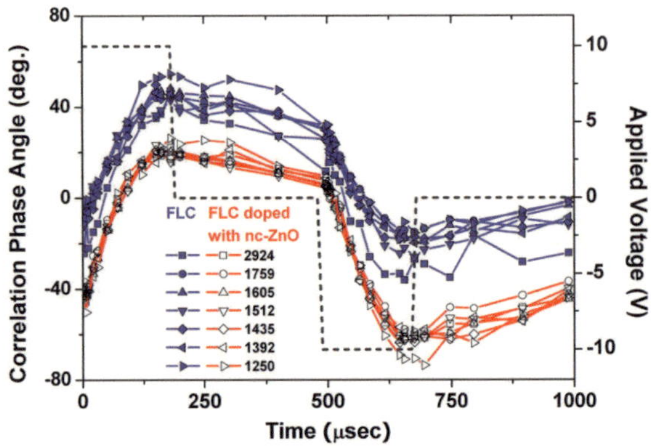

Fig. 7. Dynamic responses of functional groups. Time courses of 2D IR correlation phase angles of IR-active molecular normal modes of SSFLC with (open symbols) or without (filled symbols) nc-ZnO doping. The dashed line denotes the bipolar square wave form of applied electric field used.

SSFLC film. Secondly, at higher temperature the autocorrelation at long delay time $G(t \to \infty)$ relaxes to a higher level than that at lower temperature. Note that the value of $G(0)$ can be related to the total number of fluctuating domains lying in the optically illuminated region by $N = G(0)/(G(0) - 1)$. Although the SSFLC cells used for this study were prepared to have single domain, the cells do not fluctuate in unison. Instead, they are divided into about 500 coarse grains. By using $G(0)$ of Fig. 8(a) and a light spot of 20 μm, we estimated each thermally fluctuating domains to have a diameter of about 1 μm.

Fig. 8(b) presents the DLS curve of doped SSFLC with 1% nc-ZnO measured at 45°C. An interesting oscillating feature can be seen in the curve, which most likely originates from an increase in elastic constant by the nc-ZnO doping. The increased elastic constant could convert the mechanical response of the SSFLC from over-damping to under-damping. We also noted that doping with 1% nc-ZnO significantly increases DLS signal, indicating that the doping may reduce the total number of fluctuating domains in the optically illuminated region.

Researchers had noted that stochastic resonances (Gammaitoni et al. (1998)) can occur in a system with a proper combination of nonlinearity and stochastic excitation. The stochastic resonances had been exploited to minimize the ill effects of noise on the system. SSFLC is essentially a nonlinear system. FLC molecules in a SSFLC device typically encounter two driving forces: one from a deterministic control field and the other a stochastic thermal excitation force. One may be curious about how the FLC molecules behave under the influences of these two forces. To answer the question, we investigated the DLS of SSFLC by using a sinusoidal bipolar waveform of varying frequencies with V_{pp}=1 V. The measured DLS curves are presented in Fig. 9. Note that without the interruption of the thermal excitation the correlated DLS response of FLC molecules to the periodic waveform shall persist forever. Instead, we found that the envelopes of the DLS curves decrease as delay time

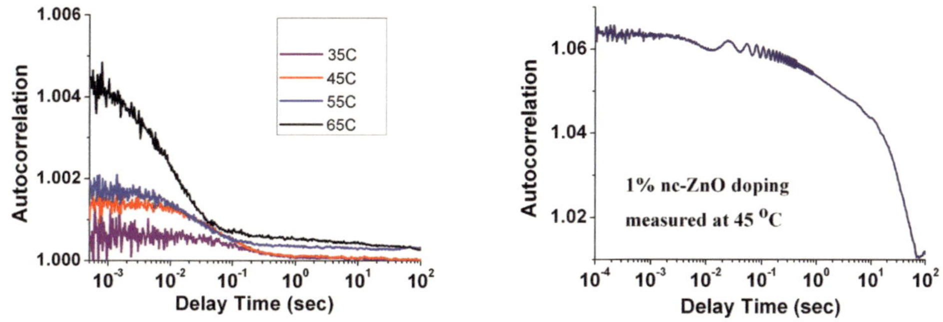

(a) (b)

Fig. 8. DLS data of SSFLC. (a) The measured autocorrelation curves of DLS from pure SSFLC at various temperatures (the curves with black, blue, red, and purple colors correspond to DLS at 65°C, 55°C, 45°C, and 35°C, respectively. (b) The DLS autocorrelation curve of SSFLC at a doping level of 1% meaured at 45°C.

increases, revealing clearly the randomizing effect of the fluctuating thermal force. Based on the rotational viscosity and spontaneous polarization of the FLC material, we expect that the FLC molecules shall not respond to an applied field with a frequency of 10 kHz or above. At such high frequencies, the FLC molecules remain essentially at the original orientation with a small random excursion caused by the thermal excitation force. Indeed, the DLS curve driven at 10 kHz (see the green-colored line in Fig. 9(a)) exhibits a flat response. However, it is surprising to find that after doping with nc-ZnO (see Fig. 9(b)), the FLC molecules become responsive even at 10 kHz.

3.4 The Model of Dynamic Light Scattering from SSFLC

To retrieve the underlying physics, we developed a stochastic model to describe the dynamic light scattering behavior of SSFLC. The model is comprised of two parts: the first part describes the dynamic light scattering process from a medium with a fluctuating index of refraction, and the second part focuses on the description of the fluctuating process of the medium. In this section, we will focus our attention on the second part of the model by reducing SSFLC to an effective medium with a moment of inertia per unit volume ρ and a rotational viscosity coefficient η. Furthermore, we assumed the SSFLC film to be well below the transition temperature of smectic C* phase. Thus, we can focus on the fluctuating behavior of the azimuthal angle φ of the SSFLC director. Thus, the equation of motion of SSFLC with thermal force $F_R(t)$ becomes (Chandrasekhar (1992))

$$\rho\frac{\partial^2\varphi(t)}{\partial t^2}+\eta\frac{\partial\varphi(t)}{\partial t}=\frac{\delta E_{free}}{\delta\varphi}+F_R(t). \tag{4}$$

We further simplified the description of elastic deformation energy of SSFLC with single elastic constant K. The total free energy per unit area can then be expressed as

$$E_{\text{free}}=d\,[\frac{1}{2}K\sin^2\theta(\frac{\partial\varphi}{\partial x})^2-P_sE\cos\varphi]. \tag{5}$$

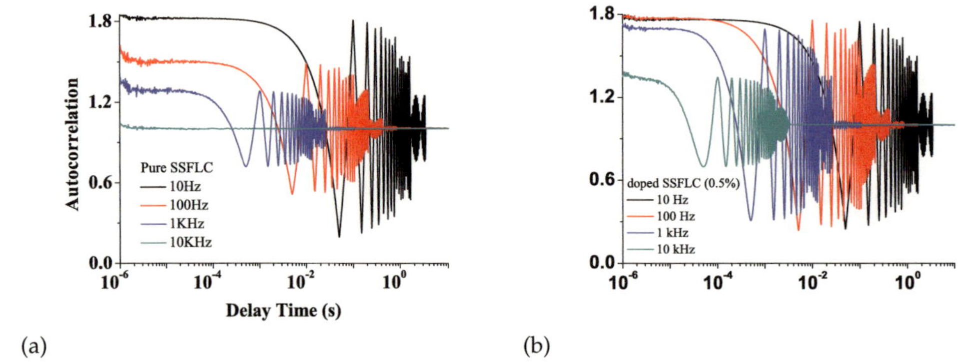

Fig. 9. Forced DLS of SSFLC. (a) The measured autocorrelation curves of DLS from pure SSFLC at various driving frequencies. (b) The DLS autocorrelation curves of SSFLC at a doping level of 0.5% meaured at 45°C.

Here d denotes the thickness of the SSFLC layer. The equation of motion in absence of external electric field finally reads

$$\rho \frac{\partial^2 \varphi(t)}{\partial t^2} + \eta \frac{\partial \varphi(t)}{\partial t} + K \sin^2\theta \frac{\partial^2 \varphi(t)}{\partial x^2} = F_R(t). \tag{6}$$

We implemented a strong anchoring boundary condition on φ by using $\varphi(0,t) = 0, \ \ \varphi(d,t) = 0$. Assuming the rubbing direction of the SSFLC to be along the z-axis and by applying the technique of the separation of variables, we can expand φ as a series of sinusoidal functions as $\varphi(x,t) = \sum_{n=1}^{\infty} \psi(t) \sin(n\pi x/d)$. We exploited a forward scattering geometry with small scattering angle to ensure only the fundamental mode of thermal excitation to be detected. This can greatly simplify Eq. (6) to yield

$$\rho \frac{\partial^2 \psi(\mathrm{t})}{\partial t^2} + \eta \frac{\partial \psi(t)}{\partial t} + \tilde{K} \psi(t) = F_R(t). \tag{7}$$

with $\tilde{K} = K \sin^2\theta (\pi/d)^2$. This is essentially a damped harmonic oscillator driven by a stochastic thermal force $F_R(t)$. The thermal force is supposed to vary extremely rapidly over the time of any observation, causing the excitation effect to be summarized by its first and second moments

$$\langle F_R(t) \rangle = 0, \ \langle F_R(t) F_R(0) \rangle = 2\eta \, k_B T \, \delta(t). \tag{8}$$

Here k_B is the Boltzmann constant and T is the temperature. The bracket notation $\langle \cdot\cdot \rangle$ denotes an average with respect to the distribution of the realizations of the stochastic variable $F_R(t)$. The important result of the second moment is well known as the Fluctuation-Dissipation Theorem (FDT) (Kubo (1966)), which relates the strength of the fluctuating force to the magnitude of the dissipation. It expresses the balance between dissipation which tends to drive any system to a completely dead state and fluctuation force which tends to keep the system alive. This balance is required to have a thermal equilibrium state at long times.

By applying Laplace transform on Eq. (7), we can derive an analytical solution of Eq. (7) as

$$\begin{array}{l}\psi(t) = \psi(0)\frac{s_1e^{s_1t}-s_2e^{s_2t}}{s_1-s_2} + \left[\psi_t(0) + \frac{\eta}{\rho}\psi(0)\right]\frac{e^{s_1t}-e^{s_2t}}{s_1-s_2} + \\ \frac{1}{\rho(s_1-s_2)}\int_0^t d\tau F_R(\tau)\left(e^{s_1(t-\tau)} - e^{s_2(t-\tau)}\right)\end{array}. \tag{9}$$

where $\psi_t(0) = \left.\frac{\partial\psi}{\partial t}\right|_{t=0}$ and $s_1 = -\frac{\eta}{2\rho} + i\sqrt{(\frac{\tilde{K}}{\rho})^2 - (\frac{\eta}{2\rho})^2}$, $s_2 = -\frac{\eta}{2\rho} - i\sqrt{(\frac{\tilde{K}}{\rho})^2 - (\frac{\eta}{2\rho})^2}$.
By using Eq. (8), we can further calculate the autocorrelation of dynamic light scattering intensity as

$$\begin{array}{l}G(t) - 1 = \langle I(t)I(0)\rangle/\langle I(0)I(0)\rangle - 1 \\ \propto \frac{1}{N}\{\left[\psi(0)\frac{s_1e^{s_1t}-s_2e^{s_2t}}{s_1-s_2} + [\psi_t(0) + \frac{\eta}{\rho}\psi(0)]\frac{e^{s_1t}-e^{s_2t}}{s_1-s_2}\right]^2 + \\ \frac{2\eta k_BT}{\rho^2(s_1-s_2)^2}\left[\frac{e^{2s_1t}-1}{2s_1} + \frac{e^{2s_2t}-1}{2s_2} + \frac{2}{s_1+s_2}(1 - e^{(s_1+s_2)t})\right]\}\end{array}. \tag{10}$$

Eq. (9) indicates that although the typical mechanical response of SSFLC is over damping with $\tilde{K}/\rho< \eta/(2\rho)$, it can become under damping ($\tilde{K}/\rho> \eta/(2\rho)$) if the elastic constant can be increased to conquer the viscous effect.

For an over-damped SSFLC, the acceleration is quickly damped by dragging force, causing the rotational kinetic energy term $\rho\partial^2\psi(t)/\partial t^2$ to be negligible in Eq. (4). In this case, the equation of motion becomes

$$\eta\frac{\partial\psi(t)}{\partial t} + \tilde{K}\psi(t) = F_R(t). \tag{11}$$

The autocorrelation of dynamic light scattering intensity is then simplified to be

$$G(t) - 1 \propto \frac{\psi^2(0)}{N}\{\psi^2(0)e^{-2t\tilde{K}/\eta} - \frac{k_BT}{\tilde{K}}(e^{-2t\tilde{K}/\eta} - 1)\}. \tag{12}$$

The autocorrelation at the short delay time limit can be found to approach a value of

$$G(0) - 1 \propto \frac{1}{N}|\psi(0)|^4 = \frac{1}{N}(\frac{k_BT}{\tilde{K}})^2. \tag{13}$$

In a liquid crystal medium, the effective elastic constant is usually proportional to the square of order parameter S as $\tilde{K} = \tilde{K}_0S^2(T)$, which decreases as temperature is increased. This leads to a higher short-time-limit of the intensity autocorrelation as temperature is increased. At long delay time, the intensity autocorrelation decays to $G(\infty) - 1 = \frac{1}{N}\left(k_BT/\tilde{K}\right)^2$, implying a higher background level at higher temperature. These asymptotic behaviors explain well the experimental observation shown in Fig. 8(a).
Both the elastic constant and rotational viscosity of the FLC decrease as temperature is increased. Eq. (12) predicts the decay time of the intensity autocorrelation function to be 0.3 sec at 35 °C, 0.2 sec at 45 °C, 0.15 sec at 55 °C, and 0.1 sec at 65 °C, respectively. By inserting the material parameters into Eq. (11), we can solve the stochastic differential equation numerically at the four temperatures. We can use the solution to calculate the intensity autocorrelation function. The calculated DLS curves are presented in Fig. 10(a), which fit to the experimental data (shown as symbols) quite well. By using a larger elastic constant and solving Eq. (7) numerically with a thermal force corresponding to a temperature of 45°C, we can calculate an intensity autocorrelation curve. The result is presented in Fig. 10(b), which successfully reproduces the oscillating feature of the measured DLS curve shown in Fig. 8(b).

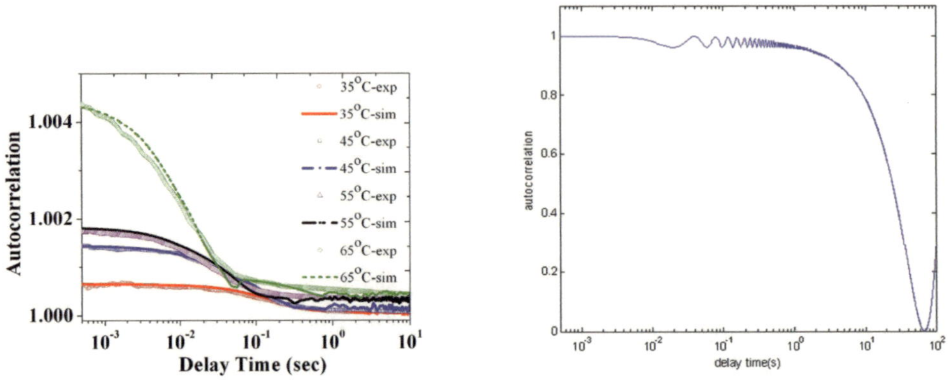

(a) (b)

Fig. 10. Simulated DLS curves of SSFLC. (a) The autocorrelation functions at a series temperature. The symbols represent the experimental data and the lines are the simulation results. The related parameters used in simulation are listed in table 1. (b) A simulation result for underdamped dynamics. Note that the oscillation behavior begins at delay time 0.02 seconds, which equals to the experiment result of SSFLC doped with 1% ZnO at 45°C.

In response to an external electric field applied along the x-direction, the equation of motion becomes

$$\eta\frac{\partial\varphi(t)}{\partial t} = -K\sin^2\theta\frac{\partial^2\varphi(t)}{\partial x^2} + P_s E\sin\varphi + F_R(t). \tag{14}$$

Assuming the SSFLC to have a bookshelf geometry with uniform profile along the x-axis, the first term in the right hand side of the equation can be neglected, which reduces the equation to

$$\eta\frac{d\varphi}{dt} = P_s E\sin\varphi + F_R(t). \tag{15}$$

Here P_s/η plays a major role in determining the response time of FLC driven by an applied electric field E. In this case, we encounter two driving forces with $P_s E\sin\varphi$ being deterministic, and the other $F_R(t)$ having a stochastic nature. It is interesting to investigate the dynamics with simulation and compare it with the experimental results. The parameters used in the simulation include the spontaneous polarization $P_s = 47$ nC/cm^2, cell thickness $d = 2\ \mu m$, and rotational viscosity $\eta = 105\, mPa\cdot$s. The magnitude of random force strength $\sqrt{2\eta\, k_B T}$ was adjusted to meet the condition of $2k_B T/\eta \simeq 0.5$, which corresponds to a temperature of about 25 °C. As shown in Fig. 11(a) for a undoped SSFLC, the simulation curve (the solid line) calculated with Eq. (15) can successfully reproduce the measured DLS curve (the open circles) measured at 100 Hz. By increasing the driving frequency to 10 kHz, the simulation predicts a flat DLS response, which agrees well with the experimental results (open circles) presented in Fig. 11(b)). The surprising result of Fig. 9(b), showing that the FLC molecules become responsive at 10 kHz after doping with nc-ZnO, can be attributed to a doping-induced increase in P_s/η. Currently we have sampled only very limited parameter space. We believe many interesting phenomena remain to be explored in SSFLC with or without nc-ZnO doping.

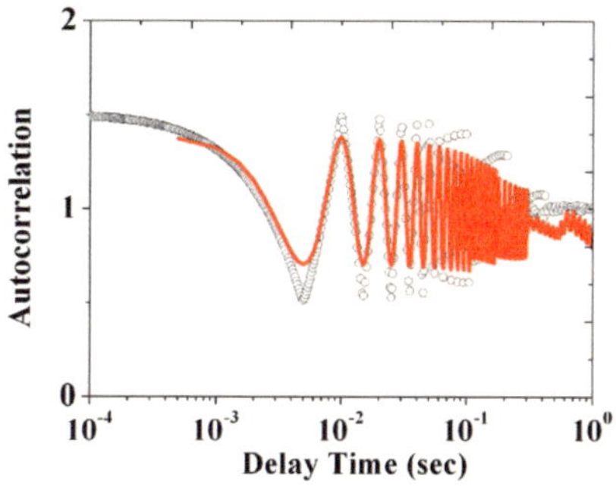

(a)

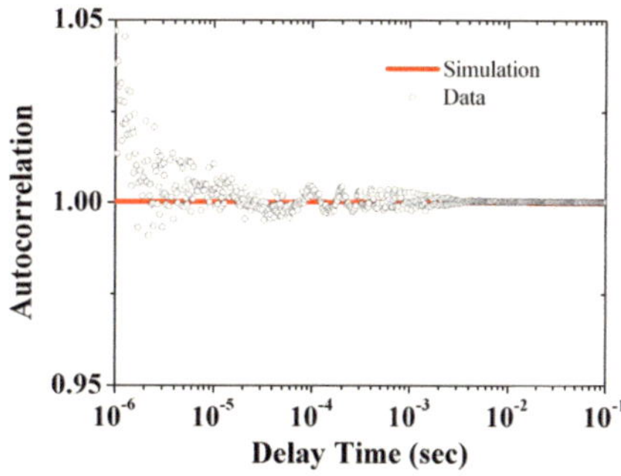

(b)

Fig. 11. Simulated curves of forced DLS from SSFLC. The comparison of experimental data (symbols) and simulated curve (solid line) of DLS at 25°C by applying a sinusoidal waveform with V_{pp}=1V and (a) f=100 Hz, (b) f=10 kHz on a SSFLC cell.

4. Importance of the Fundamental Mode of Molecular Orientation Fluctuation in Liquid Crystal

Why LC molecules under thermal excitation fluctuate collectively in orientation and how important to investigate the forward DLS from the resulting orientational fluctuation with small scattering angle? The answers to these questions may lie in the fundamental concept of symmetry and broken symmetry. In Fig. 12, we present a schematic diagram to summarize the essential concepts to be discussed.

Typically, a crystal belongs to one of the 230 space groups, which have both the positional and translational order of constituent atoms or ions. The most important behavior of a condensed-phase material is a collective excitation of its constituent elements. One type of the collective excitations in a crystal is phonon, in which each constituent nuclei oscillates coherently with its neighbors at the same frequency. The oscillatory excitations in a crystal are under-damped, resulting from the positionally ordered configurations that reduce the chance of lattice nuclei to bump into each other and enable the lattice to oscillate with a price of relatively low energy.

For materials with lower symmetry, they can be either positionally or orientationally disordered. Liquid crystals are positionally disordered, but have an orientational order. The orientational order reflects the degree of the molecular tendency to align to a specific direction. In LCs, the analogy to phonon in a crystal is orientational fluctuations of LC molecules. The collective excitations in LCs are typically over-damped, originating from the positional disorder of LC constituents. This leads to that inertial forces are much smaller than the viscous force. FDT describes the relation between the molecular fluctuations and their energy dissipations in equilibrium.

The dispersion relations for phonons in a crystal and orientational fluctuations in LC offer further insight into the underlying physics. A zero-frequency mode can exist in both material systems. The long-wavelength limit of acoustic phonon modes in a crystal always renders

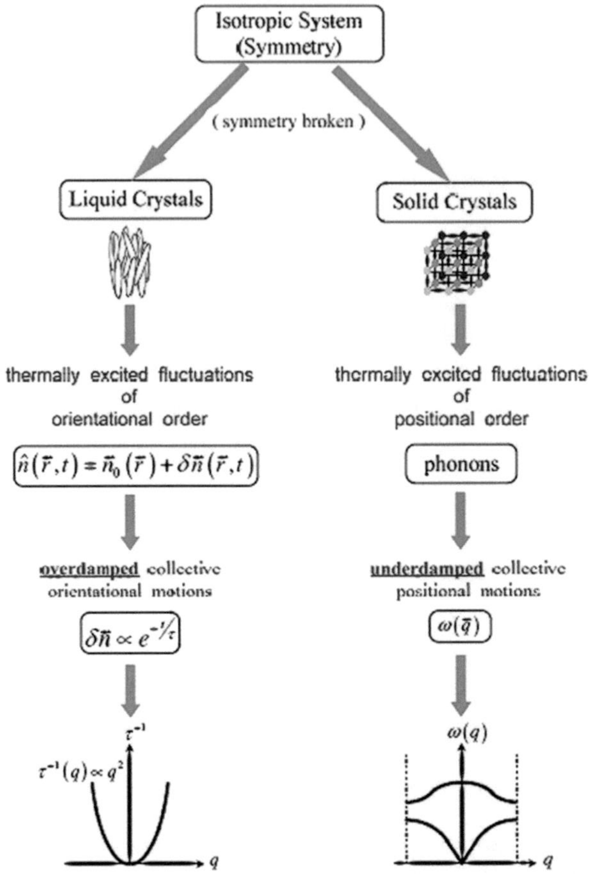

Fig. 12. Schematic showing the comparison of positionalal order in a crystal and orientational order in a nematic liquid crystal. The two figures at the bottom are the dispersion relations $\omega(q)$, which is the oscillating frequency of the phonons with $\vec{q}$ denoting the wavevector of the collective excitation. $\vec{n}_0$ is the average orientation of the LC director, and $\delta\vec{n}$ denotes a fluctuation of the director with a relaxation time of τ.

into the zero-frequency mode, whereas the zero-frequency mode of LC originates from the long-wavelength limit of orientational fluctuation. However, the q-dependent behaviors are quite different. The acoustic phonon of a crystal exhibits a linear dispersion of $\omega(q) \propto q$. The dispersion relation in LC is parabolic $\tau^{-1} \propto q^2$, a characteristic property of an over-damping system.

The existence of the zero-frequency mode in fact originates from the spontaneous breaking of the symmetry in a system. In the context of Goldstone's theorem (Strocchi (2008)), which states if the ground state of a many-body system has a broken symmetry, a gapless branch of collective excitations must exist in the system to restore the lost symmetry. It is this zero-frequency mode that makes all the neighboring LC molecules to change the orientations coherently when external perturbations such as thermal forces are exerted on them. Forward

DLS with small scattering angle offers us a unique opportunity to probe deeply into the zero-frequency mode of the orientational fluctuation.

5. Conclusion

In this chapter, we reviewed our recent advances on the ordered structure and dynamic responses of ferroelectric liquid crystal to an external deterministic force and stochastic thermal excitation. We studied how the ordered structure and dynamic responses were tailored by doping the FLC with ZnO nanoparticles. Combined capacitance-voltage measurement, two-dimensional infrared absorption correlation technique, and dynamic light scattering were shown to be valuable to yield clear picture for the complex system. Specifically, the CV characterization method revealed that doping FLC with nc-ZnO can improve the electro-optical response by increasing the spontaneous polarization and reducing the coercive voltage of the beneficial dipolar species while minimizing the polarization of the species with counter effects. Our 2D IR correlation data showed that at the submolecular level the ZnO nanocrystals were dispersed uniformly into the SSFLC medium. This uniform dispersion yields stronger correlations among the IR-active modes. The resulting stronger correlations lead to a more concerted reorientation dynamics. Dipolar interaction of a ZnO nanodot with surrounding C=O groups of FLC molecules was proposed to produce a doping-induced binding effect with an energy stability of about 1000 J/m3. The doping-induced stability also revealed in the dynamic light scattering study. Fluctuations in the optical response of a medium have traditionally been regarded as noise. We found that fluctuations in DLS can be used as a source of information for the complex system. By comparing the theoretical and experimental results of DLS, we found that an increase in elastic constant and spontaneous polarization by nc-ZnO doping can convert the mechanical response of the SSFLC from over-damping to under-damping. We offered a general discussion on the collective excitation of molecular orientation in SSFLC based on the Goldstone's theorem and argued that the zero-frequency mode may exist in the system to restore the lost symmetry. Based on the theoretical and experimental studies, we concluded that nc-ZnO doping into FLC can become a simple yet effective way to tailor the field-induced switching properties of an existing FLC.

This research was supported by the National Science Council of the Republic of China under grant NSC 97-2112-M-009-006-MY3. The author also likes to thank his graduate students Liu S. Li, Meng M. Cheng, and Bin K. Chen for their contribution and devotion to this research.

^oC	η(mPas.s)	θ (^{o})	P_s(nC/cm2)	$\tilde{K}$ (N/m2)	$\tau = (\eta/2\tilde{K})$ (s)
35	105	14	47	0.17	0.31
45	60	13	39	0.15	0.20
55	35	12	32	0.12	0.15
65	20	10	25	0.10	0.10

Table 1. The Related Parameters of FELIX 017/100 FLC used for the simulation

6. References

Berne, B. J. & Pecora, R. (2000). *Dynamic Light Scattering with Application to Chemistry, Biology and Physics*, Dover Publications, Mineola and New York.

Blinov, L. M. & Chigrinov, V. G. (1993). *Electrooptic Effects in Liquid Crystal Materials*, Springer-Verlag, ISBN-10: 0387947086, Berlin.

Chandrasekhar, S. (1992). *Liquid Crystal*, Cambridge University Press, Cambridge,United Kingdom.

Cheon, C. I.; Li, L.; Glushchenko, A.; West, J. L.; Reznikov, Y.; Kim, J. S. & Kim, D. H. (2005). Electro-Optics of Liquid Crystals Doped with Ferroelectric Nano-Powder, *Society for Information Display Digest of Technical Papers*, Vol. 36, pp. 1471-1473

Clark, N. A. & Lagerwall, S. T. (1980). Submicrosecond bistable electro-optic switching in liquid crystals. *Appl. Phys. Lett.*, Vol. 36, No. , page numbers (899-901).

Demus, D.; Goodby, J.; Gray, G. W.; Spiess, H.-W.; Vill, V. & Lagerwall, S. T. (1998). Chiral Smectic Liquid Crystals: Ferroelectric Liquid Crystals, In: *Handbook of Liquid Crystals Vol 2B*, Demus, D.; Goodby, J.; Gray, G. W.; Spiess, H. W. & Vill, V. (Eds.), page numbers (515-664), WILEY-VCH Verlag GmbH, New York.

Gammaitoni, L.; Hanggi, P.; Jung, P. & Marchesoni, F. (1998). Stochastic Resonance. *Reviews of Modern Physics*, Vol. 70, No. 1, page numbers (223-287).

Goyal, D. J.; Agashe, C.; Takwale, M. G.; Bhide, V. G.; Mahamuni, S. & Kulkarni, S. K. (1993). Stochastic Resonance. *J. Mater. Res.* , Vol. 8, No. , page numbers (1052-1054).

Hegmann, T; Qi, H. & Marx, V. M.(2007). Nanoparticles in Liquid Crystals: Synthesis, Self-Assembly, Defect Formation and Potential Applications. *Journal of Inorganic and Organometallic Polymers and Materials*, Vol. 17, No. 3, page numbers (483-508).

Huang, J. Y. & Shih, W. T. (2006). Probing field-induced submolecular motions in a ferroelectric liquid crystal mixture with time-resolved two-dimensional infrared spectroscopy. *J. Phys. C* , Vol. 18, No. 32, page numbers (7593-7603).

Huang, J. Y.; Li, L. S. & Chen, M. C. (2008). Probing Molecular Binding Effect from Zinc Oxide Nanocrystal Doping in Surface-Stabilized Ferroelectric Liquid Crystal with Two-Dimensional Infrared Correlation Technique. *J. Phys. Chem. C*, Vol. 112, No. 14, page numbers (5410-5415).

Hung, F. R. & Bale, S. (2009). Faceted nanoparticles in a nematic liquid crystal: defect structures and potentials of mean force. *Molecular Simulation*, Vol. 35, No. 10-11, page numbers (822ąV834).

Kaur, S.; Singh, S. P.; Biradar, A. M.; Choudhary, A. & Sreenivas, K. (2007). Enhanced electro-optical properties in gold nanoparticles doped ferroelectric liquid crystals. *Appl. Phys. Lett.*, Vol. 91, No. 2, page numbers (023120).

Keesom, W. H. (1921). *Phys. Z.*, Vol. 22, No. , page numbers (129).

Kobayashi, S. (2009). Nanoparticle-Embedded Liquid Crystal Display Devices. *EKISHO*, Vol. 13, No. , page numbers (241-249).

Kubo, R. (1966). The fluctuation-dissipation theorem. *Rep. Prog. Phys.*, Vol. 29, No. , page numbers (255-).

Lagerwall, S. T. (1999). *Ferroelectric and Antiferroelectric Liquid Crystals*, Wiley-VCH, Weinheim.

Lapointe, C. P.; Mason, T. G. & Smalyukh, I. I. (2009). Shape-Controlled Colloidal Interactions in Nematic Liquid Crystals. *Science*, Vol. 326, No. , page numbers (1083-1086).

Li, F.; Buchnev, O.; Cheon, C. I.; Glushchenko, A.; Reshetnyak, V.; Reznikov, Y.; Sluckin, T. J. & West, J. L. (2006). Orientational Coupling Amplification in Ferroelectric Nematic Colloids. *Phys. Rev. Lett.*, Vol. 97, No. 14, page numbers (147801-4).

Li, L. S. & Huang, J. Y. (2009). Tailoring switching properties of dipolar species in ferroelectric liquid crystal with ZnO nanoparticles. *J. Phys. D: Appl. Phys.*, Vol. 42, No. , page numbers (125413-125417).

Magnasco, V.; Battezzati, M.; Rapallo, A. & Costa, C. (2006). Keesom coefficients in gases. *Chem. Phys. Lett.* , Vol. 428, No.4-6, page numbers (231-235).

Mallia, A.; Vemula, P. K.; John, G.; Kumar, A.; Pulickel, M. & Ajayan, P. (2007). In Situ Synthesis and Assembly of Gold Nanoparticles Embedded in Glass-Forming Liquid Crystals. *Angewandte Chemie International Edition*, Vol. 46, No. , page numbers (3269-3274).

Meulenkamp, E. A. (1998). Synthesis and Growth of ZnO Nanoparticles. *J. Phys. Chem. B*, Vol. 102, No. , page numbers (5566-5572).

Morita, S. (2004). 2Dshige. *Kwansei-Gakuin University*, `http://scitech.ksc.kwansei.ac.jp/_ozaki/`.

Nann, T. & Schneider, J. (2004). . *Chem. Phys. Lett.*, Vol. 384, No. , page numbers (150-).

Noda, I. (1993). Two-Dimensional Correlation Method Applicable to Infrared, Raman and Other Types of Spectroscopy. *Appl. Spectrosc.*, Vol. 47, No. , page numbers (1329-1336).

Ouskova, E.; Buchnev, O.; Reshetnyak, V.; Reznikov, Y. & Kresse, H. (2003). Dielectric relaxation spectroscopy of a nematic liquid crystal doped with ferroelectric Sn2P2S6 nanoparticles. *Liquid Crystals*, Vol. 30, No. 10, page numbers (1235-1239).

Ozaki, Y. & Noda, I. (2000). *Two-Dimensional Correlation Spectroscopy, AIP Conf. Proc. Vol 508*, AIP, Melville, New York.

Reznikov, Y.; Buchnev, O.; Tereshchenko, O.; Reshetnyak, V.; Glushchenko, A. & West, J. (2003). *Appl. Phys. Lett.*, Vol. 82, No. , page numbers (1917-).

Shim, M. & Guyot-Sionnest, P. (1999). Permanent Dipole Moment and Charges in Colloidal Quantum Dots. *J. Chem. Phys.*, Vol. 111, No. , page numbers (6955-6964).

Shiraishi, Y., Toshima, N., Maeda, K., Yoshikawa, H., Xu, J. & Kobayashi, S. (2002). . *Appl. Phys. Lett.*, Vol. 81, No. , page numbers (2845-2847).

Strocchi, F. (2008). *Breaking of Continuous Symmetries. Goldstone's Theorem*, Symmetry Breaking, Lecture Notes in Physics Vol. 643, Springer, New York.

14

Organic nanocrystals for nanomedicine and biophotonics

Koichi Baba[1]*, Hitoshi Kasai[2,3], Kohji Nishida[1] and Hachiro Nakanishi[2]
1. Osaka University, 2. Tohoku University, 3. PRESTO, Japan Science and Technology
Japan

1. Introduction

In this chapter, we will describe the advanced methodology for opening the new gate of nanomedicine and bioimaging using organic nanocrystals formulation, which is free from organic solvent and nano carrier. The organic nanocrystals aqueous dispersion is prepared by our originally developed technique, the reprecipitation method.

The importance of the contribution of nanotechnology to biomedical application has been greatly increased in few decades. Especially the current scientific endeavours on the researches of nanotechnology based nanomedicine and bioimaging have been giving us the novel and important information of biological events *in vitro* and *in vivo*. When focusing bioimaging and nanomedicine, we definitely treat the fluorescent dyes and drugs. However, currently used numerous organic compounds including drugs and fluorescent dye probes are hydrophobic in nature. Therefore, for their using in drug administration to human and animals or in dye staining imaging in living cells *in vitro* and *in vivo*, organic solvents are widely used for their increasing water solubility. However, organic solvent has not only the problems of causing systemic toxicity and cytotoxic activity, but also disturb obtaining accurately the biological experimental data, which may be modified by the influence of organic solvent. Many researchers have been making the scientific endeavour to clear these problems. Using nano carrier such as liposome, nanocapsule, nanosphere, dendrimer, emulsion, and surfactant have concluded successfully in some part of increasing water solubility of drugs and dyes. However, these carries still suffer from systemic toxicity and cytotoxic activity caused by their own. Additionally, the nano carriers have disadvantage of low amount loading of drugs and dyes in carrier and of their leakage.

To overcome these problems, recently the organic solvent and carrier free method for photodynamic cancer therapy and for bioimaging of living cells *in vitro* using organic nanocrystals are reported including our group (Baba et al., 2007 and 2009). Organic nanocrystals were prepared by the reprecipitation method (details are described in the flowing section 2). Employing these anticancer drug and fluorescent dye nanocrystals are superior in anticancer function and excellent fluorescent cell imaging, which are at least comparable with conventional manner. The crystals are most densely packed structure of molecules and nanocrystals are too. Therefore, the amounts of molecules in nanocrystals, which consist of 100% of molecules, are higher than that of the same size of nano carriers, and nanocrystals are free from leakage problems. The details of using organic nanocrystals for nanomedicine and bioimaging are described in the section 3. Future perspective of organic nanocrystals in nanomedicine and bioimaging are described in the section 4. Until now, to our knowledge, the scientific researches dealing with pure organic nanocrystals for nanomedicine and bioimaging are quite less. In this chapter, we will introduce recent researches of organic nanocrystals in biomedical applications.

2. The reprecipitation method, which is the preparation method of organic nanocrystals

The reprecipitation method is bottom-up type of preparing technique of organic nanocrystals (Kasai et al., 1992), which is the solvent exchange method from good solvent to poor solvent. For example of the preparation, first the targeting hydrophobic compound for nanocrystallization is dissolved in good solvent acetone (in mM order), then the acetone solution (200 μl) is speedy injected into vigorously stirred poor solvent water (10 ml) using microsyringe. Then the sudden exchanges of solubility from good to poor cause the precipitation, and when the concentration, temperature, and kinds of solvent are adequately selected, nanocrystals are stably dispersed in water, with usually giving the negative charged ζ-potential on the surface of nanocrystals. The ζ-potential helps nanocrystals for preventing their aggregation because of the created balanced stability by negative charged repulsion force of them. The schematic images of the reprecipitation method are shown in Figure 1. The reprecipitation method is excellent for precise controls of their size and morphology compared with that of the other conventional top-down procedure such as milling method. The typical size control range is from several tens nanometres to several micrometers (Fig. 2). The imaging pictures of water dispersion of nanocrystals are shown in Figure 3. The transparent and solution-like appearances are because of the quite low light

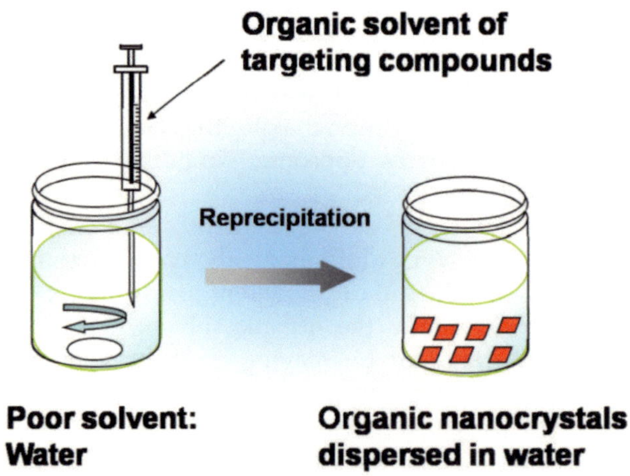

Fig. 1. The scheme of the reprecipitation method.

scattering originated in nano-sized crystals. These dispersions have both liquid and crystal characteristics. Therefore, we can directly measure the physiochemical properties of crystals as the liquid dispersions, or we can fabricate the nanocrystals-layered structure by collecting these nanocrystals from dispersion (Masuhara et al, 2001). Until now, our scientific endeavour revealed the size-controlled and size-dependent optoelectrical properties of functional organic nanocrystals such as nonlinear optical materials (Nakanishi & Katagi, 1998), п-conjugate functional dyes (Kasai et al., 1996), organic pigment for colour filter of liquid crystal display (Miyashita et al., 2008), and fullerene (Masuhara et al., 2009). Recently, numerous researches focusing on the optoelectrical property of organic nanocrystals have been reported by several groups (Zhao et al., 2008, An et al., 2004). However, the applications of organic nanocrystals on biomedical researchers are quite less (Sin et al., 2006, Bwambok et al., 2009), instead of their radical increasing importance. Therefore, recent our interest and challenging topics are applying organic nanocrystals technology toward the applications of drugs for medical treatments and fluorescent probe for bioimaging.

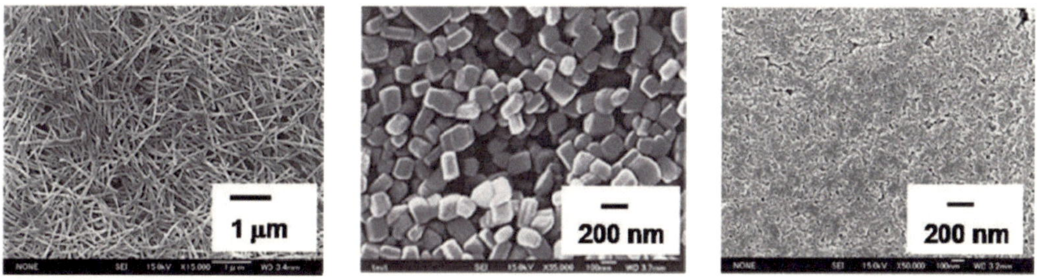

Fig. 2. Size control of organic nanocrystals in same compound.

Fig. 3. Imaging pictures of organic nanocrystals water dispersion.

3. The recent achievements of organic nanocrystals in nanomedicine and biophotonics

3.1 Organic nanocrystals used for nanomedicine

Recently, as with developing the nanoparticles preparation technology, several reports are mentioning about the fabrication technique of drug nanocrystals. However, almost all the reports are concerning about the drug formulation in industrial interest (Kipp, 2004). As our knowledge, quite few reports mentioning about the scientific aspects of using pure organic nanocrystals forms of drugs. Prof. Prasad groups firstly demonstrated the using of photosensitizing anticancer drugs in photodynamic cancer therapy (Baba et al., 2007). The concept of this study based on how to increase the water solubility of hydrophobic drugs without using solubilising agents such as surfactants and nanocarrier, which cause cytotoxicity, and also to exhibit the high efficacy of the drug.

Not only the photosensitizing drugs, but currently used many of pharmaceutical agents including various drugs are hydrophobic in nature (about 60% of drug products). Therefore, special formulations are required to make their aqueous dispersion for delivery of these drugs. Usually, surfactants or nanocarrier based delivery vehicles are used. In the cancer therapy treatment, once systemic administration of drugs are carried out, these nanocarriers including drugs are gradually taken up by tumour tissues according to the concept of the "enhanced permeability and retention effect" (Maeda et al., 2006), which rely on the characteristics of tumour tissues in trapping and retaining circulating nanoparticles because of their leaky stricture. The nanocarriers are member of such as liposomes, polymeric micelles, oil dispersions micelles, polymeric nanoparticles, drug encapsulated polymer complexes. However, such carriers tend to increase the systemic toxicity caused by own. Therefore, there is increasing interest in the developing the novel type of drug formulation and delivery approaches without auditioning any solubilising agents such as surfactants and/or nanocarriers. One novel method proposed for this was using the reprecipitation method. Though the organic nanocrystals formations of hydrophobic compounds have been well studied in optoelectrical materials, there was no report of using this method for drug delivery.

To demonstrate the concept, the nanocrystal formulation of the hydrophobic drug of photodynamic therapy was selected, and its efficacy was compared with the conventional surfactant-supported formulation. Photodynamic therapy is a promising approach for curing several types of cancers as well as some dermatological and ophthalmic diseases such as age meditated macular degeneration. The main advantage of photodynamic therapy compared with the conventional cancer chemotherapy is in the localized treatment using selective light exposure to the tumour tissue. Usually, photodynamic therapy is carried out by systemic administration of photosensitizing drugs, followed by exposing the light to the tumour tissue. Basically, visible or near-infrared light are selected, and after the light is exposed, the photosensitizing molecules transfers their excited state energy to oxygen molecular in the surroundings, and as the result, reactive oxygen are formed, so-called singlet oxygen. The singlet oxygen has ability to distract the structure of cells, and cell deaths are induced in tumour tissue. Though the destruction process requires the combination of photosensitizing, light, and oxygen, there is enough advantage for photodynamic therapy in selective distraction of tumour tissues. However, one of the major problems of photodynamic therapy is in the poor water solubility of photosensitizing drugs, thus making their stable formulation for systemic administration are highly required. One of the overcoming these problems are using solubilising agent such as nanocarrier. However, these nanocarriers often have the problem of drug leakage from carriers, resulting in fewer efficacies. Furthermore, rejection reactions caused by nanocarriers are one of the anxiety problems. Therefore, the ideal formulation for safe and for good efficacy photodynamic therapy requires the solubilising agent free method. Thus, in this time, as new method for the delivery of water insoluble drugs, which is free from using additional solubilising agents was reported. This new drug formulation was fabricated by the reprecipitation method. The resulting nanocrystals were monodispersed and taking stably dispersion, with diameter about 100 nm.

In this research, there was an interesting finding that the fluorescence and photodynamic activity of the drug nanocrystals were initially quenched in aqueous media because of the peculiarity of crystal structure; however both recovered under *in vitro* and *in vivo* conditions with the treatment of serum, resulting in creating molecular form. The recovery of fluorescence and photodynamic activities were verified by confocal fluorescent microscopy observation of cells and cellular phototoxicity assay. Efficacy of the nanocrystal formulation *in vitro* as well as *in vivo* was found to be comparable with that of the same drug formulated in the conventional surfactant delivery vehicle. This study was first example of the use of organic nanocrystals prepared by the reprecipitation method.

3.2 Organic nanocrystals used for bioimaging

Fluorescence microscopy observation is one of the most broadly utilized imaging techniques in biomedical researches, which allows the noninvasive imaging of cells and tissues with molecular specificity. These imaging requires fluorescent dyes to enter cells and tissues before visualization. Currently used dyes include classes of the coumarin, rhodamine, fluorescein, and carbocyanine, as well as their derivatives. 3,3′-Dioctadecyloxacarbocyanine perchlorate [DiO: DiOC18(3)] is the long-chain dialkylcarbocyanine dye, and commonly used for visualizing the anterograde and retrograde neuronal tracers in living cells. This lipophilic carbocyanine is also employed for many other applications, including the tracking of cell migration and lipid diffusion in membranes through fluorescence recovery after photobleaching (Gordon et al., 1995), the labeling of lipoproteins (Lohne et al., 1995), and cytotoxicity assays (Johann et al., 1995). Nevertheless, the hydrophobic nature of these dyes require, in many cases, organic solvents [e.g. dimethylsulfoxide (DMSO) and dimethylformamide] and surfactants for successful cell imaging with increased water solubility of dyes. However, unfortunately, organic solvents and surfactants themselves tend to increase cytotoxicity *in vitro* and *in vivo*. On the other hand, the direct applications of micronized crystals have also been investigated, although such crystals are probably not small enough to allow the diffusion of dyes into cells, thus imaging efficency is poor.

Furthremore, not only DiO dyes, but also many of other fluorescent dyes are hydrophobic in nature, including perylene, which is widely studied hydrophobic material as functioning with high quantum yield in organic electroluminescence devices. Although using perylene is potentially useful for bioimaging, the preparation of aqueous dispersions of perylene dyes requires special formulation techniques, similar to those of DiO. One of the very useful technique for dispersing hydrophobic compounds in water is the "reprecipitation method", which has been used for the nanocrystallization of organic optoelectronic materials exhibiting size-dependent optical properties. Recently, by using this reprecpitation mehod, we demonstrated an organic solvent free bioimaging method employing nanocrystals of

hydrophobic fluorescent dye, and their applications to *in vitro* were evaluated by confocal fluorescent laser microscopy observation (Baba et al., 2009). The fluorescent dyes that we used were DiO and perylene.
The aqueous dispersions of DiO and perylene prepared using the reprecipitation method have been dialyzed for 6 h to remove acetone, thereby resulting in organic solvent–free dispersions were obtained (Fig. 4). The resulting particle sizes and morphologies were less than 150 nm for the DiO nanocrystals and 155 nm for the perylene nanocrystals, which was revealed by scanning electron microscopy observation, and these values correspond reasonably well with those determined through dynamic light scattering measurements (Fig. 5). The average ζ-potentials of the DiO and perylene nanocrystals were ca. +36 and –10 mV, respectively. The relatively high positive surface charge of the DiO nanocrystals, presumably derived from the positively charged molecular structure, imparted this water dispersion with excellent stability. In the case of the perylene nanocrystals, the factors causing their negative surface potential remains unclear, although negative charging of organic nanocrystals in water is well known empirically to inhibit particle aggregation. The aqueous dispersion of perylene nanocrystals was stable for more than 3 months. The crystallinity of their prepared DiO and perylene nanocrystals were confirmed by powder X-ray diffraction analysis.
The obtained UV–Vis absorption and fluorescence emission spectra of these nanocrystals are significantly different from that of their solution form, namely we observed suppression and broadening in the absorption spectra and decreases in emission intensities. The decreases in fluorescence intensities are well-known aggregation effects for fluorescent molecules. In our case, this phenomenon was due to the low solubility of the hydrophobic dyes, with corresponding aggregation effects in aqueous system. A decreased fluorescent yield will induce a decreased imaging efficiency.

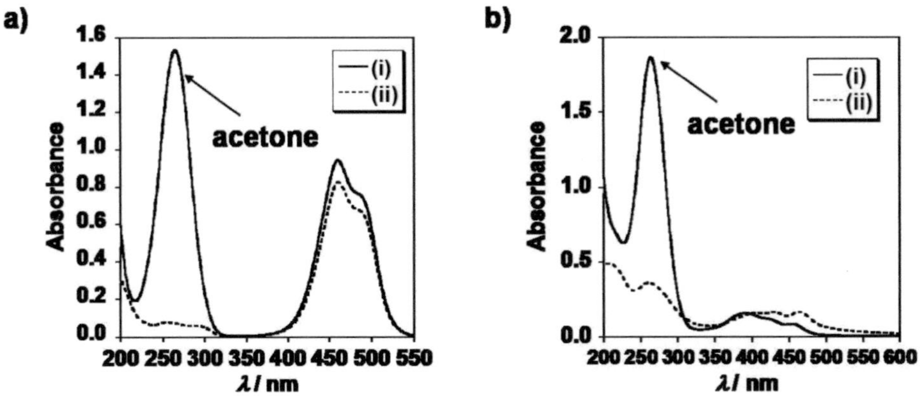

Fig. 4. Absorption spectra of water dispersions of (a) DiO nanocrystals and (b) perylene nanocrystals. (i) before and (ii) after dialysis. The disappearences of the absorbance of acetone (at λ_{max} = 265 nm) were confirmed.

a)

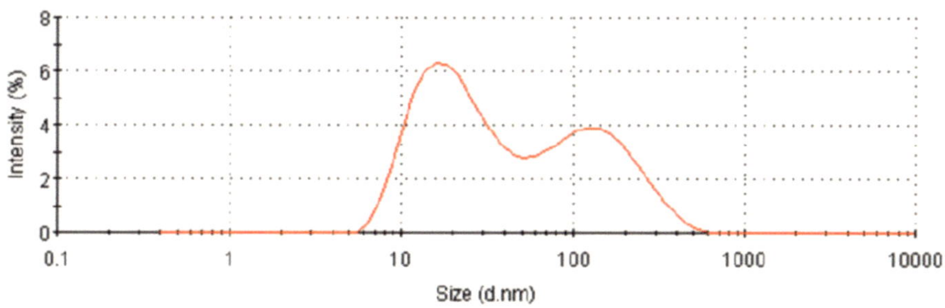

b)

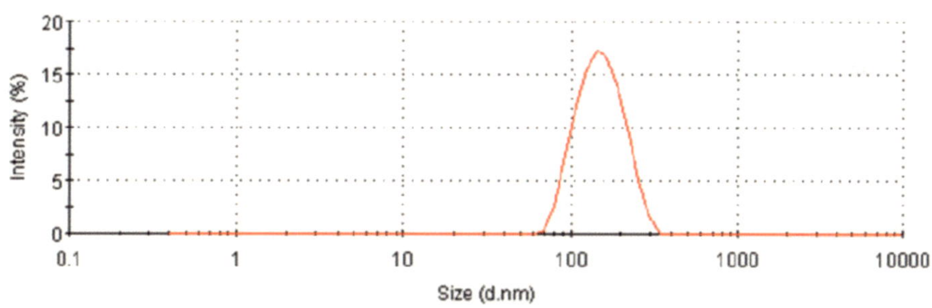

Fig. 5. Size distributions of (a) DiO and (b) perylene nanocrystals dispersed in water. The peak sizes of the DiO particles were 22 nm (intensity: 58%) and 146 nm (intensity: 42%); the peak size of the perylene particles was 155 nm (intensity: 100%; i.e., monodispersity).

However, there was interesting finding that the fluorescence of the DiO and perylene nanocrystals were recovered by elapsed time under *in vitro* conditions in the presence of 10% fetal bovine serum. It seemed that this recovery in fluorescence in the presence of fetal bovine serum induced the DiO and perylene nanocrystals to reasonably good soluble. Although the exact mechanism for the improved solubility is not cleared yet, but we consider that it may rely on the affinity interactions between the hydrophobic moieties of the proteins and the hydrophobic surfaces of the nanocrystals, which cause the dissolution of nanocrystals, resulting in converting nanocrystals into the molecular forms of the dyes, thereby, leading to fluorescence recovery.

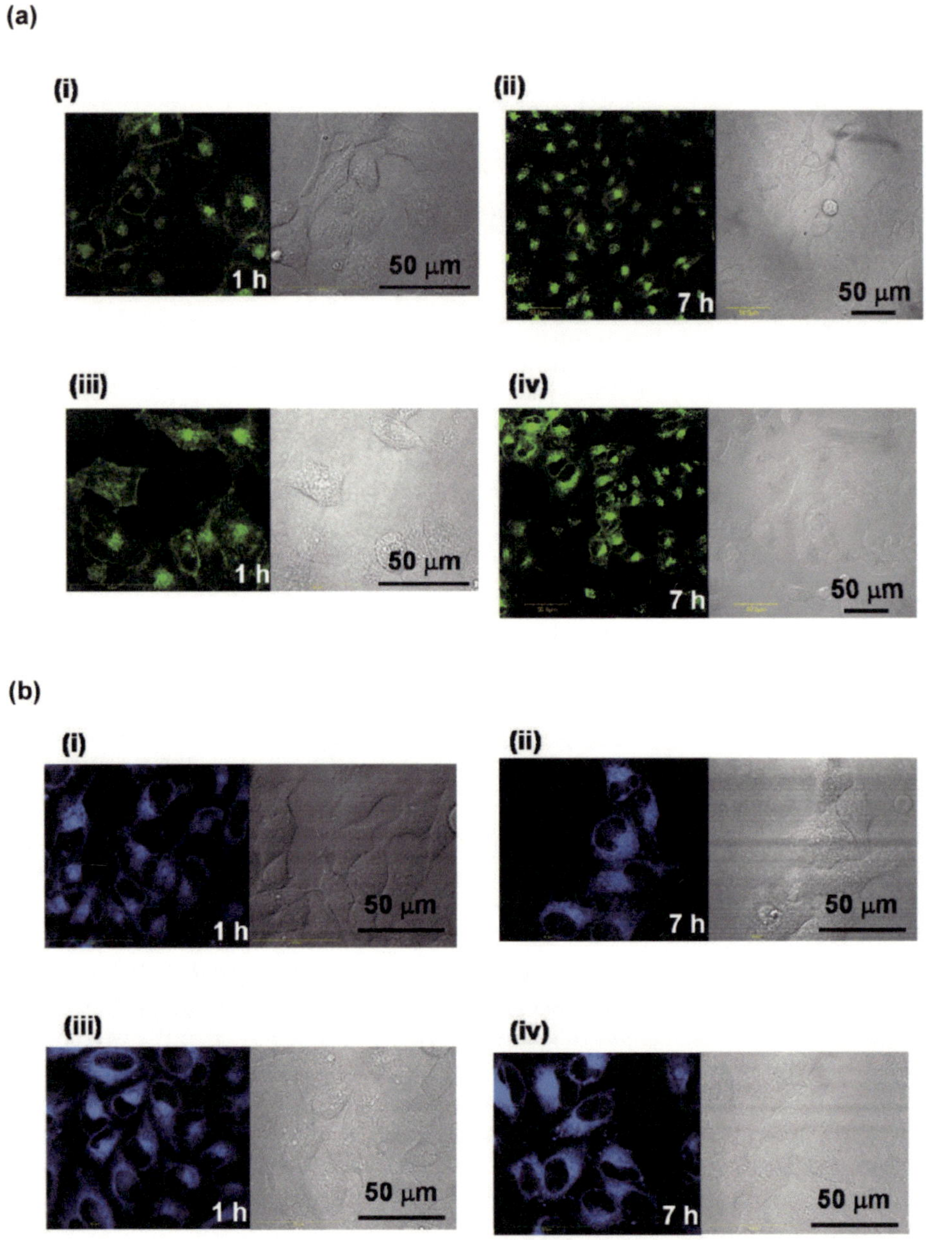

Fig. 6. *In vitro* fluorescent confocal images of cells recorded 1 h and 7 hrs after incubation with (i), (ii) nanocrystals of (a) DiO and (b) perylene, and (iii), (iv) dyes solution in DMSO of (a) DiO and (b) perylene.

To evalute if the organic nanocrystals were useful for the bioimaging of living cells *in vitro*, we performed confocal fluorescence imaging of tumor cells using the DiO and perylene nanocrystal formulations (Fig. 6). Although the fluorescence signals in the nanocrystals were initially weak, it increased after cellular uptake (1 h) as similar to the behavior of nanocrystals in the fetal bovine serum-containing medium in bothe case of perylene nanocrystals and DiO nanocrytals. Compared with the using of solution form of dyes disolved in DMSO, where exhibiting higher fluorescence signals from at the initial stage of dye dosing in culture medium, nanocrystals revealed comparable levels of cellular uptake by elapsed time. Therefore, the long-term cellular uptake of the DiO and perylene dyes in their nanocrystal formulations were similar to that of the dyes in their solution formulations. The DiO dyes specifically stained the cell membrane; the perylene dyes stained the cytoplasm.

On the other hand, we investigated whether nanocrystal formation was really necessary for efficient cell imaging, especially remarking on the crystal size. Aqueous dispersions of microcrystals, with the average sizes were in the micrometer range, were prepared through sonication of the bulk crystals in water. The concentrations of these dispersions were adjusted to be the same as those of the nanocrystals. Figure 7 displays scanning electron microscopy images of the DiO and perylene microcrystals used for cell imaging. In both cases, their sizes were on the order of several to a few tens of micrometers. In microcrystals, no significant recovery in fluorescence occurred in the cell culture medium containing 10% fetal bovin serum during 6 hrs after the addition of the DiO microcrystals, and whereas a slight recovery of fluorescence in the case of the perylene microcrystals was observed. Presumably, one of the main reasons for the significantly lower fluorescence recoveries of the microcrystals was in their size. When the size of a crystal increases, its surface area per unit volume decreases, as a result, its interactions with the serum components become weak, and resulting in decreased solubility. Figure 8a presents the confocal microscopy images we obtained when applying the DiO microcrystals. Interestingly, only the surrounding areas where the microcrystals were attached to the cells

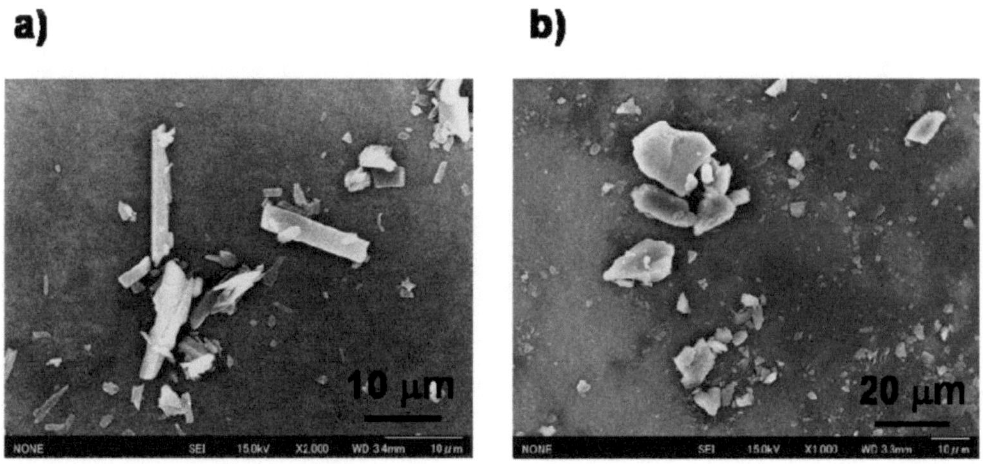

Fig. 7. Scanning electron microscopy images of (a) DiO and (b) perylene microcrystals.

displayed fluorescence (e.g., yellow circle in Figure 8a). The direct application of DiO microcrystals to fluorescence imaging of cells is a commonly used technique. Nevertheless, we observed no fluorescence from those cells not presenting any attached DiO microcrystals, meaning that this imaging technique is poorly efficient. Figure 8b reveals that, in the case of the added perylene microcrystals, the fluorescence imaging was less efficient than that obtained using the corresponding nanocrystal formulation (Fig. 6b, ii) Clearly, the low solubility of the perylene microcrystals in the serum led to poor imaging. Additionally, we also imaged cells lacking added dyes as a control, and we observed no fluorescence signals under the corresponding measurement conditions (Fig. 8c). Thus, the fluorescence images we obtained using nanocrystals, dyes in DMSO treatments, and microcrystals were not the result of autofluorescence of the cells.

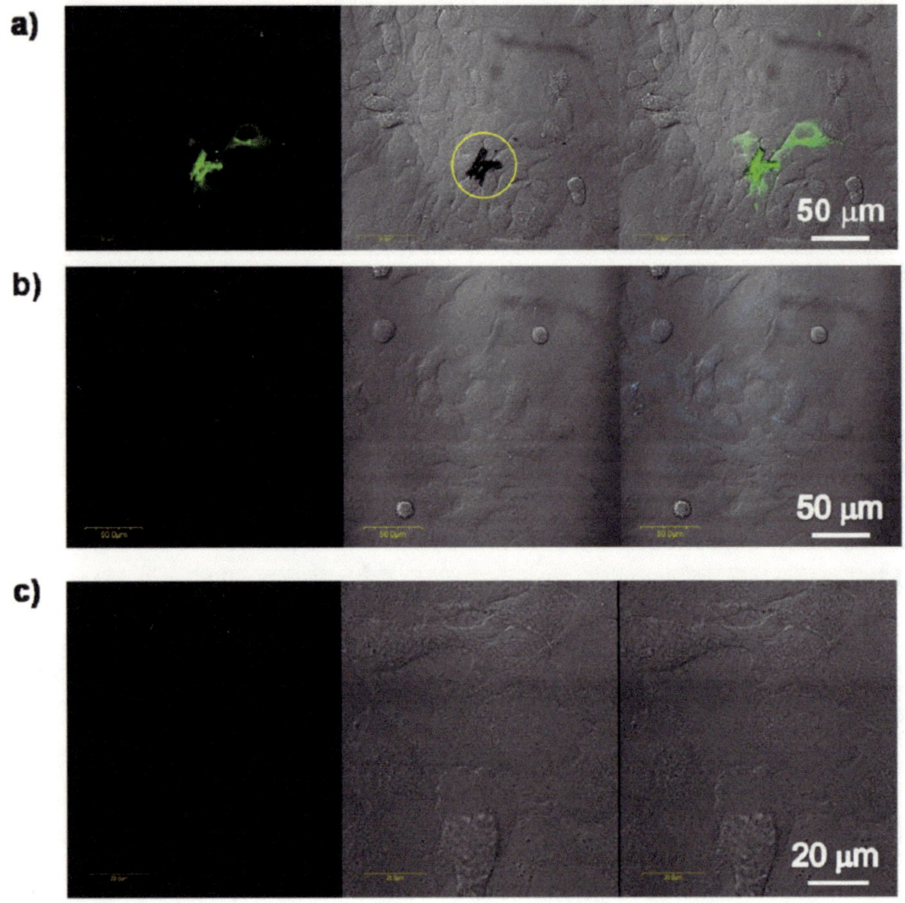

Fig. 8. Fluorescent confocal images of (a) DiO microcrystals, (b) perylene microcrystals, and (c) control: the pictures of fluorescent images (left), tranceperant images (middle), and overlaped images of the left and middle (right). The yellow circle in (a: middle) represents the microcrystals. These fluorescent images were obtained after 7 hrs incubation.

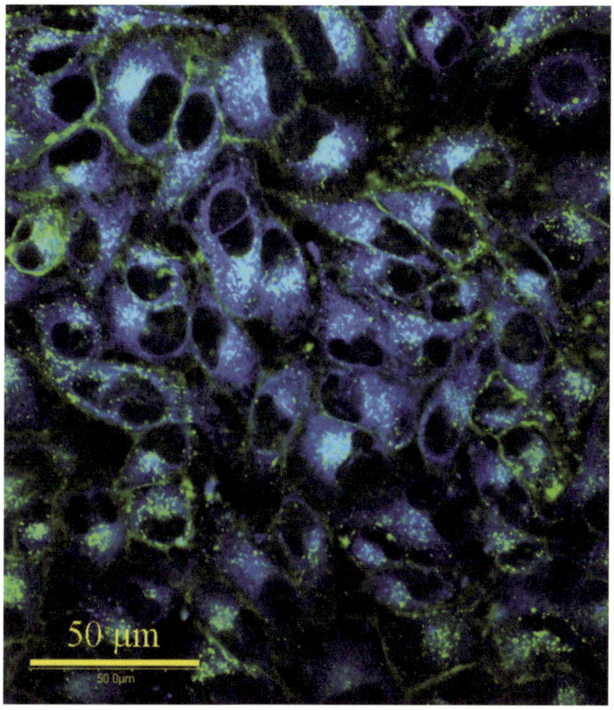

Fig. 9. Double staining images of cells using DiO and perylene nanocrystals. The green and blue fluorescent colours come from DiO and perylene, respectively.

When we investigated double staining imaging of both the DiO and perylene nanocrystals in cells in culture medium we initially detected each fluorescence signal individually by confocal fluorescent laser microscopy observation, and Figure 9 presents the over laped pictures of them. DiO resulted in significant membrane staining (Fig. 9: green color), whereas perylene stained in mostly cytoplasmic (Fig. 9: blue color). One of the reason for this behavior might be due to differences in ζ-potentials of nanocrystals/dyes, namely the DiO particles had a positive ζ-potential (+36 mV), which may have aided in their trapping on the slightly negatively charged cell membrane. We also confirmed that the fluorescence obtained were not the result of autofluorescence of the cells. Although the detail mechanisms of endocytosis of the nanocrystals/dyes remained unclear, we were successful in recording a double-staining cell images using the formulation of organic nanocrystals.

3.3 Application of the reprecipitation method toward dye doped polymeric nanoparticles

We will introuduce the other approach for fabricating bioimaging tools using the reprecipitation mehtod, where the fluorescent dye doped biodegradable polymer nanoparticles were demonstrated (Baba et al., 2005). The polymer used was poly (D,L-lactide-co-glycolide), and the dyes used were infrared emitting dyes and two photon excitable fluorescent dyes. Current technologies in fluorescent imaging have advanced significantly in developing for multiprobe applications in the study of biological events. Especially, for further advanced fluorescent imaging capability, organic fluorophores absorbing and emitting in near infrared region have been developed. Such near infrared dyes have the advantage for providing the better signals avoiding the autofluorescence caused by ultra violet or visible light. Secondly, tissue penetration of near infrared lighg is great due to low absorption of excitation light in tissue. However, many of these dyes have less solubility and emissions in aqueous systems, making them undiserable for biological applications. The infrared emitting dyes used as demonstration shows fluorescence around 1.1 to 1.35 μm in organic solvents, but was significantly quenched in aqueous media. Whereas, two photon excitable fluorescent dye has a polar D-π-A structure (Lin et al., 2004), in which the π-system is end-capped by an electron donor (D) and an electron acceptor (A), having one of the most effective molecular models for both second- and third-order nonlinear optical materials, thus such dyes have great potential for two photon imaging. However, as with these dyes lacks the solubility properties for the application of biological systems. The nanoparticle technology had the ability to encapsulate such hydrophobic dyes into some matrix for producing the aqueous dispersion, enabling functionality in biological systems. For example, the fabrication of dye-doped polymeric nanoparticle water dispersions were quite effective method for bioimaging of such hydrophobic fluorescent dyes. The biodegradable poly (D,L-lactide-co-glycolide) were useful for preparing particles, which have been well investigated as a drug and gene delivery vehicle (Jain, 2000). The applying of the reprecipitation method, which is simple and surfactant-free fabrication method, for the incorporation of hydrophobic, near infrared emitting dyes and two photon excitable emitting dyes into poly (D,L-lactide-co-glycolide) nanoparticles resulted in creating an aqueous dispersion of nanoparticle, with their sizes of less than 100 nm. Then, the optical properties of dye solution, dye nanocrystals water dispersion, and dye doped polymeric nanoparticle water dispersion for these two dyes were investigated, respectively. For infrared emitting dyes, the fluorescence from dye solution was in the spectral range of ~1.1 to 1.35 μm. Whereas, in the case of dye nanocrystal water dispersion, even at the same concentration of dye solution, the emission was not noticeable. This was because of the fluorescence quenching, estimated to be including collisional quenching, static quenching, excited state reactions, electron transfer and energy transfer. The main reason for dye fluorescence quenching in water dispersion might comes from the crystallization of dye, as it is well known for several kinds of organic molecules. Similar changes in absorption were usually associated with the appearance of the molecular *H*-aggregates, which were nonradiative. However, when fabricating the dye doped poly (D,L-lactide-co-glycolide) nanoparticle water dispersion, it exhibitted fluorescence in the aqueous environment. A series of systematic studys of the relationship between fluorescence and micelle microemulsion in water environment has been reported. Our results also indicated that dye was successfully encapsulated in the polymer matrix. Thus, crystallization of dyes were prevented, and also dyes were protected from contact with the water environment, thus fluorescence quenching was avoided. In the case of two photon excitable emitting dyes, optical properties were investigated especially forcusing on dye nanocrystals and dye doped polymeric nanoparticles. The size of dye nanocrystals was 35 ± 9 nm, and the sizes of these polymeric nanoparticles were controlled as ca. 36 ± 8 nm, 50 ± 11 nm and 96 ± 19 nm, respectively, using the reprecipitation mehod. According to the procedure of the reprecipitation mehod, when the initial concentration of polymer solution increased, the size of polymeric particles also increased. There have been several reports on fabrication of poly (D,L-lactide-co-glycolide) particles with sizes from several hundreds nanometer to few tens micrometer. However, polymeric nanoparticles less than 100 nm in size were succesuffly prepared. Even though, two photon excitable fluorescent dye nanocrystals showed fluorescence in aqueous media, the fluorescence intensity of dye-doped polymeric nanoparticles was higher than that of the dye nanocrystals. Furthermore, we observed the increasing fluorescence intensity of dye doped polymeric nanoparticles, with an increasing in their particle size. The role of the polymer for influencing the fluorescence intensity should be similar to the case of infrared emitting dyes. Namely, when two photon excitable fluorescent dyes were surrounded by the polymer matrix, the crystallization of dyes and contact of dyes with the water environment were prevented, resulting in increasing fluorescent intensity. The reason for increasing the fluorescence intensity with an increasing the polymeric particle size was that when the ratio of polymer to dye increased, correspondingly, number of dyes individually embedded in the polymer matrix increased without interaction between dyes. The uptake of two photon excitable fluorescent dye nanocrystals and their dyes-doped polymeric nanoparticles by cancer cells in culture media were investigated. Confocal laser microscopy of two-photon imaging revealed a difference in the fluorescence intensity and staining pattern of fluorescence in cells incubated with dye nanocrystals (34 ± 11 nm in size) and the dye-doped polymeric nanoparticle (32 ± 10 nm in size), respectively. The fluorescence intensity obtained from cells incubated with dye-doped polymeric nanoparticles were about ten times higher than that of dye nanocrystals. These results suggested that the coating of fluorescence dye by adequate polymer was an important methodology for efficient cell imaging. We can say that not only nanocrystals formation, but alos dye-doped polymeric nanoparticles prepared by the reprecipitation mehod are effective way for efficient bioimaging.

4. Future perspective of organic nanocrystals in nanomedicine and biophotonics

4.1 Organic nanocrystals for biophotonics

Herein, we will report the new type anti-photo breaching fluorescent organic nanocrystals used in bioimaging, which have the potential to overcome the problems of quantum dots and molecular probe. Currently, using quantum dot and molecular probe for bioimaging and nanomedicine are numerously reported. Quantum dots have been applied for detection and imaging in several areas in the life sciences, ranging from microarray technologies to fluorescence *in situ* hybridization to *in vitro* imaging (Medintz et al., 2005). Despite many superior optical properties, such as size-tuneable absorption and emission, extremely broad and intense absorption enabling a unique flexibility in excitation, high fluorescence quantum yields even in the NIR wavelengths and large two-photon action cross-section as compared to organic dyes, the solutions for using quantum dots have so far been individual ones. Mainly, quantum dots need surface modification by complicated procedure to increase the water solubility and to avoid the cytotoxic caused by leaking of heavy metal ions. Additionally, we need patient to find a solution to the challenges of their particular experimental system against the benefits of the advanced spectroscopic features of quantum dots. On the other hand, the optical property of the molecular probe depend on the electronic transitions involved and can be fine-tuned by elaborate design strategies if the structure-property relationship is known for the given class of dye (Mason, 1999). The emission of organic dyes typically originates either from an optical transition delocalized over the whole chromospheres or form intramolecular charge transfer transitions. The majority of common fluorophores, such as fluorescein, rhodamines, and most cyanines, are resonant dyes that are characterized by slightly structured, comparatively narrow absorption and emission bands that often mirror each other, a small solvent polarity-insensitive Stokes shift, high molar absorption coefficients, and niderate-to-high fluorescence quantum yields. The major problems of these dyes are both less water soluble in nature, thus need organic synthesis making hydrophilic salts form or using solubilising agents, and the weak resistance against photo breaching. We recently developed the new type of fluorescent organic nanocrystals for bioimaging. These organic nanocrystals have advantage in high water solubility, cytotoxicities-free caused by heavy metal ions, and quite strong against photo breaching. The organic nanocrystals were class of organic pigments, quinacridone deliverities. The organic pigments are very rigid compound, which have strong light resistance and weather fastness, thus have been used for several applications such as colour filter and coating materials. Therefore, making water dispersion of organic pigments is quite difficult because of their poor water solubility. However, we found that if applying the reprecipitation method, fine dispersion of quinacridone nanocrystals were obtained (Ujiiye-Ishii et al., 2006). Under the confocal laser fluorescent microscopy observation after dosing quinacridone nanocrytals in cell culture medium, we have been successfully obtained fine and tough fluorescent imaging during nearly 15 min in laser irradiation; despite the output power of laser source was nearly 100%. One can see that almost no fluorescent breaching was observed in the imaging pictures at the beginning and the ending of the observation (Fig. 10). Despite the merit of strong light resistance originated from crystal structure, the fluorescent intensity of quanicridone nanocrystals were not strong like such as fluorescein because of the quenching effect caused by crystallization. Thus, we are now further investigating to find the new approach that satisfies both high fluorescent intensity and light resistance. We believe that these kinds of right resistance fluorescent organic nanocrystals are candidate for new type of bioimaging tools in near future.

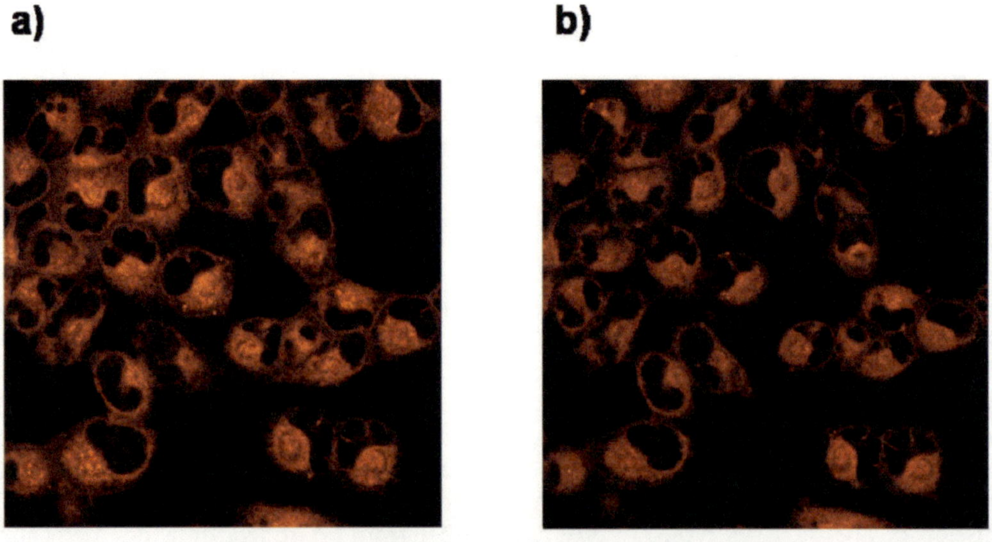

Fig. 10. Fluorescent confocal images of cells stained by quinacridone nanocrystals. The fluorescent images observed at the (a) beginning stage and (b) ending stage.

4.2 The size effect of nanocrystals for nanomedicine and biophotonics

The size is the important factor in biological events especially for cellular uptake, systemic circulation in the case of systemic administration, and solubility in living organism, which affect the efficacy of drugs and efficient fluorescent cell imaging. Actually, we found that hydrophobic nanocrystals that had poor solubility in water had good solubility in serum component, and which is peculiar to nano-sized crystals and not to more than micro-sized crystals. These findings give us the very important information for considering the strategy in nanomedicine and bioimaging for the selection of efficient and appropriate approaches. For the advanced applications, one of the strategies we can select for using organic nanocrystals in bioimaging and nanomedicine is taking the step of designing molecular structure at first, which have the hydrophobic characteristics and the sight specific targeting moiety. If the fluorescent dyes and drugs are hydrophobic, their stable organic nanocrystals aqueous dispersions are well prepared by the reprecipitation method. Upon, taking administration of these nanocrystals *in vivo* and *in vitro*, the nanocrystals will start dissolving in living organism, with gradually generating the drugs/dyes activity by taking molecular formulation. If the drugs and dyes that have site specific moiety, the drugs and dyes can bind to the desired part such as affected site and cells with minimally dosing with specifically. The dissolving kinetics, disposition, and body distribution of nanocrystals will be basically aligned by controlling the size of particles. These favourable peculiarities will be achieved by without applying any external solubilising agents such as organic solvents, surfactants, and nanocarriers. This nanocrystal technology in the fields of biology should be applied for not only limited scientific researches, but also for medical treatment in near future. This simplest nanocrystal approach, which we have demonstrated, will be mostly close way toward the clinical approach, which holds down the risk caused by carries and the cost caused by manufacturing. We can say that the nanocrystals aqueous dispersion or their dried powder will be useful for drugs/diagnostic fluorescent agents in such as administrations of oral, intravascular, inhalational, transdermal, nasotracheal, and ocular.

5. Conclusion

In this chapter we described three topics; first, we explained how to prepare the organic nanocrystals in aqueous dispersion system using the reprecipitation method. Second, we referred the recent our achievements of organic nanocrystals in nanomedicine and biophotonics. Third, we remarked the future direction of organic nanocrystals in nanomedicine and biophotonics. We believe that our organic nanocrystals technology, recent results, and ideas will be helpful especially for biochemist, biophysicist, nanoscientist, and medical scientist for tremendous advances in their specialized fields and in their multidisciplinary.

6. References

An, B. K. (2004). Strongly fluorescent organogel system comprising fibrillar self-assembly of a trifluoromethyl-based cyanostilbene derivative. *J. Am. Chem. Soc.*, 126, 10232-10233

Baba, K. (2005). Infrared emitting dye and/or two photon excitable fluorescent dye encapsulated in biodegradable polymer nanoparticles for bioimaging. *Materials Research Society Symposium Proceedings*, 845, 209-214

Baba, K. (2007). New Method for Delivering a Hydrophobic Drug for Photodynamic Therapy Using Pure Nanocrystal Form of the Drug. *Mol. Pharm.*, 4, 289-297

Baba, K. (2009). Organic solvent-free fluorescence confocal imaging of living cells using pure nanocrystal forms of fluorescent dyes. *Jpn. J. Appl. Phys.*, 48, 117002/1-117002/4

Bwambok, D. K. (2009). Near-infrared fluorescent nanoGUMBOS for biomedical imaging. *ACS Nano* 3, 3854-3860

Gordon, G. W. (1995). Analysis of simulated and experimental fluorescence recovery after photobleaching. Data for two diffusing components. *Biophys. J.*, 68, 766-778

Jain, R. A. (2000). The manufacturing techniques of various drug loaded biodegradable poly(lactide-co-glycolide) (PLGA) devices. *Biomaterials*, 21, 2475-2490

Johann, S. (1995). A versatile flow cytometry-based assay for the determination of short- and long-term natural killer cell activity. *J. Immunol. Methods*, 185, 209-216

Kasai, H. (1992). A novel preparation method of organic microcrystals. *Jpn. J. Appl. Phys.*, 31, L1132-L1134

Kasai, H. (1996). Size-dependent colors and luminescences of organic microcrystals. *Jpn. J. Appl. Phys.*, 35, L221-L223

Kipp, J. E. (2004). The role of solid nanoparticle technology in the parenteral delivery of poorly water-soluble drugs. Int. J. Pharm., 284, 109-122

Lin, T.-C. (2004). Degenerate nonlinear absorption and optical power limiting properties of asymmetrically substituted stilbenoid chromophores. *J. Mater. Chem.*, 14, 982-991

Lohne, K. (1995). Standardization of a flow cytometric method for measurement of low-density lipoprotein receptor activity on blood mononuclear cells. *Cytometry.*, 20, 290-295

Maeda, H. (2006). The EPR effect and polymeric drugs: A paradigm shift for cancer chemotherapy in the 21st century. *Adv. Polym. Sci.*, 193, 103−121

Mason, W. T. (1999). *Fluorescent and luminescent probes for biological activity 2nd edn.* Academic press, London

Masuhara, A. (2001). Hetero-multilayered thin films made up of polydiacetylene microcrystals and metal fine particles. *J. Macromol. Sci. Pure Appl. Chem.*, A38, 1371-1382

Masuhara, A., (2009). Fullerene fine crystals with unique shapes and controlled size. *Jpn. J. Appl. Phys.*, 48, 050206/1-050206/3

Medintz, I. L. (2005). Quantum dot bioconjugates for imaging, labelling and sensing. *Nat. Mater.*, 4, 435-446

Miyashita, Y. (2008). A new production process of organic pigment nanocrystals. *Mol. Cryst. Liq. Cryst.*, 492, 268-274

Nakanishi, H. & Katagi, H. (1998). Microcrystals of polydiacetylene derivatives and their linear and nonlinear optical properties. *Supramol. Sci.*, 5, 289-295

Sin, K. K. (2006). A highly sensitive fluorescent immunoassay based on avidin-labeled nanocrystals. *Anal. Bioanal. Chem.*, 384, 638-644

Ujiiye-Ishii, K. (2006). Mass-Production of Pigment Nanocrystals by the Reprecipitation Method and their Encapsulation. *Mol. Cryst. Liq. Cryst.* 445, 177-183

Zhao, S. Y., (2008). Low-Dimensional Nanomaterials Based on Small Organic Molecules: Preparation and Optoelectronic Properties. *Adv. Mater.*, 20, 2859-2876